PEROVSKITE PHOTOVOLTAICS

PEROVSKITE PHOTOVOLTAICS

Basic to Advanced Concepts and Implementation

Edited by

SABU THOMAS

International and Interuniversity Centre for Nanoscience and Nanotechnology, School of Chemical Sciences, Mahatma Gandhi University, Kottayam, Kerala, India

APARNA THANKAPPAN

International and Interuniversity Centre for Nanoscience and Nanotechnology, Mahatma Gandhi University, Kottayam, Kerala, India

Baselius College, Kottayam, Kerala, India

ELSEVIER

ACADEMIC PRESS

An imprint of Elsevier

Academic Press is an imprint of Elsevier
125 London Wall, London EC2Y 5AS, United Kingdom
525 B Street, Suite 1650, San Diego, CA 92101, United States
50 Hampshire Street, 5th Floor, Cambridge, MA 02139, United States
The Boulevard, Langford Lane, Kidlington, Oxford OX5 1GB, United Kingdom

British Library Cataloguing-in-Publication Data

A catalogue record for this book is available from the British Library

Library of Congress Cataloging-in-Publication Data

A catalog record for this book is available from the Library of Congress

ISBN: 978-0-12-812915-9

For Information on all Academic Press publications
visit our website at https://www.elsevier.com/books-and-journals

Publisher: Joe Hayton
Acquisition Editor: Lisa Reading
Editorial Project Manager: Katie Chan
Production Project Manager: Kamesh Ramajogi
Cover Designer: Miles Hitchen

Typeset by MPS Limited, Chennai, India

CONTENTS

List of Contributors *xi*
Preface *xv*

1. Introduction: Why Perovskite and Perovskite Solar Cells? 1
Sarat Kumar Sahoo, Balamurugan Manoharan and Narendiran Sivakumar

1.1 Introduction 1
1.2 Present Photovoltaic Technologies 2
1.3 Perovskite 6
1.4 History of Perovskite Solar Cell 7
1.5 Classification of Perovskite System 8
1.6 Characteristic Features of Perovskite Solar Cell 10
1.7 Advantages of Perovskite Solar Cell 12
1.8 Challenges of Perovskite Solar Cell 14
1.9 Properties and Applications 17
1.10 Future Trends of Perovskite Solar Cell 20
1.11 Summary 21
References 22

2. Organometal Lead Halide Perovskite 25
Javeed Akhtar, Muhammad Aamir and Muhammad Sher

2.1 Introduction 25
2.2 Metal Halide Perovskites 25
2.3 Composition of Perovskites 26
2.4 Nature of Bonding in Organometal Lead Halide Perovskites 27
2.5 Distortion of Perovskite Geometry 30
2.6 Optical Properties of Perovskites 32
2.7 Coordination Chemistry of Lead Halide Perovskite Precursors 34
2.8 Fabrication of Perovskites 36
2.9 Stability of Hybrid Perovskites 38
References 39

3. Evolution of Perovskite Solar Cells 43
Suneth C. Watthage, Zhaoning Song, Adam B. Phillips and Michael J. Heben

3.1 Introduction 43
3.2 Evolution of the Device Architecture 45

3.3 Evolution of the Film Deposition Methods 60
3.4 Advanced Device Engineering Techniques 67
3.5 Outlook and Summary 78
Acknowledgments 79
References 79

4. Perovskite Solar Cell Architectures **89**

Vishesh Manjunath, Ramya Krishna, Suresh Maniarasu, Easwaramoorthi Ramasamy, Sakthivel Shanmugasundaram and Ganapathy Veerappan

4.1 Introduction 89
4.2 Mesoporous Architecture 91
4.3 Planar Architecture 99
4.4 Mesosuperstructured Architecture 109
4.5 Miscellaneous Architectures 112
4.6 Summary 115
References 118

5. Hybrid Inorganic Organic Perovskites: A Low-Cost-Efficient Optoelectronic Material **123**

Diptonil Banerjee and Kalyan Kumar Chattopadhyay

5.1 Basic Crystal Structure 123
5.2 Symmetry and Phase 127
5.3 Basic Synthesis 134
5.4 Optoelectronic Properties 141
5.5 Applications 147
5.6 Conclusion 157
Acknowledgments 157
References 158

6. Ion Migration in Hybrid Perovskites: Evolving Understanding of a Dynamic Phenomenon **163**

Christie L.C. Ellis, Emily Smith, Hamza Javaid, Gabrielle Berns and Dhandapani Venkataraman

6.1 Introduction 163
6.2 Background on HOIPs 164
6.3 Long Timescale Phenomena in HOIPs 166
6.4 Qualitative Evidence of Ion Transport 168
6.5 Computational Studies of Ion Transport 171

6.6 Direct Techniques for Studying Ion Diffusion 174
6.7 Tracer Experiments 174
6.8 Secondary Ion Mass Spectrometry 176
6.9 Nuclear Magnetic Resonance 177
6.10 Complicating Factors in Measurement of Ion Diffusion in HOIPs 179
6.11 Impedance Spectroscopy: A Promising Indirect Method to Probe Ion Transport 180
6.12 EIS Equivalent Circuit Models 181
6.13 Measured Diffusion Constants in HOIPs 186
6.14 Unanswered Questions on the Nature of Ion Transport in HOIPs 188
References 190

7. Low-Dimensional Perovskites 197

P.C. Reshmi Varma

7.1 Introduction 197
7.2 Low-Dimensional Material Synthesis 200
7.3 Applications 211
7.4 Conclusion 225
References 226

8. Fabrication and Life Time of Perovskite Solar Cells 231

P. Mathan Kumar, Abinash Das, Lalu Seban and Ranjith G. Nair

8.1 Introduction 231
8.2 Fabrication of Perovskite Solar Cells 234
8.3 Performance of Perovskite Solar Cells 263
8.4 Conclusion 277
Acknowledgments 278
References 278

9. Hole Conductor—Free Perovskite Solar Cells 289

Suresh Maniarasu, Vishesh Manjunath, Easwaramoorthi Ramasamy and Ganapathy Veerappan

9.1 Introduction 289
9.2 Summary 318
Acknowledgments 318
References 318
Further Reading 321

10. Heterojunction Perovskite Solar Cells **323**
Vidya Raj
10.1 Fabrication Techniques of Perovskite Solar Cells 326
10.2 Solar Cell Device Architectures 332
10.3 Types of Perovskite Solar Cells 334
10.4 Summary 337
References 337

11. Flexible Perovskite Solar Cells **341**
Suresh Maniarasu, Vishesh Manjunath, Ganapathy Veerappan and Easwaramoorthi Ramasamy
11.1 Introduction 341
11.2 Device Architecture 343
11.3 Manufacturing of Flexible PSCs 358
11.4 Stability of Flexible PSCs 360
11.5 Modern Applications of Flexible PSCs 365
11.6 Summary 367
References 369

12. Fabrication of Semitransparent Perovskite Solar Cells **373**
Jayeeta Chattopadhyay and Tara Sankar Pathak
12.1 Introduction 373
12.2 CNT and Graphene-Based Perovskite Solar Cells 376
12.3 Silver Nanowires 380
12.4 Conclusions 383
References 383

13. Commercialization of Large-Scale Perovskite Solar Energy Technology and Scaling-Up Issues **387**
Madhulika Bhati and Radhika Rai
13.1 Introduction 387
13.2 PSC Architecture 392
13.3 Landscape of PSC: Bibliometric Analysis 397
13.4 PV Technology Status 405
13.5 Technology Focus of Top Countries 406
13.6 Short-Term Industrialization Prospects 412
13.7 Japan 414
13.8 South Korea 414

13.9 China 414
13.10 Stability 429
13.11 Hysteresis 432
13.12 Cost Competitiveness 433
13.13 Commercialization Issues of PSC 434
13.14 Efficiency and Variability 436
13.15 Discussion and Conclusion 441
References 442

14. Perovskite Photovoltaics: Life Cycle Assessment 447

Minu Mohan

14.1 Introduction 447
14.2 Perovskite 448
14.3 Life Cycle Assessment 454
14.4 Direct Perovskite Solar Cell with TiO_2 as Electron Transporting Layer 457
14.5 Direct Perovskite Solar Cell with ZnO as Electron Transporting Layer 468
14.6 Inverted Perovskite Solar Cell with PEDOT:PSS as Hole Transporting Layer 469
14.7 Tin-Based Perovskite Solar Cells 471
14.8 Cost-Effective Perovskite Solar Cell 474
14.9 Conclusions 476
References 477

Index *481*

LIST OF CONTRIBUTORS

Javeed Akhtar
Department of Chemistry, Materials Lab, Mirpur University of Science & Technology (MUST), Mirpur Azad Jammu & Kashmir, Pakistan

Muhammad Aamir
Allama Iqbal Open University, Islamabad, Pakistan

Diptonil Banerjee
Indian Institute of Engineering Science and Technology, Howrah, West Bengal, India

Gabrielle Berns
University of Massachusetts Amherst, Amherst, MA, United States

Madhulika Bhati
National Institute of Science, Technology and Development Studies [CSIR-NISTADS], New Delhi, India

Kalyan Kumar Chattopadhyay
Jadavpur University, Kolkata, West Bengal, India

Jayeeta Chattopadhyay
Amity University Jharkhand, Ranchi, Jharkhand, India

Abinash Das
Solar Energy Materials Research and Testing Laboratory (SMaRT Lab), National Institute of Technology Silchar, Cachar, Assam, India

Christie L.C. Ellis
University of Massachusetts Amherst, Amherst, MA, United States

Michael J. Heben
University of Toledo, Toledo, OH, United States

Hamza Javaid
University of Massachusetts Amherst, Amherst, MA, United States

Ramya Krishna
International Advanced Research Centre for Powder, Metallurgy and New Materials (ARCI), Hyderabad, Telangana, India

Suresh Maniarasu
International Advanced Research Centre for Powder, Metallurgy and New Materials (ARCI), Hyderabad, Telangana, India

Vishesh Manjunath
International Advanced Research Centre for Powder, Metallurgy and New Materials (ARCI), Hyderabad, Telangana, India

Balamurugan Manoharan
VIT University, Vellore, Tamil Nadu, India

P. Mathan Kumar
Solar Energy Materials Research and Testing Laboratory (SMaRT Lab), National Institute of Technology Silchar, Cachar, Assam, India

Minu Mohan
Indian Institute of Science Education and Research-Thiruvananthapuram (IISER-TVM), Thiruvananthapuram, Kerala, India

Ranjith G. Nair
Solar Energy Materials Research and Testing Laboratory (SMaRT Lab), National Institute of Technology Silchar, Cachar, Assam, India

Tara Sankar Pathak
Surendra Institute of Engineering and Management, Siliguri, West Bengal, India

Adam B. Phillips
University of Toledo, Toledo, OH, United States

Radhika Rai
National Institute of Science, Technology and Development Studies [CSIR-NISTADS], New Delhi, India

Vidya Raj
IIT Bombay, Powai, Mumbai, India

Easwaramoorthi Ramasamy
International Advanced Research Centre for Powder, Metallurgy and New Materials (ARCI), Hyderabad, Telangana, India

P.C. Reshmi Varma
Indian Institute of Science Education and Research Thiruvananthapuram, Thiruvananthapuram, Kerala, India

Sarat Kumar Sahoo
VIT University, Vellore, Tamil Nadu, India

Lalu Seban
Solar Energy Materials Research and Testing Laboratory (SMaRT Lab), National Institute of Technology Silchar, Cachar, Assam, India

Sakthivel Shanmugasundaram
International Advanced Research Centre for Powder, Metallurgy and New Materials (ARCI), Hyderabad, Telangana, India

Muhammad Sher
Allama Iqbal Open University, Islamabad, Pakistan

Narendiran Sivakumar
VIT University, Vellore, Tamil Nadu, India

Emily Smith
University of Massachusetts Amherst, Amherst, MA, United States

Zhaoning Song
University of Toledo, Toledo, OH, United States

Ganapathy Veerappan
International Advanced Research Centre for Powder, Metallurgy and New Materials (ARCI), Hyderabad, Telangana, India

Dhandapani Venkataraman
University of Massachusetts Amherst, Amherst, MA, United States

Suneth C. Watthage
University of Toledo, Toledo, OH, United States

PREFACE

As the world is becoming more progressive in technology and economy, more energy is being consumed to keep up with the development and the demand on energy boomed over the past years. Solar has become one of the fastest growing renewable energy sources. It provides an excellent solution to the issue of our diminishing finite resources. Solar also provides energy "security" because it is harvested from our most abundant resources, the sun. The diminution of petroleum resources in this century and raising consciousness of environmental change caused by the burning of fossil fuels make nations and public to reconsider the importance of exploring renewable energy sources.

According to WEC (World Energy Council), world critical uncertainties in 2017 include energy and commodity prices, climate framework, and electric storage. Solar energy is the most abundant permanent energy resource on earth and it is available for use in its direct and indirect forms. The supply of the energy from the Sun to the Earth is 3×10^{24} joules a year. Even if only 0.1% of this energy can be converted at an efficiency of only 10%, it would be four times larger than the total world's electricity generating capacity of about 5000 GW. For this reason solar energy will be a viable option for energy as long as the sun exists. Solar energy represents the highest growth rate during this time period and continues to grow as technology increases and photovoltaic (PV) cells become more efficient at capturing solar energy. So experts who work with solar technology will continue to expand and become a vital resource to society.

We discuss the importance of perovskite PV cells, which exhibit high fractions of radiative recombination, with apparent long-carrier lifetimes. They have better radiative efficiencies than many other types such as dye-sensitized and organic solar cells. The name "perovskite solar cell" is derived from the ABX3 crystal (perovskite) structure of the metal-organic halides. These cells are commonly known as hybrid organic–inorganic perovskite solar cells. A number of approaches have been made for developing highly efficient perovskite solar cells. PV devices employing hybrid organic–inorganic lead halides currently display efficiencies in excess of 20%. But still there exists some questions regarding the potential environmental impact of lead content in the material and scaling up of the material to a useful commercial size. Due to the sensitivity to moisture their long-term stability is questionable and may be a big challenge for outdoor applications. But there is still a lot of work to be done to make a competitive PV device.

As we know the research on solar technology is growing very swiftly. So it gets impossible to cover all the aspects of the field. We propose to bring out a book that deals with the emergence of perovskite PVs, which is one of the biggest challenges and opportunities of the present century. To achieve broader development in this cornerstone technology, new PV devices need to be generated with the use of low cost, nontoxic, earth abundant materials using environmentally friendly and scalable processes. Scientists show a great interest in this area by developing organic-lead halide and flexible perovskite solar cells. Many excellent books are already published that highlights different PVs such as organic PVs, polymer PVs, etc. However, to our knowledge, there is no single book that consolidates the information on perovskite including recent progress solar cells. Moreover, in the past few years, tremendous advances have been witnessed in the experimental and theoretical studies on PVs. The purpose of this book is to serve as an academic level text that brings out perovskite from basic to advances and subsequently focus on the Materials Science and Research for sustaining energy.

This book is mainly related to R&D researchers, PhD students, or other technical persons in renewable energy in particular on solar PVs in advanced levels and scientific knowledge. This book covers new information on the use of the perovskite in the fabrication of solar PVs. This is very advantageous to the users. It will be very helpful as a foundation especially for R&D researchers, PhD students, or other technical persons who works on solar PVs in advanced levels and scientific knowledge. We hope that more advanced researchers also find it useful when consulting it for different aspects in this field of research and hopefully as a source of stimulation. It is advantageous to students from chemistry, physics, and material science backgrounds who wish to expand their knowledge on the subject concerned; as well as postdocs with previous experience in related fields. This book is also helpful to solar cell researchers who focused in a specific area to complete their areas of knowledge.

The goal of this book is to cover perovskite from basic to advances. This book will be a very valuable reference source for graduates and postgraduates, engineers, and research scholars. Leading researchers in the field of solar cells from industries, academics, government, and private research institutions across the globe will contribute to this book.

I am grateful to Dr. Sabu Thomas, Director of International and Interuniversity Centre for Nanoscience and Nanotechnology and Professor of School of Chemical Science, Mahatma Gandhi University, Kottayam for initiating and editing this book. This book would not have been possible without the support of my family, they deserve special thanks.

Aparna Thankappan

CHAPTER 1

Introduction: Why Perovskite and Perovskite Solar Cells?

Sarat Kumar Sahoo, Balamurugan Manoharan and Narendiran Sivakumar
VIT University, Vellore, Tamil Nadu, India

1.1 INTRODUCTION

In recent years there is a great attention in the development of renewable energy sources due to the depletion of fossil fuels and burning of those fuels causes the environment pollution. Renewable energy sources are pollution-free energies and they contribute to reducing the greenhouse effects. Solar, Wind, Tidal, biomass are the different types of renewable energy sources utilized to generate electricity. Out of these four sources, solar energy is one of the most prominent alternative sources of energy [1]. The earth obtains 2.9×10^{15} kW of energy every day in the form of electromagnetic radiation from the sun, which is about 100 times the total energy consumption of the world in a year. In 1954 silicon solar cell was invented with 4% efficiency and later it was increased up to 11%. In 1977 the cost of solar power was $77/watt and later it was reduced up to 80 cents/watt. In future the demand for the solar energy will be enhanced rapidly throughout the world. The global generation of electricity in 2005 was 17,450 TWh out of which hydro plant contributed 16%, coal plant generated 40%, gas plant generated 20%, nuclear plant generated 16%, oil plant generated 7%, and only 2% contribution from renewable energy sources (hydro, solar, wind, geothermal, etc.). The resources of renewable energy are certainly available to mankind throughout the world [2]. The renewable energy sources are available in a wide range and also plentiful in nature. At present 13.5% of the world energy demand has been met by the renewable energy sector. The sector of renewable energy especially solar energy is now increasing quicker than the development of other energy in the overall market of energy. Solar energy sources have been switched for electricity generation that offers beneficiary

Perovskite Photovoltaics
DOI: https://doi.org/10.1016/B978-0-12-812915-9.00001-0

management policies from the financial as well as the establishment of the regulatory commissions in environment theme of view. Based on the Global Status Report of Renewables 2015, global installed capacity of solar photovoltaic (PV) is 3.7 GW in 2004 and it is increased up to 177 GW in 2014. Due to the increase in the capacity of PV the dependence of grid power is reduced and therefore the solar energy is evolved as an alternative to conventional energy. PV systems are mostly rated for domestic, industrial, and commercial applications from few kilowatts to several megawatts in rooftop mounted or building-incorporated systems. Nowadays, most of PV systems are combined with grid because for stand-alone systems, batteries are essential to store the energy, which will increase the cost of the entire system [3]. The government has announced subsidies to those who are generating power on their own and also injected the generated excess power into the grid, therefore people now shifted their focus toward the PV system.

1.2 PRESENT PHOTOVOLTAIC TECHNOLOGIES

PV technology is used to convert light energy (photons) into electrical energy. The advantages of PV system are no pollution, no greenhouse gas emission, and it is available in nature abundantly. PV technologies are broadly classified into three categories, which are wafer-based technology, silicon thin-film hybrid technology, and thin-film technology [4]. The complete list of classification of PV technology is shown in Fig. 1.1.

1.2.1 Wafer-Based Technology

The most preferred technology in PV market is the wafer-based technology and more than 85% of the world PV market uses this type of cells. Crystalline silicon is one of the forms of wafer-based technology. Crystalline silicon is also called as the first-generation PV technology where the wafer-based crystalline silicon is used as the active material.

Crystalline silicon is further classified into monocrystalline and polycrystalline silicon. The theoretical efficiency of single-junction solar cells are of 27% and around 25% of practical efficiency has been achieved.

1.2.1.1 Monocrystalline Silicon Solar Cell

The crystal structure of monocrystalline silicon is homogenous, which means the lattice parameter, electronic properties, and the orientation remains constant throughout the process. To improve the power

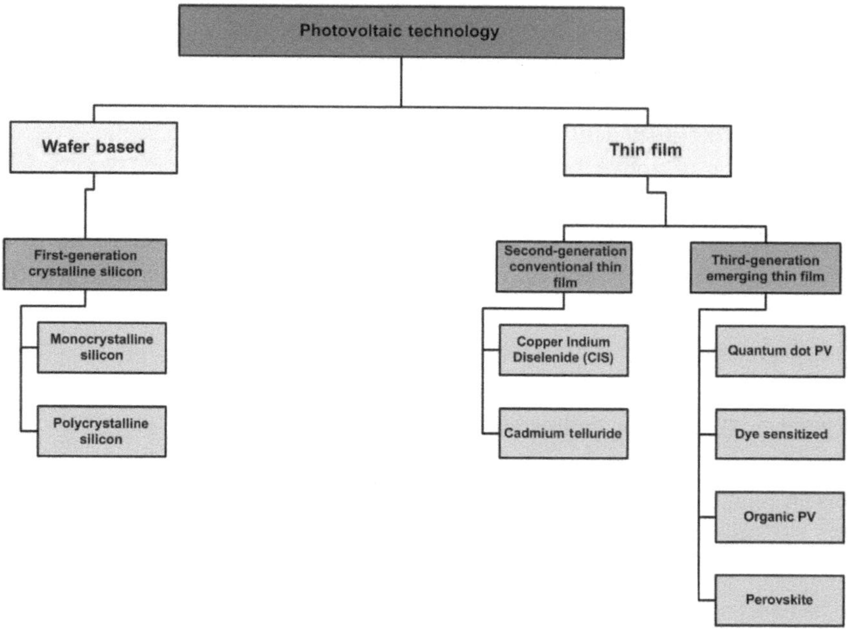

Figure 1.1 Classification of photovoltaic technology.

conversion efficiency crystal structure solar cell has been used in this technology. Monocrystalline silicon requires more expensive wafers compared to other technologies and also able to produce more energy. This type of silicon is used in rooftop and space applications.

1.2.1.2 Polycrystalline Silicon Solar Cell

Polycrystalline silicon is composed of a number of small crystals of low-grade silicon, which results in low cost and efficiency when compared to monocrystalline silicon. Polycrystalline silicon is the key technology for the manufacture of conventional silicon-based solar cells. More than half of the global supply of PV technology has been done through polycrystalline cell in 2006.

1.2.2 Thin-Film Technology

Thin-film solar cell comes under the category of the second-generation solar cell, which can be formed by combining one or more layers of plastic, glass, or metal. It is the most used technology for commercial applications throughout the world [5]. Thin-film solar cells are always very cheap compared to silicon solar cells and the efficiency is also very less but significant research has been carried out over the years to improve the performance of the thin-film solar cells.

1.2.2.1 Convention Thin-Film Technology

The amount of active material used in thin-film technology is very less compared to silicon. The light absorption capacity of the thin film is 100 times efficient than silicon that allows the material to absorb very few microns of thickness [6]. Copper Indium Gallium Diselenide (CIGS), Cadmium Telluride (CdTe), and amorphous silicon (a-Si) are some of the materials used in the commercial thin-film technologies.

1.2.2.1.1 Cadmium Telluride Solar Cell

The theoretical efficiency of CdTe is very high among the semiconductors available today for single-junction configuration. CdTe PV technology has installed worldwide capacity of 15 GW because of its thermal and chemical stability results in the enormous increase of efficiency compared to other technologies. It is ideally suited for utility-scale applications to reduce the cost of electricity per watt [7].

1.2.2.1.2 Cadmium Indium Diselenide Solar Cell

Similar to CdTe, Cadmium Indium Diselenide (CIS) also has very high absorption coefficient and high efficiency because of its wide spectrum and the cost of CIS PV technology is very low compared to other technologies. The power conversion efficiency of CIS has already crossed over 20%. CIS PV technology is ideally suited for building integrated PV applications because of their flexible property and lightweight.

1.2.2.2 Emerging Thin-Film Generation

Emerging thin-film generation comes under the category of third-generation solar cells most of the materials in this category have not yet been commercially applied and most of the materials are still in research or under development phase. In emerging thin-film technologies the production process of solar cell has been carried out through different steps namely active material synthesis and deposition [5]. But in conventional thin-film technology all the process has been done in a single step. Therefore emerging thin-film solar cells used in flexible applications and lightweight materials. This technology includes Quantum Dot, Dye Sensitized, Organic PV, and perovskite.

1.2.2.2.1 Quantum Dot Solar Cell

For the design of quantum dot solar cell this technology uses quantum dot for the absorption of PV material. Therefore the materials like silicon, CdTe, CIS can be replaced by quantum dot and it has a high bandgap that is used to tune different range of energy levels [8]. Even though the efficiency of the quantum dot as low as 9% it has attracted the researchers because of its versatile property, low cost, and lightweight.

1.2.2.2.2 Dye-Sensitized Solar Cell

Dye-sensitized solar cell (DSSC) is produced based on the semiconductor formed between the electrolyte, photosensitized anode and a photoelectrochemical system. Even though the theoretical power conversion efficiency of DSSC is very less due to its performance ratio and price it can be used as a viable option for an electric generator for grid integration [6]. The stability of DSSC under illumination is limited and the absorption coefficient is low in the infrared region are the key challenges in this technology.

1.2.2.2.3 Organic Solar Cell

Organic solar cell uses organic polymers to absorb light and transport the charge into electricity from sunlight through PV effect. The production cost of an organic solar cell is low because the molecules that are used in this process are very cheap. The main drawbacks of the organic solar cells are low stability, low efficiency, and low strength when compared to silicon cells.

1.2.2.2.4 Perovskite Solar Cell

Perovskite solar cell (PSC) includes the perovskite-structured material as an active layer based on the solution processed by tin or halide. It is a very attractive option for commercial applications because this type of cells is very cheap during the scale-up process [9].

The merits and demerits of different solar cell technologies are listed in Table 1.1.

Perovskite solar cell has emerged as one of the most standout cell in terms of efficiency. Therefore the researchers have shown tremendous interest in Perovskite solar cell. Flexibility, lightweight, and semitransparency are some of the valuable proportion of perovskite [10]. Therefore in this chapter PSC characteristic, application, and challenges have been briefly discussed.

Table 1.1 Comparison of solar cells

Solar cell type	Merits	Demerits
Wafer-based solar cell	• Tested technology • Easy fabrication process • Reliable • Inexpensive • Plentiful and nontoxic • High-conversion efficiency	• Absorption coefficient is low • Demand often exceeds supply • Efficiency decreases when temperature increases
Thin film–based solar cell	• Easy fabrication for high volume • More efficient when temperature is very high • Flexible and lightweight	• The power conversion efficiency is low • Toxic materials are used • Higher cost • Stability is very low • Lifetime is short

1.3 PEROVSKITE

Perovskite is a mineral composed of calcium titanium oxide ($CaTiO_2$) discovered by a Russian scientist, Gustav Rose, in 1839 and the research was further carried out by Russian mineralogist Lev Perovski therefore this mineral was named as Perovskite. The crystal of the mineral $CaTiO_3$ (XIIA2 + VIB4 + X2-3) is called as perovskite structure [11].

To absorb light the perovskite structure uses compounds and stoichiometry. Researchers remain uncertain why positive and negative charges produced by photoexcitation in these types of cells reach their particular electrodes so well. Yet the significant rate of development caused PV scientists in industry and academia who operated other cells to switch to perovskites or at minimum enhance the issue to their first list.

Material scientists have shown enormous interest in perovskite because it is abundantly available in nature. The perovskite mineral structure is shown in Fig. 1.2. In addition to that, it also leads the research in the exploration of superconductivity, magneto resistance, ionic conductivity, and gathering of dielectric properties, which are of significant importance in telecommunication and microelectronics.

Perovskite mineral has the capacity to absorb light and it utilizes less than 1 μm of material to seizure the similar quantity of sunlight compared to other solar cells. Perovskite is a semiconductor, which is used to transport the electric charge whenever the light hits the material. In the

Figure 1.2 Perovskite mineral structure.

United Kingdom, Oxford University physicists have found that the perovskite has been used as the replacement for thin-film solar cells [12].

A cell with the perovskite crystal structure usually consists of an organic group tin or lead and a halogen is called PSC [7].

The working principle of PSC is similar to that of thin-film solar cell. Methylammonium lead halide is the most prominently used and is one of the most commonly used perovskite cells. The first structure of PSC is based on the dye-sensitized solar cell where the TiO_2 layer is placed and the organic materials are deposited over the layer. In the PSC the perovskite will act only as the light absorbent and later it will also transport the charge as carrier for the mobility of electrons.

1.4 HISTORY OF PEROVSKITE SOLAR CELL

The first instance of perovskite mineral has been incorporated as a solar cell in 2009 by Miyasaka based on the architecture of DSSC with a thin layer of perovskite cell on titanium oxide (TiO_2). The power conversion efficiency is only 3.8% and the stability of the cell is very low because of the presence of a liquid electrolyte.

By using the similar dye-sensitized concept Park has been improved the power conversion efficiency from 3.8% to 6.5% in 2011 by using the quantum dots for the nanocrystalline material.

Henry Snaith and Mike Lee from Oxford University found out that to transport the electron instead of TiO_2 layer solid state 2, 2′, 7, 7′-Tetrakis[N, N-di(4-methoxyphenyl)amino]-9,9′-spirobifluorene

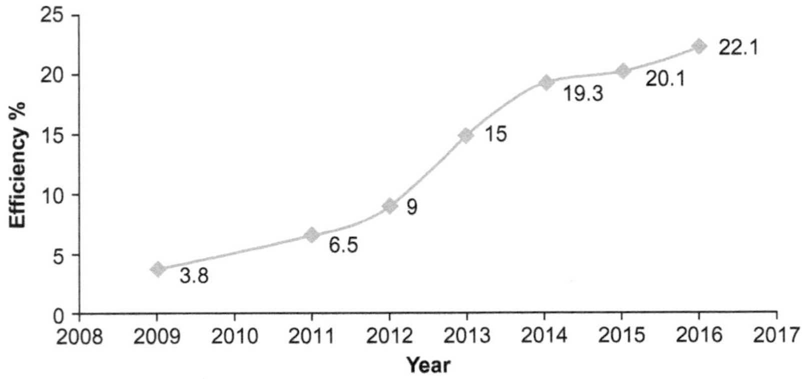

Figure 1.3 Efficiency chart of PSC.

(spiro-OMeTAD) is used to increase the stability and efficiency of perovskite from 6.5% to above 10% in 2012.

A number of developments have been seen on both the sensitized and planar architectures in 2013. By using a two-step solution processing the efficiency of the sensitized architecture exceeding 15% based on the deposition technique demonstrated by Burschka. To achieve more than 15% efficiency in planar architecture was possible by fabricating the planar solar cell by thermal evaporation and this concept was explained by Liu. Docompo has also shown the possibility of fabricating the PSC in organic cell structure with the whole concentration at the bottom layer and the electron configuration at the top layer of the thin film.

A researcher named Yang Yang from the University of California, Los Angeles has achieved 19.1% efficiency by suppressing the recombination of carriers and inject into the transport layer using planar architecture. A researcher from the Korea University of Science and Technology has reported 20.1% efficiency in November 2014. In December 2015 researchers from the Swiss Federal Institute of Technology in Lausanne have reported 21% efficiency of PSC [13].

The efficiency of PSC has reached 22.1% as of March 2016 has been reported by the researchers from Korea University of Science and Technology and Uslan National Institute of Science and Technology. The efficiency chart of PSC has been shown in Fig. 1.3.

1.5 CLASSIFICATION OF PEROVSKITE SYSTEM

The perovskite system has been classified into two broad categories, which are inorganic oxide perovskite system and halide perovskite system (Fig 1.4).

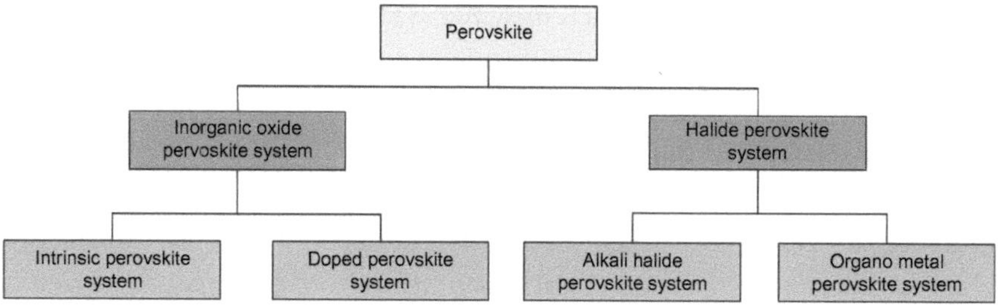

Figure 1.4 Classification of perovskite.

1.5.1 Inorganic Perovskite System

PSC can be produced by methylammonium oxide which has the high open-circuit voltage and the thickness range of the material can be well absorbed with the carrier-based diffusion length.

The inorganic perovskite system can be further classified into intrinsic and doped perovskite system.

1.5.1.1 Intrinsic Perovskite System

The power conversion efficiency of the intrinsic perovskite system is high when compared to other systems. Because of the high-conversion efficiency the stability of the material is very high and long lifetime of more than 72 hours.

1.5.1.2 Doped Perovskite System

The properties of PSC will depend upon the crystal structure of perovskite compound. Therefore doping the perovskite by using lead or halide will increase the power conversion efficiency, stability, and also increase the diffusion length of ions [14]. Therefore the wavelength absorption limit of the compound was improved.

1.5.2 Halide Perovskite System

Halide perovskite material is one of the key factors in the area of heterostructure-based optoelectronics and microelectronics. This type of material will provide a very high performance with low cost. Halide perovskite system can be further divided into alkali halide perovskite and organometal perovskite system.

1.5.2.1 Alkali Halide Perovskite System

This type of perovskite has low energy loss, which will lead to increase in power conversion efficiency and high open circuit for improved response

[15]. Alkali halide perovskite cell provides wider bandgap, which will provide cost-effective solution for the modules that have high efficiency.

1.5.2.2 Organometal Perovskite System

Organometal PSC is one of the innovative technology developed in recent years because of its light absorption coefficient and direct bandgap, which will increase the carrier mobility and power conversion efficiency of the system [16]. The detailed explanation of organometal perovskite system will be given in Chapter 2, Organometal Lead Halide Perovskite.

1.6 CHARACTERISTIC FEATURES OF PEROVSKITE SOLAR CELL

In PSC the oxides introduced inside perovskite structure demonstrate trademark highlight of their physical conduct that knows how to be pertinent for the comprehension of the mantle material. Misshape or deformations occur in perovskite structure and the four unique mechanisms are enclosed as follows:

- The tilt of octahedral complexes.
- Off-centering of the octahedrally coordinated cations.
- Distortion of the octahedral cage.
- Off-centering of the 12-fold coordinated site.

These misshape usually lead to structural inabilities related to phase transition, extremely sluggish kinetic behavior even leading to pseudo-glasses and polaronic transport. Metal halide perovskite sun powered cells have made quick increments in energy conversion efficiency, enhancing from 4% low efficiency in 2010 to an efficiency of 22% in 2016. As a result of the high-retention coefficient a thickness of just around 500 nm is expected to retain solar energy. In July 2015 noteworthy obstacles were that the biggest PSC was just the extent of a fingernail and that they debased rapidly in moist situations.

Methylammonium lead halide perovskites ($CH_3NH_3PbX_3$, X = I, Cl, Br) have pulled in a large amount of consideration for PV devices attributable to having a high light absorption coefficient and high carrier mobility. Also the perovskite components can be blended from reasonable sources, demonstrating high performance and cheap solar cells are attainable. For the most part, perovskite-based cells incorporate mesoporous

Table 1.2 Characteristics of solar cell

Characteristics	CdTe	CIGS	c-Si	Perovskite
Raw material cost	Low	Medium	Low	Low
Finished material cost	Low	High	High	Low
Fabrication cost	Medium	Medium	High	Low
Energy payback period	Medium	High	High	Low
LCOE	Medium	High	High	Low
Efficiency	Medium	Medium	High	High

LCOE, levelized cost of electricity.

Table 1.3 Bandgap, open-circuit voltage, and energy loss of solar cell

Solar cell material	Bandgap (eV)	Open-circuit voltage (eV)	Energy loss (eV)
Silicon	1.12	0.75	0.37
Gallium arsenide	1.43	1.12	0.31
Copper Indium Gallium Selenide (CIGS)	1.15	0.74	0.41
Cadmium telluride	1.49	0.90	0.59
Amorphous silicon	1.55	0.89	0.66
Perovskite	1.55	1.07	0.48

metal oxides as a framework of perovskite layer. The manufacturing procedure of perovskite-based cells including mesoporous metal oxides require high-temperature handle around 500°C and it is hard to accomplish flexible solar cells.

Several researchers researched planer sort perovskite-based cells that barred mesoporous metal oxides.

So the fabrication of perovskite-oriented solar cells under a low-temperature process got to be distinctly conceivable, empowering us to manufacture flexible cells [17]. The qualities of a solar cell with its characteristics are displayed in Table 1.2 [18].

Disregarding the way that the characteristics of PSCs under low illuminance condition have not been examined particularly, it is ordinary that open-circuit voltage will not diminish seriously even under a low illuminance from a device structure of perovskite-based cells. The bandgap, energy loss, and open-circuit voltage of solar cells are listed in Table 1.3. From the table one can clearly see that the PSC has a very high bandgap, and high open-circuit voltage, and low energy loss.

1.7 ADVANTAGES OF PEROVSKITE SOLAR CELL

Solution processing is one of the key advantage of PSC [19] due to their good solubility therefore certain low-cost processing techniques such as spray coating, spin coating, screen coating, and thermal evaporation are discussed in this section.

1.7.1 Spray Coating

Spray pyrolysis is used for the fabrication of PSC where the material is deposited on the substrate by spraying the solution [20]. The coating layer is designed in such a way that all the unstable material is removed at the working temperature. This type of technology is very effective because all the properties of the material are dependent on temperature [21]. During high temperature the power conversion efficiency of the material can be increased. The merits of the spray coating process are low cost and all the layers are of uniform thickness and highly diluted solutions are used. The parameters are very difficult to control during the process is the drawback of this method [22].

1.7.2 Spin Coating

One of the most suitable, well-known fast processing techniques for fabrication of PSC is spin coating, which is widely used in the area of organic electronics [23]. The process of spin coating involves the deposition of solution on the substrate by rapid rotation and drying is very consequent, which will result in the formation PSC [24]. The layer thickness is controlled by the spinning velocity, temperature, and the solution viscosity. The merits of spin coating method are process is very fast, low cost, and deposition of multilayered materials. Coating material loss, compatibility, and sensitive to environmental conditions are some of the drawbacks.

1.7.3 Dip Coating

Dip coating is one of the most effective processes for the production of PSC. The solution substrate is immersed in the solution for effective formation of the material [25]. Once the material is deposited then the substrate can be removed by evaporation which will result the thickness of the layer are unique [26]. The major forces that are used for the dip coating process are force of inertia, viscous drag, gravitational force, and surface tension [27]. The merits of dip coating are low cost and layer thickness can be easily adjusted. The drawbacks of dip coating are process is slow and it has the ability to block the screen, which will create major impact in the final product.

1.7.4 Thermal Evaporation

Even though this method is not included in the solution processing technique it will play a major role in the fabrication of PSC. The material is directly deposited in the vapor phase for the production of perovskite. There are two most used techniques of thermal evaporation for the production of PSCs, single-layer deposition technique and multilayer deposition technique. Single-layer deposition involves the process of the inorganic perovskite deposition material is evaporated which is placed on the sheet of metal. Applying large current on the metal sheet results in the increase in temperature for the effective production and there is no need for postproduction process of material because of the evaporation process [28]. Multilayer deposition technique involves the evaporation of both the inorganic and organic salt and annealing process is required for postproduction process [29]. The merits of thermal evaporation process are fixed pattern and uniform layer of thickness, and monitoring of the whole process is simple. The demerits of this method are the production process is slow and the material arrangement is complex and multicomponent materials are very difficult to synthesis (Table 1.4).

The major advantages of the PSC can be summarized as follows:

- Perovskite-based cells have the capacity to expand the productivity and reduce the cost of solar energy.
- Perovskite PV in reality cling to guarantee for high efficiencies, and in addition low-potential material and decreased handling costs.
- A major preferred standpoint of perovskite PV have over traditional sun-based innovation is that they can respond to different wavelengths of light, which permits them to change over a greater amount of the daylight that converts them into electricity.
- Perovskite PV basically began with the leap forward research in 2012, and in the space of a couple of years, they have figured out how to accomplish power conversion efficiencies equivalent to technologies and materials that have been used.
- The remarkable increase in efficiency is considerable and extraordinary and seems to be continuing in laboratories around the world.
- Another key advantage is the energy utilization, or the amount of a photon's energy lost in the conversion process from light to electricity.

Table 1.4 Merits and demerits of different processing techniques

Techniques	Merits	Demerits
Spray coating	• Highly diluted solutions are used • Low cost • Layer thickness are unique	• Process parameters are very difficult to control
Spin coating	• Low cost • Processing is very fast • Film thickness can be easily adjusted • Multilayer deposition	• Layer thickness are nonuniform • No fixed pattern • Coating material loss
Dip coating	• Low cost • Layer thickness can be easily adjusted	• Process is slow • Blocking of screen
Thermal evaporation	• Fixed pattern • Thickness layers are unique • Monitoring process becomes easy	• Process is slow • Complex arrangement • Multicomponent material are very difficult to synthesis

For standard excitonic-based, organic-based solar cells photon's energy loss can be as high as 50% of the absorbed energy. Though perovskite-based solar cells demonstrate a loss that is far lower.

- Perovskite-based PVs are quickly moving toward a similar stage of photon energy utilization as the present leading monolithic crystalline technologies like GaAs and silicon.
- They also have the prospective for much reduced processing charge.
- Perovskite-based powered cells offer adaptability, semitransparency, tailored form factors, lightweight, etc. Actually electronics designers and researchers are sure that such attributes will expose numerous new applications for solar-based cells.

1.8 CHALLENGES OF PEROVSKITE SOLAR CELL

The major challenges of PSC are:

- Degradation
- Stability
- Material toxicity

It needs to be addressed before it is commercialized for a wider purpose.

1.8.1 Degradation

The major challenge in perovskite PV technology is degradation. The effectiveness of the perovskite-oriented cell diminishes because of degradation. Perovskite powered cells still face various difficulties sooner than they can be executed on an extensive level [30]. The components of the PV degrade rapidly in the perovskite cell architecture and charge extraction. Likewise the efficiency of the perovskite-based cell, although moderately at peak, is dubious. To decide the efficiency of a sunlight-based cell, an $I-V$ curve shows the current versus the voltage is calculated, and the maximum power point is found. This power is evaluated with the incoming light to decide the power conversion efficiency. In perovskite cells still the $I-V$ curve change depending on how rapidly the voltage is varied and whether the voltage is scanned from high to low or low to high [31]. Hence the correct performance of a perovskite PV in a long-standing practical situation is still unclear. Fig. 1.5 displays the open-circuit voltage compared to the bandgap for a scope of technologies that the perovskites contend [12]. For standard excitonic-based, organic-based solar cells, this misfortune can be as high as 50% of the consumed energy. Nonetheless, for perovskite-based solar cells, the misfortune is far less.

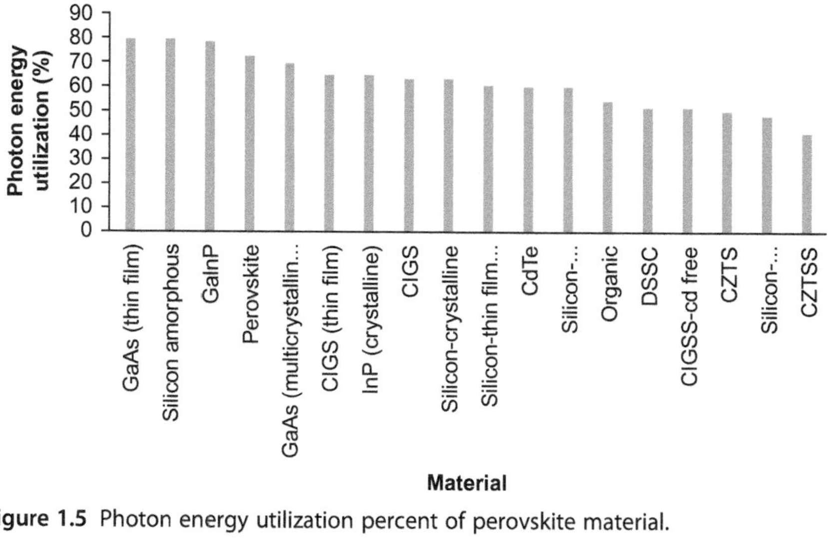

Figure 1.5 Photon energy utilization percent of perovskite material.

1.8.2 Stability

Perovskite-based powered cells are quick moving toward a similar stage of photon energy utilization as the present leading monolithic crystalline technologies, e.g., GaAs and silicon. In addition, they additionally have the possibility for less preparing cost. As of now the main significant obscure in the area of perovskite research is the stability of devices over their operational lifetime. Even though lifetime studies of actual devices are narrow, research into the stability of these films has exposed that there are several reaction pathways leading to degradation that involves water, oxygen, and even the diffusion of electrode materials. Current leading research is determined upon reproducing the high power conversion efficiencies with the adding up of stabilizing agents such as Rubidium and Cesium.

1.8.3 Material Toxicity

The real issue up till now to be completely addressed is the use of lead in perovskite components. Perovskite utilizes dangerous lead, which will prompt to environmental concern and also material degradation. This high-efficiency perovskite causes high moisture in the air. Moisture will bring about a domino impact of decomposition, in view of this issue the PV property of the perovskite material will lose. Although it is used in much lesser quantities than that which is now present in either lead- or cadmium-based batteries, the occurrence of lead in products for commercial use is challenging.

There is potential for a lead alternative to be utilized as a part of PSCs (e.g, tin perovskites), yet the power conversion efficiency of such devices is behind lead-based devices. At long last, there has been little discussion on the optical thickness of these materials, which in spite of the fact that it is superior to silicon, is still lesser than other active materials.

Thus the perovskite devices want thicker light-harvesting layers that bring about some fabrication confinements. These confinements apply especially to solution processed devices where making such thick layers with high uniformity can be complicated.

In the course of recent years the improvements in precursor material mixtures for the fabrication of PSCs have provoked a huge increment in power conversion efficiency. An innovative advancement has been made in the improvement in processing techniques used. In the past, vacuum-based techniques presented the highest efficiency devices; however,

Table 1.5 Hazardous exposures of lead and its consequences

Toxicity of lead dosage	Consequences
100–150 μg/dL exposure for less than 14 days	Kidney and brain damage to adults
15–30 μg/dL exposure for less than 14 days	For middle-aged people blood pressure is increased
10–15 μg/dL exposure for less than 14 days	Mental ability and birth weight are reduced of infants for pregnant women
15–20 μg/dL exposure for more than 14 days	Growth rate is reduced for the children

recently, advancements in a solution-based deposition during the use of solvent quenching techniques have moved the record-breaking devices to solution-based processing. The hazardous exposure of lead and its consequences provided by the United States Environmental Protection Agency in 2012 is shown in Table 1.5. The hazardous exposure of 100–150 μg/dL lead for less than 14 days will damage the kidney and brain to adults. The hazardous exposure of 15–300 μg/dL lead for less than 14 days will increase in the blood pressure for the middle-aged persons. 10–15 μg/dL hazardous exposure of lead for less than 14 days will reduce the mental ability and birth weight of infants of pregnant women. 10–15 μg/dL exposure for less than 14 days will reduce the growth rate of children.

To empower a really low cost-per-watt will require perovskite-oriented cells to have the abundantly proclaimed trio of high efficiency, long lifetimes, and low manufacturing costs.

1.9 PROPERTIES AND APPLICATIONS

The properties and utilizations of essential perovskite phase mixed-metal oxides are discussed in this section. The manufacturing cost of perovskite is very cheap because the solution is directly deposited on the material but for encapsulation process cost may increase. The material properties of perovskites are high absorption coefficient and charge carrier transport is very high and high open-circuit voltage. More studies are required to determine the stability of perovskite cell. The real use of materials utilized as a part of perovskite is based on their intrinsic dielectric, ferroelectric, piezoelectric, and pyroelectric properties of importance in electronic applications, e.g., electromechanical devices, transducers, capacitors, actuators, high-k dielectrics, dynamic RAM, field-effect transistors, and logic circuitry.

All ferroelectric materials have pyroelectric and piezoelectric potential for future use based on these properties. For example, ferroelectric materials can have their polarization reversed (switched) for memory applications at or near room temperature.

1.9.1 Multilayer Ceramic Capacitor

The drive utilized for reducing size in electronic components has built up to the improvement of ceramic capacitors, particularly the multilayer ceramic capacitor (MLCC), utilizing ceramic dielectric materials with the most elevated permittivity values. More difficult multilayer configurations have additionally been devised to satisfy the developing technology requirement for increased capacitance within an ever smaller confinement space.

In recent past, MLCCs having a capacitance of 1–100 μF have been created with interior Ni electrodes, which are made out of at least 500 or more laminated thin dielectric layers of approximately 2 μm. Another region attracting in a good deal of interest has been the utilization of ferroelectric thin film as nonvolatile memories. Current static and dynamic memory chips lose information when the power to the memory is intruded, results in retainable loss of information.

1.9.2 Storage and Switching Devices

Magnetic storage facilities provide a solution to this issue, but the need for fast, lightweight memories with low power consumption integrated into silicon chips is important for some applications. These materials would also be unaffected by radiation and magnetic fields. There are two key designs for a nonvolatile memory, which utilize ferroelectric materials, namely lateral configuration and vertical configuration.

One of the main features of these designs is to read the memory cell, it must be polled, which can invariably erase the information the cell contained. This drawback can be overcome by designing a system, so as to subsequently rewrite the information back into the memory once the cell has been initially read.

Another application currently under investigation is the use of ferroelectric materials as electro-optic switching devices for optical computers. This thought is based on the skill of a ferroelectric material to change its refractive index under an applied field. The response of this material for electro-optic applications can be judged by many factors such as film thickness, the wavelength of the light, and the properties of the electrodes. Another application for thin films would be as read/write optical storage devices. In this case the material, such as lead lanthanum zirconate titanate (PLZT), is switched between its ferroelectric state and its antiferroelectric state upon the application of a light beam, while the film is simultaneously under the control of an applied field.

The signal-to-noise ratio, which is normally poor for thin films, can be considerably improved in these particular thin PLZT film systems ($\sim 0.5\ \mu m$) under conditions that optimize interference effects. This type of optical storage device is somewhat unaffected by the presence of magnetic fields and thus, can be considered as the optical equivalent of magnetic hard disks in computers.

The properties and applications of important perovskite phase mixed-metal oxides are shown in Table 1.6. The dielectric property of $BaTiO_3$ will be used in the application of capacitors and sensors. The pyroelectric property of $(Ba, Sr)TiO_3$ will be used in the application of pyrodetector.

Table 1.6 Properties and applications of perovskite materials

Materials	Properties	Applications
$BaTiO_3$	Dielectric	Capacitor, sensor
$(Ba, Sr)TiO_3$	Pyroelectric	Pyrodetector
$PbTiO_3$	Pyroelectric	Pyrodetector
	Piezoelectric	Acoustic transducer
$Pb(Zr, Ti)O_3$	Dielectric	Nonvolatile memory
	Dielectric	Pyrodetector
	Piezoelectric	Surface acoustic wave device, substrate
	Electro-optic	Waveguide device
$(PB, la)(Zr, Ti)O_3$	Pyroelectric	Pyrodetector
	Electro-optic	Waveguide device, optical memory display
$LinbO_3$	Piezoelectric	Pyrodetector, surface acoustic wave device
$LinbO_3/Ti$	Electro-optic	Waveguide device, optic modulator, second harmonic generation
$K(Ta, Nb)O_3$	Pyroelectric	Pyrodetector
	Electro-optic	Waveguide device, frequency doubler
$Pb(Mg_{1/3}Nb_{2/3})O_3$	Dielectric	Memory, capacitor

The pyroelectric and piezoelectric property of $PbTiO_3$ will be used for pyrodetector and acoustic transducer, respectively. The dielectric and piezoelectric property of Pb(Zr, Ti)O_3 will be used for the application of nonvolatile memory, pyrodetector, and surface acoustic device, respectively. The pyroelectric and electric optic property of (PB, la)(Zr, Ti)O_3 material will be used in pyrodetector and waveguide device applications, respectively. The piezoelectric property of $LinbO_3$ is used for pyrodetector and surface acoustic wave device application. The electro-optic property of $LinbO_3$/Ti will be used for the optic modulator and waveguide applications. The pyroelectric property of K(Ta, Nb)O_3 is used for the application of pyrodetector. The dielectric property of Pb($Mg_{1/3}Nb_{2/3}$) O_3 is used for the memory and capacitor application.

1.10 FUTURE TRENDS OF PEROVSKITE SOLAR CELL

Work on sunlight-based cells utilizing perovskite material has progressed quickly accordingly of the materials fantastic light absorption, charge carrier mobility's, and lifetimes, resulting in high device efficiencies with significant chances to realize a low cost, industry-scalable technology [32]. This potential for low cost and scalability requires overcoming boundaries related with stability and ecological compatibility. Nonetheless, if these worries are tended to, a perovskite-based technology holds transformational potential for quick terawatt-scale sunlight-based deployment. The fundamental materials properties have additionally sparked interest for utilizing the hybrid perovskite semiconductors in a more extensive class of energy applications that traverse traditional electronic and optical systems [33].

In this section the future trends of PSC are discussed based on the commercial and research aspect of PSC.

1.10.1 Perovskite Commercialization

Perovskite-structured material is the active layer in PSC. Most commonly, lead or tin-based halide material is used for hybrid organic or inorganic solution. Stability and efficiency are the two parameters will play a major role in the commercialization of perovskite. Within 7 years of research the efficiency of the material has increased from 3.5% to 22.1%, which making them a quickly propelling technology and a hot debated issue in the solar cell research [34]. Perovskite sunlight-based cells are also predicted to be very cheap during the scale-up process making them an extremely smart choice for commercialization.

PSCs have yet to reach the market, but progress toward commercialization seems promising, and many analysts believe that in 2–3 years perovskite PVs will emerge as a true alternative to silicon-based panels.

1.10.2 Research and Development of Perovskite Solar Cell

A major breakthrough in the stability of PSC has been proposed by Dyesol an Australian PV researcher and PSC designer. Dyesol found the critical breakthrough on small perovskite powered cells with 10% of efficient strip cells showing less than 10% virtual degradation when visible to continuous light dowsing for more than 1000 hours. Dyesol was also granted a $0.5 million in addition to the original grant from the Australian Renewable Energy Agency (ARENA) to popularize a creative, high-efficient perovskite powered cell.

The thin film production site of Bosch Solar in Germany, to establish a fabrication with pilot-scale capacity for perovskite wafers in November 2016, by the United Kingdom-based Oxford PV researchers. Their objective is to make the PSC technology to industry standard wafer size.

1.11 SUMMARY

- At present 13.5% of the world energy demand has been met by renewable energy sector. The rapid growth in the renewable energy technology will lead to the further research in solar cell.
- Solar cells are classified into three categories. Crystalline silicon will come under the first-generation solar cell. Second-generation solar cell includes the conventional thin-film technology and emerging thin-film technology is categorized as third-generation solar cell.
- Crystalline silicon is the most used solar cell throughout the world for commercial applications.
- Most of the solar cells in the emerging thin-film category are still in research and development phase.
- Perovskite is emerged as an alternative to silicon-based solar cell because of its characteristics and low cost.
- Unlike other solar cells perovskite material does not need an electric field to produce the electric current.

- The efficiency of PSC is increased very rapidly from 3.8% in 2008 to 22.1% in 2016 and still the works have been carried out to improve the power conversion efficiency.
- The effective utilization level of photon energy in PSC is very high compared to another solar cell.
- The superiorities of PSC are high absorption coefficient, high stability, high charge carrier mobilities, High diffusion length and very high value of open-circuit voltage.
- The major challenges of PSC are degradation, stability, and material toxicity.
- PSCs have yet to reach the market but progress toward commercialization seems promising, and many analysts believe that in 2–3 years perovskite will emerge as a true alternative to silicon-based panels.

REFERENCES

[1] M. Balamurugan, S.K. Sahoo, S. Sukchai, Application of soft computing methods for grid connected PV system: a technological and status review, Renewable Sustainable Energy Rev. 75 (2016) 1493–1508.
[2] S.K. Sahoo, Renewable and sustainable energy reviews solar photovoltaic energy progress in India: a review, Renewable Sustainable Energy Rev. 59 (2016) 927–939. Available at: http://dx.doi.org/10.1016/j.rser.2016.01.049.
[3] S.K. Sahoo, et al., Energy efficiency in India: Achievements, challenges and legality, Energy Policy 88 (2016) 495–503.
[4] T. Kirchartz, et al., Classification of solar cells according to mechanisms of charge separation and charge collection, Phys. Chem. Chem. Phys. 17 (6) (2015) 4007–4014. Available at: https://doi.org/10.1039/C4CP05174B.
[5] M. Edoff, Thin film solar cells: research in an industrial perspective, Ambio 41 (Suppl. 2) (2012) 112–118.
[6] A.C. Mayer, et al., Polymer-based solar cells, Mater. Today 10 (11) (2007) 28–33. Available at: https://doi.org/10.1016/S1369-7021(07)70276-6.
[7] V.M. Fthenakis, Life cycle impact analysis of cadmium in CdTe PV production, Renewable Sustainable Energy Rev. 8 (4) (2004) 303–334.
[8] J.-H. Im, et al., 6.5% Efficient perovskite quantum-dot-sensitized solar cell, Nanoscale 3 (10) (2011) 4088. Available at: http://xlink.rsc.org/?DOI=c1nr10867k.
[9] M. Petrović, V. Chellappan, S. Ramakrishna, Perovskites: solar cells & engineering applications - materials and device developments, Sol. Energy 122 (2015) 678–699.
[10] Feng, W., 2014. Perovskite solar cells. pp. 1–6. Available at: https://www.nrel.gov/pv/perovskite-solar-cells.html.
[11] S. Patwardhan, et al., Introducing perovskite solar cells to undergraduates, J. Phys. Chem. Lett. 6 (2) (2015) 251–255.
[12] Ossila, 2015. Perovskites and perovskite solar cells: an introduction. Ossila, pp. 1–8. Available at: https://www.ossila.com/pages/perovskites-and-perovskite-solar-cells-an-introduction.
[13] W.E.I. Sha, et al., The efficiency limit of CH3NH3PbI3 perovskite solar cells, Appl. Phys. Lett. 106 (22) (2015) 221104. Available at: http://scitation.aip.org/content/aip/journal/apl/106/22/10.1063/1.4922150.

[14] T. Oku, et al., Effects of Cl addition to Sb-doped perovskite-type CH3NH3PbI3 photovoltaic devices, Metals 6 (7) (2016) 147. Available at: http://www.mdpi.com/2075-4701/6/7/147.
[15] Y.-H. Kim, H. Cho, T.-W. Lee, Metal halide perovskite light emitters, Proc. Natl. Acad. Sci. U. S. A. 113 (42) (2016) 11694–11702. Available at: http://www.ncbi.nlm.nih.gov/pubmed/27679844.
[16] I.E. Castelli, et al., Bandgap calculations and trends of organometal halide perovskites, APL Mater. 2 (8) (2014). Available at: https://doi.org/10.1063/1.4893495.
[17] Stasiulionis, J., 2015. Life cycle assessment of perovskite solar cells and comparison to silicon solar cells. p. 56. Available at: http://studentarbeten.chalmers.se/publication/218943-life-cycle-assessment-of-perovskite-solar-cells-and-comparison-to-silicon-solar-cells.
[18] Averdung, F., 2017. Pioneering technology for solar power generation. Available at: www.oxfordpv.com.
[19] D.B. Mitzi, Organic-inorganic perovskites containing trivalent metal halide layers: the templating influence of the organic cation layer, Inorg. Chem. 39 (26) (2000) 6107–6113.
[20] J. Yang, et al., Solution stability of active materials for organic photovoltaics, Sol. Energy 113 (2015) 181–188. Available at: https://doi.org/10.1016/j.solener.2015.01.001.
[21] E.N. Hoggan, et al., Spin coating of photoresists using liquid carbon dioxide, Synthesis (2004) 2113–2122.
[22] C.R. Michel, E.R. López, H.R. Zea, Synthesis of GdCo1-xCuxO3-δ (x = 0, 0.15, 0.30) perovskites by ultrasonic spray pyrolysis, Mater. Res. Bull. 41 (1) (2006) 209–216.
[23] B. Suarez, et al., Recombination study of combined halides (Cl, Br, I) perovskite solar cells, J. Phys. Chem. Lett. 5 (10) (2014) 1628–1635.
[24] Y. Deng, et al., Vividly colorful hybrid perovskite solar cells by doctor-blade coating with perovskite photonic nanostructures, Mater. Horiz. 2 (6) (2015) 578–583. Available at: http://xlink.rsc.org/?DOI=C5MH00126A.
[25] J. Gurauskis, et al., Processing of thin film ceramic membranes for oxygen separation, J. Eur. Ceram. Soc. 32 (3) (2012) 649–655. Available at: https://doi.org/10.1016/j.jeurceramsoc.2011.10.009.
[26] M.R. Othman, J. Kim, 2008. Permeability and separability of oxygen across perovskite—alumina membrane prepared from a sol–gel procedure. Ind. Eng. Chem. Res. 47(9) 3000–3007.
[27] F.C. Krebs, et al., Large area plastic solar cell modulesMater. Sci. Eng. B 138 (2) (2007) 106–111. Available at: https://doi.org/10.1016/j.mseb.2006.06.008%5Cn. Available from: https://reddog.rmu.edu/login?url=. Available from: http://search.ebscohost.com/login.aspx?direct=true&db=aph&AN=24218177&site=ehost-live&scope=site.
[28] M. Liu, M.B. Johnston, H.J. Snaith, Efficient planar heterojunction perovskite solar cells by vapour deposition, Nature 501 (7467) (2013) 395–398. Available at: http://www.ncbi.nlm.nih.gov/pubmed/24025775.
[29] K. Chondroudis, D.B. Mitzi, P. Brock, Effect of thermal annealing on the optical and morphological properties of (AETH)PbX4 (X = Br, I) perovskite films prepared using single source thermal ablation, Chem. Mater. 12 (1) (2000) 169–175.
[30] Z. Lin, N.-G. Park, G. Li, Themed issue on perovskite solar cells, J. Mater. Chem. A 3 (17) (2015) 8924–8925. Available at: http://pubs.rsc.org/en/content/articlehtml/2015/ta/c5ta90079d.
[31] M.I. Ahmed, A. Habib, S.S. Javaid, Perovskite solar cells: potentials, challenges, and opportunities, Int. J. Photoenergy 2015 (2015), 13 pages.

[32] J. Wei, et al., Potentials and challenges towards application of perovskite solar cells, Sci. China Mater. 59 (August) (2016) 1–10. Available at: http://link.springer.com/10.1007/s40843-016-5082-4.
[33] P.P. Boix, et al., Current progress and future perspectives for organic/inorganic perovskite solar cells, Mater. Today 17 (1) (2014) 16–23. Available at: https://doi.org/10.1016/j.mattod.2013.12.002.
[34] N.K. Elumalai, et al., Perovskite solar cells: progress and advancements, Energies 9 (11) (2016).

CHAPTER 2

Organometal Lead Halide Perovskite

Javeed Akhtar[1], Muhammad Aamir[2] and Muhammad Sher[2]
[1]Department of Chemistry, Materials Lab, Mirpur University of Science & Technology (MUST), Mirpur Azad, Jammu & Kashmir, Pakistan
[2]Allama Iqbal Open University, Islamabad, Pakistan

2.1 INTRODUCTION

The chemistry of organometal lead halides has been established since early 20th century. In 1970, for the first time, Weber [1] has reported the synthesis and physical properties of $CH_3NH_3PbX_3$ (X = Cl, Br, I) organometal lead halide perovskite. Today, the organometal lead halide perovskites have been emerged as the promising candidate as light harvesters in the solar cell with the first solar cell reported in 2009 [2].

The photovoltaic applications of organometal halide perovskite have been the subject of many reviews [3]. In this chapter, the basics of organometal halide perovskites including synthesis and unique properties of organolead halide perovskites will be discussed.

2.2 METAL HALIDE PEROVSKITES

The basic structure of the material can be described as ABX_3, where A is a monovalent cation, B is a smaller metal divalent (lead) cation, and X is a halide anion. The B cations are coordinated with X anions forming octahedrons that are connected with each other forming cubic voids [4]. The A site cations have two functions, fill the void generated between the connected octahedral and neutralize the charge of the octahedron networks.

There are two major classes of perovskites based on the kind of A, B, and X atoms [5]. One is inorganic perovskite in which all the bonded atoms are inorganic in nature ($CsPbBr_3$) as shown in Fig. 2.1A.

The replacement of the inorganic A site cation with an organic cation leads to the formation of organic–inorganic hybrid perovskites, the general

Perovskite Photovoltaics
DOI: https://doi.org/10.1016/B978-0-12-812915-9.00002-2

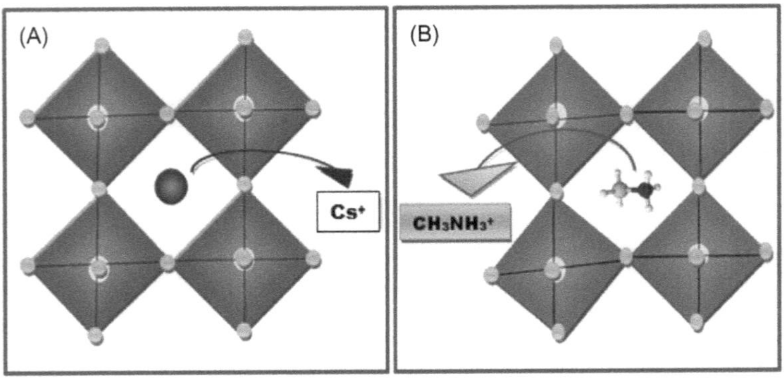

Figure 2.1 General structure of (A) inorganic perovskites and (B) hybrid perovskite.

structure is presented in Fig. 2.1B. The incorporation of an organic cation within the perovskite geometry reduces the symmetry [6,7]. Owing to the limited space in the voids of the octahedron, small-sized organic cations having three or less C–C/C–N bonds can be incorporated within the voids of perovskites to make a three dimensional (3D) geometry. However, the A cation should be larger sized compared to the B metal cation but its size is limited enough to fit in 12 folded coordinated voids of octahedral inorganic framework.

Lower dimensional (2D, 1D) perovskites are conceptually derived from the three dimensional ABX_3 perovskites by slicing or cutting the inorganic 3D network. If the size of organic cation is too large, then 3D perovskites are converted to lower dimensions by the breakdown of inorganic sheets into half sandwiching by the bilayers (bifunctional) or monolayers (nonfunctional) organic cations giving 2D or 1D hybrid perovskites [8] as shown in Fig. 2.2. The variety of organic cations of much larger or complex structures can be used as there is no more a confined cage in the octahedral geometry of the layered perovskites [9].

2.3 COMPOSITION OF PEROVSKITES

The charge balancing in the formal stoichiometry of perovskites (ABX_3) can be attained in various ways. There are two major types of perovskites based on constituents, metal oxide perovskites, and metal halide perovskites.

In metal oxide perovskites, the valence of two metal cations should be equal to six. The perovskites can be of I–V–O_3, II–IV–O_3, and III–III–O_3 composition having group I, II, III, and IV member

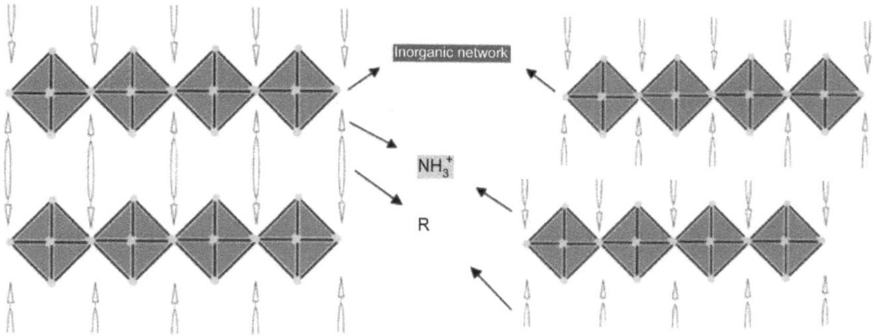

Figure 2.2 Schematic of the layered perovskites.

elements ($KTaO_3$, $CaCrO_3$, and $GdFeO_3$). The anion in the perovskites can be substituted in order to produce variety of materials like oxyhalides, oxynitrides, sulfide, and selenides [7,10]. Metal halide perovskites have composition $I-II-VII_3$, where two cations must have a valance equal to three [11]. It is known that low ionization potentials are anticipated for halide perovskites compared to metal oxide perovskites [12,13].

2.4 NATURE OF BONDING IN ORGANOMETAL LEAD HALIDE PEROVSKITES

The chemical bonding in organometal lead halide perovskites can be divided in to three components. These materials are consider to be organic–inorganic, not organometallic compounds as there is no direct chemical bond between carbon and lead according to the IUPAC definition [14].

2.4.1 Lead Halide Octahedral Cage

The bonding in a PbX_3^- anionic framework is mixed ionic/covalent interactions (heteropolar). In the case of organo lead halide perovskite, the composition of $CH_3NH_3PbI_3$ perovskites shows that the formal oxidation state of Pb (+2) and I (−1) are a good estimation of the chemical specie. However, the quantification of partial charge is ill defined due to the collective nature of the periodic electronic wavefunction [15]. The value of born effective charge of Pb can exceed four in the organo lead halide perovskites due to high ionicity [16].

In $CH_3NH_3PbI_3$ perovskites, lead has sp^3d^2 hybridization with octahedral framework involving 6 s^2 $6p^o$ $5d^o$ orbitals of lead and $5p^6$ orbitals of iodide ions forming the electronic band structure. The upper valence bond is formed by 5p orbital of iodine and 6 s and 6p orbitals of lead form the lower conduction band [17].

2.4.2 Intermolecular Interaction

The A site cations are familiarized in voids of the inorganic framework just as $CH_3NH_3^+$ cations are in the inorganic cages of PbI_6^- to form $CH_3NH_3PbI_3$. These $CH_3NH_3^+$ cations are 6 °A apart from one another in the network. There are a variety of organic cations that have been used in perovskites. These organic cations are polar and their polarization depends upon their geometry. The methyl part in $CH_3NH_3^+$ is polarized due to the ammonium ion and shows a permanent electrical dipole of 2.29 D, whereas, formamidinium ion has electrical dipole 0.21 D and ammonium ion due to its spherical structure has 0.00 D. The increase and decrease in dipole can be obtained by substituting the organic cations, e.g., mono, di, or tri fluorination of methylammonium increases the dipole from 5.35, 6.08, and 6.58 D, respectively [18]. These permanent dipoles interact with each other as well as with an external magnetic field. The value of dipole interactions is different for stationary and rotating dipolar molecules. For stationary molecules in voids the dipolar interactions tails off as $1/r^3$ while in rotating molecules it decreases and become $1/r^6$ [14].

The organic cations in organo lead halide perovskites is surrounded by the polarizable inorganic network have orientation disorder and polarization at room temperature that leads to a variety of behavior in the presence and absence of an electric field such as paraelectric, ferroelectric, superparaelectric, and antiferroelectric as shown in Fig. 2.3 [7].

The ammonium cation has a zero-dipole owing to spherical with small (0.3 KJ/mol) rotational barrier in the pseudo-cubic geometry of lead halides perovskites with low lattice electronic polarization (8 $\mu C/cm^2$). Therefore, the ammonium ion rotates inside the void and do not have dipolar interactions with nearby dipoles. While, other organic cations have molecular dipoles with high rotational barriers compared to the ammonium ion. The $CF_3NH_3^+$ has a maximum dipole (6.58 D) with high rotational barrier and strongest neighboring dipole interaction; it means this dipolar molecule has high lattice polarization (48 $\mu C/cm^2$) [19].

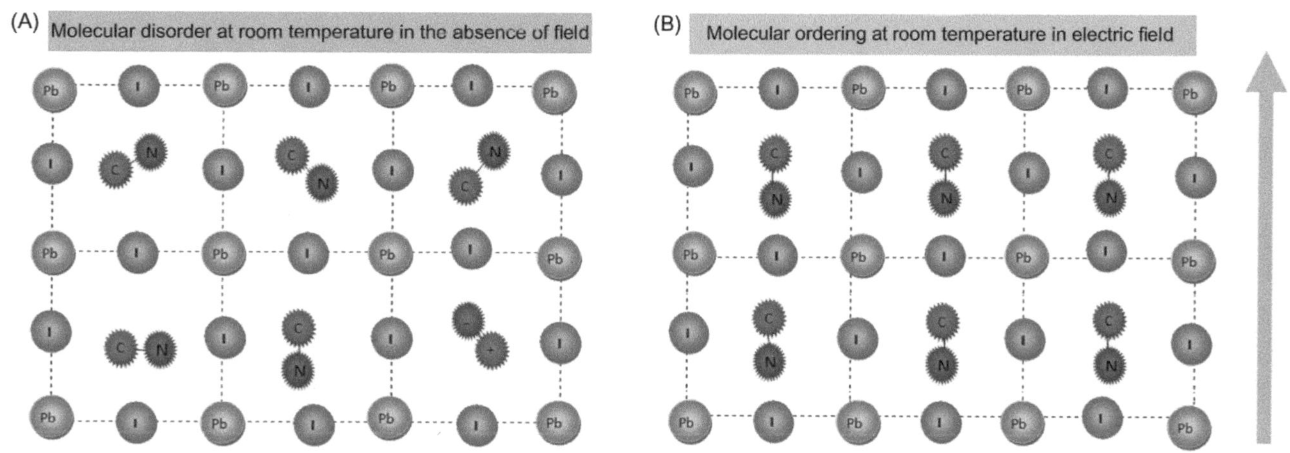

Figure 2.3 Arrangement of the ordering of molecular dipoles in the absence and presence of an external electric field.

2.4.3 Molecule Octahedral Framework Interaction

The leading force of attraction between the organic cation and inorganic framework is electrostatic in nature. There is strong electrostatic potential holding the molecule at its lattice site in the inorganic frameworks. In addition to the charge in the organic cation, there are also nonpolar parts which become polarized due to the polarization of organic cations. Therefore, charge–dipole interaction appears between the organic dipole and the inorganic framework which depends upon the molecular orientation. The dipolar molecules also polarize the I^- ions in the framework which leads to the induced–dipole interaction (Debye forces) that can be used to correlate the molecular orientation and deformation of the inorganic framework [7,14]. The polarization of the lattice has two advantages that are related to internal electric field for photovoltaic operations, enhanced charge separation and improved carrier lifetimes [20] and open circuit voltages above the band gap of the material [19].

2.5 DISTORTION OF PEROVSKITE GEOMETRY

The perovskites have cubic geometry which is its identity. The distortion in geometry from ideal structure plays a crucial role to tune physical properties of the organo lead metal perovskites [21–23] and it depends upon three variables (a) Jahn Teller Effect (b) Size Effect (c) Change in Composition of ABX_3.

2.5.1 Jahn Teller Effect

The octahedral framework of metal halide perovskites geometry can be elongated or compressed because of the Jahn Teller distortion of z ligand bonds [24]. The molecules having a degenerated electronic state, will distort its octahedral geometry in order to remove degeneracy. As a result lowering of the geometry takes place and gives the tetrahedral geometry [25]. The perovskites with a B atom having d^1, d^2, d^4, d^5, d^7, d^9 show Jahn Teller distortion, especially perovskites of Cu^{+2} and Mn^{+3} show significant Jahn Teller effects forming a tetragonal framework by the elongation of two bonds out of six bonds [4]. Jahn Teller effect leads to the distortion of octahedral framework in order to increase the stability of the compound [26]. Larger sized A site cations ($t > 1$) can increase the distortion of the structure by applying chemical pressure [27]. Polyakov et al. [28]. have reported the synthesis of $(C_6H_5CH_2CH_2NH_3)_2CuCl_4$

layered perovskite having Jahn Teller effects showing ferromagnetic and ferroelectric properties. Jahn Teller distortions provide more flexibility to the perovskite structure particularly in layered structures offering wider structural tunablility [29].

2.5.2 Size Effect

The geometrical relationship for the ideal cubic geometry of perovskites is as follows:

$$a = (r_A + r_x) = \sqrt{2}\,(r_B + r_x)$$

where 'a' is called the cell axis, r_A, r_B, r_x are the radii of A cation, B cation, and X anion. The distortion and stability of perovskites can be explained on the basis of ratio between two cell lengths, called the goldsmith tolerance (t) factor with formula [30].

$$t = \frac{(r_A + r_x)}{\sqrt{2}(r_B + r_x)}$$

Cubic perovskites can exist in range of $0.89 < t < 1$. If the A ion is smaller than $t < 1$, resulting a tilting of the octahedral geometry to fill the space giving orthorhombic and rhombohedra variants. If $t > 1$ which is due to large A sized cation give the hexagonal variants with closed packed structures [31]. $MAPbI_3$ has a distorted perovskite structure with a tetragonal phase (Fig. 2.4A), while $MAPbBr_3$ has a cubic perovskite

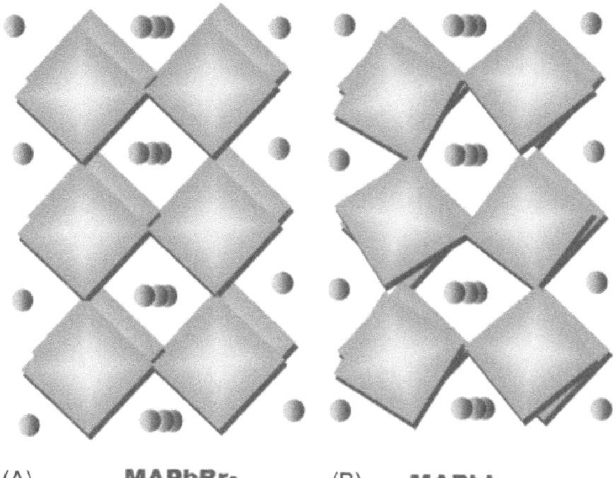

Figure 2.4 (A) Cubic perovskite structure of $MAPbBr_3$ at room temperature (B) Distorted tetragonal perovskite structure of $MAPbI_3$ at room temperature.

(Fig. 2.4B) structure at room temperature. This difference in phase is due to differences in the ionic radius of I^- (2.2 Å) and Br^- (1.96 Å) ions. It can be observed that smaller X anion leads to stable cubic geometry [32].

2.5.3 Change in Composition of ABX_3

The perovskites structure can be distorted with change in composition of X in oxidizing and reducing environment, which leads to change in oxidation state of B cation like in SrFeOx perovskite oxygen content can be varied between 2.5 and 3 [33].

2.6 OPTICAL PROPERTIES OF PEROVSKITES

Band gap has vital importance in photovoltaic related applications [34]. Hybrid 3D perovskites have direct band gap low band gap such as $CH_3NH_3PbI_3$ is a strong absorbance in the visible region and has high absorbance coefficients shows band gap of about 1.55 eV [35]. The band gap in hybrid perovskites can be tuned by tuning the geometry of perovskites. The tuning of geometry can be done by changing temperature and substituting A, B, and X components of perovskite. The band gap of $CH_3NH_3PbI_3$ hybrid perovskites decreases from 1.61 eV to 1.55 eV with decrease in temperature from 300 K to 150 K [36].

In the perovskites of stoichiometry, I–II–VII, the substitution of A site organic cations does not directly affect the electronic structure of organic–inorganic perovskites, infact it changes the crystal geometry. The organic cations regulate the band gap by changing the bond distance between metal and halide [37,38]. The smaller molecular cation decreases the band gap. For $CH_3NH_3PbI_3$ the band gap is 1.55 eV, for $CH_3NH_3^+$ cation whereas band gap reduces to 0.3 eV when $CH_3NH_3^+$ is replaced by NH_4^+ cation in the perovskites [39]. Similarly, the band gap of $HPbI_3$ less than 0.3 eV due to the smallest cation (H^+) as shown in Fig. 2.5 [7].

However, too small A site in perovskites leads to instability by tilting of octahedral networks which can change the electronic properties. The NH_4^+ and H^+ cations from the layered perovskites due to small size to fit in the perovskites inorganic networks and produces the deformation in geometry and reduces the band gap [40,41]. The shape of A site cations also deforms the geometry and help in tuning the band gap of perovskites. The formamidinium are asymmetric molecular ions, produces the distortion of geometry, and provides a fundamental reason for band gap tuning [42]. On the other hand, large A sized cations also forms layered

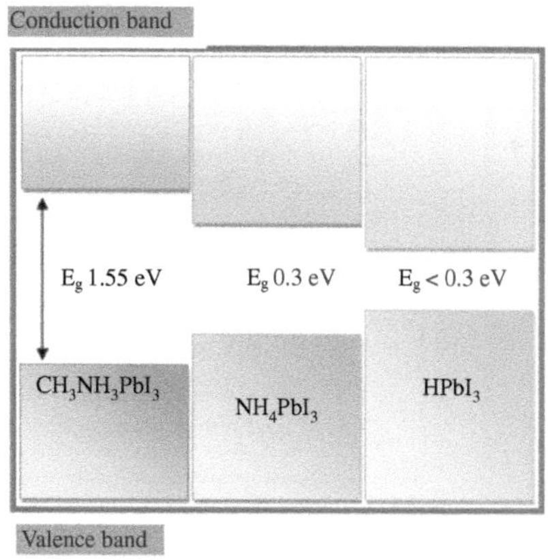

Figure 2.5 Band gap of perovskites with variation based on A site cations.

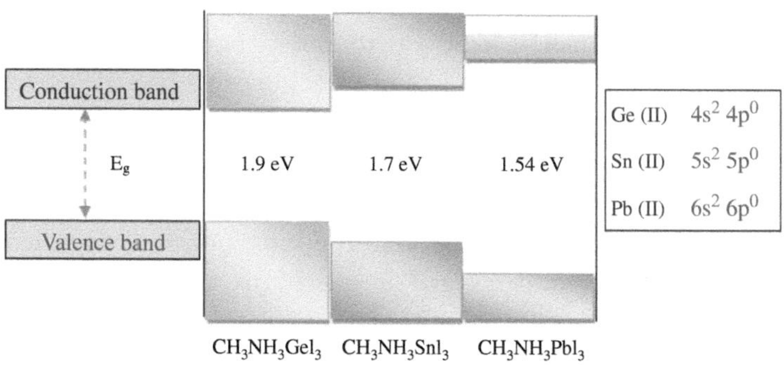

Figure 2.6 Effect of B site substitution on the band gap of perovskites.

perovskites (2D,1D) with high band gap as compared to 3D perovskites [43]. Based on above discussion, it can be concluded that size and shape of A site organic cations tunes the band gap of hybrid perovskites.

On the other hand, B site metal cations in the perovskite directly amend the conduction band. Therefore, it has been observed that the band gap of the perovskites increases when metal on the B site of perovskite lies higher in group of periodic table [44]. It has been reported that the replacement of lead (Pb) in $MAPbI_3$ by an isovalent tin (Sn) shows increase in band gap from 1.55 eV to 1.7 eV as shown in Fig. 2.6 [44,45]. Similarly, germanium can also be substituted but these metals (Sn, Gn, etc.) oxidized to IV oxidation state and leads to instability in octahedral framework [14]. Figure 2.7 describes the effect of group 14 metals in hybrid perovskites.

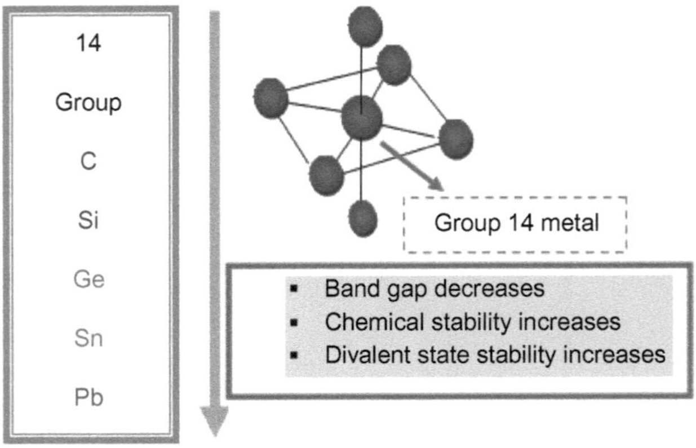

Figure 2.7 Effects of Group 14 metals in hybrid perovskites.

However, in the hybrid halide perovskites, the mixed isovalent metal halide hybrid perovskites have been reported with band gap low as compared to singe metal halide hybrid perovskites. For $CH_3NH_3Sn_{1-x}Pb_xI_3$ where x = 0 to 1, the band gap reaches to 1.18 eV, depending upon the composition of perovskites [46]. The anions (X sites) in the organic–inorganic perovskites are more effective candidates that control the band gap. The substitution of halide ions produces more shifts in the band gap of perovskites as compared to the substitution of organic cations [38,47]. The change in band gap is related to the electronic state of anions, the valence band composition changes from 3p to 4p to 5p for Cl, Br, and I, respectively. Therefore, electron binding energies decreases from Cl to I. In $CH_3NH_3PbX_3$, the change in X site from I to Cl varying the band gap from 1.55 eV to 3.0 eV as presented in Fig. 2.8. The mixed halide-based hybrid perovskites are the potential candidates for band gap tuning. Changing the I–Br ratio tune the band gap with increasing thermal stability because of their comparable ionic sizes [32]. On the other hand, the change in I–Cl ratio in perovskites does not produces significant change in band gap due to large difference in size ratio [48].

2.7 COORDINATION CHEMISTRY OF LEAD HALIDE PEROVSKITE PRECURSORS

The complexation of lead with halide ions results in complexes called plumbates which serve as precursors for the synthesis of lead halide perovskite. During the synthesis, solvents like DMF and DMSO used, donate

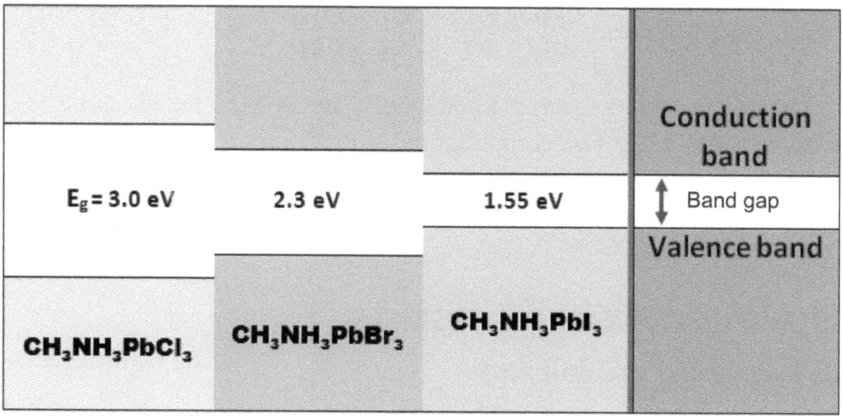

Figure 2.8 Band gap tuning of perovskites by halide substitution.

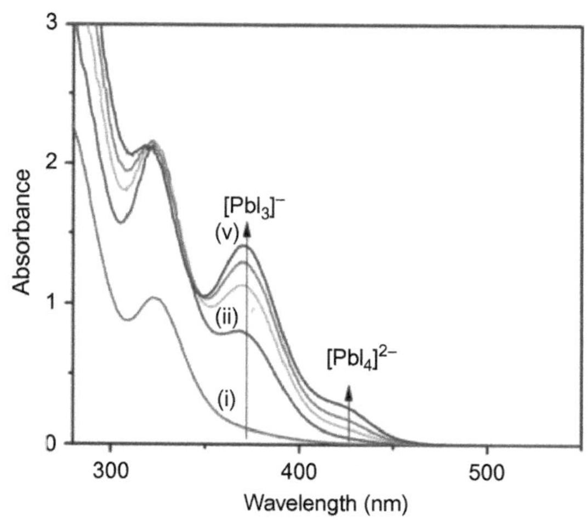

Figure 2.9 UV-Vis absorption spectra of PbI_2 solution in DMF (a) with increasing concentration of Methylaamonium iodide from 6 mM (b) to 24 mM (e). This figure is added with the copy right permission.

electron pair to form a Lewis adduct with Pb^{2+}. The formed adducts result in a colorless solution; however, the excess of iodide ions replace the coordinated solvent molecules and convert the solution into dark yellow color. The replacement of solvent ligands with I^- ions results in the formation of PbI^{3-} and PbI_4^{2-} complexes which show lower energy charge transfer absorption band [49]. The absorption of complex depends upon I^- ion concentration which enables to estimate the complexation constants (Fig. 2.9).

$$\mathbf{PbI_2 + I^- \leftarrow K_1 \rightarrow PbI_3^-}$$
$$\mathbf{PbI_3^- + I^- \leftarrow K_2 \rightarrow PbI_4^-}$$

The complexation constant for equilibria are 54 M^{-1} for K_1 and 6 M^{-1} for K_2 in DMF [49]. The complexation of Pb^{2+} ions is independent of the nature of counter ion like similar structural features were observed when CH_3NH_3I was replaced with KI.

2.8 FABRICATION OF PEROVSKITES

2.8.1 Perovskite Thin Films

The path to fabricate the thin films of perovskites from precursor solution dictate physical and optoelectronic properties of material [50]. The dramatically different properties of perovskites appeared when the stoichiometric ratio of CH_3NH_3I:PbI_3 was varied from 3:1 to 1:1 in the precursor ratio. The absorption spectra of $CH_3NH_3PbI_3$ film deposited from 3:1 ratio, prior to annealing show the spectral feature comparable to iodoplumbate complexes observed in the solution, which suggest that initial film composed iodoplumbate complexes with various degree of coordination. The annealing of the thin films provides an energy to evaporate CH_3NH_3I and gradually transforms solid state sample to perovskite. However, it was observed that the perovskite derived off from the stochiometric solution does not yield the same iodoplumbate finger print, which suggest that the fundamentally different pathway compared to solid state precursors [51,52]. The solid-state precursors of perovskite show large lattice spacing in the intermediate which is due to the formation of $\left(CH_3NH_3^+\right)_2\left(PbI_3^-\right)_2 \cdot (DMF)_2$ intermediate. However, annealing of the sample at 150 C leads to concomitant rise in tetragonal phase peaks [53].

On the contrary, utilization of equimolar antecedent solution brings about prompt crystallization upon expulsion of DMF, causing spheritic development with expansive regions of the substrate uncovered. The solid-state precursors with use of 3:1 ratio solution impede perovskite crystallization, yielding thin films with immeasurably enhanced surface coverage and homogeneity. The formation of solid state intermediate provides a wide possibility to tune perovskite thin film growths [54].

The use of different lead salts containing halides or other spectator counterions provide solid state precursors for perovskite that undergo transformation at different temperatures and rates to yield thin film with distinct grain size, surface roughness, and pinhole densities. Solvent also plays an important role in the fabrication of solution processed thin films. Coordinating solvent not only shows additional complexity but also provides an opportunity to explore solvent–solute interaction. DMSO is a coordinating solvent both as monodentate ligand through either atom of S–O or through hydrogen bonding and will serve as a model system for discussion. Other solvents like DMF and gamma-butyrolactone are also coordinating solvent and effect the film quality and morphology. In the case of solid state $PbI_2(DMSO)_2$ compound, the DMSO intercalate between the layers of PbI_2. The incorporation of DMSO between the layers of PbI_2 expand the PbI_2 lattice relative to neat PbI_2 and results in an effective preloading of the PbI_2 lattice which is more favorable to intermolecular exchange reactions to produce perovskites. The expanded lattice of PbI_2-DMSO adduct yield thicker films compared to neat PbI_2 under same working conditions. These as-deposited films were then rapidly converted to films with minimal volume expansion upon contacting isopropyl solution of organic ammonium halide salt precursors. To achieve this conversion, the PbI_2 and DMSO must be in 1:1 ratio, otherwise temperature above 100°C is required to remove DMSO [53–55].

The high ordered structure and slow growth rate gave a clear view of the crystallization process in the single crystals of the plumbate complexes that is otherwise challenge in thin films of perovskites [56]. Perovskite single crystals are grown by diffusing antisolvent into the solution to loss solubility or upon cooling of saturated solution. In addition, perovskites have shown a retrograde solubility which is rare phenomenon which is only displayed by few other materials [57,58]. The retrograde solubility is strongly influenced by nature of solvents [59]. For example, $CH_3NH_3PbCl_3$ loss solubility on heating in DMSO or mixture of DMSO/DMF. Similarly, $CH_3NH_3PbBr_3$ and $CH_3NH_3PbI_3$ show retrograde solubility in DMF and gamma- butyrolactone, respectively [60]. Since the presence of specific solvent and plumbate complexes are necessary for retrograde solubility of perovskite, the complexation between plumbates and DMSO retard the crystallization results in the conversion of amorphous film crystallization to perovskite after the annealing at the temperature sufficient to evaporate coordinated solvent molecules [61]. Similarly, the dissociation of these complexes in the solution provides a pathway for single crystal formation. The solvent and plumbate complexes dissociate at elevated temperature to begin retrograde solubility.

2.9 STABILITY OF HYBRID PEROVSKITES

The degradation of hybrid perovskites is the major problem in the solid state solar cells. The moisture, oxygen, temperature, and UV radiation affects the stability of hybrid perovskites. Especially, $CH_3NH_3PbI_3$ is sensitive to water and hydrolyzed in the presence of moisture. Similarly, bleaching of perovskites takes place in the presence of ammonia [62]. Figure 2.10 shows the decomposition of hybrid perovskites in moisture which was investigated by Walsh and coworkers and suggests that simple acid base reaction is involved in degradation. They proposed that only single water molecule which acts as Lewis base, mingle with $CH_3NH_3PbI_3$ (hybrid perovskites), take one proton from ammonium and convert the perovskites into an intermediate $\left[\left(CH_3NH_3^+\right)n\text{-}1(CH_3NH_2)PbI_3\right][H_3O]$ which then converted to HI, CH_3NH_2, and finally into PbI_2. Only single water molecule is sufficient to degrade the perovskites but to dissolve the HI and CH_3NH_2 byproducts, excess of water is required.

In the solar cell devices, the traces of water lead to the partial degradation of perovskites. The $CH_3NH_3PbI_3$ started to decompose at humidity of 55%, with the change in color from dark brown to yellow. The decomposition of perovskites becomes rapid in light as compared to dark and their absorbance peak appears at 500 nm, that match with the absorbance of PbI_2 [63].

Niu et al. have also proposed a mechanism for the degradation of perovskites in the presence of moisture, oxygen, and UV radiation. The $CH_3NH_3PbI_3$ on exposure to moisture decomposes to CH_3NH_2, PbI_2, and HI [64]. It was also observed that 2D perovskite $(PEA)_2(MA)_2[Pb_3I_{10}]$ $\left(PEA = C_6H_5(CH_2)_2\ NH_3^+,\ MA = CH_3NH_3^+\right)$ is more stable in ambient condition than 3D $CH_3NH_3PbI_3$ [65] (Fig. 2.11). On the basis of proposed mechanism, it was thought that the aprotic organic cations as such as tetramethylammonium, $(Me)4N^+$, may be chemically stable [7].

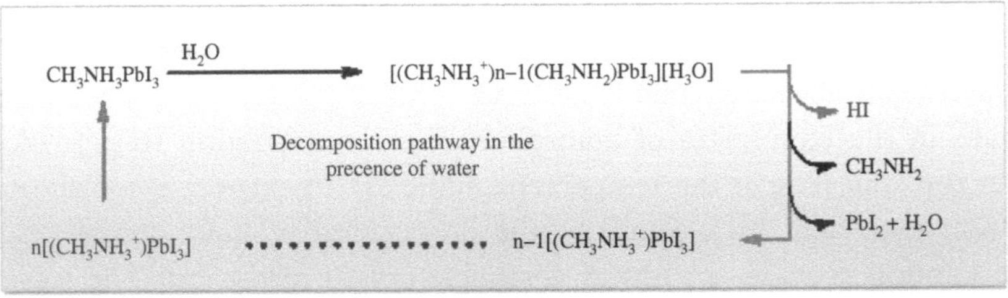

Figure 2.10 Proposed mechanism for decomposition of perovskites in the presence of water at room temperature.

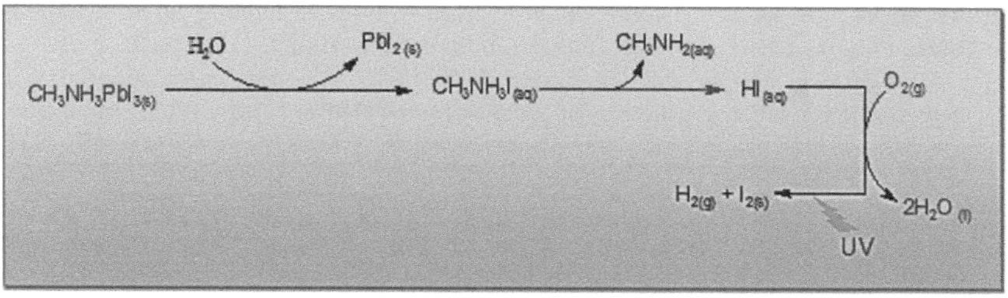

Figure 2.11 Proposed mechanism for the degradation of perovskites.

REFERENCES

[1] D. Weber, CH3NH3SnBrxI3-x (x = 0-3), ein Sn (II)-System mit kubischer Perowskitstruktur/CH3NH3SnBrxI3-x (x = 0-3), a Sn (II)-system with cubic perovskite structure, Z. Naturforsch. B 33 (8) (1978) 862–865.

[2] A. Kojima, K. Teshima, Y. Shirai, T. Miyasaka, Organometal halide perovskites as visible-light sensitizers for photovoltaic cells, J. Am. Chem. Soc. 131 (17) (2009) 6050–6051.

[3] S.D. Stranks, P.K. Nayak, W. Zhang, T. Stergiopoulos, H.J. Snaith, Formation of thin films of organic–inorganic perovskites for high-efficiency solar cells, Angewandte Chem. Intl. Ed. 54 (11) (2015) 3240–3248.

[4] P. Zolfaghari, G. De Wijs, R. De Groot, The electronic structure of organic–inorganic hybrid compounds: (NH4) 2CuCl4, (CH3NH3) 2CuCl4 and (C2H5NH3) 2CuCl4, J. Phys.: Condens. Matter. 25 (29) (2013) 295502.

[5] A. Glazer, The classification of tilted octahedra in perovskites, Acta Crystallogr. Sect. B: Struct. Crystallogr. Cryst. Chem. 28 (11) (1972) 3384–3392.

[6] I.E. Castelli, T. Olsen, S. Datta, D.D. Landis, S. Dahl, K.S. Thygesen, et al., Computational screening of perovskite metal oxides for optimal solar light capture, Energy Environ. Sci. 5 (2) (2012) 5814–5819.

[7] J.M. Frost, K.T. Butler, F. Brivio, C.H. Hendon, M. van Schilfgaarde, A. Walsh, Atomistic origins of high-performance in hybrid halide perovskite solar cells, Nano Lett. 14 (5) (2014) 2584–2590.

[8] Z. Cheng, J. Lin, Layered organic–inorganic hybrid perovskites: structure, optical properties, film preparation, patterning and templating engineering, Cryst. Eng. Comm. 12 (10) (2010) 2646–2662.

[9] C. Kagan, D. Mitzi, C. Dimitrakopoulos, Organic-inorganic hybrid materials as semiconducting channels in thin-film field-effect transistors, Science 286 (5441) (1999) 945–947.

[10] Y.-Y. Sun, Chalcogenide perovskites for photovoltaics, Nano Lett. 15 (1) (2015) 581–585.

[11] N.-G. Park, Perovskite solar cells: an emerging photovoltaic technology, Mater. Today. 18 (2) (2015) 65–72.

[12] A. Walsh, D.O. Scanlon, S. Chen, X. Gong, S.H. Wei, Self-regulation mechanism for charged point defects in hybrid halide perovskites, Angew. Chem. 54 (6) (2015) 1791–1794.

[13] D.O. Scanlon, C.W. Dunnill, J. Buckeridge, S.A. Shevlin, A.J. Logsdail, S.M. Woodley, et al., Band alignment of rutile and anatase TiO2, Nat. Mater. 12 (9) (2013) 798–801.

[14] A. Walsh, Principles of chemical bonding and band gap engineering in hybrid organic–inorganic halide perovskites, J. Phys. Chem. C 119 (11) (2015) 5755–5760.

[15] M. Jansen, U. Wedig, A piece of the picture—misunderstanding of chemical concepts, Angew. Chem. Int. Ed. 47 (52) (2008) 10026–10029.
[16] J. Brgoch, A.J. Lehner, M. Chabinyc, R. Seshadri, Ab initio calculations of band gaps and absolute band positions of polymorphs of RbPbI3 and CsPbI3: implications for main-group halide perovskite photovoltaics, J. Phys. Chem. C 118 (48) (2014) 27721–27727.
[17] C. Rao, A. Cheetham, A. Thirumurugan, Hybrid inorganic–organic materials: a new family in condensed matter physics, J. Phys. Conden. Matter. 20 (8) (2008) 083202.
[18] J.M. Frost, K.T. Butler, A. Walsh, Molecular ferroelectric contributions to anomalous hysteresis in hybrid perovskite solar cells, APL Mater. 2 (8) (2014) 081506.
[19] I. Grinberg, D.V. West, M. Torres, G. Gou, D.M. Stein, L. Wu, et al., Perovskite oxides for visible-light-absorbing ferroelectric and photovoltaic materials, Nature 503 (7477) (2013) 509–512.
[20] S. Yang, J. Seidel, S. Byrnes, P. Shafer, C.-H. Yang, M. Rossell, et al., Above-bandgap voltages from ferroelectric photovoltaic devices, Nat. Nanotechnol. 5 (2) (2010) 143–147.
[21] I. Koutselas, P. Bampoulis, E. Maratou, T. Evagelinou, G. Pagona, G. Papavassiliou, Some unconventional organic – inorganic hybrid low-dimensional semiconductors and related light-emitting devices, J. Phys. Chem. C 115 (17) (2011) 8475–8483.
[22] G.C. Papavassiliou, I. Koutselas, Structural, optical and related properties of some natural three-and lower-dimensional semiconductor systems, Synth. Met. 71 (1) (1995) 1713–1714.
[23] G.C. Papavassiliou, G.A. Mousdis, I. Koutselas, Some new organic–inorganic hybrid semiconductors based on metal halide units: structural, optical and related properties, Adv. Mater. Opt. Electron. 9 (6) (1999) 265–271.
[24] B.J. Kennedy, C.J. Howard, B.C. Chakoumakos, Phase transitions in perovskite at elevated temperatures — a powder neutron diffraction study, J. Phys. Condens. Matter. 11 (6) (1999) 1479.
[25] J. Alonso, M. Martinez-Lope, M. Casais, M. Fernandez-Diaz, Evolution of the Jahn-Teller distortion of MnO6 octahedra in RMnO3 perovskites (R = Pr, Nd, Dy, Tb, Ho, Er, Y): a neutron diffraction study, Inorgan. Chem. 39 (5) (2000) 917–923.
[26] J. Kanamori, Crystal distortion in magnetic compounds, J. Appl. Phys. 31 (5) (1960) S14–S23.
[27] M.W. Lufaso, P.M. Woodward, Jahn–Teller distortions, cation ordering and octahedral tilting in perovskites, Acta Crystallogr. Sect. B: Struct. Sci. 60 (1) (2004) 10–20.
[28] A.O. Polyakov, A.H. Arkenbout, J. Baas, G.R. Blake, A. Meetsma, A. Caretta, et al., Coexisting ferromagnetic and ferroelectric order in a CuCl4-based organic–inorganic hybrid, Chem. Mater. 24 (1) (2011) 133–139.
[29] D. Cortecchia, H.A. Dewi, J. Yin, A. Bruno, S. Chen, T. Baikie, et al., Lead-free MA2CuCl x Br4−x hybrid perovskites, Inorgan. Chem. 55 (3) (2016) 1044–1052.
[30] N.-G. Park, Organometal perovskite light absorbers toward a 20% efficiency low-cost solid-state mesoscopic solar cell, J. Phys. Chem. Lett. 4 (15) (2013) 2423–2429.
[31] M. Johnsson, P. Lemmens, Crystallography and chemistry of perovskites, Handbook of Magnetism and Advanced Magnetic Materials, Helmut Kronmueller, eds., John Wiley & Sons, Hoboken, NJ. 2007, 1–2912.
[32] J.H. Noh, S.H. Im, J.H. Heo, T.N. Mandal, S.I. Seok, Chemical management for colorful, efficient, and stable inorganic–organic hybrid nanostructured solar cells, Nano Lett. 13 (4) (2013) 1764–1769.
[33] M. Johnsson, P. Lemmens, Perovskites and thin films—crystallography and chemistry, J. Phys.: Condens. Matter. 20 (26) (2008) 264001.

[34] H. Salehi, N. Shahtahmasebi, S. Hosseini, Band structure of tetragonal $BaTiO_3$, Eur. Phys. J. B-Condens. Matt. Compl. Syst. 32 (2) (2003) 177–180.
[35] H.-S. Kim, C.-R. Lee, J.-H. Im, K.-B. Lee, T. Moehl, A. Marchioro, et al., Lead iodide perovskite sensitized all-solid-state submicron thin film mesoscopic solar cell with efficiency exceeding 9%, Sci. Rep. 2 (2012) 591.
[36] Y. Yamada, T. Nakamura, M. Endo, A. Wakamiya, Y. Kanemitsu, Near-band-edge optical responses of solution-processed organic–inorganic hybrid perovskite CH3NH3PbI3 on mesoporous TiO_2 electrodes, Appl. Phys. Exp. 7 (3) (2014) 032302.
[37] D.B. Mitzi, Solution-processed inorganic semiconductors, J. Mater. Chem. 14 (15) (2004) 2355–2365.
[38] G.E. Eperon, S.D. Stranks, C. Menelaou, M.B. Johnston, L.M. Herz, H.J. Snaith, Formamidinium lead trihalide: a broadly tunable perovskite for efficient planar heterojunction solar cells, Energy Environ. Sci. 7 (3) (2014) 982–988.
[39] F. Brivio, K.T. Butler, A. Walsh, M. Van Schilfgaarde, Relativistic quasiparticle self-consistent electronic structure of hybrid halide perovskite photovoltaic absorbers, Phys. Rev. B 89 (15) (2014) 155204.
[40] G. Kieslich, S. Sun, A.K. Cheetham, Solid-state principles applied to organic–inorganic perovskites: new tricks for an old dog, Chem. Sci. 5 (12) (2014) 4712–4715.
[41] F. Wang, H. Yu, H. Xu, N. Zhao, HPbI3: A new precursor compound for highly efficient solution-processed perovskite solar cells, Adv. Funct. Mater. 25 (7) (2015) 1120–1126.
[42] M.R. Filip, G.E. Eperon, H.J. Snaith, F. Giustino, Steric engineering of metal-halide perovskites with tunable optical band gaps, Nat. Commun. 5 (2014) 5757.
[43] D. Mitzi, S. Wang, C. Feild, C. Chess, A. Guloy, Conducting layered organic-inorganic halides containing $<110>$-oriented perovskite sheets, Science 267 (5203) (1995) 1473–1476.
[44] N.K. Noel, S.D. Stranks, A. Abate, C. Wehrenfennig, S. Guarnera, A.-A. Haghighirad, et al., Lead-free organic–inorganic tin halide perovskites for photovoltaic applications, Energy Environ. Sci. 7 (9) (2014) 3061–3068.
[45] J. Feng, B. Xiao, Effective masses and electronic and optical properties of nontoxic MASnX3 (X = Cl, Br, and I) perovskite structures as solar cell absorber: a theoretical study using HSE06, J. Phys. Chem. C 118 (34) (2014) 19655–19660.
[46] C.C. Stoumpos, C.D. Malliakas, M.G. Kanatzidis, Semiconducting tin and lead iodide perovskites with organic cations: phase transitions, high mobilities, and near-infrared photoluminescent properties, Inorgan. Chem. 52 (15) (2013) 9019–9038.
[47] F. Brivio, A.B. Walker, A. Walsh, Structural and electronic properties of hybrid perovskites for high-efficiency thin-film photovoltaics from first-principles, APL Mater. 1 (4) (2013) 042111.
[48] S. Colella, E. Mosconi, P. Fedeli, A. Listorti, F. Gazza, F. Orlandi, et al., MAPbI3-xCl x mixed halide perovskite for hybrid solar cells: the role of chloride as dopant on the transport and structural properties, Chem. Mater. 25 (22) (2013) 4613–4618.
[49] K.G. Stamplecoskie, J.S. Manser, P.V. Kamat, Dual nature of the excited state in organic–inorganic lead halide perovskites, Energy Environ. Sci. 8 (1) (2015) 208–215.
[50] G. Grancini, S. Marras, M. Prato, C. Giannini, C. Quarti, F. De Angelis, et al., The impact of the crystallization processes on the structural and optical properties of hybrid perovskite films for photovoltaics, J. Phys. Chem. Lett. 5 (21) (2014) 3836–3842.
[51] J.S. Manser, P.V. Kamat, Band filling with free charge carriers in organometal halide perovskites, Nat. Photon. 8 (9) (2014) 737–743.

[52] S. Yang, Y.C. Zheng, Y. Hou, X. Chen, Y. Chen, Y. Wang, et al., Formation mechanism of freestanding CH3NH3PbI3 functional crystals: in situ transformation vs dissolution–crystallization, Chem. Mater. 26 (23) (2014) 6705–6710.

[53] Y. Guo, K. Shoyama, W. Sato, Y. Matsuo, K. Inoue, K. Harano, et al., Chemical pathways connecting lead (II) iodide and perovskite via polymeric plumbate (II) fiber, J. Am. Chem. Soc. 137 (50) (2015) 15907–15914.

[54] J.S. Manser, B. Reid, P.V. Kamat, Evolution of organic–inorganic lead halide perovskite from solid-state iodoplumbate complexes, J. Phys. Chem. C 119 (30) (2015) 17065–17073.

[55] K. Yan, M. Long, T. Zhang, Z. Wei, H. Chen, S. Yang, et al., Hybrid halide perovskite solar cell precursors: colloidal chemistry and coordination engineering behind device processing for high efficiency, J. Am. Chem. Soc. 137 (13) (2015) 4460–4468.

[56] D. Shi, V. Adinolfi, R. Comin, M. Yuan, E. Alarousu, A. Buin, et al., Low trap-state density and long carrier diffusion in organolead trihalide perovskite single crystals, Science 347 (6221) (2015) 519–522.

[57] D.W. de Quilettes, S.M. Vorpahl, S.D. Stranks, H. Nagaoka, G.E. Eperon, M.E. Ziffer, et al., Impact of microstructure on local carrier lifetime in perovskite solar cells, Science 348 (6235) (2015) 683–686.

[58] C.G. Bischak, E.M. Sanehira, J.T. Precht, J.M. Luther, N.S. Ginsberg, Heterogeneous charge carrier dynamics in organic–inorganic hybrid materials: Nanoscale lateral and depth-dependent variation of recombination rates in methylammonium lead halide perovskite thin films, Nano Lett. 15 (7) (2015) 4799–4807.

[59] Y. Dang, Y. Liu, Y. Sun, D. Yuan, X. Liu, W. Lu, et al., Bulk crystal growth of hybrid perovskite material CH 3 NH 3 PbI 3, Cryst. Eng. Comm. 17 (3) (2015) 665–670.

[60] M.I. Saidaminov, A.L. Abdelhady, G. Maculan, O.M. Bakr, Retrograde solubility of formamidinium and methylammonium lead halide perovskites enabling rapid single crystal growth, Chem. Commun. 51 (100) (2015) 17658–17661.

[61] M.I. Saidaminov, A.L. Abdelhady, B. Murali, E. Alarousu, V.M. Burlakov, W. Peng, et al., High-quality bulk hybrid perovskite single crystals within minutes by inverse temperature crystallization, Nat., Commun. 6 (2015) 7586.

[62] G. Niu, X. Guo, L. Wang, Review of recent progress in chemical stability of perovskite solar cells, J. Mater. Chem. A 3 (17) (2015) 8970–8980.

[63] J.A. Christians, P.A. Miranda Herrera, P.V. Kamat, Transformation of the excited state and photovoltaic efficiency of CH3NH3PbI3 perovskite upon controlled exposure to humidified air, J. Am. Chem. Soc. 137 (4) (2015) 1530–1538.

[64] G. Niu, W. Li, F. Meng, L. Wang, H. Dong, Y. Qiu, Study on the stability of CH 3 NH 3 PbI 3 films and the effect of post-modification by aluminum oxide in all-solid-state hybrid solar cells, J. Mater. Chem. A 2 (3) (2014) 705–710.

[65] I.C. Smith, E.T. Hoke, D. Solis-Ibarra, M.D. McGehee, H.I. Karunadasa, A. Layered, Hybrid perovskite solar-cell absorber with enhanced moisture stability, Angew. Chem. 126 (42) (2014) 11414–11417.

CHAPTER 3

Evolution of Perovskite Solar Cells

Suneth C. Watthage, Zhaoning Song, Adam B. Phillips and Michael J. Heben
University of Toledo, Toledo, OH, United States

3.1 INTRODUCTION

Perovskite is a naturally occurring mineral of calcium titanate with a chemical formula of $CaTiO_3$. This mineral was first discovered by German mineralogist Gustav Rose in 1839 and was named in honor of the Russian mineralogist Lev Perovski (1792–1856) [1]. Generally, materials which have the same crystal structure of $CaTiO_3$ are referred to as perovskite materials. These perovskite materials typically have a cubic or tetragonal crystal structure with the stoichiometry of AMX_3, where both A and M are metal cations and X is an anion (Fig. 3.1). Each cation M is octahedrally coordinated with the anions X to form the basic building block of the perovskite structure. These MX_6 octahedra are connected in a three-dimensional corner-sharing configuration, with the cation A embedded in the space formed between adjacent MX_6 octahedron to neutralize the charge of the structure [1].

Over the past few decades, the inorganic perovskite oxides (e.g., $CaTiO_3$, $BaTiO_3$, $LaMnO_3$, etc.) and halides ($CsSnI_3$, $CsPbI_3$, etc.) have been extensively studied due to their versatile applications in optics [2], magnetics [3], electronics [4], and superconductors [5]. In recent years, another group of perovskite materials called the hybrid organic–inorganic perovskites (HOIPs), which comprise an organic ammonium cation A ($CH_3NH_3^+$, $NH = CHNH_3^+$, etc.), a divalent metal cation B (Pb^{2+}, Sn^{2+}, Cd^{2+}, Fe^{2+}, Co^{2+}, Cu^{2+}, etc.), and halide ions X (Cl^-, Br^-, or I^-), has attracted substantial research interests. Since the 1970s, significant research has been conducted on this family of materials to better understand their structural versatility and diversity in optical, magnetic, and electronic properties [6–8]. In the early 1990s, Sn- and Pb-based HOIPs were of great interest to materials scientists for fabricating field-effect

Perovskite Photovoltaics
DOI: https://doi.org/10.1016/B978-0-12-812915-9.00003-4

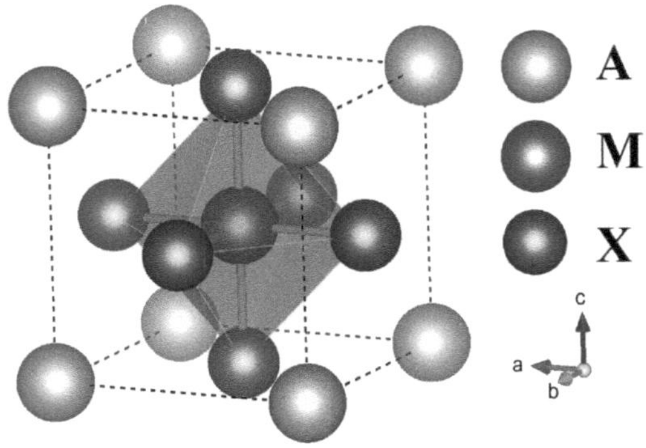

Figure 3.1 Diagram of cubic perovskite crystal structure.

transistors [9] and light-emitting diodes [10] due to their high charge carrier mobilities and strong excitonic transitions. These excellent optoelectronic properties also make the HOIP materials ideal for thin-film photovoltaic (PV) devices; however, their applications as solar cells were not considered.

The first HOIP-based PV device was reported by Kojima et al. in 2009 [11], who used methylammonium lead iodide ($CH_3NH_3PbI_3$, $MAPbI_3$) and methylammonium lead bromide ($CH_3NH_3PbBr_3$, $MAPbBr_3$) as the sensitizers to fabricate dye-sensitized solar cells (DSSCs) with a liquid electrolyte [11]. Because of a low power conversation efficiencies (PCEs) of ~3% and poor device stability, HOIPs as light absorbing materials received little attention. In 2012, the liquid electrolyte was replaced with a solid-state hole transport material (HTM), and the PCE of $MAPbI_3$-based solar cells increased to ~10% [12,13]. These revolutionary findings opened a new era of the emerging perovskite solar cells (PSCs) and spread the so-called "perovskite fever" all around the world [14–16]. Thanks to the tremendous efforts in the past few years, PSCs have rapidly progressed and reached 22.1% PCE [17]. Fig. 3.2 shows some major milestones in the progress of PSCs, including the device evolution from dye-sensitized to planar structure and the advancements in the solvent and compositional engineering techniques. This successful progress also stimulated a great interest in this emerging PV technology, as the annual number of articles published with the topic of PSCs has increased from a single digit number in 2012 to ~2300 in 2016. This chapter traces this remarkable evolution mainly from the perspectives of device architecture and material deposition method by focusing on the early developments of the materials, notable devices with excellent

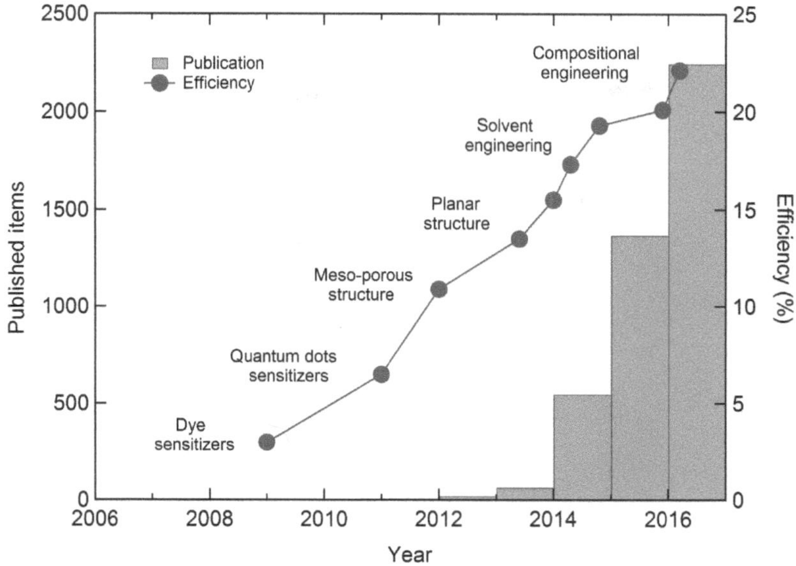

Figure 3.2 Record efficiency and numbers of publications with the topic on perovskite solar cells (PSCs) from 2009 to 2016. The publication data is adapted from ISI web of science.

performance, technical advances in preparation methods, and the future directions of the field.

3.2 EVOLUTION OF THE DEVICE ARCHITECTURE

Advances in the design of device architecture are one of the most important factors that drove the evolution of PSCs, especially in the early stage of the technical development. In 2009, PSCs were first demonstrated based on the design of the liquid electrolyte DSSC configuration (Fig. 3.3A). This device structure, inherited from the conventional DSSCs, has intrinsic material compatibility issues associated with the use of liquid electrolyte that limit the stability of the devices. Advances in solid-state HTM as a replacement for the liquid electrolyte led to the development of the mesoscopic (Fig. 3.3B) and meso-superstructured (Fig. 3.3C) architectures used in 2012. Based on these two preliminary structures, the so-called "regular" structure (Fig. 3.3D) constructed with the HOIP materials penetrating into a thin mesoporous metal oxide layer and forming a capping layer on the top was developed during 2013–2014. Since then, the regular structure has been widely used to

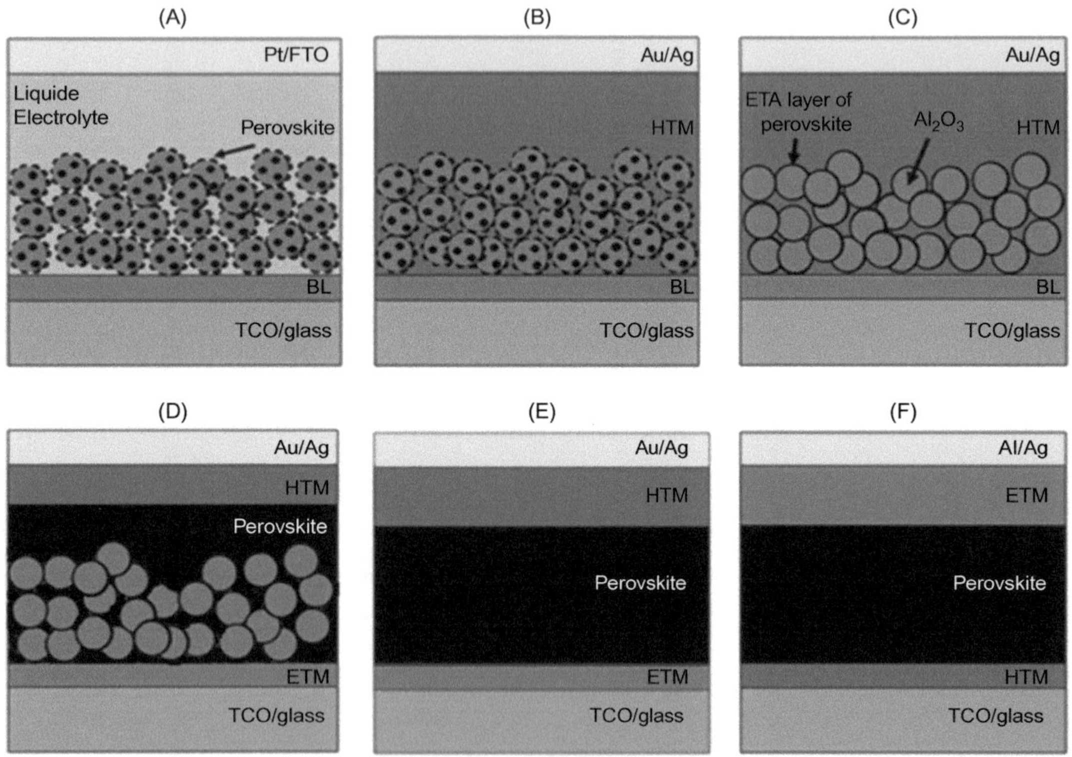

Figure 3.3 Schematics of perovskite solar cell (PSC) architectures: (A) liquid electrolyte DSSC structure, (B) solid-state mesoscopic structure, (C) meso-superstructured structure, (D) the regular structure, (E) planar n—i—p heterojunction structure, and (F) planar p—i—n heterojunction structure (inverted planar structure). *DSSC*, dye-sensitized solar cells.

fabricate high-efficiency PSCs. This success has attracted the experts of inorganic thin-films and organic PV materials to the field of PSC research, resulting in the development of the planar heterojunction structures in the n–i–p (Fig. 3.3E) and p–i–n (Fig. 3.3F) configurations. The planar p–i–n structure, with the film stack deposited in the opposite sequence of the regular structure, is often referred as the inverted structure.

3.2.1 Liquid Electrolyte DSSC Structure

PSCs were initially fabricated in the conventional DSSC structure (Fig. 3.4A), which is comprised of a glass substrate/transparent conducting oxide (TCO), a thin compact TiO_2 (c-TiO_2) hole blocking layer, a several micron thick mesoporous TiO_2 (mp-TiO_2) layer sensitized with molecular dyes, a liquid electrolyte, and a counter electrode (Pt/TCO/glass) [18]. Fig. 3.4B shows the working principle of DSSC, which is based on photoexcitation of the sensitizer dye. When the incident photons are absorbed by the photosensitizers coating on the surface of TiO_2, electrons are excited from the ground state to the excited state. The photoexcitation results in the injection of electrons into the conduction band of mp-TiO_2, and then to the front electrode (anode). This process leaves the photosensitizers in an oxidized state, which is subsequently reduced back to the ground state by an injected electron from the reductants (e.g., I^-) in the liquid electrolyte. After the electron donation, the deactivated reductants, i.e., the oxidants (e.g., I_3^-), diffuse to the metal counter electrode and are regenerated by electrons from the cathode. This complete electron flow creates a photogenerated current in a DSSC. The

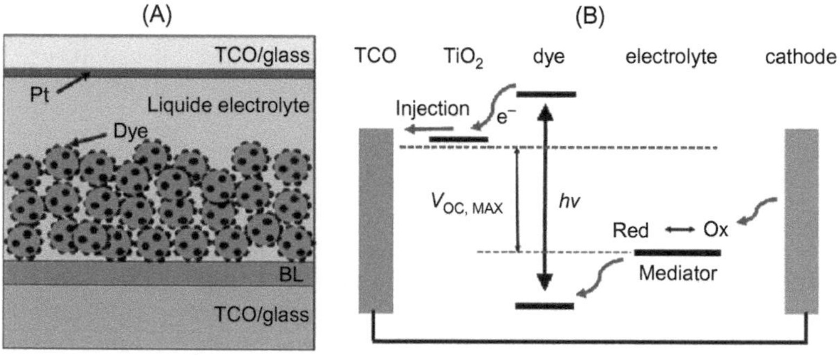

Figure 3.4 (A) Schematic diagram and (B) working mechanism of the conventional DSSC. *DSSC*, dye-sensitized solar cells.

energy difference between the Fermi level of TiO_2 and the redox potential of the mediator in the electrolyte determines the maximum value of V_{OC} [18,19].

In 2009, Kojima et al. replaced the organic dye in the DSSC structure with $MAPbI_3$ or $MAPbBr_3$ perovskite (Fig. 3.5A) by spin-coating a precursor solution containing equimolar methylammonium and lead halide, MAX and PbX_2 (X = I or Br), onto the mp-TiO_2/c-TiO_2/fluorine-doped tin oxide (FTO) substrate [11]. The best $MAPbI_3$ device demonstrated a PCE of 3.8%, with an open circuit voltage (V_{OC}) of 0.61 V, a short circuit current density (J_{SC}) of 11.0 mA/cm^2, and a fill factor (FF) of 57%. In comparison, the champion $MAPbBr_3$ device had a PCE of 3.2%, with V_{OC} = 0.96 V, J_{SC} = 5.57 mA/cm^2, and FF = 59%. The higher J_{SC} of the $MAPbI_3$ device is a result of photoconversion of longer wavelengths light (550–800 nm) (Fig. 3.5B) due to a lower bandgap (1.5 eV) relative to $MAPbBr_3$ (2.3 eV), but the V_{OC} of the $MAPbI_3$ devices is lower due to a smaller energy level difference between the conduction band of the perovskite absorber and the redox potential of the X^{3-}/X^- (X = I or Br) mediator (Fig. 3.5C). Although this work demonstrated the potential of using the HOIP materials in PV applications, it did not receive much attention due to the relatively low efficiency and instability of the devices.

Two years later, Im et al. fabricated perovskite-sensitized DSSCs using 2–3 nm $MAPbI_3$ quantum dots (QDs) to coat the 3.6 μm thick mp-TiO_2 layer (Fig. 3.6A) [19]. Their champion device shows an improved

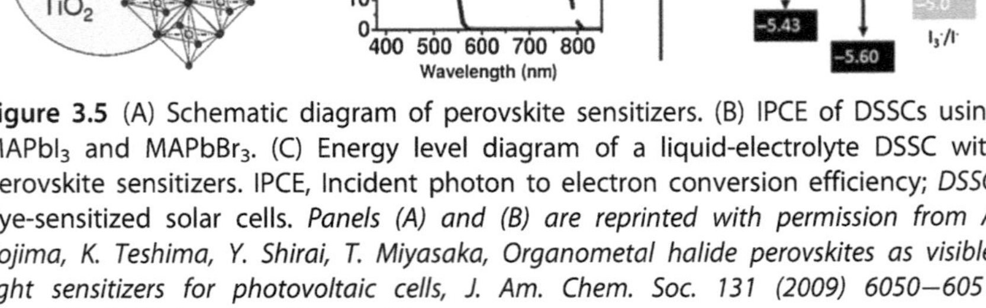

Figure 3.5 (A) Schematic diagram of perovskite sensitizers. (B) IPCE of DSSCs using $MAPbI_3$ and $MAPbBr_3$. (C) Energy level diagram of a liquid-electrolyte DSSC with perovskite sensitizers. IPCE, Incident photon to electron conversion efficiency; *DSSC*, dye-sensitized solar cells. *Panels (A) and (B) are reprinted with permission from A. Kojima, K. Teshima, Y. Shirai, T. Miyasaka, Organometal halide perovskites as visible-light sensitizers for photovoltaic cells, J. Am. Chem. Soc. 131 (2009) 6050–6051. Copyright 2009 American Chemical Society.*

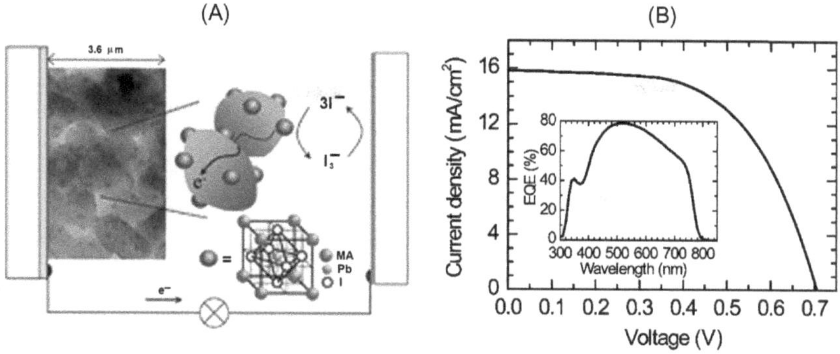

Figure 3.6 (A) Schematic of a dye-sensitized solar cells (DSSC) with the perovskite quantum dots (QDs). (B) Current density–voltage (*J–V*) and external quantum efficiency (EQE) curves of the device. *Reprinted with permission from J.-H. Im, C.-R. Lee, J.-W. Lee, S.-W. Park, N.-G. Park, 6.5% efficient perovskite quantum-dot-sensitized solar cell, Nanoscale 3 (2011) 4088–4093. Copyright 2011, Royal Society of Chemistry.*

PCE of 6.54%, with $V_{OC} = 0.706$ V, $J_{SC} = 15.82$ mA/cm^2, and FF = 58.6% (Fig. 3.6B). The J_{SC} enhancement is due to the improved light conversion in the $MAPbI_3$ QDs, which shows the maximum external quantum efficiency of ~80% at 530 nm. The increased V_{OC} is likely due to the thicker mp-TiO_2 that prevents recombination at the perovskite/FTO interface. This work revealed the high light absorption coefficient of the HOIP materials, which is one of the key factors for fabricating high-efficiency solar cells. Still, instability of these devices limits their practical application—the device performance degraded ~80% in 10 min due to the dissolution of the perovskite QDs in the liquid electrolyte.

3.2.2 Solid-State Mesoscopic Structure

To overcome the instability issue associated with the liquid electrolyte, Kim et al. in 2012 developed the first solid-state PSC by replacing the liquid electrolyte with the solid HTM 2,2′,7,7′-tetrakis-(*N*,*N*-di-*p*-methoxyphenyl-amine)-9,9′-spirobifluorene (Spiro-MeOTAD) that was successfully used in solid-state DSSC [13]. This device configuration (Fig. 3.3B) was named the mesoscopic structure because it consists of a thin mp-TiO_2 layer similar to that used in the liquid electrolyte perovskite-sensitized solar cells, although the thickness of the mp-TiO_2 layer was reduced from ~3 to ~0.6 μm. In this device configuration, part of the Spiro-MeOTAD HTM penetrates into the pores of the mp-TiO_2 layer, making direct contact with the perovskite sensitizers. The remaining HTM material forms a dense capping layer that covers the mp-TiO_2, preventing shunts between the electron transport materials (ETMs) and the back contact (Fig. 3.7A).

The working principle of this device configuration differs from the conventional DSSCs. In contrast to the electrolyte that suffered from low charge carrier mobility, the solid-state Spiro-OMeTAD allows holes to move more efficiently. The champion device demonstrated a PCE of 9.7%, with $V_{OC} = 0.89$ V, $J_{SC} = 17.6$ mA/cm^2, and FF = 62% (Fig. 3.7B). Replacing the liquid electrolyte with the solid HTM significantly improved V_{OC} from 0.71 to 0.89 V due to the better alignment of the highest occupied molecular orbital (HOMO) level of the

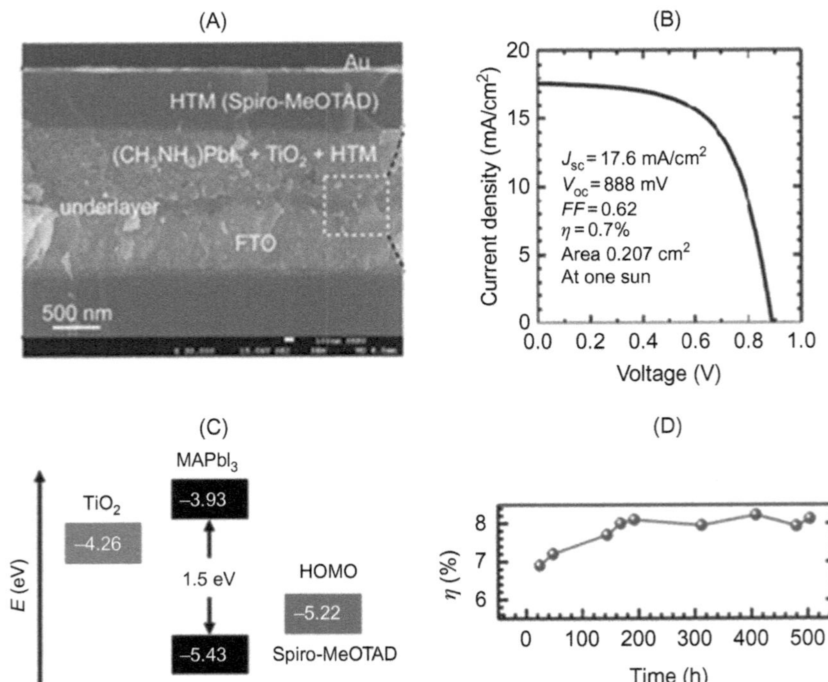

Figure 3.7 (A) Cross-sectional scanning electron microscope (SEM) image, (B) *J–V* characteristic, (C) energy level diagram, and (D) stability of a perovskite solar cell (PSC) in solid-state mesoscopic structure. *Panels (A), (B), and (D) are reprinted with permission from H.-S. Kim, C.-R. Lee, J.-H. Im, K.-B. Lee, T. Moehl, A. Marchioro, et al., Lead iodide perovskite sensitized all-solid-state submicron thin film mesoscopic solar cell with efficiency exceeding 9%, Sci. Rep. 2 (2012) 591. Copyright 2012, Nature Publishing Group.*

Spiro-MeOTAD (−5.22 eV) with the valance band of $MAPbI_3$ (−5.43 eV), enabling more efficient hole extraction from the perovskite absorber and widening the hole Fermi energy splitting under illumination (Fig. 3.7C). In addition to the PCE improvement, the device stability was also dramatically enhanced. Some devices even exhibited the enhanced PCEs after 500 h of operation (Fig. 3.7D).

3.2.3 Mesosuperstructured Structure

Analogous to the mesoscopic configuration, Snaith et al. invented the meso-superstructured structure by replacing the mp-TiO_2 layer with the insulating Al_2O_3 [12]. This device configuration has almost the same structure as in the mesoscopic structure (Fig. 3.8A), but the perovskite materials formed a continuous extremely thin absorber (ETA) layer on the surface of the porous metal oxide (Fig. 3.8B). Leijtens et al. investigated a mixed halide $CH_3NH_3PbI_{3-x}Cl_x$, instead of the pure $MAPbI_3$ perovskite as the absorber material, and studied the electronic properties of the meso-superstructured perovskite films [20]. The champion device with the porous Al_2O_3 layer achieved a PCE of 10.9%, with $V_{OC} = 0.98$ V, $J_{SC} = 17.8$ mA/cm^2, and FF = 63%, while the best control device made with mp-TiO_2 demonstrated a PCE of 7.6% with $V_{OC} = 0.80$ V, $J_{SC} = 17.8$ mA/cm^2, and FF = 53%. Interestingly, a

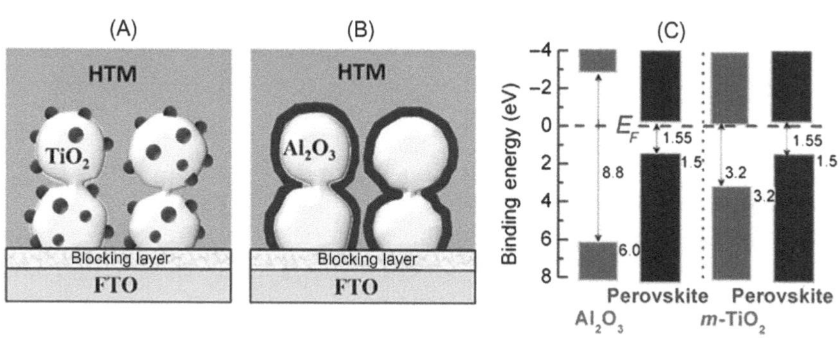

Figure 3.8 Schematic diagram of (A) mesoporous and (B) meso-superstructured perovskite solar cells (PSCs). (C) Measured binding energy of perovskite, Al_2O_3/perovskite, and mp-TiO_2/perovskite. *(A) and (B) are reprinted with permission from H.S. Kim, S.H. Im, N.G. Park, Organolead halide perovskite: new horizons in solar cell research, J. Phys. Chem. C 118 (2014) 5615–5625. Copyright 2014 American Chemical Society. (C) is reprinted with permission from T. Leijtens, S.D. Stranks, G.E. Eperon, R. Lindblad, E.M.J. Johansson, I.J. McPherson, et al., Electronic properties of meso-superstructured and planar organometal halide perovskite films: charge trapping, photodoping, and carrier mobility. ACS Nano 8 (2014) 7147–7155. Copyright 2014 American Chemical Society.*

significantly higher V_{OC} value of ~1.1 V was achieved in some devices with the Al_2O_3 layer. This low V_{OC} deficiency, i.e., the difference between E_g/e of the absorber material and V_{OC} of the corresponding devices, indicates low nonradiative recombination and long carrier diffusion lengths of the mixed halide perovskites. These excellent optoelectronic properties that favor the PV application were confirmed by optical and charge transport measurements [21,22].

The charge transport mechanism of the meso-superstructured cells differs from that of the mesoscopic devices in one critical way. Unlike the case of mp-TiO_2, injection of photoexcited electrons from the perovskite absorber into Al_2O_3 is energetically unfavorable due to the higher conduction band of Al_2O_3 relative to $MAPbI_{3-x}Cl_x$ (Fig. 3.8C) [20,23]. Moreover, Al_2O_3 is electrically insulating as indicated by the poor performance of the DSSC with the N719 dye and mesoporous Al_2O_3 [12]. Therefore, in this device, the porous Al_2O_3 layer merely serves as a mesoscale scaffold for the perovskite absorber. Electrons have to transport to the c-TiO_2 layer through the ETA layer of perovskites, which is enabled by the long diffusion lengths of both electrons and holes in the perovskite materials [21,22]. These good electric properties suggested PSCs could be fabricated without the mesoporous layer. While Snaith et al. only achieved a PCE of 1.8% for their preliminary planar junction device [12], high PCE devices were realized in the following years after the advances in deposition techniques, interface passivation, compositional control, and the understanding of the perovskite materials (see Section 3.4 for the details).

3.2.4 The Regular Structure

The success of mesosuperstructured cells motivated many researchers in the related field to study these exciting materials. The breakthrough was enabled when researcher discovered the low trap state density and long carrier diffusion length in HOIPs [21,22], indicating thicker perovskite films instead of ETA could be used in the PSCs. Increasing the absorber thickness can improve the absorption of long wavelength photons and prevent the direct shunts between the electrodes. Consequently, the device structure was developed based on the meso-superstructure configuration but with a thicker and dense perovskite layer (Fig. 3.3D). This structure was widely adopted by researchers all around the world for fabricating high-efficiency PSCs and thus is called the "regular" structure due to its popularity.

The regular device structure was originated from the PSCs with "pillared structures." In 2013, Heo et al. reported this structure with $MAPbI_3$ completely filled the pores of the mp-TiO_2 and formed individual $MAPbI_3$ pillars (Fig. 3.9A) [24]. These perovskite pillars were covered by a thin polytriarylamine (PTAA) HTM and Au electrode to complete the device. The PCE of the best cell was 12%, with $V_{OC} = 0.997$ V, $J_{SC} = 16.5$ mA/cm^2, and FF = 72.7%. Heo et al. found that devices that had a mp-TiO_2 layer of ~600 nm exhibited the best incident photon-to-electron conversion efficiency (Fig. 3.9B) and device performance (Fig. 3.9C) due to the formation of a perovskite overlayer. However, this overlayer suffered from a rough and discontinuous morphology, limiting the improvement of device performance.

The further progress of the regular structure was propelled by the advances in the perovskite film deposition techniques [1], allowing thicker perovskite films to be deposited on a thinner mp-TiO_2 layer. Burschka et al. reported PSCs with the best PCE of 15% in the regular configuration by fabricating a smooth and thin perovskite capping layer (~50 nm) on the top of a ~300 nm mp-TiO_2 layer (Fig. 3.10A) using the two-step sequential deposition method (see Section 3.3.2) [25]. The efficiency enhancement, especially the improved J_{SC} (20.0 mA/cm^2) (Fig. 3.10D), was mainly due to the formation of a dense and uniform perovskite layer. The same deposition technique was then used by Im et al. to prepare PSCs on a thinner mp-TiO_2 layer (~100 nm), resulting in the increased thickness of the perovskite capping layer (~150 nm) (Fig. 3.10B) [26], and improved the J_{SC} (21.64 mA/cm^2), V_{OC} (1.056 V), and efficiency (17%) of the devices (Fig. 3.10E). By employing the Li-doped TiO_2 and a thicker perovskite capping layer (~300 nm) (Fig. 3.10C), Giordano et al. achieved a PCE of 19.3%, with $V_{OC} = 1.114$ V, $J_{SC} = 23.0$ mA/cm^2, and FF = 74% (Fig. 3.10F) [27]. To date, the state-of-the-art PSCs with PCEs >20% were predominantly fabricated in the regular configuration [1].

3.2.5 Planar n–i–p Heterojunction Structure

The planar n–i–p heterojunction structure (Fig. 3.3E) was developed as the mesoporous oxide layer of the regular structure gets thinner that they eventually were removed. Analogous to inorganic thin-film solar cells, the structure is comprised of a TCO cathode, an n-type ETM, an intrinsic perovskite layer, a p-type HTM, and a metal anode. Due to the

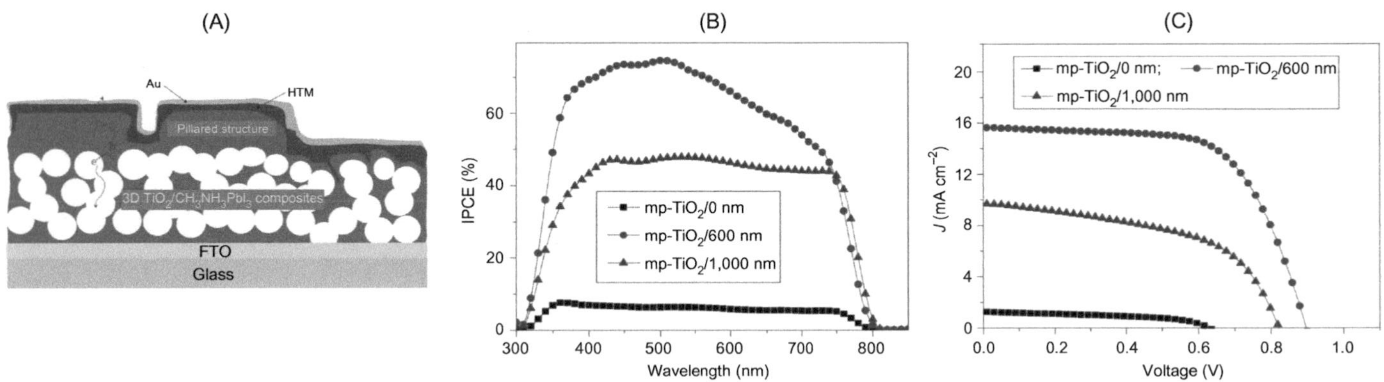

Figure 3.9 (A) Schematic illustration, (B) Incident photon-to-electron conversion efficiency (IPCE) curves, and (C) *J*–*V* characteristics of the perovskite solar cells (PSCs) with different mp-TiO_2 thickness. *Reprinted with permission from J.H. Heo, S.H. Im, J.H. Noh, T.N. Mandal, C.-S. Lim, J.A. Chang, et al., Efficient inorganic–organic hybrid heterojunction solar cells containing perovskite compound and polymeric hole conductors, Nat. Photonics 7 (2013) 486–491. Copyright 2013, Nature Publishing Group.*

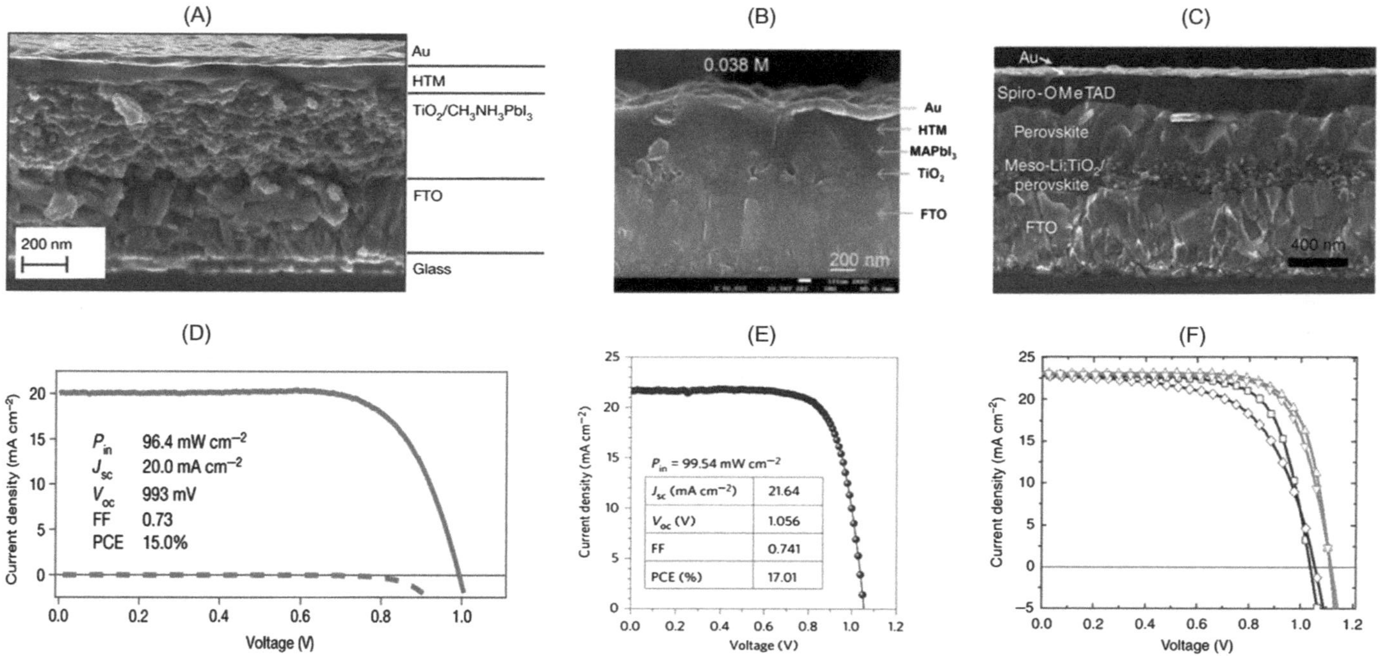

Figure 3.10 (A—C) Cross-sectional scanning electron microscope (SEM) image and (D—F) corresponding *J—V* curves of perovskite solar cells (PSCs) with different perovskite capping layers. *Panels (A) and (D) are reprinted with permission from J. Burschka, N. Pellet, S.-J. Moon, R. Humphry-Baker, P. Gao, M.K. Nazeeruddin, et al., Sequential deposition as a route to high-performance perovskite-sensitized solar cells, Nature 499 (2013) 316—319. Copyright 2013, Nature Publishing Group. Panels (B) and (E) are reprinted with permission from J.-H. Im, I.-H. Jang, N. Pellet, M. Grätzel, N.-G. Park, Growth of $CH_3NH_3PbI_3$ cuboids with controlled size for high-efficiency perovskite solar cells, Nat. Nano. 9 (2014) 927—932. Copyright 2014, Nature Publishing Group. Panels (C) and (F) are reprinted with permission from F. Giordano, A. Abate, J.P. Correa Baena, M. Saliba, T. Matsui, S.H. Im, et al., Enhanced electronic properties in mesoporous TiO_2 via lithium doping for high-efficiency perovskite solar cells, Nat. Commun. 7 (2016) 10379. Copyright 2015 Nature Publishing Group.*

low exciton bounding energy and long diffusion lengths of charge carriers, photoexcited electrons and holes have sufficiently long lifetimes to transport to the interfaces with the charge selective layers (HTM and ETM). In 2012, Snaith et al. made the first attempt on planar heterojunction PSCs using $MAPbI_{3-x}Cl_x$ but only achieved a PCE of 1.8% [12], likely due to the incomplete film coverage that led to shunts in the devices. Efforts to increase the PCE by optimizing the processing conditions, including annealing temperature and film thicknesses, resulted in the values of 4.5% [28] in 2013 and 11.4% [29] in 2014. Although the preliminary p–i–n type planar devices seemed to be comprised of smooth and uniform films in the devices (Fig. 3.11A), the perovskite

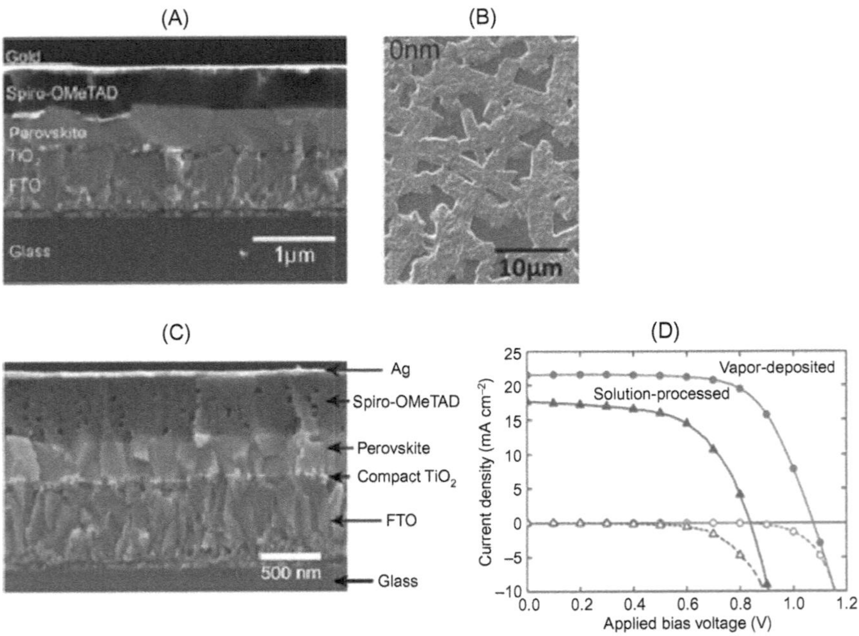

Figure 3.11 (A) Cross-sectional and (B) surface SEM images of a PSC in the planar p–i–n configuration prepared by solution-based deposition. (C) Cross-sectional scanning electron microscope (SEM) image of a perovskite solar cells (PCSs) prepared by the vapor-based deposition. (D) *J–V* characteristics of the planar p–i–n type PSCs. *Panels (A) and (B) are reprinted with permission from G.E. Eperon, V.M. Burlakov, P. Docampo, A. Goriely, H.J. Snaith, Morphological control for high performance, solution-processed planar heterojunction perovskite solar cells, Adv. Funct. Mater. 24 (2014) 151–157. Copyright 2013 WILEY-VCH Verlag GmbH & Co. KGaA, Weinheim. Panels (C) and (D) are reprinted with permission from M. Liu, M.B. Johnston, H.J. Snaith, Efficient planar heterojunction perovskite solar cells by vapour deposition, Nature 501 (2013) 395–398. Copyright 2013, Nature Publishing Group.*

films were porous and had a high surface defect density (Fig. 3.11B). To improve the smoothness of the perovskite film, Snaith et al. developed a dual-source coevaporation of $PbCl_2$ and MAI technique (see Section 3.3.4) to prepare a uniform $MAPbI_{3-x}Cl_x$ layer on the c-TiO_2 layer in 2013 (Fig. 3.11C) [30]. The champion device achieved a PCE of 15%, with $V_{OC} = 1.07$ V, $J_{SC} = 21.5$ mA/cm^2, and FF = 0.67 (Fig. 3.11D). Thanks to the advances in the film deposition techniques (see Section 3.4), the present state-of-the-art planar PSCs prepared by solution-based deposition methods yields PCEs of ~20% [31–33], with a V_{OC} of up to 1.214 V [33]. Additionally, these devices exhibited stable performance under operating conditions over hundreds of hours [34]. These advancements in efficiency and stability have made the planar device structure a promising candidate for the commercialized manufacturing of PSCs.

3.2.6 Inverted Planar p–i–n Heterojunction Structure

While the regular and planar n–i–p structures progressed, the planar p–i–n heterojunction structure originally used in the organic PV devices was adopted by the PSC community as an alternative. Because the planar p–i–n heterojunction structure has an opposite sequence of HTM and ETM than the regular configuration, it is often referred as the "inverted" structure (Fig. 3.3F). The p–i–n structure is comprised of a TCO anode, a p-type HTM, an intrinsic perovskite layer, an n-type ETM, and a metal cathode. In 2013, Jeng et al. reported the prototype of planar p–i–n PSC with the structure indium tin oxide (ITO)/PEDOT:PSS/$MAPbI_3$/C_{60}/bathocuproine (BCP)/Al [35], where they categorized the $MAPbI_3$/C_{60} as a planar heterojunction and BCP and polypolystyrene sulfonate (PEDOT:PSS) as the ETM and HTM (Fig. 3.12A). The preliminary device yielded a PCE of 1.6%, which improved to 3.9% when phenyl-C_{61}-butyric acid methyl ester ($PC_{61}BM$) was used instead of C_{60}. The increased efficiency was attributed to the larger energy offset between the HOMO of $MAPbI_3$ (donor) and the lowest unoccupied molecular orbital of $PC_{61}BM$ (acceptor) and more efficient electron injection into the Al cathode (Fig. 3.12B).

The efficiency of PSCs in the planar p–i–n structure has rapidly progressed. Sun et al. applied the solution-processed $PC_{61}BM$ to the p–i–n type PSC and improved the PCE to 7.4% [36]. Snaith et al. prepared the

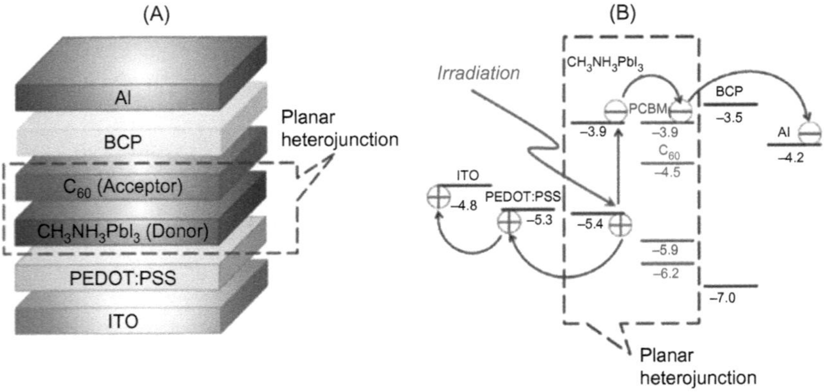

Figure 3.12 (A) Structure and (B) energy level diagram of a perovskite solar cell (PSC) in the planar p–i–n heterojunction structure. *Reprinted with permission from J.-Y. Jeng, Y.-F. Chiang, M.-H. Lee, S.-R. Peng, T.-F. Guo, P. Chen, et al., $CH_3NH_3PbI_3$ perovskite/fullerene planar-heterojunction hybrid solar cells, Adv. Mater. 25 (2013) 3727–3732. Copyright 2013 WILEY-VCH Verlag GmbH & Co. KGaA, Weinheim.*

planar PSC in the structure FTO/PEDOT:PSS/$MAPbI_{3-x}Cl_x$/$PC_{61}BM$/TiO_x/Al and obtained a PCE of 9.8% [37]. You et al. further improved the device efficiency to 11.5% by optimizing the perovskite deposition [38]. Malinkiewicz et al. employed the coevaporation technique for preparing $MAPbI_3$ and achieved a 12.0% PCE champion cell [39]. Wang et al. introduced a double fullerene layer consisting of C_{60} and $PC_{61}BM$ to passivate the perovskite film [40] and demonstrated a 12.2% PCE device with a record high FF of 80.1%.

While the early work established the standard configuration of the p–i–n planar structure (Fig. 3.13A), the further progress of the p–i–n planar PSCs was due to advances in perovskite fabrication techniques (see Section 3.4) [41,42]. Nie et al. fabricated mm-sized perovskite grains using the hot casting deposition technique and achieved a PCE of 17.7% [43]. Meanwhile, Huang et al. used nonwetting HTM and ETM to assist the preparation of large-size perovskite grains (Fig. 3.13B and C) [44,45], achieving the champion PCE of 19.5%, with $J_{SC} = 22.6\ mA/cm^2$, $V_{OC} = 1.07$ V, and FF = 80.6% (Fig. 3.13D) [45]. These state-of-the-art planar p–i–n devices not only exhibited excellent photon-to-electron conversion rate (Fig. 3.13E) but also demonstrated enhanced device stability maintaining ~90% PCE after 30 days of exposure to the ambient condition (Fig. 3.13F).

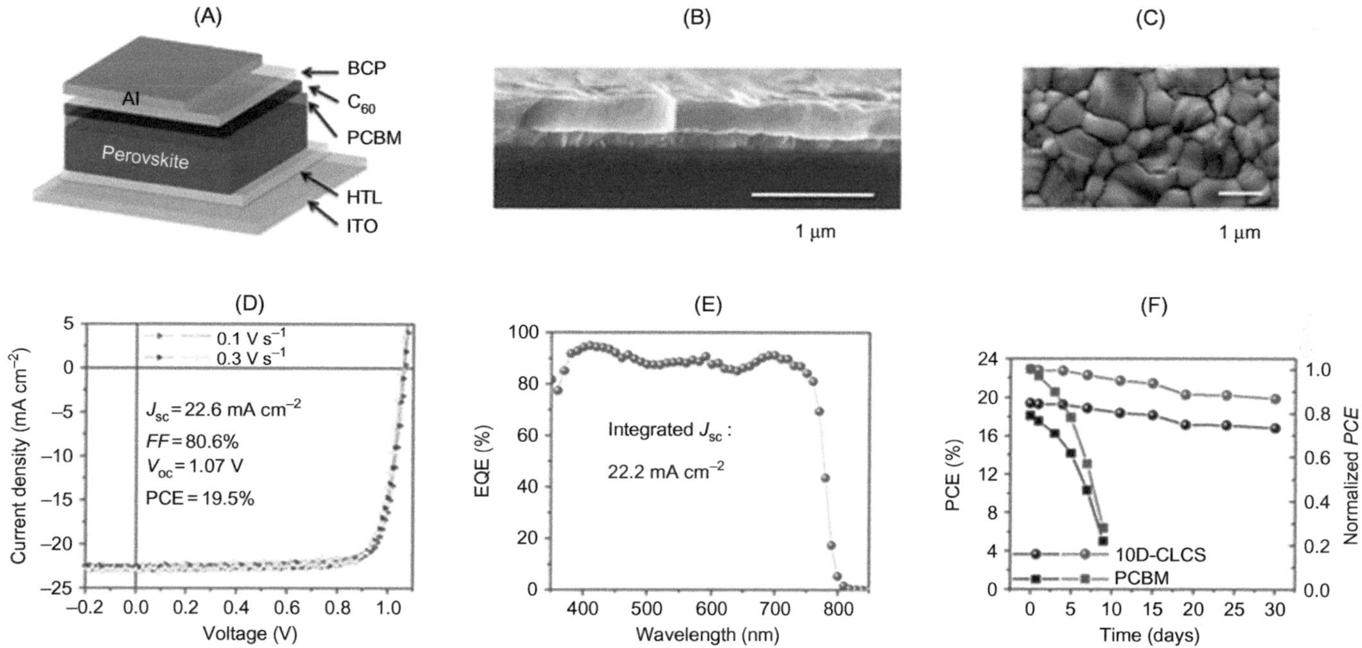

Figure 3.13 (A) Structure of a perovskite solar cell (PSC) in the planar p–i–n structure. (B) Cross-sectional and (C) surface scanning electron microscope (SEM) images of the perovskite film deposited on a Indium tin oxide (ITO) substrate. (D) *J–V* characteristic, (E) external quantum efficiency (EQE), and (F) stability measurement of a PSC in the planar p–i–n structure. *Panels (A)–(C) are reprinted with permission from C. Bi, Q. Wang, Y. Shao, Y. Yuan, Z. Xiao, J. Huang, Non-wetting surface-driven high-aspect-ratio crystalline grain growth for efficient hybrid perovskite solar cells, Nat. Commun. 6 (2015) 7747. Copyright 2015, Nature Publishing Group. Panels (D)–(F) are reprinted with permission from Y. Bai, Q. Dong, Y. Shao, Y. Deng, Q. Wang, L. Shen, et al., Enhancing stability and efficiency of perovskite solar cells with crosslinkable silane-functionalized and doped fullerene, Nat. Commun. 7 (2016) 12806. Copyright 2016, Nature Publishing Group.*

3.3 EVOLUTION OF THE FILM DEPOSITION METHODS

Advances in the design of device architecture drove the evolution of PSCs, especially in the early stage (2009–2013) of the technical evolution. As the technology progressed, preparation of the perovskite layer became essential for fabricating high-efficiency devices. Generally, delicate control of the stoichiometry, crystallographic phase, and grain structure of the HOIP material is required to fabricate high-efficiency PSCs. Because these properties can be controlled by the preparation methods, substantial research into the development of film deposition techniques has occurred. During the evolution of PSCs, four primary deposition approaches, single-step solution process (Fig. 3.14A) [12], two-step solution process (Fig. 3.14B) [25], vapor assisted solution process (Fig. 3.14C) [46], and thermal evaporation process (Fig. 3.14D) [30], were developed to fabricate the perovskite films.

3.3.1 Single-Step Solution Process

The single-step solution deposition was first used to fabricate PSCs due to the ease of processing [11–13] and is now the most widely adopted processing technique for preparing the perovskite films. In this process, the precursor solution is prepared by dissolving organic halide MAX and metal halide PbX_2 in an aprotic polar solvent, such as γ-butyrolactone (GBL), dimethyl formamide (DMF), dimethyl sulfoxide (DMSO), *N*-2-methyl pyrrolidone, or a mixture of them. A wide range of precursor MAX to PbX_2 ratios has been used to fabricate high-efficiency PV devices [47]. For example, high PCE $MAPbI_3$ devices were prepared using a MAI to PbI_2 ratio of 0.8:1 (MAI-poor) [48] and 3:1 (MAI-rich) [49]. To deposit a perovskite film using the single-step solution process, the precursor solution is spin-coated on a substrate followed by a postdeposition annealing at temperatures ranging from 80 to 150°C (Fig. 3.14A) [1,50]. The appropriate selection of postdeposition temperature and time based on the precursor composition is crucial for fabricating high-quality perovskite films [29,47,51]. Additionally, the environmental conditions (e.g., oxygen and humidity), morphology of the substrates, and other processing parameters (rotational speed and time) can also influence the uniformity, crystallinity, phase purity, surface morphology, and interface properties of the perovskite films.

Advancements of the single-step solution deposition method started with understanding the effect of substrate morphology and annealing

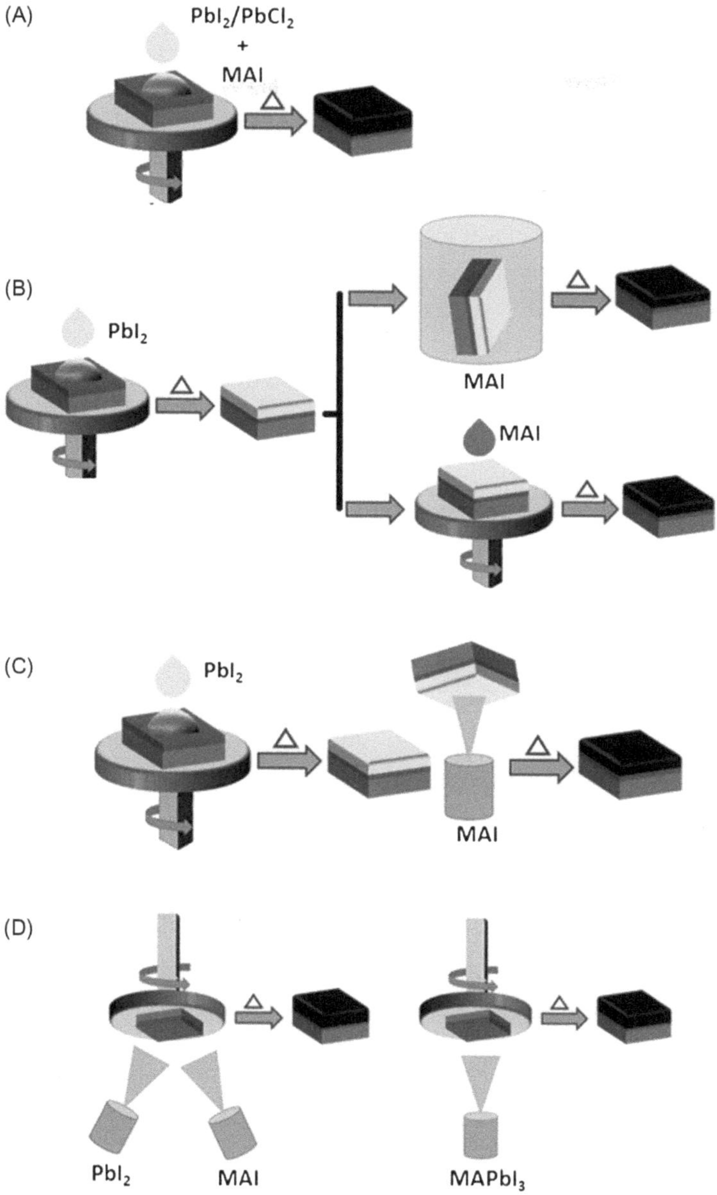

Figure 3.14 Preparation methods for perovskite absorber: (A) single-step solution process, (B) two-step solution process, (C) vapor assisted solution process, and (D) thermal vapor process.

temperature on the surface morphology of the perovskite films. Snaith et al. showed that the perovskite capping layer coverage increased with decreasing the mesoporous Al_2O_3 thickness [29]. However, poor surface coverage was observed with absence of the mesoporous Al_2O_3 layer, indicating that the mesoporous scaffold was important to form a conformal perovskite coating. They further investigated the morphology of perovskite films on a c-TiO_2 layer annealed at different temperatures and found that increasing the annealing temperature led to less surface coverage of the perovskite films [29]. Therefore, temperature that is sufficiently high to crystalize the perovskite film while preserving a good surface coverage is recommended for the film preparation. Dualeh et al. investigated the effects of postdeposition annealing on the surface morphology of $MAPbI_3$ films deposited on a mp-TiO_2 layer [52]. The best devices were achieved using temperatures in the range of 80–100°C, due to a smooth surface morphology and good phase purity.

Clearly, the empirical optimization of various deposition parameters can lead to high-efficiency PSCs. However, without the development of rational synthetic guidelines, further advances would be difficult to achieve. To address this need, a recent work systematically studied the impact of the precursor composition and reaction temperature on the $MAPbI_3$ formation using the single-step deposition method and proposed a near-equilibrium phase diagram for the perovskite formation (Fig. 3.15)

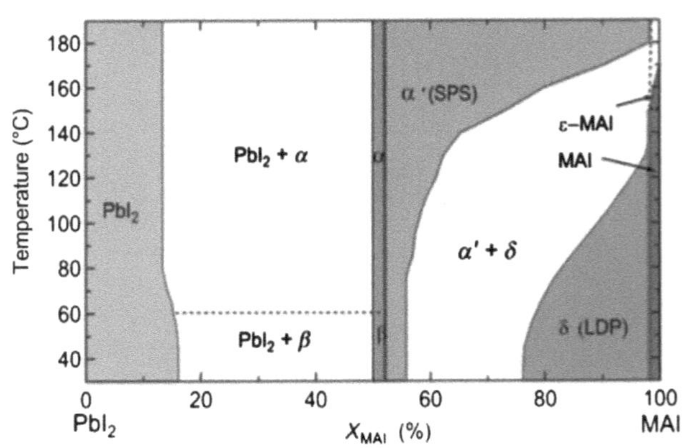

Figure 3.15 Proposed phase diagram for $MAPbI_3$. *Reprinted with permission from Z. Song, S.C. Watthage, A.B. Phillips, B.L. Tompkins, R.J. Ellingson, M.J. Heben, Impact of processing temperature and composition on the formation of methylammonium lead iodide perovskites, Chem. Mater. 27 (2015) 4612–4619. Copyright 2015 American Chemical Society.*

[47]. Although the pure tetragonal $MAPbI_3$ perovskite phase (α) occurs in a narrow compositional space where the composition is close to the stoichiometric value [$MAI/(MAI + PbI_2) = 0.5$], the benign nature of the stacked perovskite sheet phase that formed in the MAI-rich region enables a wider composition range for high-efficiency device fabrication. The result is consistent with early work, in which the off-stoichiometry MAI to PbI_2 ratio (3:1, $X_{MAI} = 75\%$) was used in the precursor solution and the films were subjected to a high temperature ($\sim 150°C$) annealing [21]. This off-stoichiometry precursor composition led to smoother morphology than using a stoichiometric precursor solution at that time.

3.3.2 Two-Step Solution Deposition

In the early stage of PSC evolution, the devices prepared by the single-step solution deposition was inhibited by the poor surface coverage of the perovskite film (Fig. 3.16A–D). To overcome this, Burschka et al. developed the two-step solution deposition method, based on the pioneering work by Mitzi et al. [8], to prepare uniform $MAPbI_3$ films on the mp-TiO_2 layer [25]. In the two-step method, a PbI_2 seed layer is first deposited and subsequently reacted with MAI in isopropanol (IPA) solution introduced by dipping or spin-coating (Fig. 3.14B) to form $MAPbI_3$.

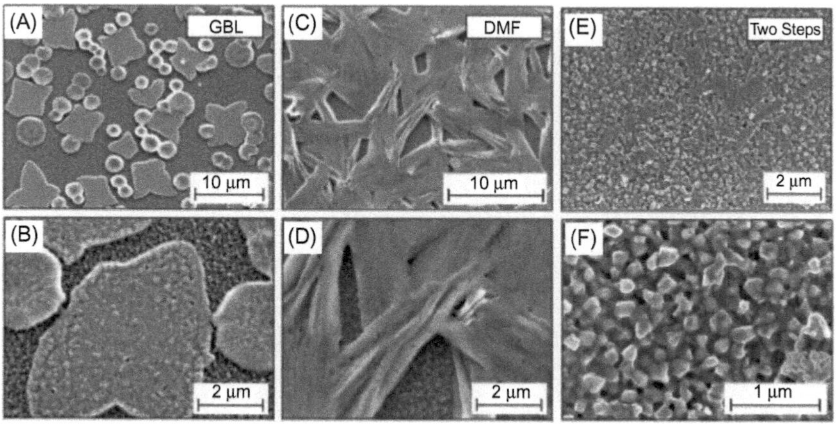

Figure 3.16 $MAPbI_3$ perovskite films prepared by the single-step solution deposition using (A), (B) γ-butyrolactone (GBL) and (C), (D) dimethyl formamide (DMF) solvents. Panels (E), (F) show the fabricated $MAPbI_3$ films using the two-step sequential deposition. *Reprinted with permission from J. Burschka, N. Pellet, S.-J. Moon, R. Humphry-Baker, P. Gao, M.K. Nazeeruddin, et al., Sequential deposition as a route to high-performance perovskite-sensitized solar cells, Nature 499 (2013) 316–319. Copyright 2013, Nature Publishing Group.*

The films prepared by the two-step method were dense and conformal (Fig. 3.16E and F). In contrast, films prepared using the single-step solution method exhibited highly porous morphologies consisting of perovskite crystallites in spherical and polygonal shapes (with GBL) (Fig. 3.16A and B) or needle shape (with DMF) (Fig. 3.16C and D).

Im et al. continued the work on the two-step deposition by investigating the relationship between grain size of perovskite and the MAI concentration [26]. They found that grain size decreased with increasing MAI concentration, with the best PCE of 17.01% achieved at the MAI concentration of 0.038 M. Due to the success in fabrication of high-efficiency PSCs in the normal device structure, the two-step solution method was subsequently applied to the planar n-i-p and p—i—n architectures [41,53,54]. However, there were several drawbacks to this method. The major one is the trade-off between perovskite grain size and surface smoothness. Surface roughness increases with increasing the size of perovskite grains, leading to a higher surface recombination and leakage current. In contrast, smooth films with small grains suffer from low carrier lifetime and short diffusion length. Another drawback is the incomplete conversion of PbI_2, especially in the planar structures [55,56]. The remaining PbI_2 not only decreases the total light absorption but also blocks the efficient charge carrier transport to the selective contacts, resulting in reduced V_{OC} and J_{SC}. In the past few years, these hurdles were overcome using advanced film deposition techniques (see Section 3.4) and devices with PCEs of >20% were fabricated using the two-step methods [57].

3.3.3 Vapor Assisted Solution Process

The vapor assisted solution process was developed as a modified two-step solution deposition where the MAI is introduced by a vapor deposition process. Chen et al. introduced this method to fabricate uniform and conformal $MAPbI_3$ films with high-phase purity (Fig. 3.17A) [46]. They deposited the PbI_2 seed layer by spin-coating, followed by exposing to the MAI vapor generated by heating the MAI powder at 150°C in an inert environment. This method effectively avoids delamination issues during the two-step solution process when the sample is removed from the MAI solution or washed by the IPA solution [41,46]. Perovskite films formed by this method showed a full conversion of PbI_2 and a conformal, uniform, and smooth film with large sized grains up to the micrometer

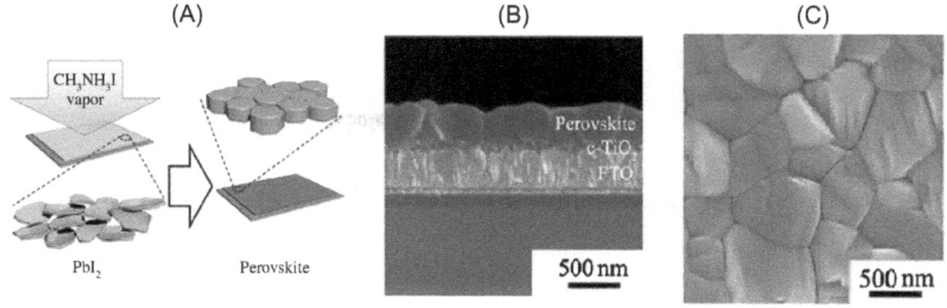

Figure 3.17 (A) Schematic illustration, (B) cross-sectional, and (C) surface scanning electron microscope (SEM) images of perovskite film prepared by vapor assisted solution process. *Reprinted with permission from Q. Chen, H. Zhou, Z. Hong, S. Luo, H.-S. Duan, H.-H. Wang, et al., Planar heterojunction perovskite solar cells via vapor-assisted solution process, J. Am. Chem. Soc. 136 (2014) 622–625. Copyright 2014 American Chemical Society.*

scale (Fig. 3.17B and C). However, the process of MAI incorporation into the PbI_2 framework is much slower than the solution deposition, needing several hours for complete conversion. This long processing time limits the practical application of this method.

3.3.4 Thermal Vapor Deposition

Thermal evaporation was used to fabricate HOIP films in the 1990s because it was widely used in the semiconductor industry [58,59]. In 2013, Snaith et al. rejuvenated this method by using the dual coevaporation to prepare the first planar heterojunction PSCs [30]. $MAPbI_{3-x}Cl_x$ films were prepared by coevaporation of $PbCl_2$ and MAI followed by annealing (Fig. 3.18A). Smooth, conformal, and uniform perovskite films were the result, and then a record PCE of 15.4% was achieved. This success stimulated development of alternative vapor deposition methods, such as the layer-by-layer vacuum evaporation [60], chemical vapor deposition (Fig. 3.18B) [61], and flash evaporation (Fig. 3.18C) [62]. Perovskite films deposited using vapor deposition are more uniform and pinhole-free compared to solution processed films [63]. This can lead to high-efficiency devices; however, due to the low thermal stability of both the precursor sources and the products, control over temperature during deposition is difficult and often results in off-stoichiometric films. Therefore, only few research groups have fabricated high-efficiency devices using this method [30,39,60,64,65].

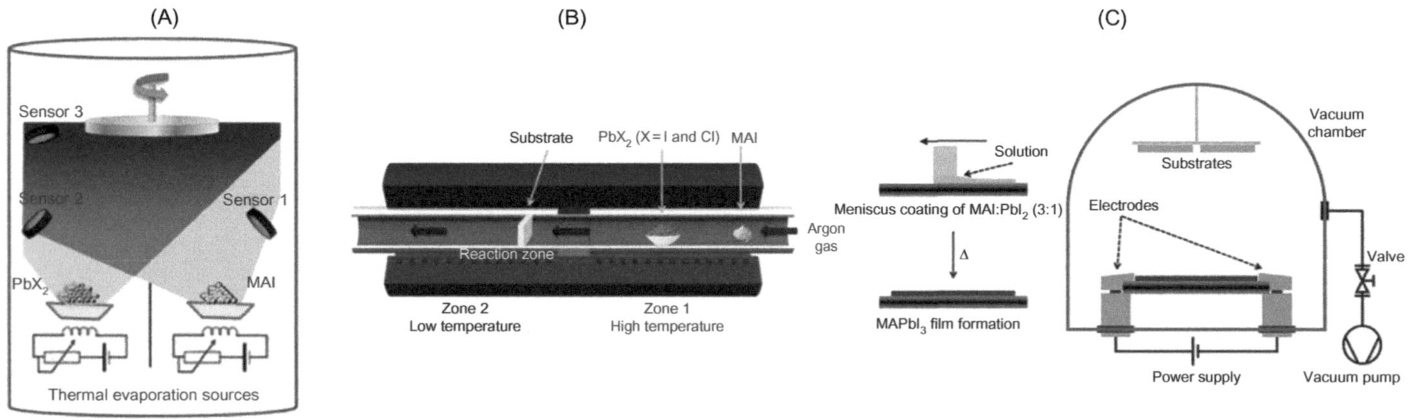

Figure 3.18 Schematic images of (A) dual source evaporation, (B) chemical vapor deposition, and (C) flash evaporation. *Panel (A) is reprinted with permission from M. Sessolo, C. Momblona, L. Gil-Escrig, H.J. Bolink, Photovoltaic devices employing vacuum-deposited perovskite layers. MRS Bull. 40 (2015) 660–666. Copyright 2015 Materials Research Society. Panel (B) is reprinted with permission from M.M. Tavakoli, L. Gu, Y. Gao, C. Reckmeier, J. He, A.L. Rogach, et al., Fabrication of efficient planar perovskite solar cells using a one-step chemical vapor deposition method, Sci. Rep. [Online], 2015, p. 14083. ⟨http://dx.doi.org/10.1038/srep14083⟩ (accessed 09/22/online). Copyright 2015 Nature Publishing Group. Panel (C) is reprinted with permission from G. Longo, L. Gil-Escrig, M.J. Degen, M. Sessolo, H.J. Bolink, Perovskite solar cells prepared by flash evaporation, Chem. Commun. 51 (2015) 7376–7378. Copyright 2015 Royal Society of Chemistry.*

3.4 ADVANCED DEVICE ENGINEERING TECHNIQUES

The rapid progress of PSCs in the early stage (2012–14) relied mainly on the innovation on the device architecture and film preparation method. During this period, the champion PCE was rapidly increased to 15%. The further advances in fabricating high-efficiency PSCs with PCEs of ~20% were enabled by some advanced device engineering techniques developed in 2014–16 to address problems associated with the primary deposition methods and improve the quality of the perovskite films [1]. These strategies, solvent engineering, deposition process engineering, contact material engineering, and band-gap engineering, yield uniform and dense perovskite films with improved optoelectronic properties and interface passivation. Consequently, the present state-of-the-art PSCs typically employed combined engineering strategies to achieve high performance and stability. In addition to single junction PSCs, research on perovskite-based tandem devices has been progressed as a promising future direction.

3.4.1 Solvent Engineering

After common device structures and film deposition methods were established, the most challenging aspect of processing PSCs became achieving a homogeneous composition and uniform thickness over large area. In the early 2014, two-step solution process was mostly used to prepare uniform and pinhole-free perovskite films because of easy process and high reproducibility. In contrast, single-step method often resulted in inconsistent film morphology with high porosity (Fig. 3.16). To address this, many researchers investigated solvent engineering techniques to improve the surface morphology of single-step solution processed perovskite films.

Jeon et al. first used DMSO as an additive to the precursor solution, improving the crystallinity and morphology of perovskite films [66]. During the single-step deposition, the intermediate phase of MAI-PbI_2-DMSO was formed via an intercalation process, retarding the crystallization kinetics of perovskite formation. The intermediate phase was then converted into $MAPbI_3$ via a heat treatment, which removes the DMSO molecules (Fig. 3.19A). This process allowed a uniform grain growth rate of perovskite crystals, leading to the formation of uniform and dense perovskite layers. By incorporating the DMSO additive, Jeon et al. demonstrated PSCs with an improved PCE of 16.5% (Fig. 3.19B) and high

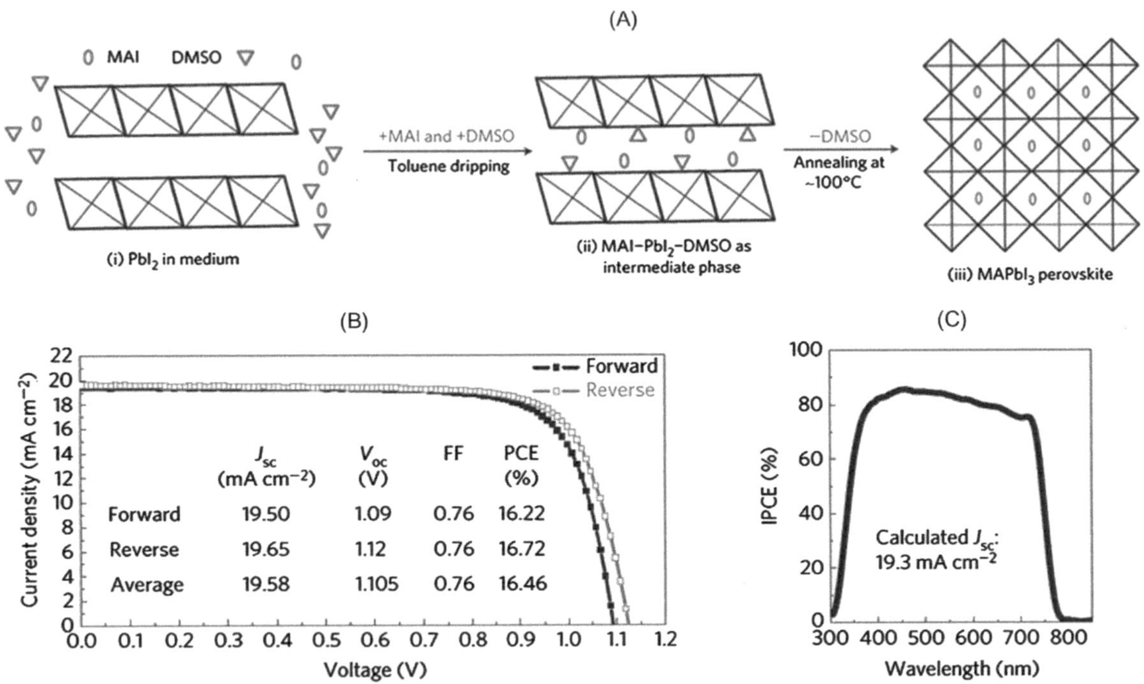

Figure 3.19 (A) Mechanism illustration of solvent engineering using dimethyl sulfoxide (DMSO). (B) J–V characteristics and (C) incident photon-to-electron conversion efficiency (IPCE) curve of perovskite solar cells (PSCs) prepared with DMSO additive. *Reprinted with permission from N.J. Jeon, J.H. Noh, Y.C. Kim, W.S. Yang, S. Ryu, S.I. Seok, Solvent engineering for high-performance inorganic–organic hybrid perovskite solar cells, Nat. Mater. 13 (2014) 897–903. Copyright 2014, Nature Publishing Group.*

photon-to-electron conversion efficiencies over the visible light spectrum (Fig. 3.19C) [66].

Besides DMSO that was added to the precursor solution for decoupling the nucleation and grain growth processes during the deposition, several other additives such as CH_3NH_3Cl [67], HI [68], MAI/I_2 [69], NH_4Cl [70], H_2O/HBr [71], 1,8-diiodooctane [72], 5-ammoniumvaleric acid [73], and phosphonic acid ammonium [74] have also been used in the single-step solution deposition to fabricate perovskite films with higher crystallinity and better surface coverage. The solvent engineering technique was also applied to the two-step solution deposition. The conventional two-step solution deposition is hampered by the incomplete conversion that leaves a dense PbI_2 layer on the ETM or HTM [75,76]. When MAI is introduced on to a dense PbI_2 seed layer, the top surface reacts with MAI and forms a dense perovskite capping layer, which hinders the MAI diffusion to the underlying PbI_2 layer [75]. Fabricating a porous PbI_2 film is an effective way to avoid the formation of dense perovskite capping layer and increase the perovskite conversion. Adding additives such as *n*-butanol [77], MAI [53], H_2O [78], or 4-tert-butylpyridine [54] in the PbI_2/DMF precursor solution, or by employing DMSO as a solvent [55], researchers can control the porosity in the PbI_2 layer. Additionally, modifying the MAI solution by incorporating small concentrations of metal ions (e.g., Zn^{2+}, Cd^{2+}, Hg^{2+}, etc.) was also used to prepare perovskite films with uniform morphology and large grains [79,80]. Despite these advancements, two-step has become less popular after the recent progress of deposition process engineering techniques.

3.4.2 Deposition Process Engineering

Jeon et al. introduced the antisolvent drop casting technique to complement the solvent engineering in the single-step solution deposition [66]. A precursor solution consisting of the equimolar MAX and PbX_2 in DMF/DMSO is span onto a substrate. During the last tens of seconds of spinning, an antisolvent to DMF, such as toluene, chloroform, chlorobenzene, diethyl ether is dripped on the substrate (Fig. 3.20A) [66,81,82]. The antisolvent addition rapidly extracts DMSO from the precursor solution, leading to the increased solute concentration in the precursor solution. Consequently, the intermediate $MAI{-}PbI_2{-}DMSO$ phase is rapidly formed, which avoids the needle- and spherical-shaped perovskite crystals that usually occur with DMF and GBL solvents, respectively.

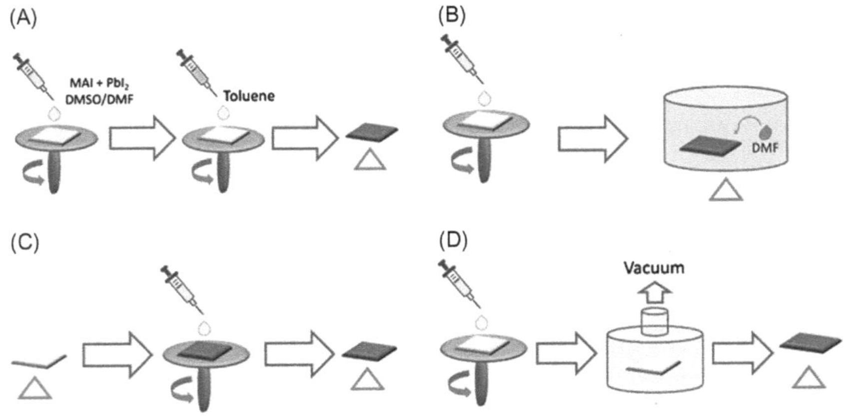

Figure 3.20 Schematics of (A) antisolvent drop casting, (B) solvent annealing process, (C) hot-casting deposition method, and (D) vacuum-flash assisted solution process.

After annealing, smooth and dense perovskite films with high-phase purity which are essential for fabricating high-efficiency PCSs are formed. Many research groups have further improved PCEs to >20% by optimizing the precursor composition and preparation recipe of this advanced processing technique [81–84].

Other modifications on the perovskite deposition and/or postdeposition processes have also been successfully applied for fabricating high-efficiency PSCs. Xiao et al. developed a solvent annealing technique for the two-step solution deposition (Fig. 3.20B) to enhance the grain size [42]. Annealing the as-deposited perovskite films with a small amount of DMF vapor enhanced the perovskite grain size from 300 nm to ~1 μm and increased PCE from 9.9% to 15.6%. Perovskite films annealed with pyridine [85] or MAI [86] vapor also demonstrated the improved photoluminescence and carrier lifetimes, indicating the formation of high-quality perovskites films that could be used to fabricate high-efficiency solar cells.

While the antisolvent technique promotes the formation of intermediate phase by delaying the crystal growth kinetics, the same results can be achieved by speeding the nucleation kinetics. Nie et al. demonstrated that the hot-casting method (Fig. 3.20C), in which crystallization of the perovskite film occurs immediately after a hot precursor solution is loaded onto the substrate hold at an elevated temperature, can be used to prepare high-quality perovskite films with millimeter-scale grains [43]. Li et al.

developed another approach called the vacuum-flash assisted solution process to replace the addition of antisolvent, (Fig. 3.20D) [87]. In this process, the spin-coated perovskite precursor film is placed in a vacuum chamber for a few seconds to remove the residual solvents more quickly. This rapid solvent extraction due to different vapor pressures of the solvents accelerates the crystallization of the perovskite intermediate phase, and the subsequent thermal annealing produces a high-quality perovskite film. The advantage of this technique over the antisolvent method is the ability to fabricate a uniform perovskite film over a large area. The champion cell fabricated with this method with a $>1\ cm^2$ aperture area achieved a maximum PCE of 20.5%, with a $V_{OC} = 1.14$ V, $J_{SC} = 23.24\ mA/cm^2$, and FF = 75.9%.

3.4.3 Contact Materials Engineering

In addition to the advances in the perovskite film preparation, the progress in ETM and HTM also played an essential role in the development of PSCs. Various materials possessing combined merits, including proper band alignment with the conduction or valence band of the perovskite absorber, efficient charge carrier transport, and low charge recombination defect densities at the interfaces, have been developed as the ETMs or HTMs. Fig. 3.21 shows the band energy levels of these commonly used ETM and HTM in comparison with HOIPs and electrodes.

Since the PSCs emerged, TiO_2 has been widely used as the compact and porous ETM in the p–i–n structure configuration [11,19]. With the evolution of perovskite technology, the thickness of the mp-TiO_2 layer, conventionally used in DSSCs, gradually decreased from $\sim 4\ \mu m$ to ~ 100 nm in the regular structure and was omitted in the planar structure [26,29–30]. This progress was enabled by the high crystallinity and long carrier lifetime of the HOIP materials. Removing the mesoporous oxide layer made the compact ETM an essential component of the devices for efficient collection of photoexcited electrons. Therefore, some dedicated control technique (e.g., atomic layer deposition (ALD)) has become popular to fabricate high-quality ETM. Besides c-TiO_2, other ETMs such as ZnO [88], SnO_2 [31–33,89,90], CdSe [91], CdS [92], TiO_2-graphene [40], and yttrium-doped TiO_2 [49] have been developed as ETMs. The use of these alternative ETMs allows PSCs to be processed without high temperature ($\sim 500°C$) treatment that is required for the anatase TiO_2. C_{60} and its derivatives such $PC_{61}BM$ or $PC_{71}BM$ are among the

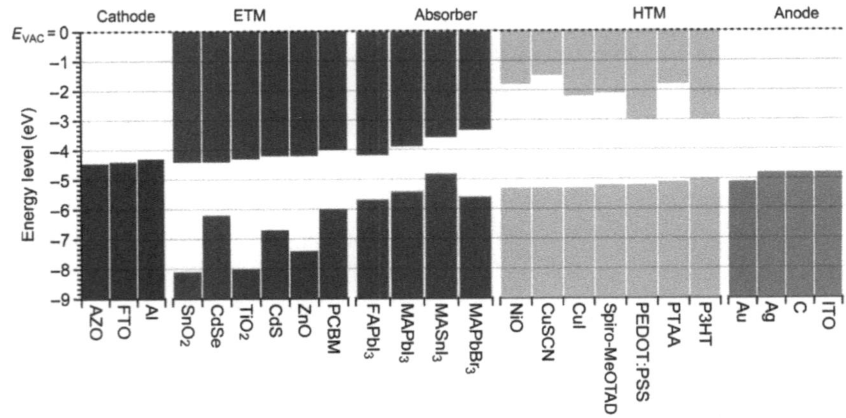

Figure 3.21 Energy levels of commonly used cathode, n-type (ETM), absorber, p-type (HTM), and anode materials in perovskite solar cells. *ETM*, electron transport materials; *HTM*, hole transport material. *Reprinted with permission from Z. Song, S.C. Watthage, A.B. Phillips, M.J. Heben, Pathways toward high-performance perovskite solar cells: review of recent advances in organo-metal halide perovskites for photovoltaic applications, J. Photonics Energy 6 (2016) 022001. Copyright 2016 Society of Photo Optical Instrumentation Engineers.*

commonly used organic ETM materials in the p–i–n device configuration [36,93]. These materials can also be processed at low temperatures, enabling the potential for roll-to-roll manufacturing on flexible substrates and incorporating with another cell to fabricate tandem devices.

HTMs are also critical for a good PV performance, especially for a high V_{OC} [94]. Both organic and inorganic HTMs have been investigated in recent years. The organic HTMs can be categorized into two groups: small molecules and conducting polymers. Spiro-MeOTAD is the most used small molecules HTM due to high device performance (>20% PCEs) [87]. However, the devices using spiro-OMeTAD typically exhibit poor stability, especially after exposure to humidity [95–97]. Conducting polymers of PTAA, P3HT [Poly(3-hexylthiophene-2,5-diyl)], and triazatruxene derivatives have also been applied as the HTMs [98]. Among them, PTAA (20.2%) [57] and triazatruxene derivatives (18.3%) [99] showed competitive PCEs to the Spiro-MeOTAD but improved stability. P3HT, on the other hand, is not good as the Spiro-MeOTAD, demonstrating only 12%–15% efficiencies [98,100]. Nonetheless, higher material cost and lack of stability of the organic HTMs are still the main hurdles for the commercialization of PSCs. One possible solution is employing cost-effective and more stable inorganic HTMs, such as NiO_x

[101−104], CuO [105], Cu_2O [105], and CuSCN [106]. Among these, the devices using NiO_x showed >17% PCE [103−104], which is promising for utilization in commercial manufacturing.

3.4.4 Band-gap Engineering of Perovskites

In the early stage of PSC progress (2009−2013), $MAPbI_3$ was used as the prototype due to its bandgap of 1.55 eV that is suitable for high-efficiency solar conversion [11,13]. As the field advanced, the composition of the perovskite AMX_3 has been altered by incorporating other monovalent cations A [formamidinium (FA^+) and Cs^+], divalent cation M (Sn^{2+}), and halide ions X (Br^- and Cl^-), which results in optical absorption tunability (Fig. 3.22) [107,108]. This versatility in optical bandgap enables various PV applications of PSCs, including single-junction devices as well as both the top and bottom cells in tandem structures [109].

The substitution of the monovalent A cation does not directly affect the electronic band structure in halide perovskites because the valance band maximum and conduction band minimum are mainly contributed

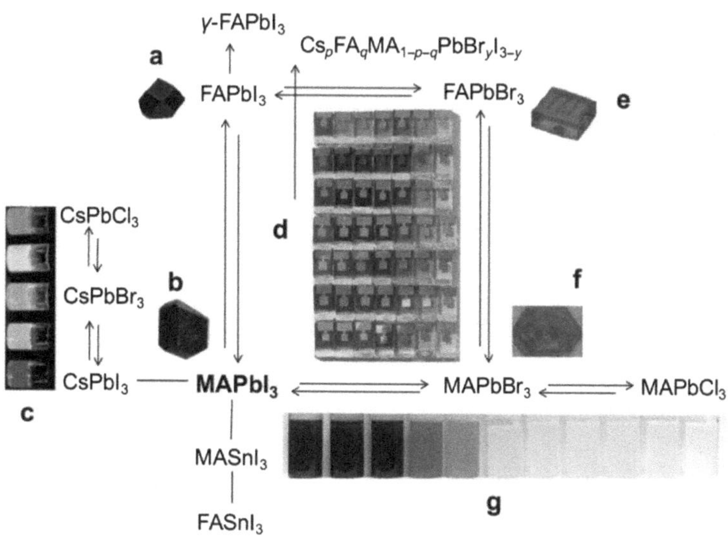

Figure 3.22 The versatility of hybrid perovskite materials $Cs_pFA_qMA_{1-p-q}PbBr_yI_{3-y}$ and their absorption tunability. *Reprinted with permission from J.-P. Correa-Baena, A. Abate, M. Saliba, W. Tress, T. Jesper Jacobsson, M. Gratzel, et al., The rapid evolution of highly efficient perovskite solar cells, Energy Environ. Sci. 10 (2017) 710−727. Copyright 2017, Royal Society of Chemistry.*

from the strong Pb-s and I-p antibonding state and Pb-p state, respectively [110]. However, the size of the A cation affects to the tolerance factor t and thus changes to the spacing of $[MX_6]^{4-}$ octahedra, altering the bandgap [15]. The A cations of FA^+, MA^+, and Cs^+ have the decreasing ionic sizes of 253, 217, and 181 pm, respectively, and thus the increasing bandgaps 1.48 eV ($FAPbI_3$), 1.52 eV ($MAPbI_3$) and 1.67 eV ($CsPbI_3$) (Fig. 3.23A) [108,111]. The $MA_xFA_{1-x}PbI_3$ mixed perovskite fabricated by engineering a mixture of FA^+ and MA^+ in the A site showed a red (black in print versions) shifted optical absorption with respect to $MAPbI_3$, extending solar energy absorption to the longer wavelengths thereby increasing the J_{SC} [112]. This mixed cation perovskite $MA_xFA_{1-x}PbI_3$ was used to fabricated PSCs with PCE >20% [57]. Additionally, the mixed cation perovskite $Cs_xFA_{1-x}PbI_3$ (Fig. 3.23B) has improved moisture stability and photoexcited carrier lifetime over $MAPbI_3$ or $FAPbI_3$ [34,113,114]. The enhanced interaction between FA^+ and I^- due to reduction of the octahedral volume increases the chemical inertness of the perovskite [113]. Moreover, incorporation of the inorganic Cs^+ stabilizes the perovskite phase of $FAPbI_3$ and reduces the defect density [114].

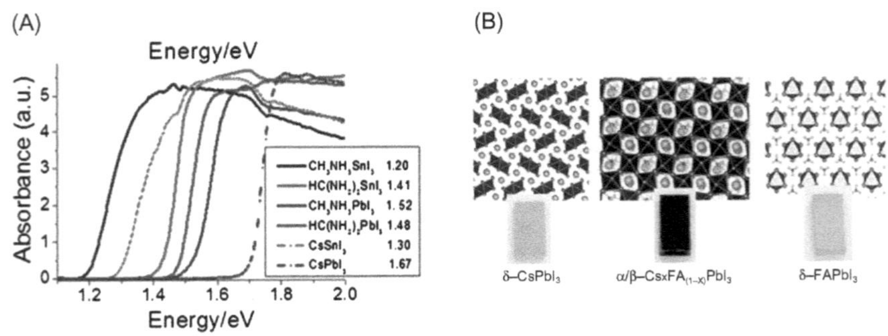

Figure 3.23 (A) UV−vis absorbance and band-gap values of $MASnI_3$, $FASnI_3$, $MAPbI_3$, $FAPbI_3$, $CsSnI_3$, and $CsPbI_3$. (B) Crystal structure and photos of $CsPbI_3$, $Cs_xFA_{1-x}PbI_3$, and $FAPbI_3$. *Panel (A) is reprinted with permission from C.C. Stoumpos, C.D. Malliakas, M.G. Kanatzidis, Semiconducting tin and lead iodide perovskites with organic cations: phase transitions, high mobilities, and near-infrared photoluminescent properties, Inorg. Chem. 52 (2013) 9019−9038. Copyright 2013 American Chemical Society. Panel (B) is reprinted with permission from C. Yi, J. Luo, S. Meloni, A. Boziki, N. Ashari-Astani, C. Gratzel, et al., Entropic stabilization of mixed A-cation ABX_3 metal halide perovskites for high performance perovskite solar cells, Energy Environ. Sci. 9 (2016) 656−662. Copyright 2016 Royal Society of Chemistry.*

The substitution of different halide anions X can be used to tune the optical bandgap of the metal halide perovskites (Fig. 3.22). $MAPb(I_{1-x}Br_x)_3$ is commonly used to fabricate high bandgap perovskite absorber layer [111,115,116]. Increasing Br content (x value) results in a continuous blue (gray in print versions) shift of the absorption band edge from 1.5 eV (830 nm) at $x = 0$ to 2.3 eV (540 nm) at $x = 1$ (Fig. 3.24A) [115–117]. In contrast to Br incorporation, which is used to alter the bandgap, the Cl mixed phase $MAPb(I_{1-x}Cl_x)_3$ shows negligible bandgap alternation to the pristine $MAPbI_3$ (Fig. 3.24B) due to the relatively low Cl concentrations (<4%) in the final product [12]. However, the charge carrier diffusion length in $MAPb(I_{1-x}Cl_x)_3$ is greater than 1 μm, which is one order of magnitude higher than the pure iodide perovskite [21], leading to the improved PV performance. Therefore, $MAPb(I_{1-x}Cl_x)_3$ was widely used to fabricate high-performance devices in the early stage (2012–2014) [12,117–119]. To date, perovskite absorbers combined mixed organic cations (FA^+ and MA^+) and halides (I^- and Br^-) have been widely used to fabricate $FA_{1-x}MA_xPb(I_{1-y}Br_y)_3$ based solar cells [120], which has achieved >20% PCE with a high degree of reproducibility [81,87]. The present champion PSC with the $FA_{1-x}MA_xPb$

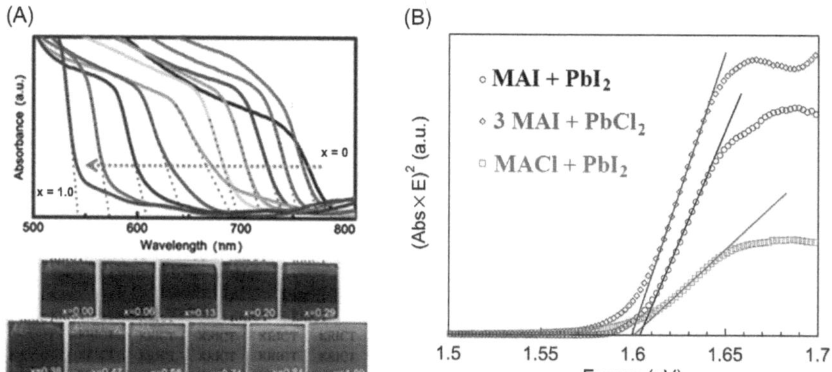

Figure 3.24 (A) Photographs and UV–vis absorption spectra of $MAPb(I_{1-x}Br_x)_3$. (B) UV–vis spectra of $MAPb(I_{1-x}Cl_x)_3$. *Panel (A) is reprinted with permission from J.H. Noh, S.H. Im, J.H. Heo, T.N. Mandal, S.I. Seok, Chemical management for colorful, efficient, and stable inorganic–organic hybrid nanostructured solar cells, Nano Lett. 13 (2013) 1764–1769. Copyright 2013 American Chemical Society. Panel (B) is reprinted with permission from S. Colella, E. Mosconi, P. Fedeli, A. Listorti, F. Gazza, F. Orlandi, et al., $MAPbI_{3-x}Cl_x$ mixed halide perovskite for hybrid solar cells: the role of chloride as dopant on the transport and structural properties, Chem. Mater. 25 (2013) 4613–4618. Copyright 2013 American Chemical Society.*

$(I_{1-y}Br_y)_3$ absorber demonstrated a PCE of 22.1%, with $V_{OC} = 1.105$, $J_{SC} = 24.97$ mA/cm^2, and FF = 80.3% [57].

The substitution of divalent cation M can also alter the bandgap of the perovskite absorber. Due to the environmental concerns with water-soluble lead based perovskites [121,122], many research efforts have been devoted to replacing Pb by nontoxic Sn [123,124]. The $MASnI_3$ perovskite has a lower bandgap (~1.2 eV) compared to its Pb counterpart (1.55 eV) and thus exhibits light absorption at the longer wavelengths, which is beneficial for a higher J_{SC} [108]. However, due to the high V_{OC} deficit, the pure Sn-based PSCs typically exhibited poor PV performance [123,125]. More recently, mixed tin−lead perovskite low-bandgap PSCs have been developed for bottom-cell tandem solar cells [126,127]. Recently, Zhao et al. achieved a 17.6% PCE, with $V_{OC} = 0.853$ V, $J_{SC} = 28.5$ mA/cm^2, and FF = 72.5% in a single junction PSC based on $(FASnI_3)_{0.6}(MAPbI_3)_{0.4}$ perovskite film [127]. Further, by connecting this with a ~1.58 eV bandgap perovskite top cell, they were able to achieve a steady-state efficiency of 21.0% in the four-terminal tandem cell. However, stability of the Sn-based PSCs is not comparable to their Pb-based counterparts [108,128] and needs to be addressed in the future research.

3.4.5 Perovskite Tandem Solar Cells

Due to their tunable bandgap that ranges of 1.2−2.3 eV and versatility in the device configurations, HOIPs are ideal candidates for tandem solar cells (Fig. 3.25) [109,111,115,116]. The $MAPb(I_xBr_{1-x})_3$ perovskites with relatively high bandgaps (1.5−2.3 eV) and high PCEs (~20%) are commonly used as the top cell for a variety of combinations of tandems, including perovskite-Si [129−131], perovskite-$CuInGaSe_2$ [132,133], and perovskite-Cu_2ZnSnS_4 [134]. The low band (1.2−1.5 eV) perovskites $(FASnI_3)_x(MAPbI_3)_{1-x}$ were developed recently as the bottom cells for the perovskite−perovskite tandems [127,135].

The perovskite tandem solar cells were first demonstrated in the four-terminal configuration (Fig. 3.25B) due to the simplicity of implementation. In this configuration, a high band-gap cell is mechanically mounted on top of a low band-gap cell. The subcells are fabricated independently, often by incompatible processes. Individual operation of the subcells is possible as well, which allows for a high combined PCE since the two cells can be independently maintained at differing maximum power

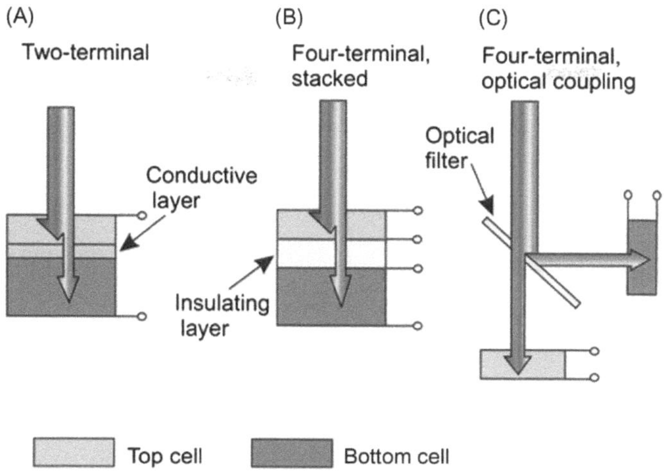

Figure 3.25 Three main configurations of tandem solar cells: (A) two-terminal monolithic, (B) four-terminal stacked, and (C) four-terminal optical coupling structures. *Reprinted with permission from Z. Yu, M. Leilaeioun, Z. Holman, Selecting tandem partners for silicon solar cells, Nat. Energy 1 (2016) 16137. Copyright 2016, Nature Publishing Group.*

points. Bailie et al. fabricated the first four-terminal perovskite tandems using the $MAPbI_3$ top cell coupled with $CuInGaSe_2$ and Si bottom cells, achieving 18.6% and 17.0% PCEs, respectively [133]. More recently, McMeekin et al. reported a high PCE of 25.2% by combining a top cell with $FA_{0.83}Cs_{0.17}Pb(I_{0.6}Br_{0.4})_3$ perovskite ($E_g = \sim 1.74$ eV) with a Silicon-heterojunction bottom cell [136]. Fu et al. demonstrated perovskite-$CuInGaSe_2$ tandems with an efficiency of 22.1% [137].

The four-terminal optical coupling tandem structure was developed as an alternative four-terminal configuration (Fig. 3.25C [138]). Instead of stacking the subcells, an optical splitter was used to split photons with short and long wavelengths to the high and low band-gap cells, respectively. Uzu et al. fabricated such tandems by coupling a $MAPbI_3$ perovskite cell and a silicon heterojunction bottom cell with a 550 nm cutoff splitter and achieved a PCE of 28% [139]. The high PCE is due to the optical splitter which allows more efficient utilization of long wavelength photons; however, the additional cost for the optical components may make it infeasible for practical application.

After the demonstration of four-terminal tandem devices and the advances in the processing techniques for HOIP materials, two-terminal monolithic perovskite-based tandems were developed. The low temperature processing allows the perovskite top cell being processed directly on the bottom cell (e.g., Si and $CuInGaSe_2$) and connected via a recombination layer. Due to the two-terminal configuration, only a single transparent conductive electrode is needed, lowering parasitic absorption in the electrode and manufacturing cost. However, it is challenging to fabricate high-efficiency devices due to the complexity of processing the PSC on a completed bottom cell and electrical and optical design of the whole tandem. At present, perovskite-Si and perovskite-$CuInGaSe_2$ tandems in this configuration have demonstrated PCEs up to 23.6% [130,131,140] and 10.9% [132], respectively.

More recently, perovskite–perovskite tandem solar cells proposed by many researchers due to a tunable bandgap were demonstrated [111,115,116]. Both two-terminal monolithically integrated and four-terminal mechanically stacked perovskite–perovskite tandems have been developed in recent years [127,135,141]. Eperon et al. fabricated both types using a $FA_{0.83}Cs_{0.17}Pb(I_{0.5}Br_{0.5})_3$ top cell (1.8 eV) and a $FA_{0.75}Cs_{0.25}Sn_{0.5}Pb_{0.5}I_3$ bottom cell (1.2 eV), demonstrating 17% and 20% PCEs, respectively [135]. This perovskite–perovskite tandem is very promising due to the potential of high PCE and low costs, which could be the future direction of PSCs.

3.5 OUTLOOK AND SUMMARY

Despite the rapid progress of PSCs, the obvious challenge that limited the commercialization of PSCs is lack of long-term and operational stability. In recent years, significant effort has been directed to understanding the operative decay mechanisms and developing appropriate solutions to improve device stability [34,142], but there is much to learn and control if PSCs are to be widely deployed. Although the underlying degradation mechanisms are far from completely understood, degradation has been attributed to humidity [143], oxygen [144], light [145], heat [146], and intrinsic instability [147]. In nearly all studies, humidity has been demonstrated as the most aggressive cause of instability in materials and devices due to the strong interaction between water molecules and perovskite materials [97]. However, isolating the device from the external environment via encapsulation procedures [148,149] can significantly slow the degradation. Additionally, PSCs employing hydrophobic back contact materials (e.g., carbon fiber) have also demonstrated enhanced performance stability in outdoor tests [150,151].

Hysteresis in the $J-V$ characteristics of devices is another important effect that must be controlled. Without a complete understanding of the origins of hysteresis, as well as methods to circumvent or control it, it will be impossible to manage the operation of PSCs. Hysteresis has received elevated attention recently, but its origin and mitigation are still under investigation [152–156]. Although there are evidences that hysteresis could be attributed to the formation and migration of ionic defects and the interfaces between the perovskite absorber and the selective contacts, a detailed understanding of the degradation and hysteresis mechanisms in PSCs is needed. If the advances in this emerging PV technology continue and the stability problem can be solved, perovskite PV would be a viable option in the PV market. Such an economic advantage without subsidies would stimulate a wider degree of PV deployment.

In summary, the solution-processed PSCs have made remarkable progress, becoming one of the most promising PV technologies of the future. High-efficiency PSCs with a diverse set of device architectures and material combinations have been successfully developed. In this chapter, we discussed the evolution of device structures, film deposition methods, and advanced device engineering techniques that have led to the unprecedented rate of device advancement. Additional research is needed to provide a better understanding of the material and device properties and address challenges such as long-term stability and hysteresis in $J-V$ characteristics. Moving forward, it is expected that more encouraging results will be reported, and PSCs will continue to be an area of interest of the PV community.

ACKNOWLEDGMENTS

This work was financially supported by the Office of Naval Research (contract no. N00014-17-1-2223) and the National Science Foundation (contract no. ECCS-1665172).

REFERENCES

[1] Z. Song, S.C. Watthage, A.B. Phillips, M.J. Heben, Pathways toward high-performance perovskite solar cells: review of recent advances in organo-metal halide perovskites for photovoltaic applications, J. Photonics Energy 6 (2016) 022001.

[2] U. Schwarz, F. Wagner, K. Syassen, H. Hillebrecht, Effect of pressure on the optical-absorption edges of CsGeBr3 and CsGeCl3, Phys. Rev. B: Condens. Matter 53 (1996) 12545–12548.

[3] R. von Helmolt, J. Wecker, B. Holzapfel, L. Schultz, K. Samwer, Giant negative magnetoresistance in perovskitelike $La_{2/3}Ba_{1/3}MnO_x$ ferromagnetic films, Phys. Rev. Lett. 71 (1993) 2331–2333.
[4] J.H. Haeni, P. Irvin, W. Chang, R. Uecker, P. Reiche, Y.L. Li, et al., Room-temperature ferroelectricity in strained $SrTiO_3$, Nature 430 (2004) 758–761.
[5] J.G. Bednorz, K.A. Müller, Perovskite-type oxides—the new approach to high-Tc superconductivity, Rev. Mod. Phys. 60 (1988) 585–600.
[6] H. Arend, W. Huber, F.H. Mischgofsky, G.K. Richter-Van Leeuwen, Layer perovskites of the $(C_nH_{2n+1}NH_3)_2MX_4$ and $NH_3(CH_2)_mNH_3MX_4$ families with M = Cd, Cu, Fe, Mn OR Pd and X = Cl OR Br: importance, solubilities and simple growth techniques, J. Cryst. Growth 43 (1978) 213–223.
[7] D.B. Mitzi, Templating and structural engineering in organic–inorganic perovskites, J. Chem. Soc., Dalton Trans. 1 (2001) 1–12.
[8] K. Liang, D.B. Mitzi, M.T. Prikas, Synthesis and characterization of organic–inorganic perovskite thin films prepared using a versatile two-step dipping technique, Chem. Mater. 10 (1998) 403–411.
[9] C.R. Kagan, D.B. Mitzi, C.D. Dimitrakopoulos, Organic–inorganic hybrid materials as semiconducting channels in thin-film field-effect transistors, Science 286 (1999) 945–947.
[10] D.B. Mitzi, K. Chondroudis, C.R. Kagan, Organic–inorganic electronics, IBM J. Res. Dev. 45 (2001) 29–45.
[11] A. Kojima, K. Teshima, Y. Shirai, T. Miyasaka, Organometal halide perovskites as visible-light sensitizers for photovoltaic cells, J. Am. Chem. Soc. 131 (2009) 6050–6051.
[12] M.M. Lee, J. Teuscher, T. Miyasaka, T.N. Murakami, H.J. Snaith, Efficient hybrid solar cells based on meso-superstructured organometal halide perovskites, Science 338 (2012) 643–647.
[13] H.-S. Kim, C.-R. Lee, J.-H. Im, K.-B. Lee, T. Moehl, A. Marchioro, et al., Lead iodide perovskite sensitized all-solid-state submicron thin film mesoscopic solar cell with efficiency exceeding 9%, Sci. Rep. 2 (2012) 591.
[14] H.J. Snaith, Perovskites: the emergence of a new era for low-cost, high-efficiency solar cells, J. Phys. Chem. Lett. 4 (2013) 3623–3630.
[15] M.A. Green, A. Ho-Baillie, H.J. Snaith, The emergence of perovskite solar cells, Nat. Photonics 8 (2014) 506–514.
[16] [No authors listed], Perovskite fever, Nat. Mater. 13 (2014) 837.
[17] NREL, Solar cell efficiency chart, 2017 ⟨https://www.nrel.gov/pv/assets/images/efficiency-chart.png⟩ (accessed March 2017).
[18] B. O'Regan, M. Gratzel, A low-cost, high-efficiency solar cell based on dye-sensitized colloidal TiO_2 films, Nature 353 (1991) 737–740.
[19] J.-H. Im, C.-R. Lee, J.-W. Lee, S.-W. Park, N.-G. Park, 6.5% efficient perovskite quantum-dot-sensitized solar cell, Nanoscale 3 (2011) 4088–4093.
[20] T. Leijtens, S.D. Stranks, G.E. Eperon, R. Lindblad, E.M.J. Johansson, I.J. McPherson, et al., Electronic properties of meso-superstructured and planar organometal halide perovskite films: charge trapping, photodoping, and carrier mobility, ACS Nano 8 (2014) 7147–7155.
[21] S.D. Stranks, G.E. Eperon, G. Grancini, C. Menelaou, M.J.P. Alcocer, T. Leijtens, et al., Electron-hole diffusion lengths exceeding 1 micrometer in an organometal trihalide perovskite absorber, Science 342 (2013) 341–344.
[22] G. Xing, N. Mathews, S. Sun, S.S. Lim, Y.M. Lam, M. Graetzel, et al., Long-range balanced electron- and hole-transport lengths in organic–inorganic $CH_3NH_3PbI_3$, Science 342 (2013) 344–347.

[23] H.S. Kim, S.H. Im, N.G. Park, Organolead halide perovskite: new horizons in solar cell research, J. Phys. Chem. C 118 (2014) 5615–5625.

[24] J.H. Heo, S.H. Im, J.H. Noh, T.N. Mandal, C.-S. Lim, J.A. Chang, et al., Efficient inorganic–organic hybrid heterojunction solar cells containing perovskite compound and polymeric hole conductors, Nat. Photonics 7 (2013) 486–491.

[25] J. Burschka, N. Pellet, S.-J. Moon, R. Humphry-Baker, P. Gao, M.K. Nazeeruddin, et al., Sequential deposition as a route to high-performance perovskite-sensitized solar cells, Nature 499 (2013) 316–319.

[26] J.-H. Im, I.-H. Jang, N. Pellet, M. Grätzel, N.-G. Park, Growth of $CH_3NH_3PbI_3$ cuboids with controlled size for high-efficiency perovskite solar cells, Nat. Nano. 9 (2014) 927–932.

[27] F. Giordano, A. Abate, J.P. Correa Baena, M. Saliba, T. Matsui, S.H. Im, et al., Enhanced electronic properties in mesoporous TiO_2 via lithium doping for high-efficiency perovskite solar cells, Nat. Commun. 7 (2016) 10379.

[28] J.M. Ball, M.M. Lee, A. Hey, H.J. Snaith, Low-temperature processed meso-superstructured to thin-film perovskite solar cells, Energy Environ. Sci. 6 (2013) 1739–1743.

[29] G.E. Eperon, V.M. Burlakov, P. Docampo, A. Goriely, H.J. Snaith, Morphological control for high performance, solution-processed planar heterojunction perovskite solar cells, Adv. Funct. Mater. 24 (2014) 151–157.

[30] M. Liu, M.B. Johnston, H.J. Snaith, Efficient planar heterojunction perovskite solar cells by vapour deposition, Nature 501 (2013) 395–398.

[31] Q. Jiang, L. Zhang, H. Wang, X. Yang, J. Meng, H. Liu, et al., Enhanced electron extraction using SnO_2 for high-efficiency planar-structure $HC(NH_2)_2PbI_3$-based perovskite solar cells, Nat. Energy 2 (2016) 16177.

[32] J.-Y. Seo, T. Matsui, J. Luo, J.-P. Correa-Baena, F. Giordano, M. Saliba, et al., Ionic liquid control crystal growth to enhance planar perovskite solar cells efficiency, Adv. Energy Mater. 6 (2016) 1600767.

[33] E.H. Anaraki, A. Kermanpur, L. Steier, K. Domanski, T. Matsui, W. Tress, et al., Highly efficient and stable planar perovskite solar cells by solution-processed tin oxide, Energy Environ. Sci. 9 (2016) 3128–3134.

[34] T. Leijtens, G.E. Eperon, N.K. Noel, S.N. Habisreutinger, A. Petrozza, H.J. Snaith, Stability of metal halide perovskite solar cells, Adv. Energy Mater. 5 (2015) 1500963.

[35] J.-Y. Jeng, Y.-F. Chiang, M.-H. Lee, S.-R. Peng, T.-F. Guo, P. Chen, et al., $CH_3NH_3PbI_3$ perovskite/fullerene planar-heterojunction hybrid solar cells, Adv. Mater. 25 (2013) 3727–3732.

[36] S. Sun, T. Salim, N. Mathews, M. Duchamp, C. Boothroyd, G. Xing, et al., The origin of high efficiency in low-temperature solution-processable bilayer organometal halide hybrid solar cells, Energy Environ. Sci. 7 (2014) 399–407.

[37] P. Docampo, J.M. Ball, M. Darwich, G.E. Eperon, H.J. Snaith, Efficient organometal trihalide perovskite planar-heterojunction solar cells on flexible polymer substrates, Nat. Commun. 4 (2013) 2761.

[38] J. You, Z. Hong, Y. Yang, Q. Chen, M. Cai, T.-B. Song, et al., Low-temperature solution-processed perovskite solar cells with high efficiency and flexibility, ACS Nano 8 (2014) 1674–1680.

[39] O. Malinkiewicz, A. Yella, Y.H. Lee, G. Minguez Espallargas, M. Graetzel, M.K. Nazeeruddin, et al., Perovskite solar cells employing organic charge-transport layers, Nat. Photonics 8 (2014) 128–132.

[40] J.T.-W. Wang, J.M. Ball, E.M. Barea, A. Abate, J.A. Alexander-Webber, J. Huang, et al., Low-temperature processed electron collection layers of graphene/TiO_2 nanocomposites in thin film perovskite solar cells, Nano Lett. 14 (2014) 724–730.

[41] Z. Xiao, C. Bi, Y. Shao, Q. Dong, Q. Wang, Y. Yuan, et al., Efficient, high yield perovskite photovoltaic devices grown by interdiffusion of solution-processed precursor stacking layers, Energy Environ. Sci. 7 (2014) 2619–2623.
[42] Z. Xiao, Q. Dong, C. Bi, Y. Shao, Y. Yuan, J. Huang, Solvent annealing of perovskite-induced crystal growth for photovoltaic-device efficiency enhancement, Adv. Mater. 26 (2014) 6503–6509.
[43] W. Nie, H. Tsai, R. Asadpour, J.-C. Blancon, A.J. Neukirch, G. Gupta, et al., High-efficiency solution-processed perovskite solar cells with millimeter-scale grains, Science 347 (2015) 522–525.
[44] C. Bi, Q. Wang, Y. Shao, Y. Yuan, Z. Xiao, J. Huang, Non-wetting surface-driven high-aspect-ratio crystalline grain growth for efficient hybrid perovskite solar cells, Nat. Commun. 6 (2015) 7747.
[45] Y. Bai, Q. Dong, Y. Shao, Y. Deng, Q. Wang, L. Shen, et al., Enhancing stability and efficiency of perovskite solar cells with crosslinkable silane-functionalized and doped fullerene, Nat. Commun. 7 (2016) 12806.
[46] Q. Chen, H. Zhou, Z. Hong, S. Luo, H.-S. Duan, H.-H. Wang, et al., Planar heterojunction perovskite solar cells via vapor-assisted solution process, J. Am. Chem. Soc. 136 (2014) 622–625.
[47] Z. Song, S.C. Watthage, A.B. Phillips, B.L. Tompkins, R.J. Ellingson, M.J. Heben, Impact of processing temperature and composition on the formation of methylammonium lead iodide perovskites, Chem. Mater. 27 (2015) 4612–4619.
[48] C. Roldan-Carmona, P. Gratia, I. Zimmermann, G. Grancini, P. Gao, M. Graetzel, et al., High efficiency methylammonium lead triiodide perovskite solar cells: the relevance of non-stoichiometric precursors, Energy Environ. Sci. 8 (2015) 3550–3556.
[49] H. Zhou, Q. Chen, G. Li, S. Luo, T.-B. Song, H.-S. Duan, et al., Interface engineering of highly efficient perovskite solar cells, Science 345 (2014) 542–546.
[50] X. Liu, W. Zhao, H. Cui, Y.A. Xie, Y. Wang, T. Xu, et al., Organic–inorganic halide perovskite based solar cells—revolutionary progress in photovoltaics, Inorg. Chem. Front. 2 (2015) 315–335.
[51] Q. Wang, Y. Shao, Q. Dong, Z. Xiao, Y. Yuan, J. Huang, Large fill-factor bilayer iodine perovskite solar cells fabricated by a low-temperature solution-process, Energy Environ. Sci. 7 (2014) 2359–2365.
[52] A. Dualeh, N. Tétreault, T. Moehl, P. Gao, M.K. Nazeeruddin, M. Grätzel, Effect of annealing temperature on film morphology of organic–inorganic hybrid perovskite solid-state solar cells, Adv. Funct. Mater. 24 (2014) 3250–3258.
[53] T. Zhang, M. Yang, Y. Zhao, K. Zhu, Controllable sequential deposition of planar $CH_3NH_3PbI_3$ perovskite films via adjustable volume expansion, Nano Lett. 15 (2015) 3959–3963.
[54] H. Zhang, J. Mao, H. He, D. Zhang, H.L. Zhu, F. Xie, et al., A smooth $CH_3NH_3PbI_3$ film via a new approach for forming the PbI_2 nanostructure together with strategically high CH_3NH_3I concentration for high efficient planar-heterojunction solar cells, Adv. Energy Mater. 5 (2015) 1501354.
[55] Y. Wu, A. Islam, X. Yang, C. Qin, J. Liu, K. Zhang, et al., Retarding the crystallization of PbI_2 for highly reproducible planar-structured perovskite solar cells via sequential deposition, Energy Environ. Sci. 7 (2014) 2934–2938.
[56] Y. Zhou, M. Yang, A.L. Vasiliev, H.F. Garces, Y. Zhao, D. Wang, et al., Growth control of compact $CH_3NH_3PbI_3$ thin films via enhanced solid-state precursor reaction for efficient planar perovskite solar cells, J. Mater. Chem. A 3 (2015) 9249–9256.
[57] W.S. Yang, J.H. Noh, N.J. Jeon, Y.C. Kim, S. Ryu, J. Seo, et al., High-performance photovoltaic perovskite layers fabricated through intramolecular exchange, Science 348 (2015) 1234–1237.

[58] M. Era, T. Hattori, T. Taira, T. Tsutsui, Self-organized growth of PbI-based layered perovskite quantum well by dual-source vapor deposition, Chem. Mater. 9 (1997) 8–10.
[59] D.B. Mitzi, M.T. Prikas, K. Chondroudis, Thin film deposition of organic–inorganic hybrid materials using a single source thermal ablation technique, Chem. Mater. 11 (1999) 542–544.
[60] C.-W. Chen, H.-W. Kang, S.-Y. Hsiao, P.-F. Yang, K.-M. Chiang, H.-W. Lin, Efficient and uniform planar-type perovskite solar cells by simple sequential vacuum deposition, Adv. Mater. 26 (2014) 6647–6652.
[61] M.M. Tavakoli, L. Gu, Y. Gao, C. Reckmeier, J. He, A.L. Rogach, et al., Fabrication of efficient planar perovskite solar cells using a one-step chemical vapor deposition method, Sci. Rep. [Online] (2015) 14083. Available from: https://doi.org/10.1038/srep14083 (accessed 09/22/online).
[62] G. Longo, L. Gil-Escrig, M.J. Degen, M. Sessolo, H.J. Bolink, Perovskite solar cells prepared by flash evaporation, Chem. Commun. 51 (2015) 7376–7378.
[63] M. Sessolo, C. Momblona, L. Gil-Escrig, H.J. Bolink, Photovoltaic devices employing vacuum-deposited perovskite layers, MRS Bull. 40 (2015) 660–666.
[64] D. Zhao, W. Ke, C.R. Grice, A.J. Cimaroli, X. Tan, M. Yang, et al., Annealing-free efficient vacuum-deposited planar perovskite solar cells with evaporated fullerenes as electron-selective layers, Nano Energy 19 (2016) 88–97.
[65] L.K. Ono, M.R. Leyden, S. Wang, Y. Qi, Organometal halide perovskite thin films and solar cells by vapor deposition, J. Mater. Chem. A 4 (2016) 6693–6713.
[66] N.J. Jeon, J.H. Noh, Y.C. Kim, W.S. Yang, S. Ryu, S.I. Seok, Solvent engineering for high-performance inorganic–organic hybrid perovskite solar cells, Nat. Mater. 13 (2014) 897–903.
[67] Y. Zhao, K. Zhu, Efficient planar perovskite solar cells based on 1.8 eV band gap $CH_3NH_3PbI_2Br$ nanosheets via thermal decomposition, J. Am. Chem. Soc. 136 (2014) 12241–12244.
[68] J.H. Heo, D.H. Song, H.J. Han, S.Y. Kim, J.H. Kim, D. Kim, et al., Planar $CH_3NH_3PbI_3$ perovskite solar cells with constant 17.2% average power conversion efficiency irrespective of the scan rate, Adv. Mater. 27 (2015) 3424–3430.
[69] H. Tsai, W. Nie, P. Cheruku, N.H. Mack, P. Xu, G. Gupta, et al., Optimizing composition and morphology for large-grain perovskite solar cells via chemical control, Chem. Mater. 27 (2015) 5570–5576.
[70] Y. Chen, Y. Zhao, Z. Liang, Non-thermal annealing fabrication of efficient planar perovskite solar cells with inclusion of NH_4Cl, Chem. Mater. 27 (2015) 1448–1451.
[71] J.H. Heo, D.H. Song, S.H. Im, Planar $CH_3NH_3PbBr_3$ hybrid solar cells with 10.4% power conversion efficiency, fabricated by controlled crystallization in the spin-coating process, Adv. Mater. 26 (2014) 8179–8183.
[72] P.-W. Liang, C.-Y. Liao, C.-C. Chueh, F. Zuo, S.T. Williams, X.-K. Xin, et al., Additive enhanced crystallization of solution-processed perovskite for highly efficient planar-heterojunction solar cells, Adv. Mater. 26 (2014) 3748–3754.
[73] A. Mei, X. Li, L. Liu, Z. Ku, T. Liu, Y. Rong, et al., A hole-conductor–free, fully printable mesoscopic perovskite solar cell with high stability, Science 345 (2014) 295–298.
[74] X. Li, M. Ibrahim Dar, C. Yi, J. Luo, M. Tschumi, S.M. Zakeeruddin, et al., Improved performance and stability of perovskite solar cells by crystal crosslinking with alkylphosphonic acid ω-ammonium chlorides, Nat. Chem. 7 (2015) 703–711.
[75] Y. Fu, F. Meng, M.B. Rowley, B.J. Thompson, M.J. Shearer, D. Ma, et al., Solution growth of single crystal methylammonium lead halide perovskite nanostructures for optoelectronic and photovoltaic applications, J. Am. Chem. Soc. 137 (2015) 5810–5818.

[76] S.C. Watthage, Z. Song, G.K. Liyanage, A.B. Phillips, M.J. Heben, Investigation on the nucleation and growth mechanisms of perovskite formation in the two-step solution process, in: 2016 IEEE 43rd Photovoltaic Specialists Conference (PVSC), 5–10 June 2016, 2016, pp. 0831–0834.
[77] W. Wu, H. Li, S. Liu, B. Zheng, Y. Xue, X. Liu, et al., Tuning PbI_2 layers by *n*-butanol additive for improving $CH_3NH_3PbI_3$ light harvesters of perovskite solar cells, RSC Adv. 6 (2016) 89609–89613.
[78] C.-G. Wu, C.-H. Chiang, Z.-L. Tseng, M.K. Nazeeruddin, A. Hagfeldt, M. Gratzel, High efficiency stable inverted perovskite solar cells without current hysteresis, Energy Environ. Sci. 8 (2015) 2725–2733.
[79] S.C. Watthage, Z. Song, N. Shrestha, A.B. Phillips, G.K. Liyanage, P.J. Roland, et al., Enhanced grain size, photoluminescence, and photoconversion efficiency with cadmium addition during the two-step growth of $CH_3NH_3PbI_3$, ACS Appl. Mater. Interfaces 9 (2016) 2334–2341.
[80] S.C. Watthage, Z. Song, N. Shrestha, A.B. Phillips, G.K. Liyanage, P.J. Roland, et al., Impact of divalent metal additives on the structural and optoelectronic properties of $CH_3NH_3PbI_3$ perovskite prepared by the two-step solution process, MRS Adv. (2017) 1–6.
[81] D. Bi, W. Tress, M.I. Dar, P. Gao, J. Luo, C. Renevier, et al., Efficient luminescent solar cells based on tailored mixed-cation perovskites, Sci. Adv. 2 (2016). 1501170.
[82] N. Ahn, D.-Y. Son, I.-H. Jang, S.M. Kang, M. Choi, N.-G. Park, Highly reproducible perovskite solar cells with average efficiency of 18.3% and best efficiency of 19.7% fabricated via Lewis base adduct of lead(II) iodide, J. Am. Chem. Soc. 137 (2015) 8696–8699.
[83] X. Zheng, B. Chen, C. Wu, S. Priya, Room temperature fabrication of $CH_3NH_3PbBr_3$ by anti-solvent assisted crystallization approach for perovskite solar cells with fast response and small $J-V$ hysteresis, Nano Energy 17 (2015) 269–278.
[84] X. Fang, Y. Wu, Y. Lu, Y. Sun, S. Zhang, J. Zhang, et al., Annealing-free perovskite films based on solvent engineering for efficient solar cells, J. Mater. Chem. C 5 (2017) 842–847.
[85] D.W. de Quilettes, S.M. Vorpahl, S.D. Stranks, H. Nagaoka, G.E. Eperon, M.E. Ziffer, et al., Impact of microstructure on local carrier lifetime in perovskite solar cells, Science 348 (2015) 683–686.
[86] B.S. Tosun, H.W. Hillhouse, Enhanced carrier lifetimes of pure iodide hybrid perovskite via vapor-equilibrated re-growth (VERG), J. Phys. Chem. Lett. 6 (2015) 2503–2508.
[87] X. Li, D. Bi, C. Yi, J.-D. Décoppet, J. Luo, S.M. Zakeeruddin, et al., A vacuum flash–assisted solution process for high-efficiency large-area perovskite solar cells, Science 353 (2016) 58–62.
[88] D. Liu, T.L. Kelly, Perovskite solar cells with a planar heterojunction structure prepared using room-temperature solution processing techniques, Nat. Photonics. 8 (2014) 133–138.
[89] W. Ke, G. Fang, Q. Liu, L. Xiong, P. Qin, H. Tao, et al., Low-temperature solution-processed tin oxide as an alternative electron transporting layer for efficient perovskite solar cells, J. Am. Chem. Soc. 137 (2015) 6730–6733.
[90] J.P. Correa Baena, L. Steier, W. Tress, M. Saliba, S. Neutzner, T. Matsui, et al., Highly efficient planar perovskite solar cells through band alignment engineering, Energy Environ. Sci. 8 (2015) 2928–2934.
[91] L. Wang, W. Fu, Z. Gu, C. Fan, X. Yang, H. Li, et al., Low temperature solution processed planar heterojunction perovskite solar cells with a CdSe nanocrystal as an electron transport/extraction layer, J. Mater. Chem. C 2 (2014) 9087–9090.

[92] J. Liu, C. Gao, L. Luo, Q. Ye, X. He, L. Ouyang, et al., Low-temperature, solution processed metal sulfide as an electron transport layer for efficient planar perovskite solar cells, J. Mater. Chem. A 3 (2015) 11750–11755.
[93] C.-H. Chiang, Z.-L. Tseng, C.-G. Wu, Planar heterojunction perovskite/PC71BM solar cells with enhanced open-circuit voltage via a (2/1)-step spin-coating process, J. Mater. Chem. A 2 (2014) 15897–15903.
[94] E.J. Juarez-Perez, M. Wußler, F. Fabregat-Santiago, K. Lakus-Wollny, E. Mankel, T. Mayer, et al., Role of the selective contacts in the performance of lead halide perovskite solar cells, J. Phys. Chem. Lett. 5 (2014) 680–685.
[95] Z. Song, A. Abate, S.C. Watthage, G.K. Liyanage, A.B. Phillips, U. Steiner, et al. In-situ observation of moisture-induced degradation of perovskite solar cells using laser-beam induced current, in: 2016 IEEE 43rd Photovoltaic Specialists Conference (PVSC), 5–10 June 2016, 2016, pp. 1202–1206.
[96] Z. Song, S.C. Watthage, A.B. Phillips, G.K. Liyanage, R.R. Khanal, B.L. Tompkins, et al., Investigation of degradation mechanisms of perovskite-based photovoltaic devices using laser beam induced current mapping, Proc. SPIE 9561 (2015) 956107.
[97] Z. Song, A. Abate, S.C. Watthage, G.K. Liyanage, A.B. Phillips, U. Steiner, et al., Perovskite solar cell stability in humid air: partially reversible phase transitions in the $PbI_2-CH_3NH_3I-H_2O$ system, Adv. Energy Mater. 6 (2016) 1600846.
[98] Y. Guo, C. Liu, K. Inoue, K. Harano, H. Tanaka, E. Nakamura, Enhancement in the efficiency of an organic–inorganic hybrid solar cell with a doped P3HT hole-transporting layer on a void-free perovskite active layer, J. Mater. Chem. A 2 (2014) 13827–13830.
[99] K. Rakstys, A. Abate, M.I. Dar, P. Gao, V. Jankauskas, G. Jacopin, et al., Triazatruxene-based hole transporting materials for highly efficient perovskite solar cells, J. Am. Chem. Soc. 137 (2015) 16172–16178.
[100] Y. Zhang, M. Elawad, Z. Yu, X. Jiang, J. Lai, L. Sun, Enhanced performance of perovskite solar cells with P3HT hole-transporting materials via molecular p-type doping, RSC Adv. 6 (2016) 108888–108895.
[101] J.-Y. Jeng, K.-C. Chen, T.-Y. Chiang, P.-Y. Lin, T.-D. Tsai, Y.-C. Chang, et al., Nickel oxide electrode interlayer in $CH_3NH_3PbI_3$ perovskite/PCBM planar-heterojunction hybrid solar cells, Adv. Mater. 26 (2014) 4107–4113.
[102] J.H. Kim, P.-W. Liang, S.T. Williams, N. Cho, C.-C. Chueh, M.S. Glaz, et al., High-performance and environmentally stable planar heterojunction perovskite solar cells based on a solution-processed copper-doped nickel oxide hole-transporting layer, Adv. Mater. 27 (2015) 695–701.
[103] J.H. Park, J. Seo, S. Park, S.S. Shin, Y.C. Kim, N.J. Jeon, et al., Efficient $CH_3NH_3PbI_3$ perovskite solar cells employing nanostructured p-type NiO electrode formed by a pulsed laser deposition, Adv. Mater. 27 (2015) 4013–4019.
[104] N. Li, Z. Zhu, C.-C. Chueh, H. Liu, B. Peng, A. Petrone, et al., Mixed cation $FA_xPEA_{1-x}PbI_3$ with enhanced phase and ambient stability toward high-performance perovskite solar cells, Adv. Energy Mater. 7 (2017) 1601307.
[105] C. Zuo, L. Ding, Solution-processed Cu_2O and CuO as hole transport materials for efficient perovskite solar cells, Small 11 (2015) 5528–5532.
[106] S. Ye, W. Sun, Y. Li, W. Yan, H. Peng, Z. Bian, et al., CuSCN-based inverted planar perovskite solar cell with an average PCE of 15.6%, Nano Lett. 15 (2015) 3723–3728.
[107] J.-P. Correa-Baena, A. Abate, M. Saliba, W. Tress, T. Jesper Jacobsson, M. Gratzel, et al., The rapid evolution of highly efficient perovskite solar cells, Energy Environ. Sci. 10 (2017) 710–727.

[108] C.C. Stoumpos, C.D. Malliakas, M.G. Kanatzidis, Semiconducting tin and lead iodide perovskites with organic cations: phase transitions, high mobilities, and near-infrared photoluminescent properties, Inorg. Chem. 52 (2013) 9019–9038.
[109] B. Chen, X. Zheng, Y. Bai, N.P. Padture, J. Huang, Progress in tandem solar cells based on hybrid organic–inorganic perovskites, Adv. Energy Mater. 7 (2017) 1602400.
[110] W.-J. Yin, J.-H. Yang, J. Kang, Y. Yan, S.-H. Wei, Halide perovskite materials for solar cells: a theoretical review, J. Mater. Chem. A 3 (2015) 8926–8942.
[111] G.E. Eperon, S.D. Stranks, C. Menelaou, M.B. Johnston, L.M. Herz, H.J. Snaith, Formamidinium lead trihalide: a broadly tunable perovskite for efficient planar heterojunction solar cells, Energy Environ. Sci. 7 (2014) 982–988.
[112] N. Pellet, P. Gao, G. Gregori, T.-Y. Yang, M.K. Nazeeruddin, J. Maier, et al., Mixed-organic-cation perovskite photovoltaics for enhanced solar-light harvesting, Angew. Chem. Int. Ed. 53 (2014) 3151–3157.
[113] J.-W. Lee, D.-H. Kim, H.-S. Kim, S.-W. Seo, S.M. Cho, N.-G. Park, Formamidinium and cesium hybridization for photo- and moisture-stable perovskite solar cell, Adv. Energy Mater. 5 (2015) 1501310.
[114] C. Yi, J. Luo, S. Meloni, A. Boziki, N. Ashari-Astani, C. Gratzel, et al., Entropic stabilization of mixed A-cation ABX3 metal halide perovskites for high performance perovskite solar cells, Energy Environ. Sci. 9 (2016) 656–662.
[115] S.A. Kulkarni, T. Baikie, P.P. Boix, N. Yantara, N. Mathews, S. Mhaisalkar, Band-gap tuning of lead halide perovskites using a sequential deposition process, J. Mater. Chem. A 2 (2014) 9221–9225.
[116] J.H. Noh, S.H. Im, J.H. Heo, T.N. Mandal, S.I. Seok, Chemical management for colorful, efficient, and stable inorganic–organic hybrid nanostructured solar cells, Nano Lett. 13 (2013) 1764–1769.
[117] E. Mosconi, A. Amat, M.K. Nazeeruddin, M. Grätzel, F. De Angelis, First-principles modeling of mixed halide organometal perovskites for photovoltaic applications, J. Mater. Chem. C 117 (2013) 13902–13913.
[118] S. Colella, E. Mosconi, P. Fedeli, A. Listorti, F. Gazza, F. Orlandi, et al., $MAPbI_{3-x}Cl_x$ mixed halide perovskite for hybrid solar cells: the role of chloride as dopant on the transport and structural properties, Chem. Mater. 25 (2013) 4613–4618.
[119] Q. Chen, H. Zhou, Y. Fang, A.Z. Stieg, T.-B. Song, H.-H. Wang, et al., The optoelectronic role of chlorine in $CH_3NH_3PbI_3$(Cl)-based perovskite solar cells, Nat. Commun. 6 (2015) 7269.
[120] N.J. Jeon, J.H. Noh, W.S. Yang, Y.C. Kim, S. Ryu, J. Seo, et al., Compositional engineering of perovskite materials for high-performance solar cells, Nature 517 (2015) 476–480.
[121] J. Gong, S.B. Darling, F. You, Perovskite photovoltaics: life-cycle assessment of energy and environmental impacts, Energy Environ. Sci. 8 (2015) 1953–1968.
[122] I. Celik, Z. Song, A.J. Cimaroli, Y. Yan, M.J. Heben, D. Apul, Life cycle assessment (LCA) of perovskite PV cells projected from lab to fab, Sol. Energy Mater. Sol. Cells 156 (2016) 157–169.
[123] N.K. Noel, S.D. Stranks, A. Abate, C. Wehrenfennig, S. Guarnera, A.-A. Haghighirad, et al., Lead-free organic–inorganic tin halide perovskites for photovoltaic applications, Energy Environ. Sci. 7 (2014) 3061–3068.
[124] F. Hao, C.C. Stoumpos, D.H. Cao, R.P.H. Chang, M.G. Kanatzidis, Lead-free solid-state organic–inorganic halide perovskite solar cells, Nat. Photonics 8 (2014) 489–494.

[125] W. Ke, C.C. Stoumpos, J.L. Logsdon, M.R. Wasielewski, Y. Yan, G. Fang, et al., TiO_2-ZnS cascade electron transport layer for efficient formamidinium tin iodide perovskite solar cells, J. Am. Chem. Soc. 138 (2016) 14998–15003.
[126] Y. Ogomi, A. Morita, S. Tsukamoto, T. Saitho, N. Fujikawa, Q. Shen, et al., $CH_3NH_3Sn_xPb_{(1-x)}I_3$ perovskite solar cells covering up to 1060 nm, J. Phys. Chem. Lett. 5 (2014) 1004–1011.
[127] D. Zhao, Y. Yu, C. Wang, W. Liao, N. Shrestha, C.R. Grice, et al., Low-bandgap mixed tin–lead iodide perovskite absorbers with long carrier lifetimes for all-perovskite tandem solar cells, Nat. Energy 2 (2017) 17018.
[128] P. Umari, E. Mosconi, F. De Angelis, Relativistic GW calculations on $CH_3NH_3PbI_3$ and $CH_3NH_3SnI_3$ perovskites for solar cell applications, Sci. Rep. 4 (2014) 4467.
[129] J. Werner, C.-H. Weng, A. Walter, L. Fesquet, J.P. Seif, S. De Wolf, et al., Efficient monolithic perovskite/silicon tandem solar cell with cell area >1 cm^2, J. Phys. Chem. Lett. 7 (2016) 161–166.
[130] S. Albrecht, M. Saliba, J.P. Correa Baena, F. Lang, L. Kegelmann, M. Mews, et al., Monolithic perovskite/silicon-heterojunction tandem solar cells processed at low temperature, Energy Environ. Sci. 9 (2016) 81–88.
[131] J.P. Mailoa, C.D. Bailie, E.C. Johlin, E.T. Hoke, A.J. Akey, W.H. Nguyen, et al., A 2-terminal perovskite/silicon multijunction solar cell enabled by a silicon tunnel junction, Appl. Phys. Lett. 106 (2015) 121105.
[132] T. Todorov, T. Gershon, O. Gunawan, Y.S. Lee, C. Sturdevant, L.-Y. Chang, et al., Monolithic perovskite-CIGS tandem solar cells via in situ band gap engineering, Adv. Energy Mater. 5 (2015) 1500799.
[133] C.D. Bailie, M.G. Christoforo, J.P. Mailoa, A.R. Bowring, E.L. Unger, W.H. Nguyen, et al., Semi-transparent perovskite solar cells for tandems with silicon and CIGS, Energy Environ. Sci. 8 (2015) 956–963.
[134] T. Todorov, T. Gershon, O. Gunawan, C. Sturdevant, S. Guha, Perovskite-kesterite monolithic tandem solar cells with high open-circuit voltage, Appl. Phys. Lett. 105 (2014) 173902.
[135] G.E. Eperon, T. Leijtens, K.A. Bush, R. Prasanna, T. Green, J.T.-W. Wang, et al., Perovskite–perovskite tandem photovoltaics with optimized band gaps, Science 354 (2016) 861–865.
[136] D.P. McMeekin, G. Sadoughi, W. Rehman, G.E. Eperon, M. Saliba, M.T. Hörantner, et al., A mixed-cation lead mixed-halide perovskite absorber for tandem solar cells, Science 351 (2016) 151–155.
[137] F. Fu, T. Feurer, T.P. Weiss, S. Pisoni, E. Avancini, C. Andres, et al., High-efficiency inverted semi-transparent planar perovskite solar cells in substrate configuration, Nat. Energy 2 (2016) 16190.
[138] Z. Yu, M. Leilaeioun, Z. Holman, Selecting tandem partners for silicon solar cells, Nat. Energy 1 (2016) 16137.
[139] H. Uzu, M. Ichikawa, M. Hino, K. Nakano, T. Meguro, J.L. Hernández, et al., High efficiency solar cells combining a perovskite and a silicon heterojunction solar cells via an optical splitting system, Appl. Phys. Lett. 106 (2015) 013506.
[140] K.A. Bush, A.F. Palmstrom, Z.J. Yu, M. Boccard, R. Cheacharoen, J.P. Mailoa, et al., 23.6%-efficient monolithic perovskite/silicon tandem solar cells with improved stability, Nat. Energy 2 (2017) 17009.
[141] Z. Yang, A. Rajagopal, C.-C. Chueh, S.B. Jo, B. Liu, T. Zhao, et al., Stable low-bandgap Pb–Sn binary perovskites for tandem solar cells, Adv. Mater. (2016) 1602696.

[142] T.A. Berhe, W.-N. Su, C.-H. Chen, C.-J. Pan, J.-H. Cheng, H.-M. Chen, et al., Organometal halide perovskite solar cells: degradation and stability, Energy Environ. Sci. 9 (2016) 323–356.
[143] Y. Han, S. Meyer, Y. Dkhissi, K. Weber, J.M. Pringle, U. Bach, et al., Degradation observations of encapsulated planar $CH_3NH_3PbI_3$ perovskite solar cells at high temperatures and humidity, J. Mater. Chem. A 3 (2015) 8139–8147.
[144] N. Aristidou, I. Sanchez-Molina, T. Chotchuangchutchaval, M. Brown, L. Martinez, T. Rath, et al., The role of oxygen in the degradation of methylammonium lead trihalide perovskite photoactive layers, Angew. Chem. Int. Ed. 54 (2015) 8208–8212.
[145] T. Leijtens, G.E. Eperon, S. Pathak, A. Abate, M.M. Lee, H.J. Snaith, Overcoming ultraviolet light instability of sensitized TiO_2 with meso-superstructured organometal tri-halide perovskite solar cells, Nat. Commun. [Online] (2013) 3885. <Go to ISI>://WOS:000329396200003 (accessed Dec).
[146] G. Divitini, S. Cacovich, F. Matteocci, L. Cinà, A. Di Carlo, C. Ducati, In situ observation of heat-induced degradation of perovskite solar cells, Nat. Energy 1 (2016) 15012.
[147] S. Wang, Y. Jiang, EmilioJ. Juarez-Perez, LuisK. Ono, Y. Qi, Accelerated degradation of methylammonium lead iodide perovskites induced by exposure to iodine vapour, Nat. Energy 2 (2016) 16195.
[148] F. Matteocci, L. Cinà, E. Lamanna, S. Cacovich, G. Divitini, P.A. Midgley, et al., Encapsulation for long-term stability enhancement of perovskite solar cells, Nano Energy 30 (2016) 162–172.
[149] Q. Dong, F. Liu, M.K. Wong, H.W. Tam, A.B. Djurišić, A. Ng, et al., Encapsulation of perovskite solar cells for high humidity conditions, ChemSusChem 9 (2016) 2597–2603.
[150] S. Gholipour, J.-P. Correa-Baena, K. Domanski, T. Matsui, L. Steier, F. Giordano, et al., Highly efficient and stable perovskite solar cells based on a low-cost carbon cloth, Adv. Energy Mater. 6 (2016) 1601116.
[151] X. Li, M. Tschumi, H. Han, S.S. Babkair, R.A. Alzubaydi, A.A. Ansari, et al., Outdoor performance and stability under elevated temperatures and long-term light soaking of triple-layer mesoporous perovskite photovoltaics, Energy Technology 3 (2015) 551–555.
[152] H.J. Snaith, A. Abate, J.M. Ball, G.E. Eperon, T. Leijtens, N.K. Noel, et al., Anomalous hysteresis in perovskite solar cells, J. Phys. Chem. Lett. 5 (2014) 1511–1515.
[153] S. van Reenen, M. Kemerink, H.J. Snaith, Modeling anomalous hysteresis in perovskite solar cells, J. Phys. Chem. Lett. 6 (2015) 3808–3814.
[154] T.-Y. Yang, G. Gregori, N. Pellet, M. Grätzel, J. Maier, The significance of ion conduction in a hybrid organic–inorganic lead-iodide-based perovskite photosensitizer, Angew. Chem. Int. Ed. 54 (2015) 7905–7910.
[155] Y. Zhang, M. Liu, G.E. Eperon, T.C. Leijtens, D. McMeekin, M. Saliba, et al., Charge selective contacts, mobile ions and anomalous hysteresis in organic–inorganic perovskite solar cells, Mater. Horiz. 2 (2015) 315–322.
[156] H. Zhang, C. Liang, Y. Zhao, M. Sun, H. Liu, J. Liang, et al., Dynamic interface charge governing the current–voltage hysteresis in perovskite solar cells, Phys. Chem. Chem. Phys. 17 (2015) 9613–9618.

CHAPTER 4

Perovskite Solar Cell Architectures

Vishesh Manjunath, Ramya Krishna, Suresh Maniarasu, Easwaramoorthi Ramasamy, Sakthivel Shanmugasundaram and Ganapathy Veerappan
International Advanced Research Centre for Powder, Metallurgy and New Materials (ARCI), Hyderabad, Telangana, India

4.1 INTRODUCTION

Kojima et al. in 2009 reported the use of organic–inorganic hybrid perovskites such as $CH_3NH_3PbBr_3$ and $CH_3NH_3PbI_3$ as light sensitizer. The device was constructed by screen printing nanocrystalline TiO_2 on a fluorine-doped tin oxide (FTO) substrate. Organic–inorganic perovskite absorbers were coated by spin coating the precursor solutions over TiO_2 layer. Platinum-coated FTO glass was used as cathode. Lithium halide and redox couple was used as an electrolyte. Power conversion efficiencies (PCEs) of 3.13% and 3.81% were obtained for $CH_3NH_3PbBr_3$ and $CH_3NH_3PbI_3$, respectively [1]. Im et al. in 2011 reported 6.54% PCE using quantum dot-sensitized solar cell based on perovskite $CH_3NH_3PbI_3$ sensitizer. $CH_3NH_3PbI_3$ quantum dots were formed on the surface of TiO_2 and sandwiched in between the electrodes and filled with liquid electrolyte. One major problem encountered in this architecture was the dissolution of perovskite quantum dots in the redox electrolyte solution in less than 10 minutes of continuous illumination [2]. To avoid the dissolution of perovskite sensitizer in the electrolyte solution, Kim et al. used solid-state electrolyte, namely, spiro-OMeTAD. PCE of 9.7% was possible with this architecture and devices retained about 80% of its initial stability for 500 hours [3]. This breakthrough in efficiency triggered the photovoltaic community. Later, Lee et al. reported low-cost and solution-processable solar cells by developing mesosuperstructured solar cell (MSSC) architecture. This architecture couples perovskite light absorbers

Perovskite Photovoltaics
DOI: https://doi.org/10.1016/B978-0-12-812915-9.00004-6

with nonelectron injecting Al_2O_3 scaffolds. The MSSC device architecture produced a PCE of around 8% [4]. In 2012, Etgaret et al. developed a perovskite solar cell (PSC) without hole-transport material (HTM), by just utilizing perovskite-sensitized TiO_2 electrode as anode and gold deposited onto that acts as a cathode. This means that in this architecture, neither solid nor liquid electrolyte was used. Moreover, the ability of perovskite light absorber to transport holes was studied and a PCE of 5.5% was achieved [5]. Snaith et al. in 2012 formulated a planar device architecture containing compact-TiO_2 layer over the FTO substrate. Perovskite absorber coated over the compact TiO_2 by solution process and vacuum deposition were compared. HTM layer was used in this architecture, and finally a thin layer of gold as cathode was thermally evaporated over the HTM [6]. The perovskite absorber layer formed by vacuum deposition process had complete coverage over the substrates without discontinuity and a PCE of 15% was achieved; on the other hand, solution-processed perovskite films had discontinuity and a PCE of 8% was achieved. Fig. 4.1 demonstrates and summarizes the evolution of PSCs [7].

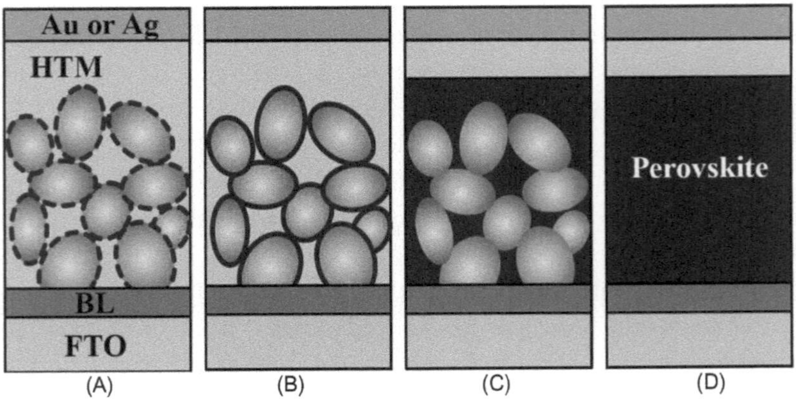

Figure 4.1 (A) Concept of sensitizing TiO_2 with perovskite nanodots. (B) Mesosuperstructured architecture with noninjecting Al_2O_3 scaffolds. (C) Mesoporous architecture with nanocrystalline TiO_2 (metal oxide). (D) Planar architecture. BL spheres in (A) and (C) represent TiO_2 and spheres in (B) represent Al_2O_3. *BL*, blocking layer. *Reproduced with permission from N.G. Park, Perovskite solar cells: an emerging photovoltaic technology. Mater. Today 18 (2) (2015) 65–72. Copyright 2015, Elsevier Publications.*

4.2 MESOPOROUS ARCHITECTURE

The carrier diffusion length (l_D) and thickness (t) of the perovskite absorber layer is the deciding factor for adopting mesoporous architecture. When the thickness of the perovskite absorber is greater than the carrier diffusion length, the generated carriers will recombine before reaching the electron transport layer (ETL). To avoid this recombination, mesoporous TiO_2 is deposited over the hole-blocking layer. The extended network of mesoporous TiO_2 allows the photogenerated electrons to flow through it and reach the anode before they recombine. Abrusci et al. functionalized mesoporous titania with self-assembled fullerene monolayer. Two device structures with $CH_3NH_3PbI_{3-x}Cl_x$ as absorber were compared. In the first device, light-absorbing hole-conductor P3HT was used, whereas in the second device transparent spiro-OMeTAD was used as HTM. The later device achieved a PCE of 11.7% compared to the former device with P3HT [8]. Further, to avoid high-temperature fabrication of TiO_2 layer, as this may hinder the fabrication of PSCs on plastic substrates, Wang et al. developed a solution-processable low-temperature route utilizing graphene–TiO_2 nanocomposites as mesoporous layer [9]. Fig. 4.2 depicts the band diagram of this cell architecture. PCE as high as 5.6% was reported with this architecture.

Burschka et al. in 2013 reported sequential deposition of perovskite precursors on mesoporous TiO_2 to fabricate $CH_3NH_3PbI_3$ absorber. In this deposition technique, a PbI_2 precursor solution was infiltrated in

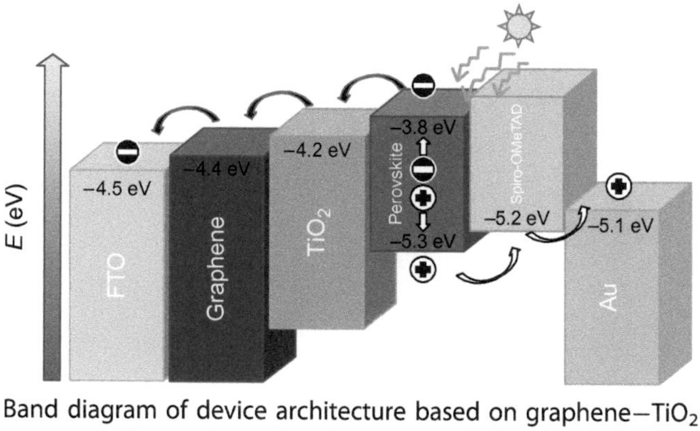

Figure 4.2 Band diagram of device architecture based on graphene–TiO_2 nanocomposite as mesoporous layers. *Reproduced with permission from J.T.W. Wang, J.M. Ball, E.M. Barea, A. Abate, J.A. Alexander-Webber, J. Huang, et al., Low-temperature processed electron collection layers of graphene/TiO_2 nanocomposites in thin film perovskite solar cells. Nano Lett. 14 (2) (2013) 724–730. Copyright 2013, ACS Publications.*

mesoporous TiO_2 by spin coating and later dried. For the champion devices, PbI_2-coated electrodes were first dipped in 2-propanol followed by dipping in a solution of CH_3NH_3I in 2-propanol. The main variable in this mesoporous architecture was the deposition of perovskite film. Moreover, complete conversion of PbI_2 into $CH_3NH_3PbI_3$ was observed with mesoporous TiO_2 as compared to flat TiO_2 films [10].

Transparency, wide bandgap, and band-edge alignment of zinc oxide (ZnO) with perovskite and FTO enable them to be used as a replacement for TiO_2. Also, employing one-dimensional structure instead of mesoporous layers would lead to better charge dynamics. Son et al. optimized the condition to grow ZnO nanorods (NRs) as ETL and compared them with TiO_2 NRs. PCE of 11.13% and 10.02% was achieved for architecture with ZnO and TiO_2 NRs, respectively [10]. But the basicity of ZnO surface causes degradation of $CH_3NH_3PbI_3$ with time. This degradation is minimal with the use of more acidic TiO_2 [11]. Further, the usage of mesoporous nickel oxide (NiO) photocathodes used for sensitizing $CH_3NH_3PbI_3$ and other dyes in dye-sensitized solar cell architecture with liquid electrolyte were studied. Time-resolved photoluminescence studies revealed high hole injection by employing a mesoporous NiO sensitizer. The main limitation with this architecture was poor light absorption of NiO photocathode [12]. Fig. 4.3 represents the schematics of $CH_3NH_3PbI_3$ sensitized on NiO photocathode.

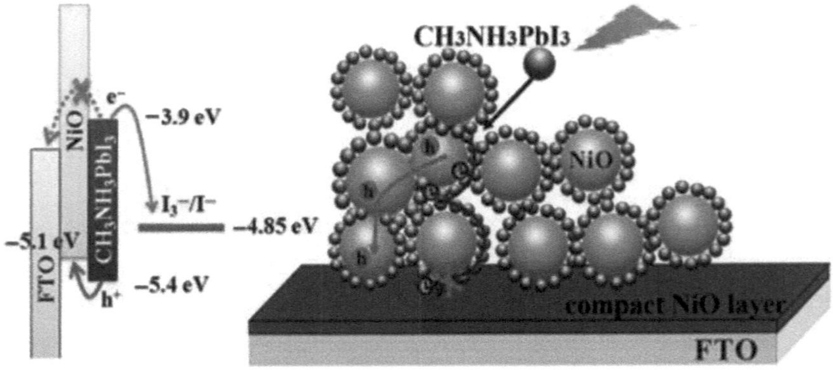

Figure 4.3 Schematic of NiO photocathode sensitized with $CH_3NH_3PbI_3$. *Reproduced with permission from H. Wang, X. Zeng, Z. Huang, W. Zhang, X. Qiao, B. Hu, et al., Boosting the photocurrent density of p-type solar cells based on organometal halide perovskite-sensitized mesoporous NiO photocathodes. ACS Appl. Mater. Interfaces 6 (15) (2014) 12609–12617. Copyright 2014, ACS Publications.*

Lindblad et al. studied the electronic structure of $TiO_2/CH_3NH_3PbI_3$ interface formed by one- and two-step deposition. They claimed that the stoichiometry and the electronic structure of the two-step are very similar to those of the one-step processed $TiO_2/CH_3NH_3PbI_3$ interface. Devices with unconverted PbI_2 in two-step process showed under stoichiometry of iodine, i.e., in the two-step process, devices showed unconverted PbI_2 [13]. Hu et al. demonstrated an approach to avoid the deposition of compact metal oxide film and replaced it with alkali salt solution. The alkali salt solution had well-aligned energy band edges with transparent conductive oxide (TCO) and perovskite. Fig. 4.4 shows the schematic, scanning electron microscopy (SEM) cross-sectional image and band diagram of the reported structure [14].

Liang et al. made use of highly reliable mature sputtering technique, which can be adopted in laboratory and industry to deposit magnetron sputtered ZnO layer. ZnO film consisted of NRs aligned with the *c*-axis.

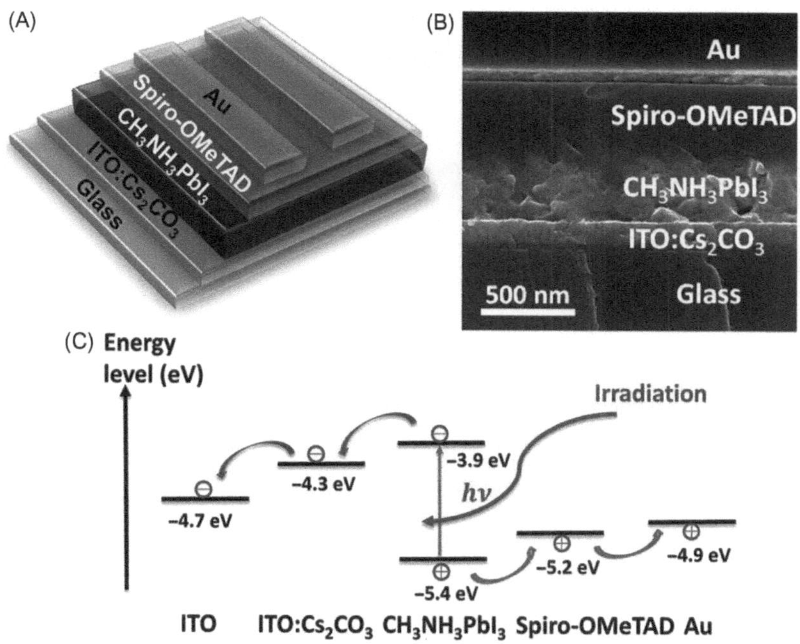

Figure 4.4 (A) Schematic of the device structure. (B) SEM cross-sectional image of a perovskite solar cell based on Cs_2CO_3-modified ITO substrate. (C) Energy-level diagram. *SEM*, scanning electron microscopy. *Reproduced with permission from Q. Hu, J. Wu, C. Jiang, T. Liu, X. Que, R. Zhu, et al., Engineering of electron-selective contact for perovskite solar cells with efficiency exceeding 15%. ACS Nano 8 (10) (2014) 10161–10167. Copyright 2014, ACS Publications.*

Fabricating these 1D ZnO layer by sputtering can be adopted for large-scale fabrication of 1D ZnO-based PSCs [15]. Yella et al. deposited rutile nanocrystalline TiO_2 by hydrolysis of $TiCl_4$ (chemical bath deposition) and achieved a PCE of around 14%. Authors also claimed that nanocrystalline rutile TiO_2 achieved better performance than the anatase TiO_2, due to improved heterojunction between nanocrystalline rutile TiO_2 and $CH_3NH_3PbI_3$ light absorber. By varying the concentration of $TiCl_4$ during low-temperature chemical bath deposition, the size of the nanoparticles increased as shown in Fig. 4.5. The reasons for the formation of rutile TiO_2 were due to usage of $TiCl_4$ as precursor. Since the FTO support also has the rutile structure, epitaxial growth was possible and required no further sintering [16].

Bera et al. developed a device architecture with mesoporous $SrTiO_3$ as electron extraction layer and achieved 25% and 17% increase in open circuit voltage and fill factor compared to the traditional mesoporous TiO_2. The perovskite absorbers coated over the mesoporous $SrTiO_3$ had increased coverage and large grains [17] (Fig. 4.6).

Oh et al. reported the use of Zn_2SnO_4 as electron-transporting layer. The effect of thickness influence on Zn_2SnO_4 layer was minimal on its

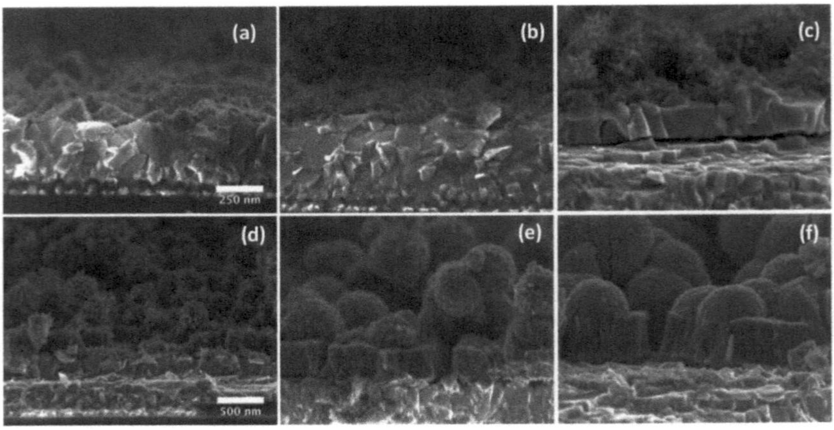

Figure 4.5 Cross-section SEM images of the TiO_2 particles formed by chemical bath deposition using different concentrations of $TiCl_4$ at (A) 100 mM, (B) 200 mM, (C) 250 mM, (D) 300 mM, (E) 400 mM, and (F) 500 mM. Panels (A)–(C) has the same scale bar of 250 nm, while the scale bar of panels (D) – (F) is 500 nm. *SEM*, scanning electron microscopy. *Reproduced with permission from A. Yella, L.P. Heiniger, P. Gao, M.K. Nazeeruddin, M. Grätzel, Nanocrystalline rutile electron extraction layer enables low-temperature solution processed perovskite photovoltaics with 13.7% efficiency. Nano Lett. 14 (5) (2014) 2591–2596. Copyright 2014, ACS Publications.*

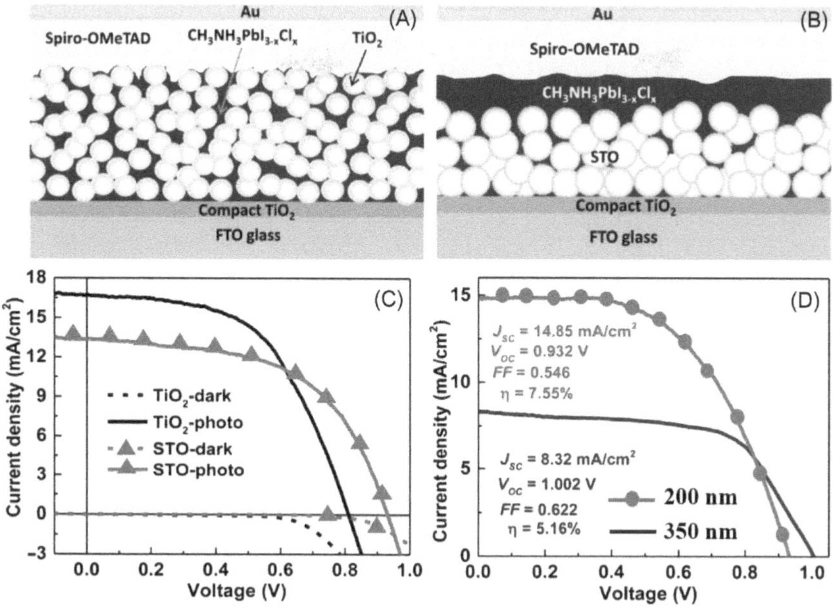

Figure 4.6 Schematic of the $CH_3NH_3PbI_{3-x}Cl_x$ solar cells with (A) meso-TiO_2 and (B) meso-$SrTiO_3$ (STO) as electron-transporting materials. (C) Average $J-V$ characteristics of a batch of 24 devices using meso-TiO_2 and meso-STO. (D) $J-V$ characteristics of two typical solar cells with the meso-STO thicknesses of 200 (-▲-) and 350 nm (—). *Reproduced with permission from A. Bera, K. Wu, A. Sheikh, E. Alarousu, O.F. Mohammed, T. Wu, Perovskite oxide $SrTiO_3$ as an efficient electron transporter for hybrid perovskite solar cells. J. Phys. Chem. C 118 (49) (2014) 28494–28501. Copyright 2014, ACS Publications.*

transmittance, and the thick layers effectively blocked the holes. Open circuit voltage and fill factor increased with increase in thickness of Zn_2SnO_4 layer but the photocurrent density decreased. Zn_2SnO_4 compact layer as compared to the conventional TiO_2 compact layer is crucial for having the Zn_2SnO_4-based PSCs worked properly due to the electron blocking effect resulting from the higher conduction band-edge position of TiO_2 compared to Zn_2SnO_4 [18].

Compact and mesoporous TiO_2 processing plays a prominent role in the device architecture. The requirement of two-step firing of compact and mesoporous TiO_2 can be eliminated by replacing it with one-step sintering. This can be done by thermal oxidation of sputtered TiO_2 film. The oxidized films are dense compared to solution-processed compact

Table 4.1 Photovoltaic parameters of the PSCs using solution coated TiO_2 (SC-TiO_2) and thermal oxidation TiO_2 (TO-TiO_2)

	V_{OC} (V)	J_{sc} (mA/cm²)	FF (%)	PCE (%)	R_S (Ω/cm²)
40 nm SC-TiO_2	1.09	18.01	0.53	12.04	5.49
50 nm SC-TiO_2	1.09	20.83	0.56	12.97	4.86
60 nm SC-TiO_2	1.09	21.08	0.58	13.47	3.97
70 nm SC-TiO_2	1.09	21.16	0.58	12.80	4.29
80 nm SC-TiO_2	1.08	20.11	0.57	12.20	4.67
5 nm TO-TiO_2	1.07	19.72	0.56	12.29	5.17
10 nm TO-TiO_2	1.08	20.66	0.60	14.59	4.32
15 nm TO-TiO_2	1.09	22.19	0.63	15.07	3.57
20 nm TO-TiO_2	1.09	21.30	0.64	14.76	3.76
25 nm TO-TiO_2	1.09	21.10	0.62	14.15	4.08

Source: Reproduced with permission W. Ke, G. Fang, J. Wang, P. Qin, H. Tao, H. Lei, et al., Perovskite solar cell with an efficient TiO_2 compact film. ACS Appl. Mater. Interfaces 6 (18) (2014), 15959–15965. Copyright 2014, ACS Publications.

films. The thermal oxidized films achieved a PCE of 15% as compared to two-step solution-processed films yielding a PCE of 13% due to decreased series and recombination resistance [19]. The optimum thickness of thermally oxidized and solution-coated TiO_2 films were also obtained. The photovoltaic parameters are shown in Table 4.1. Usages of sacrificial templates are also one of the methods to deposit well-aligned crack-free mesoporous TiO_2 [20].

1-D nanostructures are more efficient electron transporters. So modification of ETL nanostructures can provide a direct pathway for electron transport and also offer open channels for pore filling with sensitizers and HTM for efficient solar cells. Hydrothermal/solvothermal synthesis is a facile approach to grow rutile TiO_2 NRs on TCO surfaces [21]. Sheng et al. developed a technique to grow epitaxial single-crystal-like 3-D TiO_2 branched nanowire arrays consisting of 1-D branch. They also notice 72% increases in surface area as compared to 1-D TiO_2 nanowire array [22]. Zhong et al. grew oriented rutile TiO_2 nanocones on FTO under nearly neutral conditions by facile hydrothermal approach. They also compared the photovoltaic parameters of TiO_2 NRs with nanocones and concluded nanocone TiO_2 has superior photovoltaic parameters [23]. $J-V$ curves and incident photon conversion efficiency (IPCE) spectra comparing TiO_2 NRs with nanocones is as shown in Fig. 4.7. Son et al. used seed layer assisted growth of ZnO NRs as ETL. The structures of these ZnO NRs deposited with different seed layers were studied.

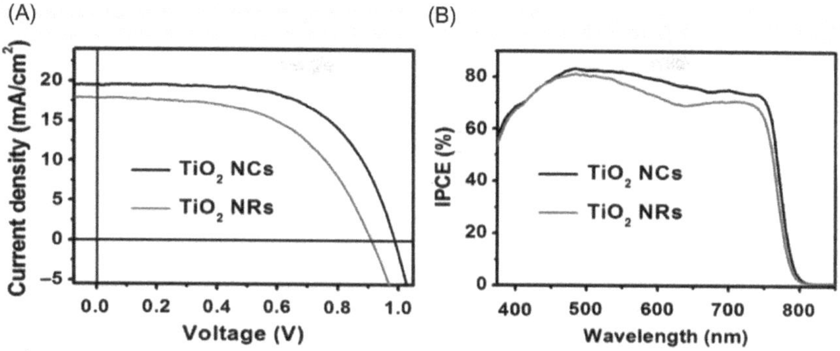

Figure 4.7 (A) *J–V* curves and (B) IPCE of the nanocones and nanorods of fixed length. *IPCE*, incident photon conversion efficiency. *Reproduced with permission from D. Zhong, B. Cai, X. Wang, Z. Yang, Y. Xing, S. Miao, et al., Synthesis of oriented TiO_2 nanocones with fast charge transfer for perovskite solar cells. Nano Energy 11 (2015) 409–418. Copyright 2015, Elsevier Publications.*

Furthermore, impedance spectroscopy results revealed the interface between perovskite and ZnO ETL acted as recombination sites and were overcome by passivizing the ZnO NRs [24].

Various techniques used to deposit TiO_2 ETL results in varied morphologies and thicknesses, resulting in varied photovoltaic parameters. Hong et al. developed a low-cost dip-coating method to deposit hole-blocking TiO_2 layer. The thickness was optimized by controlling the concentration of the dipping solution [25]. Dip-coating method also enables large-scale fabrication of such ETLs. Fig. 4.8 depicts the *J–V* curves comparing spin coating and dip-coating methods for deposition of TiO_2. The stability of PSCs depends on interface between perovskite and ETLs. Perovskite absorber layer morphology, electronic and physical interface between perovskites and ETL is highly influenced by the roughness and structure of ETL [26].

WO_x is also one of the candidates, which can fulfill the role of electron transporting in PSCs. Amorphous WO_x can be prepared via low-temperature approach. ETL layer composed of WO_x and TiO_2 are utilized to make PSC and were compared, and their corresponding photovoltaic curves are given in Fig. 4.9 [27]. Due to the superior conductivity of WO_x over TiO_2 phototelectrodes, they showed much better photovoltaic performance. Ke et al. reported the use of low-temperature solution-processed SnO_2 as ETL. The nanocrystalline SnO_2 possessed antireflection and good carrier dynamics properties. The band edges of SnO_2 ETL were

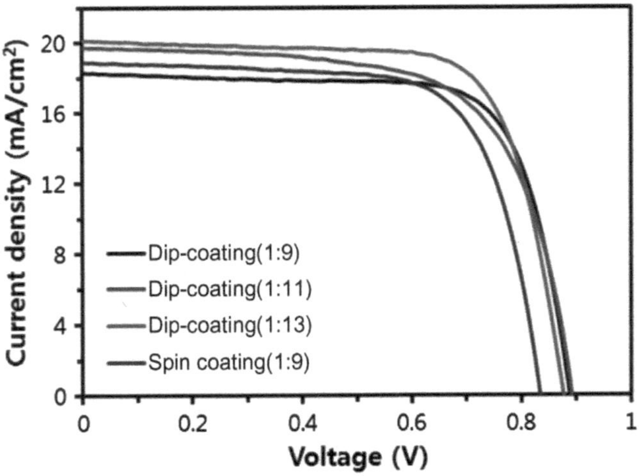

Figure 4.8 *I–V* test results of perovskite solar cells based on different compact-TiO_2 layers. *Reproduced with permission from S. Hong, A. Han, E.C. Lee, K.W. Ko, J.H. Park, H. J. Song, et al., A facile and low-cost fabrication of TiO_2 compact layer for efficient perovskite solar cells. Curr. Appl. Phys. 15 (5) (2015) 574–579. Copyright 2015, Elsevier Publications.*

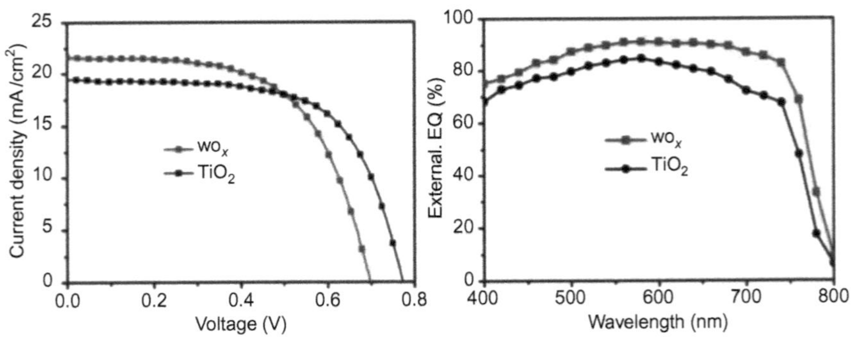

Figure 4.9 *J* – *V* curves and the IPCE of PSCs based on WO_X or TiO_2. *IPCE*, incident photon conversion efficiency. *Reproduced with permission from K. Wang, Y. Shi, Q. Dong, Y. Li, S. Wang, X. Yu, et al., Low-temperature and solution-processed amorphous WO_X as electron-selective layer for perovskite solar cells. J. Phys. Lett. 6 (5) (2015) 755–759. Copyright 2015, ACS Publications.*

well aligned with TCO and perovskite absorber as shown in Fig. 4.10. This low-temperature coating of SnO_2 ETL is feasible in large-scale roll-to-roll production of PSCs [28].

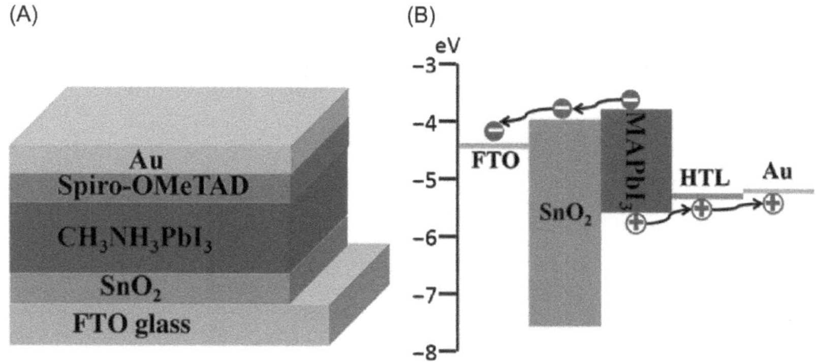

Figure 4.10 (A) Schematics of SnO_2-based device and (B) band diagram of the complete device. *Reproduced with permission from W. Ke, G. Fang, Q. Liu, L. Xiong, P. Qin, H. Tao, et al., Low-temperature solution-processed tin oxide as an alternative electron transporting layer for efficient perovskite solar cells. J. Am. Chem. Soc. 137 (21) (2015) 6730–6733. Copyright 2015, ACS Publications.*

Fabrication of PSCs with two-dimensional TiO_2 nanosheet thin film as the compact layer by electrophoretic deposition technique followed by mesoporous TiO_2 layer as ETL was reported [29]. By using these titania nanosheets, hysteresis observed in mesoscopic devices was minimized. Surface treatments of these nanocrystalline TiO_2 will enhance the performance of the device. Usually, $TiCl_4$ and UV (O_3) are used in this regard [30]. Nagaoka et al. incorporated zirconium into the TiO_2 electrodes. Zr-incorporated TiO_2 electron-transporting layers had increased carrier lifetime and higher charge densities compared to pristine TiO_2. Zr incorporation into titania also minimized the hysteresis effect and overall performance of the device increased. Usage of these electrodes and further passivizing them with pyridine resulted in reduced defect sites at perovskite ETL interface [31]. Jiang et al. reported compact titania-free fully printable mesoscopic PSCs. In these cells, the need for hole conductors can also be eliminated [32]. The device architecture, cross-sectional SEM image, and band diagram of such cells is as shown in Fig. 4.11.

4.3 PLANAR ARCHITECTURE

In planar device architecture, the charge diffusion length (l_D) of the carriers is greater than the thickness (t) of the perovskite absorber layer ($l_D > t$). In this case, the need for nanocrystalline TiO_2 becomes nonexistent [33] (Fig. 4.12; [34]).

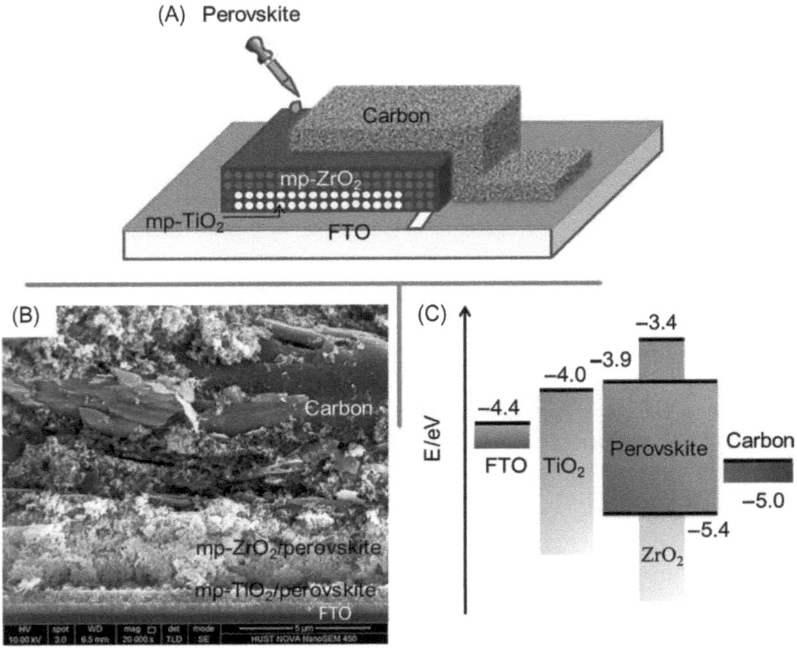

Figure 4.11 (A) Schematic structure, (B) the actual cross-sectional SEM image, and (C) the energy-level diagram of compact-layer free, hole-conductor free fully printable Mesoscopic perovskite solar cell (MPSC). *SEM*, scanning electron microscopy. *Reproduced with permission from X. Jiang, Y. Xiong, A. Mei, Y. Rong, Y. Hu, L. Hong, et al., Efficient compact-layer-free, hole-conductor-free, fully printable mesoscopic perovskite solar cell. J. Phys. Lett. 7 (20) (2016) 4142–4146. Copyright 2016, ACS Publications.*

Figure 4.12 Mesoscopic perovskite solar cell with mesoporous TiO_2 layer and planar structure without a mesoporous TiO_2 layer. *Reproduced with permission from Y. Shi, Y. Xing, Y. Li, Q. Dong, K. Wang, Y. Du, et al., $CH_3NH_3PbI_3$ and $CH_3NH_3PbI_{3-X}Cl_X$ in planar or mesoporous perovskite solar cells: comprehensive insight into the dependence of performance on architecture. J. Phys. Chem. C 119 (28) (2015) 15868–15873. Copyright 2015, ACS Publications.*

The complete device structure usually comprises TCO/ETL/perovskite absorber/electron blocking layer/metal cathode. Snaith et al. in 2012 formulated a planar device architecture containing compact-TiO_2 layer over the FTO substrate. Planar architecture devices were fabricated by both vapor and solution processes [6]. Chen et al. reported a vapor-assisted solution process to fabricate planar architecture PSC device. The resulting perovskite films have improved surface roughness and large grain size. These devices achieved a PCE of ~12% [35]. Zhu et al. used $CH_3NH_3PbI_2Br$ nanosheets over the compact-TiO_2 layer and achieved efficiency around ~10% [36]. In planar device architecture, films produced by spin coating usually show a dendritic grain morphology giving many gaps in the film, resulting in poor coverage of the substrate and thus low PCE. These dendritic grain morphologies can be engineered by solvent engineering by allowing the solution to be supersaturated to form a large number of nucleation sites. Huang et al. reported gas-assisted solution technique to grow uniform reproducible films. The schematic of fabrication and corresponding SEM image is shown in Fig. 4.13 [37] and achieved PCE of ~16% with their new technique.

Wojciechowski et al. through spectroscopic measurements found out that electron transfer from the perovskite to the TiO_2 in the standard planar junction cells is slow. Hence, they functionalized compact-TiO_2 layer as shown in Fig. 4.14. By immersing hot TiO_2 substrates in the C_{60}-SAM solution (0.1 mg/mL in chlorobenzene, filtered with 0.2 μm

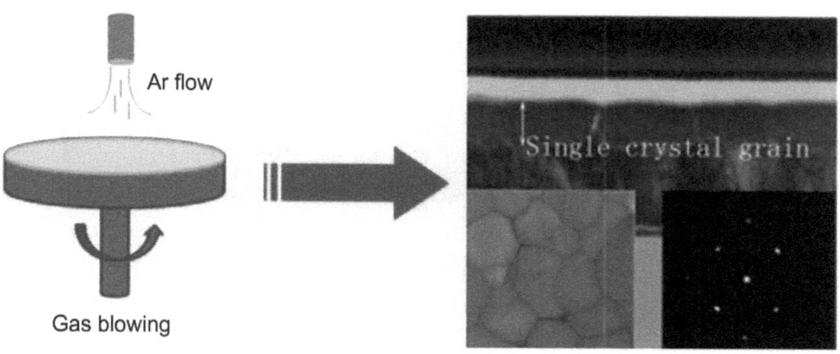

Figure 4.13 Schematic of the gas-blowing technique and corresponding cross-sectional SEM image of the perovskite film obtained. *SEM*, scanning electron microscopy. *Reproduced with permission from F. Huang, Y. Dkhissi, W. Huang, M. Xiao, I. Benesperi, S. Rubanov, et al., Gas-assisted preparation of lead iodide perovskite films consisting of a monolayer of single crystalline grains for high efficiency planar solar cells. Nano Energy 10 (2014) 10–18. Copyright 2014, Elsevier Publications.*

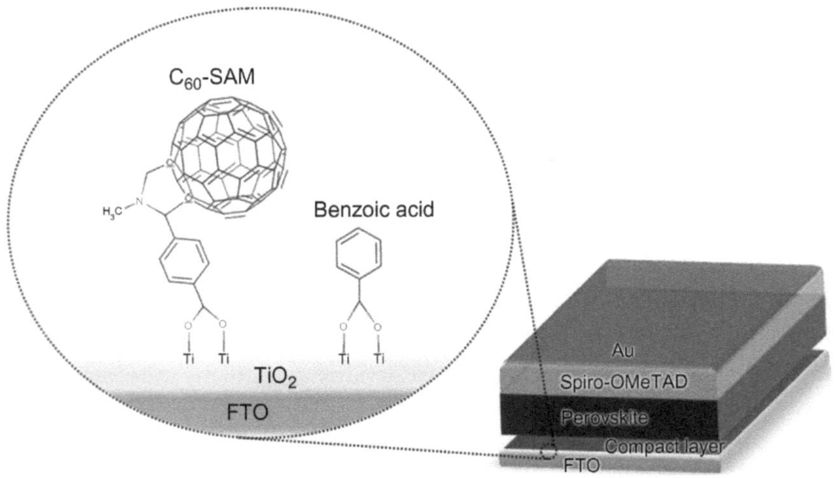

Figure 4.14 Schematic of device structure and in the inset binding of SAMs to TiO_2 surface. *Reproduced with permission from K. Wojciechowski, S.D. Stranks, A. Abate, G. Sadoughi, A. Sadhanala, N. Kopidakis, et al., Heterojunction modification for highly efficient organic–inorganic perovskite solar cells. ACS Nano 8 (12) (2014) 12701–12709. Copyright 2014, ACS Publications.*

Polytetrafluoroethylene (PTFE) filter) for 24–36 hours, followed by rinsing with chlorobenzene and drying for 10 minutes at 120°C, they were able to functionalize compact-TiO_2. This functionalization of compact-TiO_2 was done as C_{60}-SAM is a good electron acceptor. By modifying the n-type contact with a self-assembled fullerene monolayer, electron transfer improved. The fullerene-modified planar architecture devices achieved up to 17.3% PCE [38].

Since the solution-processed TiO_2 has low electrical conductivity when used in planar device architecture. Liu et al. reported the use of $PC_{61}BM$ with solution-processed TiO_2 as ETL. Preferable electron extraction and transportation of $PC_{61}BM$ at the cathode side tremendously boosted short-circuit current density, fill factor, and enhanced PCE. The wettability of perovskite on $PC_{61}BM$ was increased by introduction of water-soluble fullerene (WS-C_{60}). The modified electron transport resulted in improvised electron extraction and transport [39]. The planar device architecture schematic is shown in Fig. 4.15.

Later Paek et al. reported the use of solution-processed $PC_{71}BM$ planar heterojunction [40]. They used inverted planar device architecture as shown in Fig. 4.16. The usage of buffer layer, surfactant layers, etc. was eliminated in this architecture. IPCE of the devices employing $PC_{61}BM$

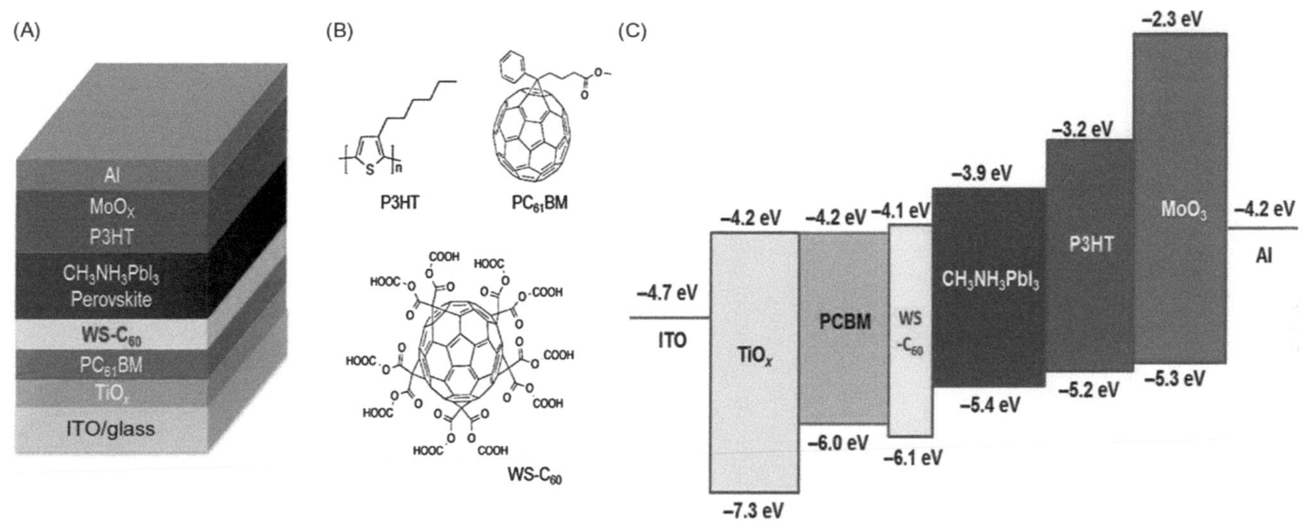

Figure 4.15 Schematic structure of the device architecture and the corresponding band diagram. *Reproduced with permission from C. Liu, K. Wang, P. Du, T. Meng, X. Yu, S.Z. Cheng, et al., High performance planar heterojunction perovskite solar cells with fullerene derivatives as the electron transport layer. ACS Appl. Mater. Interfaces 7 (2) (2015) 1153–1159. Copyright 2015, ACS Publications.*

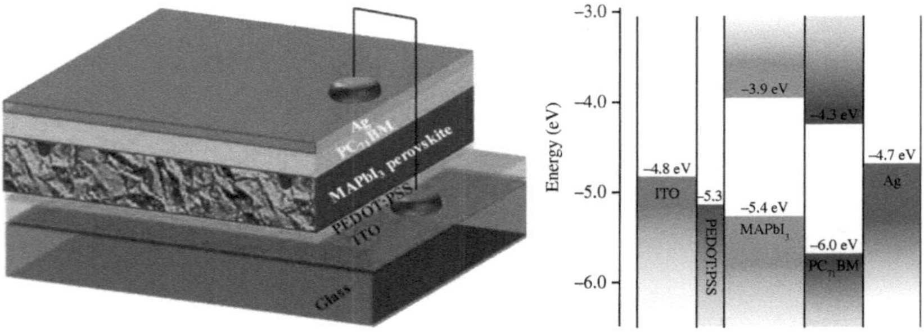

Figure 4.16 Schematic of device architecture and the corresponding band diagram. *Reproduced with permission from S. Paek, N. Cho, H. Choi, H. Jeong, J.S. Lim, J.Y. Hwang, et al., Improved external quantum efficiency from solution-processed (CH3NH3) PbI3 perovskite/PC71BM planar heterojunction for high efficiency hybrid solar cells. J. Phys. Chem. C 118 (45) (2014) 25899–25905. Copyright 2014, ACS Publications.*

and $PC_{71}BM$ were compared, with the later showing increased IPCE. The surface morphologies of the devices were also studied by varying the solvents used for dissolving the precursor salts. The mixed solvent of γ-butyrolactone and dimethyl sulfoxide showed good surface coverage and hence improved photovoltaic performance.

Liu et al. integrated a wide bandgap DOR3T-TBDT small molecule as electron donor in bulk heterojunction (BHJ) films with perovskite absorber in the planar configuration as shown in Fig. 4.17. Increased short current density of $21.2\ mA/cm^2$ was achieved with devices employing BHJ as compared to devices employing traditional HTMs ($19.3\ mA/cm^2$). Integrated devices exhibit a photoresponse spanning the whole visible and infrared spectrum. These integrated devices showed a new way to harvest more light to make perovskite-based solar cells more efficient. PCE of $\sim 14\%$ was achieved in this type of device architecture [41].

Kang et al. reported spinning speed, drying, and stepwise ramp annealing are critical for obtaining optimized planar hybrid PSCs with device architecture composed of ITO/PEDOT:PSS/$CH_3NH_3PbI_{3-X}Cl_X$/PCBM/Al and showed photovoltaic performance up to 11.8% [42]. The pictorial representation of different stages of nucleation and growth is shown in Fig. 4.18. Nucleation and growth of the perovskite absorber is highly dependent on spinning and annealing parameters.

Xia et al. reported inverted planar architecture with usage of (amphiphilic surfactant) oleamide-doped PCBM. About 5% doping of oleamide resulted in good coverage of PCBM over the perovskite absorber layer. Oleamide doping may have increased the wettability of PCBM over the

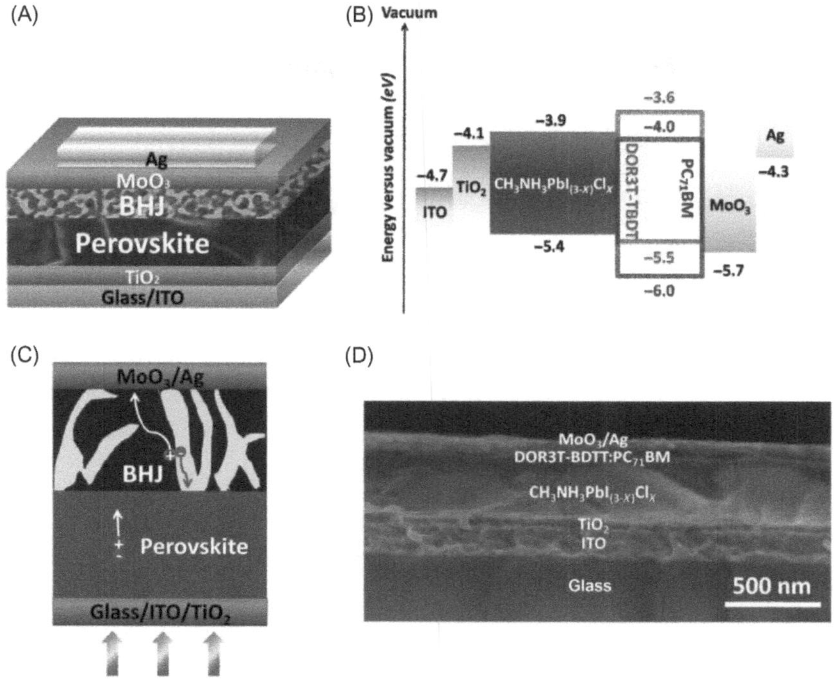

Figure 4.17 (A) Schematic structure of the integrated perovskite/BHJ device. (B) Schematic energy-level diagrams of the integrated perovskite/(DOR3T-TBDT:$PC_{71}BM$) device. (C) Charge generation and transport mechanism of the integrated device. (D) SEM image of cross-sectional structure of the integrated device. *SEM*, scanning electron microscopy. *Reproduced with permission from Y. Liu, Z. Hong, Q. Chen, W. Chang, H. Zhou, T.B. Song, et al., Integrated perovskite/bulk-heterojunction toward efficient solar cells. Nano Lett. 15 (1) (2014) 662–668. Copyright 2014, ACS Publications.*

perovskite absorber. Due to this large surface coverage, increase in short current density and fill factor was obtained. Impedance spectroscopy results revealed low interface recombination at the PCBM/perovskite interface [43] (Fig. 4.19).

Li et al. achieved a PCE of ~17% in planar device architecture based on low pressure, reduced temperature vapor annealing. They also stated that by controlling temperature and vapor pressure of CH_3NH_3I vapors, controlled intercalation can be achieved to form pinhole-free high-quality perovskite films in planar architectures [44] (Fig. 4.20).

Gu et al. developed a novel planar architecture employing CdS NR array as ETL [45]. Simple hydrothermal methods were used to grow the CdS NRs and posttreatments namely UV-ozone and thermal annealing

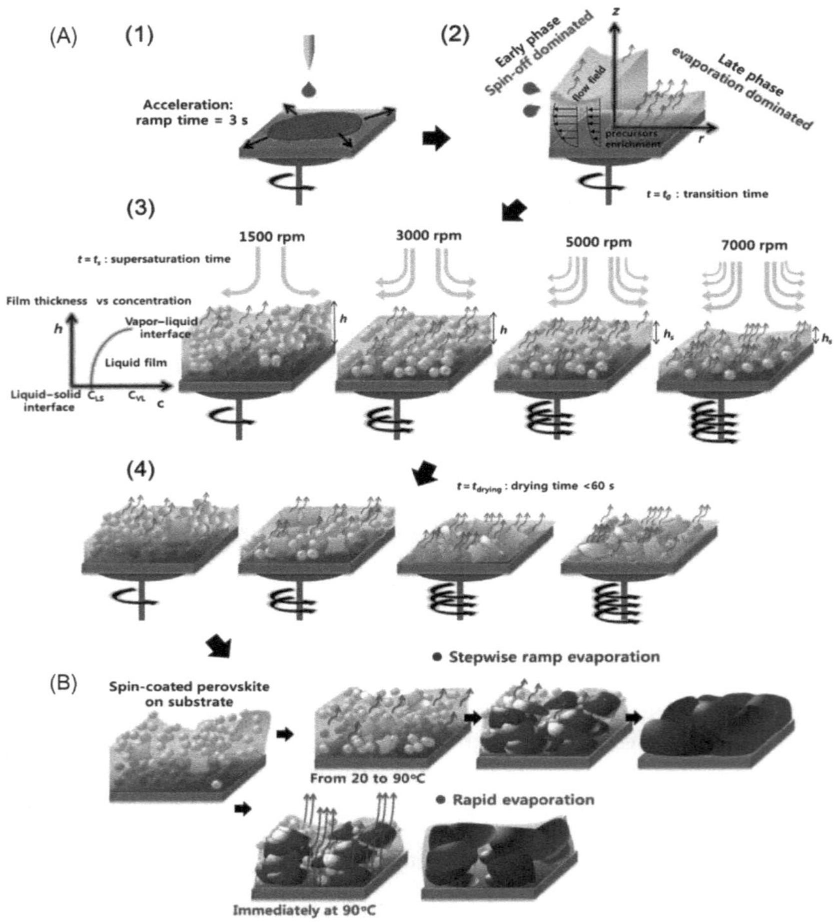

Figure 4.18 (A) Stages: (1) the precursor solution is placed onto the substrate; (2) spin coating showing the dominate processes in the early and late stages, followed by reaching a transition thickness; (3) the solution becomes supersaturated with a thickness; (4) nucleation and growth occur during the drying time. (B) The samples are subjected to stepwise ramp and rapid thermal annealing. *Reproduced with permission from R. Kang, J.E. Kim, J.S. Yeo, S. Lee, Y.J. Jeon, D.Y. Kim, Optimized organometal halide perovskite planar hybrid solar cells via control of solvent evaporation rate. J. Phys. Chem. C 118 (46) (2014) 26513–26520. Copyright 2014, ACS Publications.*

were used to modify the surface of CdS layer. The schematic of the device architecture is shown in Fig. 4.21.

Jin and Chumanov reported planar device architecture without HTL. Electrochemically exfoliated graphite was directly used as the top electrode over the perovskite film. Devices with this planar architecture

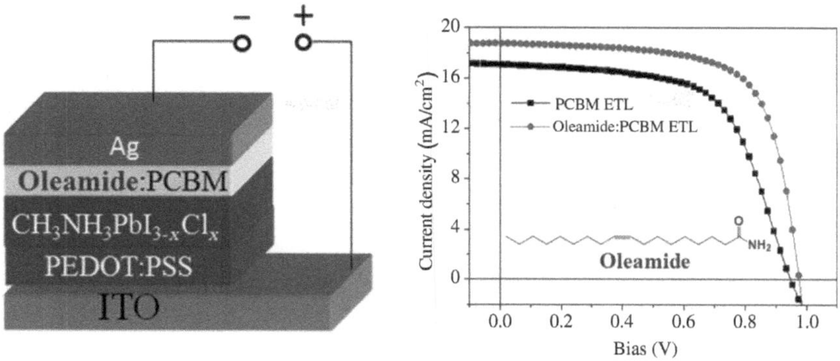

Figure 4.19 Schematics of device architecture and the corresponding *J–V* curve. *Reproduced with permission from F. Xia, Q. Wu, P. Zhou, Y. Li, X. Chen, Q. Liu, et al., Efficiency enhancement of inverted structure perovskite solar cells via oleamide doping of PCBM electron transport layer. ACS Appl. Mater. Interfaces 7 (24) (2015) 13659–13665. Copyright 2015, ACS Publications.*

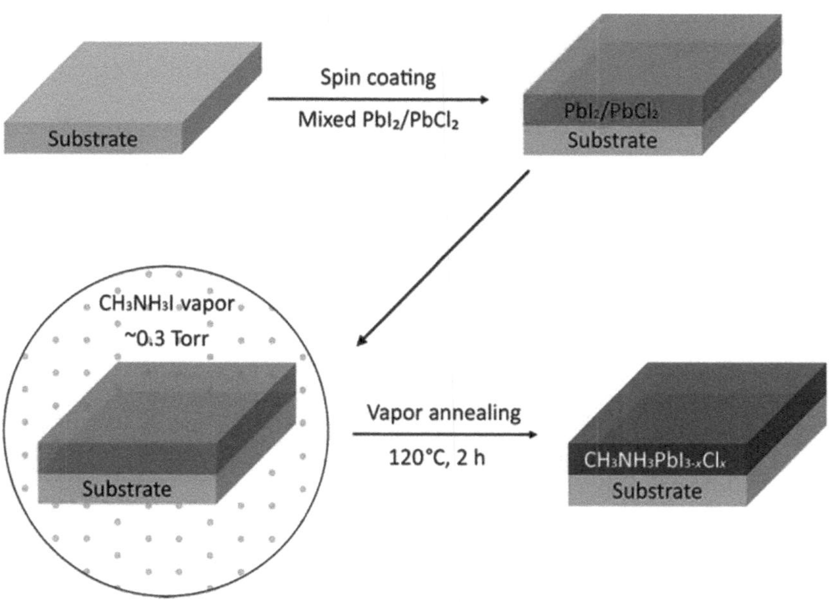

Figure 4.20 Schematics of the low-pressure vapor-assisted process. *Reproduced with permission from Y. Li, J.K. Cooper, R. Buonsanti, C. Giannini, Y. Liu, F.M. Toma, et al., Fabrication of planar heterojunction perovskite solar cells by controlled low-pressure vapor annealing. J. Phys. Lett. 6 (3) (2015) 493–499. Copyright 2015, ACS Publications.*

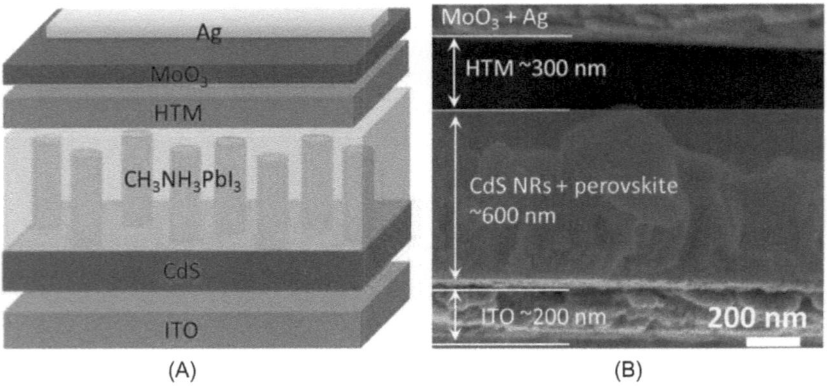

Figure 4.21 (A) Planar heterostructure hybrid solar cell device configuration (ITO/CdS NRs/$CH_3NH_3PbI_3$/HTM/MoO_3/Ag), (B) cross-sectional SEM image of the hybrid solar cell with UV-ozone. *SEM*, scanning electron microscopy. *Reproduced with permission from Z. Gu, F. Chen, X. Zhang, Y. Liu, C. Fan, G. Wu, et al., Novel planar heterostructure perovskite solar cells with CdS nanorods array as electron transport layer. Sol. Energy Mater. Sol. Cells 140 (2015) 396–404. Copyright 2015, Elsevier Publications.*

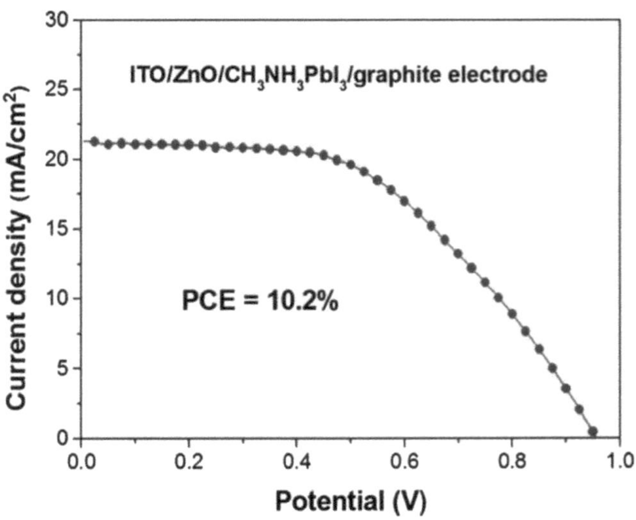

Figure 4.22 *J–V* curve of the device. *Reproduced with permission from Y. Jin, G. Chumanov, Solution-processed planar perovskite solar cell without a hole transport layer. ACS Appl. Mater. Interfaces 7 (22) (2015) 12015–12021. Copyright 2015, ACS Publications.*

achieved a PCE of ~10% [46]. Device consisted of ITO/ZnO/$CH_3NH_3PbI_3$/graphite electrode (Fig. 4.22).

4.4 MESOSUPERSTRUCTURED ARCHITECTURE

For the first time Lee et al. reported low-cost and solution-processable solar cells by developing MSSC architecture. This architecture couples perovskite light absorbers with nonelectron injecting Al_2O_3 scaffolds. The MSSC device architecture produced a PCE of around 8% [4]. Using Al_2O_3 scaffold instead of mesoporous TiO_2 is advantageous in a few ways, i.e., it eliminates high-temperature sintering and the devices with insulating Al_2O_3 scaffolds yield high open circuit voltage. Edri et al. report the MSSC architecture with organic–inorganic lead bromide perovskite also acts as a light absorber [47]. The device architecture is shown in Fig. 4.23.

Ball et al. reported a PCE of ~12% with MSSC device architecture as shown in Fig. 4.24. A solution-processable route was developed with $CH_3NH_3PbI_{3-X}Cl_X$ as perovskite light absorber [48]. The metal oxide scaffold was fabricated at temperatures below 150°C.

Annealing of the perovskite films has direct effect on the grain evolution and morphology of the film. Saliba et al. reported short rapid thermal annealing at 130°C leads to the growth of large micron-sized

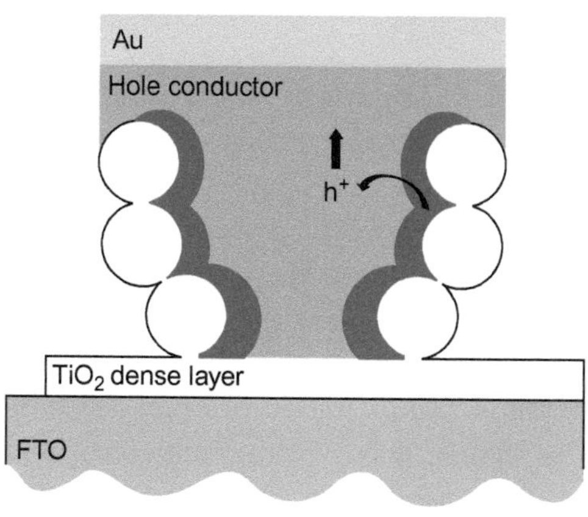

Figure 4.23 Schematics of MSSC device architecture with Al_2O_3 scaffold and $CH_3NH_3PbBr_3$ as perovskite light absorber. *MSSC*, mesosuperstructured solar cell. *Reproduced with permission from E. Edri, S. Kirmayer, D. Cahen, G. Hodes, High open-circuit voltage solar cells based on organic–inorganic lead bromide perovskite. J. Phys. Lett. 4 (6) (2013) 897–902. Copyright 2013, ACS Publications.*

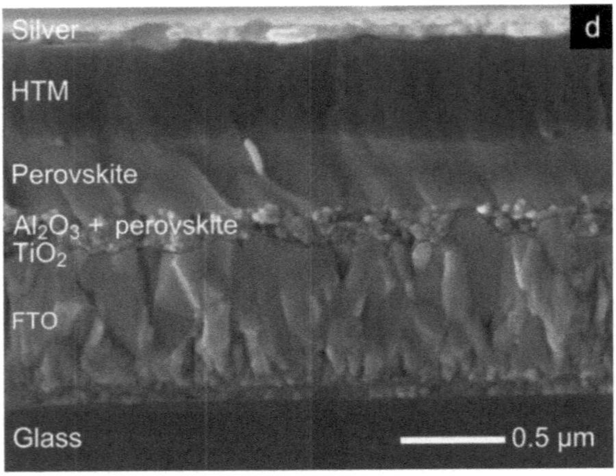

Figure 4.24 Cross-sectional SEM image of the reported device. *SEM*, scanning electron microscopy. *Reproduced with permission from J.M. Ball, M.M. Lee, A. Hey, H.J. Snaith, Low-temperature processed meso-superstructured to thin-film perovskite solar cells. Energy Environ. Sci. 6 (6) (2013) 1739–1743. Copyright 2013, The Royal Society of Chemistry.*

textured perovskite domains and improves the short-circuit current and PCE [49]. Short rapid annealing at 130°C helped in the growth of large perovskite and at the same time resulted in formation of 100–1000 nm polycrystalline perovskite domains during dwelling at 100°C. These results were compared between planar and MSSC device architecture with Al_2O_3 scaffolds. The device architecture used to in this study is shown in Fig. 4.25.

Anaya et al. studied the optical response of the MSSC device, both theoretically and experimentally. In theoretical studies the optical effects that arise from the presence of all layers comprising the photovoltaic device were considered. Theoretical model served to describe the optical response, i.e., reflectance, transmittance and absorbance of each layer. The same were compared with the fabricated MSSC devices [50]. Guarnera et al. worked on the long-term stability of MSSC PSCs and found out that the metal migration from cathode through the hole transporter and then through the perovskite is one of the influential factors degrading the device [33]. To overcome this, Al_2O_3 nanoparticles were used as the buffer layer between perovskite and top electrode. Hole transporter was infiltrated into these buffer layers. The buffer layer prevented the metal migration, and overall stability of the device was improved (Fig. 4.26; [51]).

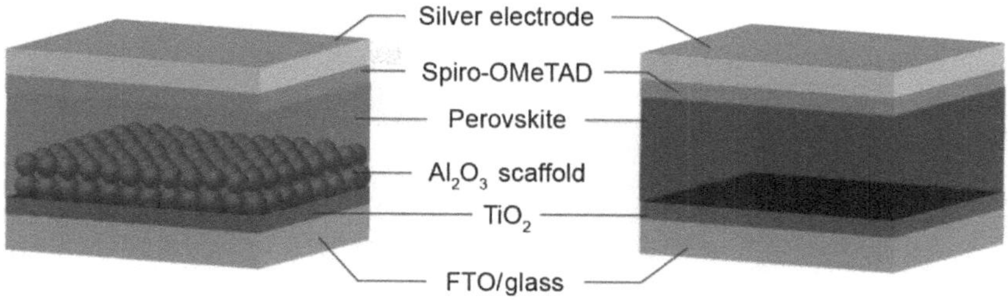

Figure 4.25 Schematic of mesosuperstructured and planar heterojunction perovskite solar cells. *Reproduced with permission from M. Saliba, K.W. Tan, H. Sai, D.T. Moore, T. Scott, W. Zhang, et al., Influence of thermal processing protocol upon the crystallization and photovoltaic performance of organic–inorganic lead trihalide perovskites. J. Phys. Chem. C 118 (30) (2014) 17171–17177. Copyright 2014, ACS Publications.*

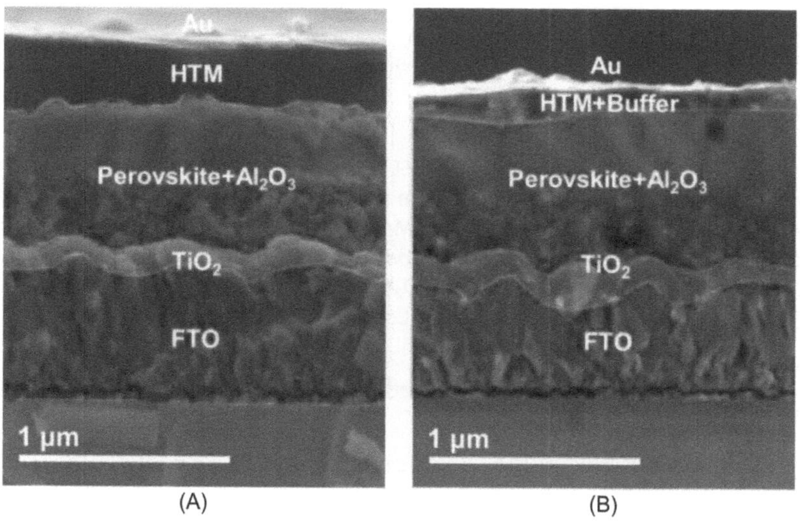

Figure 4.26 (A) SEM image of a standard MSSC (unbuffered cell); (b) SEM image of a MSSC with an Al_2O_3 buffer layer (buffered cell). *SEM*, scanning electron microscopy; *MSSC*, mesosuperstructured solar cell. *Reproduced with permission from S. Guarnera, A. Abate, W. Zhang, J.M. Foster, G. Richardson, A. Petrozza, et al., Improving the long-term stability of perovskite solar cells with a porous Al_2O_3 buffer layer. J. Phys. Lett. 6 (3) (2015) 432–437. Copyright 2015, ACS Publications.*

Liu et al. developed PSCs with efficiency exceeding 13% based on TiO_2/ZrO_2/carbon triple layer device architecture by controlling the critical parameters, such as thickness of spacer layer (ZrO_2), pore-filling and morphology of $MAPbI_3$ perovskite, temperature of substrate and

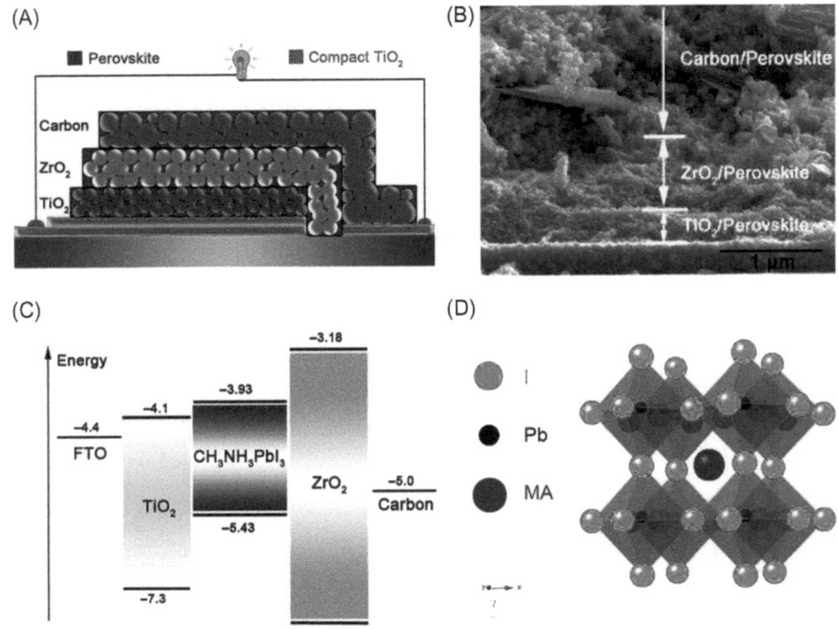

Figure 4.27 (A) Schematic diagram and SEM images (B) of the cross-section of the hole-conductor free mesoscopic perovskite solar cell; (C) energy band diagram of the device, and (D) the crystal structure of $MAPbI_3$ perovskite. *SEM*, scanning electron microscopy. *Reproduced with permission from T. Liu, L. Liu, M. Hu, Y. Yang, L. Zhang, A. Mei, et al., Critical parameters in TiO_2/ZrO_2/carbon-based mesoscopic perovskite solar cell. J. Power Sources 293 (2015) 533–538. Copyright 2015, Elsevier Publications.*

precursors. These cells showed an efficiency of 10.3% [52]. The device architecture is shown in Fig. 4.27.

Cheng et al. reported the incorporation of SiO_2 layer into the HTM-free PSCs by spin coating a SiO_2 paste onto the TiO_2 scaffold layer. This SiO_2 layer can act as an insulating layer and effectively inhibit the charge recombination between the TiO_2 layer and carbon electrode [53]. Mejía Escobar et al. implemented an electron collection bilayer composed of a thin layer of zirconia coated with titania, over the TCO film. The device architecture is shown in Fig. 4.28 [54].

4.5 MISCELLANEOUS ARCHITECTURES

Since the stability of perovskite absorber reduces with time due to degradation of these absorbers when exposed to atmospheric moisture, to overcome this, Kang et al. coated antireflective lotus leaf-inspired hierarchical

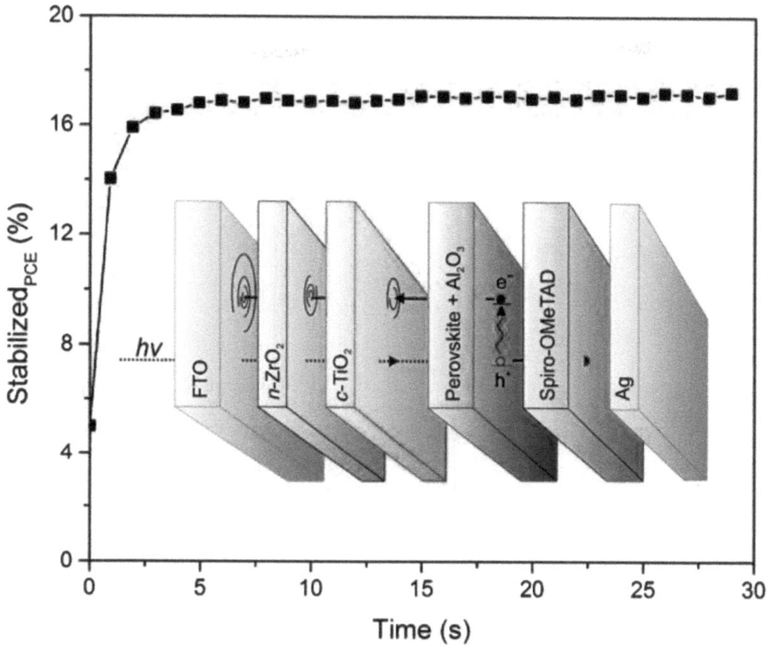

Figure 4.28 Stabilized power conversion efficiency and schematics of device. *Reproduced with permission from M.A. Mejía Escobar, S. Pathak, J. Liu, H.J. Snaith, F. Jaramillo, ZrO$_2$/TiO$_2$ electron collection layer for efficient meso-superstructured hybrid perovskite solar cells. ACS Appl. Mater. Interfaces 9 (3) (2017) 2342–2349. Copyright 2017, ACS Publications.*

pyramidal arrays with self-cleaning characteristics on top of the conventional PSCs. The PCE of $CH_3NH_3PbI_3$PSC was improved from 13.12% to 14.01% due to the antireflection effect arose from bifunctional polydimethylsiloxane (PDMS) film usage in PSCs. Moreover, the PDMS film adhered to the PSC demonstrated excellent superhydrophobicity [55]. The reflectance and transmittance of PDMS samples were low and high compared to bare samples as shown in Fig. 4.29. The cross-sectional SEM images confirmed the formation of pyramidal structures and the schematics illustrated the mechanism of antireflection. These PDMS coatings were studied for their self-cleaning and hydrophobic nature. The usage of patterned PDMS coating showed excellent hydrophobicity and self-cleaning properties as shown in Fig. 4.30.

Hwang et al. reported the passivation of PSC by spin coating hydrophobic Teflon polymer [56]. This simple spin coating has a good potential to hinder the moisture from contacting the perovskite absorber. The

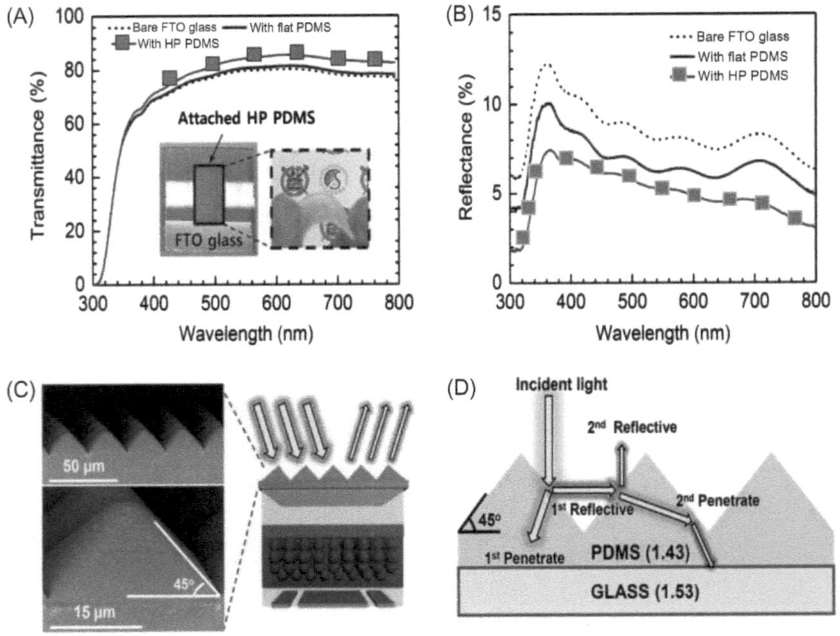

Figure 4.29 (A and B) Transmittance and reflectance spectra of FTO glass with three different cases, bare FTO, flat PDMS, and HP PDMS. The inset photographs represent the fabricated flexible HP PDMS film attached on a FTO glass, showing the visual effect of antireflection. (c) SEM images of the HP arrays with a 45° angle with the bottom edge side. The HP PDMS film serves as an antireflection layer on the perovskite solar cell device. (d) Theoretical scheme of the antireflection mechanism on the HP PDMS film. The size and color of arrows indicate the relative magnitude and direction of each light. *FTO*, fluorine-doped tin oxide; *HP*, hierarchical pyramid; *PDMS*, polydimethylsiloxane; *SEM*, scanning electron microscopy. *Reproduced with permission from S.M. Kang, N. Ahn, J.W. Lee, M. Choi, N.G. Park, Water-repellent perovskite solar cell. J. Mater. Chem. A 2 (47) (2014) 20017–20021. Copyright 2014, RSC Publications.*

hydrophobicity and nonreactivity of Teflon polymer with the perovskite has enabled this device architecture. The contact angle of the water droplet over the bare PSC had a contact angle of 53° while Teflon polymer passivized PSC had increased contact angle of 118°. This increase in contact angle is attributed to hydrophobicity of Teflon polymer (Fig. 4.31).

Kim et al. reported improvement in the stability of inverted PSCs against water and high operating temperature (100°C). This was achieved by integrating an ultrathin amorphous oxide electron extraction layer via atomic layer deposition. These unencapsulated inverted devices exhibit a stable operation over at least 10 hours when subjected to high thermal

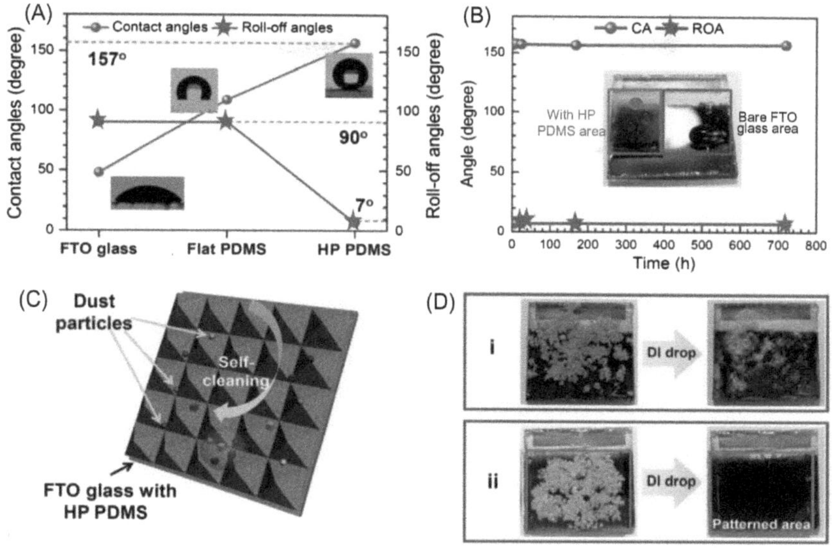

Figure 4.30 (A) Static CAs (—) and ROAs (-★-) of DI water on the three different surfaces. Microscope images of a droplet on each surface are shown in the insets. (B) Time-dependent CA and ROA of the HP PDMS film measured over a time period of 1 month. Inset photograph shows the degree of reflectivity and superhydrophobicity for the different areas with and without an HP PDMS film on the same perovskite solar cell. Water was colored with red dye to demonstrate visually. (C) Schematic illustration of the self-cleaning effect on the HP PDMS surface. (D) Self-cleaning experiments for the perovskite solar cell devices without (i) and with (ii) the HP PDMS film. *CA*, contact anglel; *DI*, deionized; *ROA*, roll-off angle; *HP*, hierarchical pyramid; *PDMS*, polydimethylsiloxane. *Reproduced with permission from S.M. Kang, N. Ahn, J.W. Lee, M. Choi, N.G. Park, Water-repellent perovskite solar cell. J. Mater. Chem. A 2 (47) (2014) 20017–20021. Copyright 2014, RSC Publications.*

stress (100°C) in ambient environments, as well as upon direct contact with a droplet of water without further encapsulation [57] (Fig. 4.32).

4.6 SUMMARY

Various device architectures evolved in PSCs have been discussed briefly in this chapter. Organic/inorganic perovskite absorbers are extensively studied in mesoscopic, planar and other device architectures. Mesoscopic architectures involve the usage of nanocrystalline particles, whereas in planar architectures the need to use nanocrystalline layer is nonexistent. In mesoscopic architecture, electron injection from perovskite absorber into

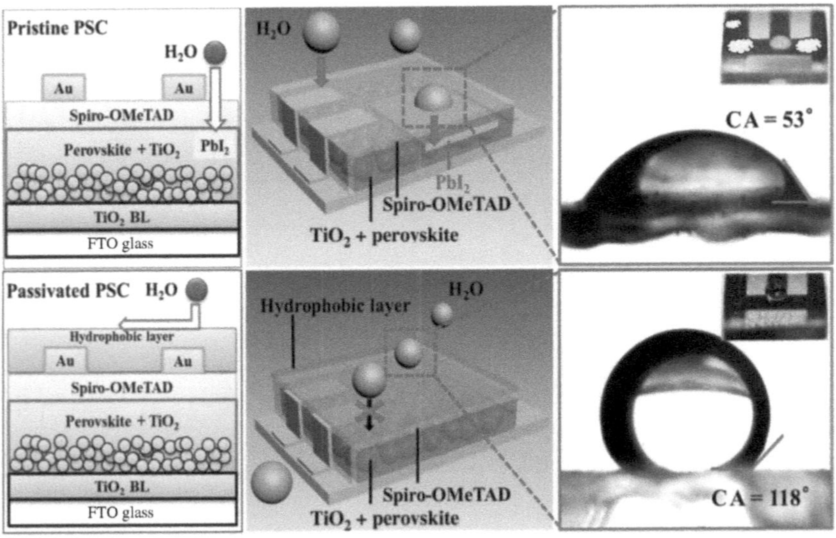

Figure 4.31 From left, schematic illustration shows the principle of hydrophobic passivation, contact angles of the untreated cell (top) and passivized cell (bottom), digital photographs of both cells showing contact angles. Also insets show water drop on cell surfaces. The water drop on untreated cell (top inset image) is caused dissolution of perovskite by water. The layer of perovskite in passivated cell does not change when the water drops on to the device. *Reproduced with permission from I. Hwang, I. Jeong, J. Lee, M.J. Ko, K. Yong, Enhancing stability of perovskite solar cells to moisture by the facile hydrophobic passivation. ACS Appl. Mater. Interfaces 7 (31) (2015) 17330–17336. Copyright 2015, ACS Publications.*

mesoporous TiO_2 is evident and in planar and mesostructured architectures ambipolar carrier transport in perovskites is exploited. Prior to adapting a device architecture, electronic properties like band-edge matching, charge carrier diffusion length, etc., optical properties such as transmittance, reflectance, absorbance, etc., physical properties like surface coverage of films, density of pin holes etc., and interfacial matching need to be considered. Further to improve the long-term stability of PSCs, the architectures employing moisture passivation, hydrophobic, self-cleaning, etc. layer is also discussed. Architectures discussed in this chapter possess variety of advantages and disadvantages leading to a trade-off between efficiency, cost, and stability. Hence careful adoption of a particular architecture becomes relevant for realization of successful and commercial PSC.

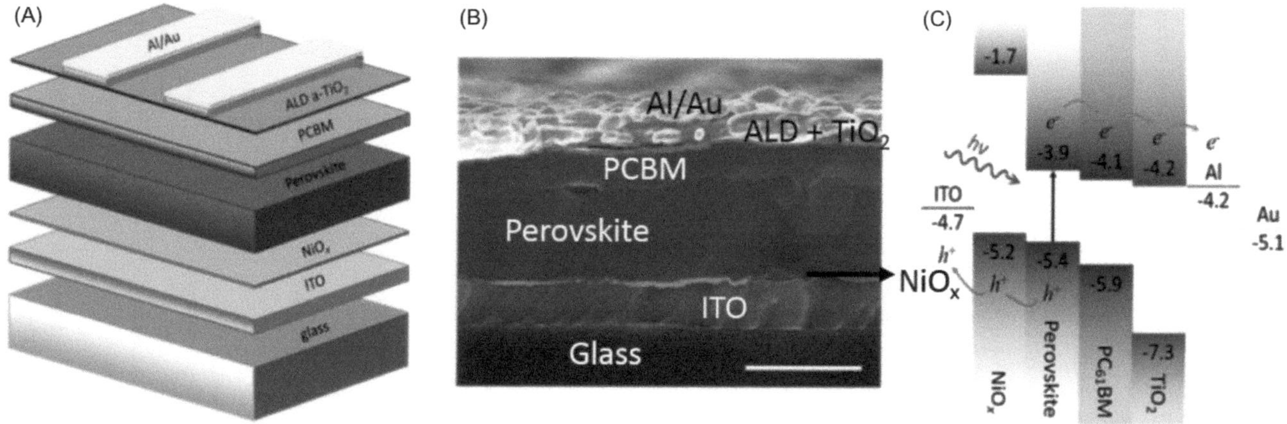

Figure 4.32 A representative (A) device architecture and (B) false color SEM image of the inverted perovskite photovoltaic device (scale bar 500 nm). (c) Idealized energy-level diagram of the corresponding device architecture. *SEM*, scanning electron microscopy. *Reproduced with permission from I.S. Kim, D.H. Cao, D.B. Buchholz, J.D. Emery, O.K. Farha, J.T. Hupp, et al., Liquid water-and heat-resistant hybrid perovskite photovoltaics via an inverted ALD oxide electron extraction layer design. Nano Lett. 16 (12) (2016) 7786–7790. Copyright 2016, ACS Publications.*

REFERENCES

[1] A. Kojima, K. Teshima, Y. Shirai, T. Miyasaka, Organometal halide perovskites as visible-light sensitizers for photovoltaic cells, J. Am. Chem. Soc. 131 (17) (2009) 6050–6051.
[2] J.H. Im, C.R. Lee, J.W. Lee, S.W. Park, N.G. Park, 6.5% efficient perovskite quantum-dot-sensitized solar cell, Nanoscale 3 (10) (2011) 4088–4093.
[3] H.S. Kim, C.R. Lee, J.H. Im, K.B. Lee, T. Moehl, A. Marchioro, et al., Lead iodide perovskite sensitized all-solid-state submicron thin film mesoscopic solar cell with efficiency exceeding 9%, Sci. Rep. 2 (2012) 591.
[4] M.M. Lee, J. Teuscher, T. Miyasaka, T.N. Murakami, H.J. Snaith, Efficient hybrid solar cells based on meso-superstructured organometal halide perovskites, Science 338 (6107) (2012) 643–647.
[5] L. Etgar, P. Gao, Z. Xue, Q. Peng, A.K. Chandiran, B. Liu, et al., Mesoscopic $CH_3NH_3PbI_3/TiO_2$ heterojunction solar cells, J. Am. Chem. Soc. 134 (42) (2012) 17396–17399.
[6] M. Liu, M.B. Johnston, H.J. Snaith, Efficient planar heterojunction perovskite solar cells by vapour deposition, Nature 501 (7467) (2013) 395–398.
[7] N.G. Park, Perovskite solar cells: an emerging photovoltaic technology, Mater. Today 18 (2) (2015) 65–72.
[8] A. Abrusci, S.D. Stranks, P. Docampo, H.L. Yip, A.K.Y. Jen, H.J. Snaith, High-performance perovskite-polymer hybrid solar cells via electronic coupling with fullerene monolayers, Nano Lett. 13 (7) (2013) 3124–3128.
[9] J.T.W. Wang, J.M. Ball, E.M. Barea, A. Abate, J.A. Alexander-Webber, J. Huang, et al., Low-temperature processed electron collection layers of graphene/TiO_2 nanocomposites in thin film perovskite solar cells, Nano Lett. 14 (2) (2013) 724–730.
[10] D.Y. Son, J.H. Im, H.S. Kim, N.G. Park, 11% efficient perovskite solar cell based on ZnO nanorods: an effective charge collection system, J. Phys. Chem. C 118 (30) (2014) 16567–16573.
[11] N.G. Park, T. Miyasaka, M. Grätzel, Organic–Inorganic Halide Perovskite Photovoltaics, Springer, 2016.
[12] H. Wang, X. Zeng, Z. Huang, W. Zhang, X. Qiao, B. Hu, et al., Boosting the photocurrent density of p-type solar cells based on organometal halide perovskite-sensitized mesoporous NiO photocathodes, ACS Appl. Mater. Interfaces 6 (15) (2014) 12609–12617.
[13] R. Lindblad, D. Bi, B.W. Park, J. Oscarsson, M. Gorgoi, H. Siegbahn, et al., Electronic structure of $TiO_2/CH_3NH_3PbI_3$ perovskite solar cell interfaces, J. Phys. Lett. 5 (4) (2014) 648–653.
[14] Q. Hu, J. Wu, C. Jiang, T. Liu, X. Que, R. Zhu, et al., Engineering of electron-selective contact for perovskite solar cells with efficiency exceeding 15%, ACS Nano 8 (10) (2014) 10161–10167.
[15] L. Liang, Z. Huang, L. Cai, W. Chen, B. Wang, K. Chen, et al., Magnetron sputtered zinc oxide nanorods as thickness-insensitive cathode interlayer for perovskite planar-heterojunction solar cells, ACS Appl. Mater. Interfaces 6 (23) (2014) 20585–20589.
[16] A. Yella, L.P. Heiniger, P. Gao, M.K. Nazeeruddin, M. Grätzel, Nanocrystalline rutile electron extraction layer enables low-temperature solution processed perovskite photovoltaics with 13.7% efficiency, Nano Lett. 14 (5) (2014) 2591–2596.
[17] A. Bera, K. Wu, A. Sheikh, E. Alarousu, O.F. Mohammed, T. Wu, Perovskite oxide $SrTiO_3$ as an efficient electron transporter for hybrid perovskite solar cells, J. Phys. Chem. C 118 (49) (2014) 28494–28501.

[18] L.S. Oh, D.H. Kim, J.A. Lee, S.S. Shin, J.W. Lee, I.J. Park, et al., Zn_2SnO_4-based photoelectrodes for organolead halide perovskite solar cells, J. Phys. Chem. C 118 (40) (2014) 22991−22994.
[19] W. Ke, G. Fang, J. Wang, P. Qin, H. Tao, H. Lei, et al., Perovskite solar cell with an efficient TiO_2 compact film, ACS Appl. Mater. Interfaces 6 (18) (2014) 15959−15965.
[20] A. Sarkar, N.J. Jeon, J.H. Noh, S.I. Seok, Well-organized mesoporous TiO_2 photoelectrodes by block copolymer-induced sol−gel assembly for inorganic−organic hybrid perovskite solar cells, J. Phys. Chem. C 118 (30) (2014) 16688−16693.
[21] X. Feng, K. Shankar, O.K. Varghese, M. Paulose, T.J. Latempa, C.A. Grimes, Vertically aligned single crystal TiO_2 nanowire arrays grown directly on transparent conducting oxide coated glass: synthesis details and applications, Nano Lett. 8 (11) (2008) 3781−3786.
[22] X. Sheng, D. He, J. Yang, K. Zhu, X. Feng, Oriented assembled TiO_2 hierarchical nanowire arrays with fast electron transport properties, Nano Lett. 14 (4) (2014) 1848−1852.
[23] D. Zhong, B. Cai, X. Wang, Z. Yang, Y. Xing, S. Miao, et al., Synthesis of oriented TiO_2 nanocones with fast charge transfer for perovskite solar cells, Nano Energy 11 (2015) 409−418.
[24] D.Y. Son, K.H. Bae, H.S. Kim, N.G. Park, Effects of seed layer on growth of ZnO nanorod and performance of perovskite solar cell, J. Phys. Chem. C 119 (19) (2015) 10321−10328.
[25] S. Hong, A. Han, E.C. Lee, K.W. Ko, J.H. Park, H.J. Song, et al., A facile and low-cost fabrication of TiO_2 compact layer for efficient perovskite solar cells, Curr. Appl. Phys. 15 (5) (2015) 574−579.
[26] J. Zhang, T. Pauporte, Effects of oxide contact layer on the preparation and properties of $CH_3NH_3PbI_3$ for perovskite solar cell application, J. Phys. Chem. C 119 (27) (2015) 14919−14928.
[27] K. Wang, Y. Shi, Q. Dong, Y. Li, S. Wang, X. Yu, et al., Low-temperature and solution-processed amorphous WO_X as electron-selective layer for perovskite solar cells, J. Phys. Lett. 6 (5) (2015) 755−759.
[28] W. Ke, G. Fang, Q. Liu, L. Xiong, P. Qin, H. Tao, et al., Low-temperature solution-processed tin oxide as an alternative electron transporting layer for efficient perovskite solar cells, J. Am. Chem. Soc. 137 (21) (2015) 6730−6733.
[29] C. Li, Y. Li, Y. Xing, Z. Zhang, X. Zhang, Z. Li, et al., Perovskite solar cell using a two-dimensional titania nanosheet thin film as the compact layer, ACS Appl. Mater. Interfaces 7 (28) (2015) 15117−15122.
[30] L. Cojocaru, S. Uchida, Y. Sanehira, J. Nakazaki, T. Kubo, H. Segawa, Surface treatment of the compact TiO_2 layer for efficient planar heterojunction perovskite solar cells, Chem. Lett. 44 (5) (2015) 674−676.
[31] H. Nagaoka, F. Ma, D.W. deQuilettes, S.M. Vorpahl, M.S. Glaz, A.E. Colbert, et al., Zr incorporation into TiO_2 electrodes reduces hysteresis and improves performance in hybrid perovskite solar cells while increasing carrier lifetimes, J. Phys. Lett. 6 (4) (2015) 669−675.
[32] X. Jiang, Y. Xiong, A. Mei, Y. Rong, Y. Hu, L. Hong, et al., Efficient compact-layer-free, hole-conductor-free, fully printable mesoscopic perovskite solar cell, J. Phys. Lett. 7 (20) (2016) 4142−4146.
[33] H.S. Jung, N.G. Park, Perovskite solar cells: from materials to devices, Small 11 (1) (2015) 10−25.
[34] Y. Shi, Y. Xing, Y. Li, Q. Dong, K. Wang, Y. Du, et al., $CH_3NH_3PbI_3$ and $CH_3NH_3PbI_{3-X}Cl_X$ in planar or mesoporous perovskite solar cells: comprehensive

insight into the dependence of performance on architecture, J. Phys. Chem. C 119 (28) (2015) 15868–15873.
[35] Q. Chen, H. Zhou, Z. Hong, S. Luo, H.S. Duan, H.H. Wang, et al., Planar heterojunction perovskite solar cells via vapor-assisted solution process, J. Am. Chem. Soc. 136 (2) (2013) 622–625.
[36] Y. Zhao, K. Zhu, Efficient planar perovskite solar cells based on 1.8 eV band gap $CH_3NH_3PbI_2Br$ nanosheets via thermal decomposition, J. Am. Chem. Soc. 136 (35) (2014) 12241–12244.
[37] F. Huang, Y. Dkhissi, W. Huang, M. Xiao, I. Benesperi, S. Rubanov, et al., Gas-assisted preparation of lead iodide perovskite films consisting of a monolayer of single crystalline grains for high efficiency planar solar cells, Nano Energy 10 (2014) 10–18.
[38] K. Wojciechowski, S.D. Stranks, A. Abate, G. Sadoughi, A. Sadhanala, N. Kopidakis, et al., Heterojunction modification for highly efficient organic–inorganic perovskite solar cells, ACS Nano 8 (12) (2014) 12701–12709.
[39] C. Liu, K. Wang, P. Du, T. Meng, X. Yu, S.Z. Cheng, et al., High performance planar heterojunction perovskite solar cells with fullerene derivatives as the electron transport layer, ACS Appl. Mater. Interfaces 7 (2) (2015) 1153–1159.
[40] S. Paek, N. Cho, H. Choi, H. Jeong, J.S. Lim, J.Y. Hwang, et al., Improved external quantum efficiency from solution-processed $(CH_3NH_3)PbI_3$ perovskite/$PC_{71}BM$ planar heterojunction for high efficiency hybrid solar cells, J. Phys. Chem. C 118 (45) (2014) 25899–25905.
[41] Y. Liu, Z. Hong, Q. Chen, W. Chang, H. Zhou, T.B. Song, et al., Integrated perovskite/bulk-heterojunction toward efficient solar cells, Nano Lett. 15 (1) (2014) 662–668.
[42] R. Kang, J.E. Kim, J.S. Yeo, S. Lee, Y.J. Jeon, D.Y. Kim, Optimized organometal halide perovskite planar hybrid solar cells via control of solvent evaporation rate, J. Phys. Chem. C 118 (46) (2014) 26513–26520.
[43] F. Xia, Q. Wu, P. Zhou, Y. Li, X. Chen, Q. Liu, et al., Efficiency enhancement of inverted structure perovskite solar cells via oleamide doping of PCBM electron transport layer, ACS Appl. Mater. Interfaces 7 (24) (2015) 13659–13665.
[44] Y. Li, J.K. Cooper, R. Buonsanti, C. Giannini, Y. Liu, F.M. Toma, et al., Fabrication of planar heterojunction perovskite solar cells by controlled low-pressure vapor annealing, J. Phys. Lett. 6 (3) (2015) 493–499.
[45] Z. Gu, F. Chen, X. Zhang, Y. Liu, C. Fan, G. Wu, et al., Novel planar heterostructure perovskite solar cells with CdS nanorods array as electron transport layer, Sol. Energy Mater. Sol. Cells 140 (2015) 396–404.
[46] Y. Jin, G. Chumanov, Solution-processed planar perovskite solar cell without a hole transport layer, ACS Appl. Mater. Interfaces 7 (22) (2015) 12015–12021.
[47] E. Edri, S. Kirmayer, D. Cahen, G. Hodes, High open-circuit voltage solar cells based on organic–inorganic lead bromide perovskite, J. Phys. Lett. 4 (6) (2013) 897–902.
[48] J.M. Ball, M.M. Lee, A. Hey, H.J. Snaith, Low-temperature processed meso-superstructured to thin-film perovskite solar cells, Energy Environ. Sci. 6 (6) (2013) 1739–1743.
[49] M. Saliba, K.W. Tan, H. Sai, D.T. Moore, T. Scott, W. Zhang, et al., Influence of thermal processing protocol upon the crystallization and photovoltaic performance of organic–inorganic lead trihalide perovskites, J. Phys. Chem. C 118 (30) (2014) 17171–17177.
[50] M. Anaya, G. Lozano, M.E. Calvo, W. Zhang, M.B. Johnston, H.J. Snaith, et al., Optical description of mesostructured organic–inorganic halide perovskite solar cells, J. Phys. Lett. 6 (1) (2014) 48–53.

[51] S. Guarnera, A. Abate, W. Zhang, J.M. Foster, G. Richardson, A. Petrozza, et al., Improving the long-term stability of perovskite solar cells with a porous Al_2O_3 buffer layer, J. Phys. Lett. 6 (3) (2015) 432–437.
[52] T. Liu, L. Liu, M. Hu, Y. Yang, L. Zhang, A. Mei, et al., Critical parameters in TiO_2/ZrO_2/carbon-based mesoscopic perovskite solar cell, J. Power Sources 293 (2015) 533–538.
[53] N. Cheng, P. Liu, S. Bai, Z. Yu, W. Liu, S.S. Guo, et al., Application of mesoporous SiO_2 layer as an insulating layer in high performance hole transport material free $CH_3NH_3PbI_3$ perovskite solar cells, J. Power Sources 321 (2016) 71–75.
[54] M.A. Mejía Escobar, S. Pathak, J. Liu, H.J. Snaith, F. Jaramillo, ZrO_2/TiO_2 electron collection layer for efficient meso-superstructured hybrid perovskite solar cells, ACS Appl. Mater. Interfaces 9 (3) (2017) 2342–2349.
[55] S.M. Kang, N. Ahn, J.W. Lee, M. Choi, N.G. Park, Water-repellent perovskite solar cell, J. Mater. Chem. A 2 (47) (2014) 20017–20021.
[56] I. Hwang, I. Jeong, J. Lee, M.J. Ko, K. Yong, Enhancing stability of perovskite solar cells to moisture by the facile hydrophobic passivation, ACS Appl. Mater. Interfaces 7 (31) (2015) 17330–17336.
[57] I.S. Kim, D.H. Cao, D.B. Buchholz, J.D. Emery, O.K. Farha, J.T. Hupp, et al., Liquid water-and heat-resistant hybrid perovskite photovoltaics via an inverted ALD oxide electron extraction layer design, Nano Lett. 16 (12) (2016) 7786–7790.

CHAPTER 5

Hybrid Inorganic Organic Perovskites: A Low-Cost-Efficient Optoelectronic Material

Diptonil Banerjee[1] and Kalyan Kumar Chattopadhyay[2]
[1]Indian Institute of Engineering Science and Technology, Howrah, West Bengal, India
[2]Jadavpur University, Kolkata, West Bengal, India

5.1 BASIC CRYSTAL STRUCTURE

Due to the simultaneous possession of different properties like optical, electronic, magnetic properties along with their tunability as well as structural tenability and good processability, hybrid organic–inorganic perovskite (HOIP) has attracted the attention of researchers both from the field of science and technology since long back.

The materials have a general structure that can be best described in general by an ABX_3 kind of formula where both A and B represent certain cation with X being the representative of anions as shown in Fig. 5.1 [1].

The ideal perovskite materials possess the highest symmetry, i.e., cubic symmetry where BX_6 octahedral network with a B–X–B bond angle of 180° shares the corners of the structures with A staying in the interstices. However, the structures can be changed to a lower symmetric phase that include tetragonal or orthorhombic by partially distorting the BX_6 octahedral.

It should be noted that these kinds of materials possess (ns^2) metals like Sn(II) or Pb(II) and thus exhibits unconventional electronic properties which can be explained if one understands the structure–property relationship.

Thus the properties as well as the structural stabilities are basically defined by the synthesis condition of the materials. The origin of the electrical and optical properties of the material and thus of all the similar

Perovskite Photovoltaics
DOI: https://doi.org/10.1016/B978-0-12-812915-9.00005-8

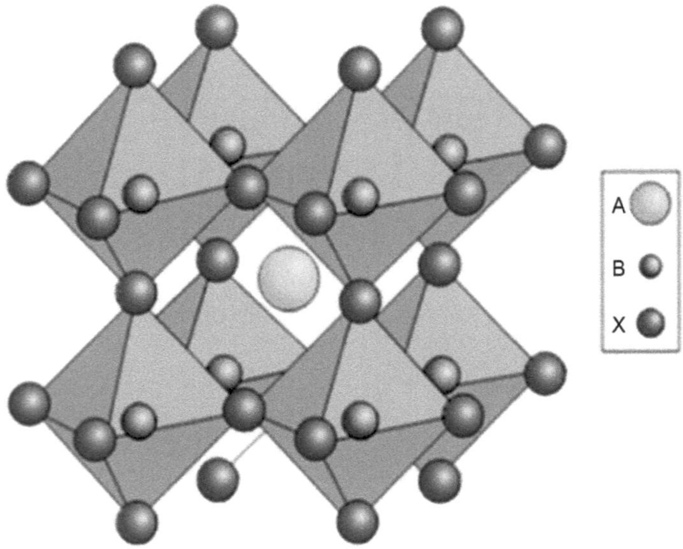

Figure 5.1 Cubic perovskite ABX_3 crystal structure. *Published with permission from J. Chen, S. Zhou, S. Jin, H. Li, T.Zhai, Crystal organometal halide perovskites with promising optoelectronic applications, J. Mater. Chem. C 4 (1) (2016) 11–27.*

materials can be very well speculated if someone understands structure in depth. Using the structural features as guides, the origin of the optical and electronic properties of similar perovskites has been clarified thus setting the basis for tuning and controlling them. The arrangement of BX_6 octahedral unit plays a crucial role for determination of the bandgap in such kind of materials as well as the transport properties shown by them. For instance the materials with less tilted cubic like arrangement show a redshift in bandgap with a profound decrease in effective mass. This behavior or rather dependency is associated with the spin–orbit coupling with the heavy metal and thus the structural and electronic degrees of freedom. The magnitude of the coupling is very much dependent on the extent of BX_6 octahedral tilting. In this point, it should be noted that among various perovskites, tin–lead perovskites are always a material with special attention due to the possibilities of bandgap to be tuned in the near-infrared region (NIR) as well as controlling the lead content in the device opening the possibilities of developing perovskite–perovskite tandem cells.

Solution-processed (one of the main means for developing the perovskite structures) materials—both the inorganic lead halide matrix and the mobile (organic) cations, such as cesium, methylammonium (MA), or formamidinium (FA)—inherently have soft structures. In a hybrid perovskite, the inorganic and organic components are interdependent; thus a change or deformation in the organic cations immediately results a subsequent effects or modifications in their inorganic counterpart (sublattice). The unusual behavior of perovskites, showing spectacularly large length of diffusions and thus low recombination, which are believed to be the main reason for the success of this class of materials due to the implication of such soft dynamical behavior and several hypotheses were made in order to explain them.

The materials are special in another aspect as the system shows dynamical behaviors that are associated with different time scale from ultrafast femtosecond scale to milliseconds or even minutes or hours. The materials can be distinguished from the other common semiconductors mainly based on the following three points: first, they show very long carrier lifetime and simultaneously a modest mobility; second, even after their low-temperature solution-processed growth, they exhibit a rather low trap density; and lastly, flexible inorganic lattice being associated with sharp diffraction and optical absorption onset [2–5].

It should be noted that the understanding the nature of carrier–phonon interaction is extremely important in order to analyze the carrier dynamics and transport processes in perovskite system. This is due to the fact that due to the deformable nature of the crystal lattice in perovskite system, shifting of ions from the equilibrium system will likely have interaction with the charge carrier thus influencing electronic as well as transport properties.

So far, the established theory predicts screening of nascent charge carriers by the formation of large polaron of couple of hundred femtoseconds time scale resulting generation of excited electrons with very high lifetime on the order of 100 ps. This is much higher compared to the carrier lifetime seen in other conventional semiconductors. Another point that should be mentioned here is that formation of protected energetic carriers need the formation of polarons that is associated with very fast dynamics (rather ultrafast) in the disordered hybrid perovskite system [6,7].

However, the energetic carriers can see the protection of long-lived band-edge carriers without any organic cation being present in the system, and thus it can be concluded that this is likely an intrinsic property of the soft perovskite inorganic lattice.

As the synthesis of the materials involves a low-temperature process, one can expect a large number of structural and chemical disorders, i.e., defects present in the perovskite materials that definitely alter the optoelectronic material. The actual influence of these defects present in the perovskite materials on the transport properties is very difficult to understand, and this lack of knowledge truly limits the final performance of the devices made using that materials. Also, the defects present limit the reproducibility of the material synthesized in different lab which is one of the key steps to make the associated technology market appealing. The confocal fluorescence lifetime imaging microscopy and scanning-probe microscopy are two of the most widely used techniques for having an idea about the origin and distribution of the heterogeneity present in the material and thus influencing the properties like local photoluminescence (PL), local electroluminescence, local conductivity, and local surface potential.

Time-of-flight mass spectrometry and density functional theoretical calculation indicated the surface states of the material can be modified by introducing certain additive or illumination to them. Also, the "photocleaning" effects repeatedly observed in perovskites can be associated with photoinduced ion migration. As mentioned before the selective adopting of different synthesis processes and additives the carrier lifetime and the PL intensities of the perovskite materials can be tailored and can thus be made approached to that of single crystal opening the way for these materials to approach the Shockley–Queisser limit [8,9]. Apart from the defects originated from synthesis condition as well as additives and illumination, another key parameter that governs the optoelectronic performance of the perovskite materials based device is the interfaces made with certain contacts. It is assumed that the dynamic interaction shown by the perovskite systems is possibly underlying the hysteresis of perovskite solar cells.

The hysteresis can be classified into two categories, namely, capacitive and noncapacitive hysteresis, both giving the capacitance as function of frequencies with or without illumination. The results show that the capacitance mainly arises due to the contacts to the perovskite layer and also due to ion migration. Capacitive interface and related hysteresis which basically signifies the ability to accommodate both ionic and electronic charges in a reversible way can be seen in perovskite/TiO_2 interface system, whereas noncapacitive hysteresis cause device degradation associated with the modification of the interface or contact reactivity caused by perovskite ionic motion [10,11].

5.2 SYMMETRY AND PHASE

As mentioned before, the B–X unit in ABX_3 type perovskite material plays very important role in determining the electronic properties of materials. However, the A ion is almost neutral to such aspects, whereas the size of A can cause a distortion in the B–X bond thus distorting the symmetry as well. In an ideal case, perovskite can be found in the highest symmetrized cubic phase associated with a pm3m space group. The crystal structure is associated with coordination number of 12 and 6 for the cations A and B, respectively, with BX_6 octahedra residing at the corner. The cubic symmetry offers the optimum electronic properties because of the high degree of ionic bonding, and thus it means that deviation from the cubic symmetry can alter the properties of the material as expected of which the first noticeable influence is the size effect. How the cubic structure can be obtained w.r.t. ionic size can be speculated by well-known Goldschmidt's tolerance factor [12] given by

$$t = \frac{R_A + R_x}{\sqrt{2}(R_B + R_x)} \tag{5.1}$$

where R_S are the ionic radii and the suffix denotes the corresponding ion (i.e., A, B, X, etc.)

It is seen that as per the criteria set, the oxide perovskites show the cubic symmetry for $0.89 < t < 1$, whereas for halide perovskite, the corresponding value is given by $0.85 < t < 1.11$ [13,14].

Conditions when there are mismatching between the ionic radii crystal symmetry can be changed resulting a drastic change in the symmetry. For $t < 1$, the B–X and A–X bonds, respectively, undergo compression and tension, resulting a rotation of octahedral in order to accommodate this induced stress. This will cause a distortion and tilting of BX_6 octahedral breaking the symmetry. On the other hand, for $t > 1$ (that may occur as result of large R_A or small R_B), more stabilized hexagonal structure is the result.

Also depending upon the size of the cation, A can influence the dimensionality, i.e., the shape of the perovskites. For instance, monovalent A cations result in a three-dimensional framework, whereas replacement of this with a large A cations results in destruction of 3D network giving a reduced dimensionality of two or three corresponding to $t \ll 1$.

The Goldschmidt's tolerance factor has been alternatively developed by Li et al. [14]. As per the rule set by them, the ratio $\mu = R_B/R_X$

directly correlates BX_6 to perovskite, and according to the assumption, $\mu > 0.442$ is associated with formation of halide perovskite, whereas below this value, BX_6 octahedron will become unstable and the formation will not occur. This assumption along with Goldschmidt's criteria provides a guideline regarding the condition for the formation of halide perovskites. However, they are not certainly the sufficient condition for predicting all structure formations within the perovskite family. Apart from the size effect, other external perturbations like heat, environment (oxidative or reductive) can also affect the structure by introducing distortion via changing the composition. The change in the valency of transition metal ion or the voids present in the sample due to different oxygen content can affect the structural stability within the material. Also, it is seen that BX_6 octahedra can be distorted by the B cation (e.g., Mn) due to residing of odd numbers electron in the bandgap energy level. This is known as Jahn–Teller distortion and associated with the spatial degeneracy [13]. Detailed electronic band structure analysis can predict that apart from the abovementioned fact, the structure has a pressure dependency also, and pressure beyond a certain value can cause redshift in the energy-band diagram [15].

Thus one will not make mistake by concluding that if one wants to tailor a specific property of the perovskite, one has to synthesize the material with a particular symmetry. Moreover, for obtaining a particular symmetry in perovskite, one should not only take care of the crystal component selection but should also take hold on the other external parameters as well.

A detailed crystallographic analysis reveals (especially based on the X-ray diffraction pattern) that perovskite can be found with the following symmetry: cubic (Pm3m), tetragonal (I4/mcm), and orthorhombic (Pnam) with the fact that generally it exhibits a cubic symmetry and that can be transformed to tetragonal and orthorhombic upon the reduction of temperature [16,17]. Important information that one obtained from transmission electron microscope (TEM) study is that perovskite can be stabilized by forming pseudocubic structure that basically results from octahedral tilting and cation rotation [18]. For instance, one has seen that in the temperature range 300–400K, the tetragonal structure of $MAPb_3$ got stabilized by forming a pseudocubic structure. It should be mentioned that unless otherwise mentioned, the symmetry of perovskite structure increases with temperature. However, when room temperature is concerned, the symmetries of $MAPbI_3$, $MAPbCl_3$, and $MAPbBr_3$ are, respectively, tetragonal, cubic, and cubic. Also, the materials having chemical formula of the form of $MAPbI_{3-x}Cl_x$ adopt a tetragonal crystal structure similar to $MAPbI_3$ [19,20]. As per the detailed study done by Kanatzidis et al. on several perovskites that include MA, FA, Pb, and Sn and found that there are three distinct phase transition that can occur during a change in temperature and they are a high temperature α phase, an intermediate temperature ß phase, and a low-temperature γ phase [21]. These results are in well agreement with that reported by other workers. Another important point that is to be remembered is that for field use of perovskite-based solar cell, the influences of pressure and temperature are two very important factors, and it is found that FA-based perovskites do not have any high temperature phase, which has undoubtedly made them suitable for using in field-based solar cells.

5.2.1 "A" Site

As mentioned earlier, the A cation directly has no influence on the electronic properties of the material. Fig. 5.2 shows the contour plot of electron density for the hybrid material $CH_3NH_3PbBr_3$ where one can see the week interaction and very small overlapping between the organic part and the octahedral unit.

Still A sites have an indirect influence on the electronic properties of perovskite through their well-known size effect by altering the extent of distortion of the B−X bond in the BX_6 octahedra. For small A cations like Cs, Rb, MA, or FA, the dimensionality remains 3. With the increase in the ionic radii R_A, the tolerance factor approaches unit resulting higher packing symmetry that in turn reduces the bandgap [23].

So far, in perovskite-based optoelectronic device (especially in solar cell), MA cation is the most widely used cation for hybrid perovskite. Use of larger cations has been proven to be useful for enhanced harvesting of light due to associated higher symmetry and smaller bandgap. Thus people has tried to replace MA with cations like ethylammonium, FA, or Cs with $R_{Cs} < R_{MA} < R_{FA} < R_{EA}$. However, it is also seen that replacement of MA with excessive large size disturbed the dimensionality and change from 3D to 2D orthorhombic structure with relatively large bandgap. Thus optimization should be done w.r.t. the cations having size (R say) between $R_{MA} < R < R_{EA}$.

From the previous discussion, it can be concluded that FA-based perovskite will give rise to higher symmetry compared to MA and thus

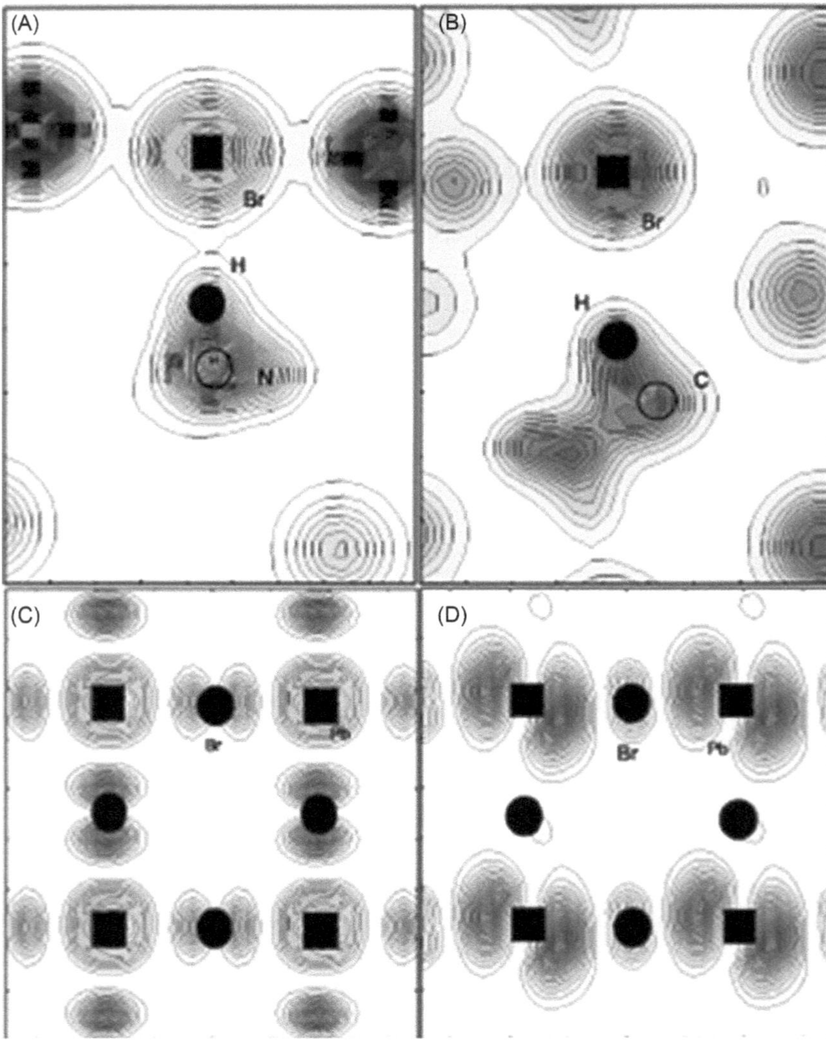

Figure 5.2 Contour plot depicting electron orbital overlap and total electron density of $CH_3NH_3PbBr_3$ for (A) the N—H—Br plane and (B) the C—H—Br plane. Electron density representing the (C) VBM and (D) CBM states. *CBM*, conduction band minimum; *VBM*, valence band maximum. *Published with permission from M.I. Dar, N. Arora, P. Gao, S. Ahmad, M. Grätzel, M.K. Nazeeruddin, Investigation regarding the role of chloride in organic—inorganic halide perovskites obtained from chloride containing precursors, Nano Lett. 14 (12) (2014) 6991—6996 [22].*

expected to give better performance. For this, FA-based perovskite has been studied extensively [24]. It should be noted that though FA shows nearly a cubic symmetry, there is an inherent distortion in the FA—I bond due to which FA ion sits little off center in the octahedron making the system trigonal [25].

Cs is another widely studied cation, which was investigated by Scaife and company since long back (even back to the early 1970s). It has been seen that Cs-based perovskite-like $CsSnI_3$ shows very strong PL intensity at room temperature in the near IR energy region. This perovskite has been reported to have four different structural phases at room temperature and more interestingly has a p-type conductivity due to the vacancy of Sn creating highly mobile hole [26]. The value of tolerance factor is less in case of $CsPbI_3$ compared to that of $MAPbI_3$ resulting a higher bandgap of 1.73 eV drifting away from the optimum value of 1.4 eV.

There is perovskite material having mixed cation $MA_xFA_{1-x}PbI_3$ that was first reported by Pellet et al. [24]. The optical properties of the material or these kinds of materials can be varied by optimizing the value of x. In this case, power conversion efficiency (PCE) as high as 14.9% has been achieved. This better result compared to pure MA was due to greater absorption of the particular spectral region giving better photocurrent. For the other mixed cation system, cesium-based perovskite offered a PCE as high as 7.68%. Another interesting system giving PCE as high 12.8% can be found in the system where MA has been partially replaced by 5-aminovaleric acid (5-AVA). This was first developed by Mei and company [27] giving a reduced number of defects and an adjacent titania layer gives higher degree of conformity. The partial substitution of MA with 5-AVA results an orthorhombic phase with significantly enhanced values of lattice parameter b and c. Apart from the high value of PCE, other notable advantage of this 5-AVA is the improved stability of the material under environmental perturbation compared to pure MA-based system.

5.2.2 "B" Site

Regarding B metal cation, the two most widely studied ions are Pb^{2+} and Sn^{2+}, whereas Ge^{2+} has not been investigated much. Lead has been proven to be superior compared to Sn giving superior stability as well as performance. It is seen that if one proceeds through periodic table from lead to germanium, the stability of the system gets reduced because of the reduced inert electron pair effect. However, we cannot ignore this way as it also corresponds to increasing of electronegativity giving lower bandgap favorable for optoelectronic operation. Thus, though it may seem to us that opting a lower atomic number group IVA element should be the best way, we cannot forget the corresponding compromise regarding the stability of the divalent metal. This can be well understood if one goes to the $MASnX_3$, the system gives an ideal bandgap of (1.2−1.4 eV) compared to $MAPbX_3$ (1.6−1.8 eV), but the first one suffers rapid degradation of inherent instability that makes them change their oxidation state from +2 to +4 upon exposure in the air-producing volatile SnI_4 compound [28]. So far the author is concerned, there is no report regarding the successful development of Ge-based hybrid perovskite as this effect is more prominent in such system. There are reports of developing lead-free perovskite like $MASnI_3$ and $MASnI_{3-x}Br_x$ giving a PCE around 5.73% which has been further enhanced to 6% under 1 sun illumination as reported by Noel et al. for their lead-free Sn-based devices [29]. It should be noted that both the Sn- and Pb-based ones have a tetragonal structure (P4mm) under ambient conditions; Sn-based materials possess a higher symmetrized α phase compared to the lower symmetrized ß phase associated with Pb-based perovskite [21]. Also Sn-based materials have further advantages of producing higher photocurrent, but their commercialization of use has been largely hindered by the stability issue discussed before.

Mixed B cations like $MASn_xPb_{1-x}I_3$ have been studied for their tenability of bandgap by varying the Sn/Pb ratio, i.e., by varying the value of x. It has been seen that for the values of x up to $x = 0.5$, the structure shows a pseudocubic (P4mm) symmetry like $MASnI_3$. However, with further increase of x, the structure changes to tetragonal (I4cm). With increasing Sn content, the system shows higher absorption and thus a reduced bandgap.

5.2.3 "X" Site

When one goes down through group VII (Cl−I), the atomic size of the halide increases producing a redshift of the energy absorption and thus a reduction in the energy occurs. This happens mainly due to the fact that with increasing atomic size, the electronegativity decreases and at some point matches well with Pb thus giving more covalent character. The halide ion in the hybrid perovskite so far is the most flexible tool to play with.

Iodide (I): So far, the iodide-based perovskites give efficiencies as high as 15% and thus is one of the most well-studied materials in the hybrid halide perovskite. This atom matches well with Pb having a similar covalent structure and thus together with Pb gives a very stable system. However, it suffers a drawback of being oxidized under suitable perturbation and the stability becomes questionable. This problem has made the study of iodine substitution and mixed halide perovskite very important.

Chlorine (Cl): Cl has probably been the most important halide used in perovskites. Still now, due to improved device performance, the incorporation of Cl and related mechanism for the enhanced optoelectronic properties is being studied most rigorously [30]. The efficiency is enhanced as Cl has shown to offer improved diffusion lengths and carrier lifetimes [19]. Another interesting point in Cl-substituted halide perovskite is the low miscibility of Cl (compared to Br) with I. This is due to the fact that Cl has higher difference in ionic radii and higher degree of covalent character with respect to Br. It is seen that pure chloride-based perovskite has a cubic structure, whereas the mixed halide perovskite has highly oriented structure though the point is still undergoing debates. X-ray diffraction (XRD) studies predict that stoichiometry of $MAPbI_{3-x}Cl_x$ to be very similar to that of $MAPbI_3$, with a maximum 3%–4% Cl-to-I ratio. Again X-ray photo electron spectroscopy (XPS) studies indicate that the Cl content in the hybrid halide is only 1%–2% suggesting possibilities of sublimation of Cl leaving remains of pure $MAPbI_3$ [31]. Another new claim is that Cl is not contained within the hybrid halide perovskite as $MAPbI_{3-x}Cl_x$, it rather exists as $PbCl_2$ serving as nucleation site in order to improve the surface coverage [22]. Moreover, it is concluded that as chlorine containing precursors have limited solubility [like $PbCl_2$ in dimethylformamide (DMF)], the nucleation of complex ion occurs that basically initializes the crystallization in the system. The crystallographic texture is generally determined by Cl that helps release of excess organic compound during rearrangement of constituent ions in the crystal-growth process. It is also evident that Cl inclusion has a direct effect on the morphology of the system that in turn governs the material property.

Bromide (Br): Inclusion of Br into the halide perovskite helps tuning the bandgap and thus the conversion efficiency. This is due to the fact that when Br is included in the I-based perovskite, a compressive stress is produced on Pb–I bond which in turn caused structural disorder giving rise to change in the bandgap. Pure Br-based device $MAPbBr_3$ gives rise to a cubic structure at room temperature, whereas upon substitution of Br by I gives a higher conversion efficiency and stability through the chemical management to tune the bandgap. It is reported by Snaith et al. that substitution of Br by I in $FAPbI_xBr_{3-x}$ produces perovskite with tunable bandgap between 1.48 and 2.23 eV and thus producing samples with variety of colors [23]. Increasing the iodide content in the structural phase of the sample changes from Br-rich cubic phase to I-rich tetragonal phase with a redshift in bandgap. The maximum efficiency they could achieve was 14.2% with $J_{sc} = 23$ mA/cm. Also, they have noticed that with iodide content, there was a considerable decease in J_{sc}.

Fluoride (F): The fluoride-based perovskite could be a very effective material regarding the optoelectronic properties due to the strong electronegativity of the fluoride ions that are capable of forming hydrogen bonding over other halides (N−H−F) [20], but its inclusion offers an unfavorable tolerance factor and high degree of lattice strain preventing light harvesting capabilities and conductivity. However, inclusion of F in terms of $(BF_4)^-$ forming a perovskite having chemical formula like $MAPbI_{3-x}(BF_4)_x$ can form a strong hydrogen bonding with MA ions and thus can prevent volatilization of MA. This inclusion also result an increase of low-frequency conductivity compared to I^-. Due to the similarity in size of BF_4^- and I^-, both the pure and substituted perovskites show identical tetragonal structure as well as similar bandgap of around 1.5 eV.

For HOIPs, A is a monovalent organic cation, whereas B is divalent cation as previously discussed, and X is halides or oxides. It was more than 20 years ago when the investigation regarding the optoelectronic properties of HOIPs was started mainly by Weber et al. [20].

Investigation of the fundamental optoelectronic properties of 3D-crystalline HOIPs began with the work of Weber et al. (Fig. 5.2), whereas almost in the same time Mitzi et al. reported the employment of HOIPs and associated 2D layered structures in different optoelectronic devices (Fig. 5.3).

Since then the hybrid perovskites specially the hybrid halide perovskites came to center of attraction of the researchers due to the fascinating properties and versatile application. Fig. 5.4 gives a statistical representation of the growth of the publications in the recent years.

5.3 BASIC SYNTHESIS

In 1978, Weber and group first time reported the synthesis of organometallic halide perovskite $CH_3NH_3PbX_3$ by mixing of different precursors

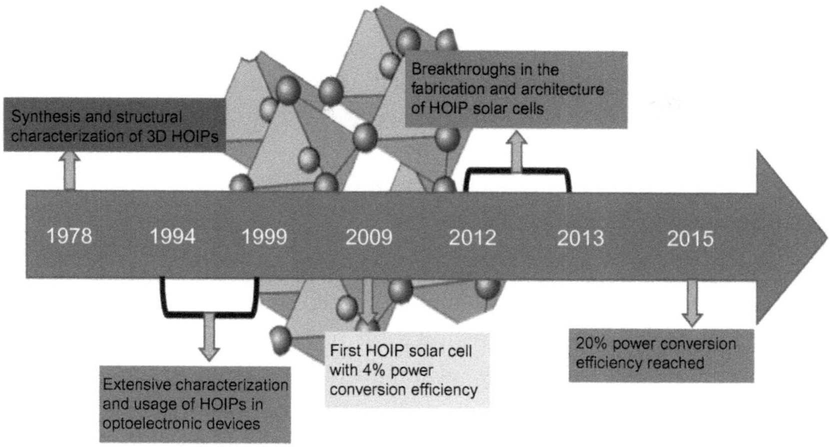

Figure 5.3 Timeline of research into the optoelectronic properties of HOIPs and important discoveries in the development of HOIP solar cells. *HOIP*, hybrid organic–inorganic perovskite. *Reproduced from T.M. Brenner, D.A. Egger, L. Kronik, G. Hodes, D Cahen, Hybrid organic–inorganic perovskites: low-cost semiconductors with intriguing charge-transport properties. Nat. Rev. Mater. 1 (2016) 15007.*

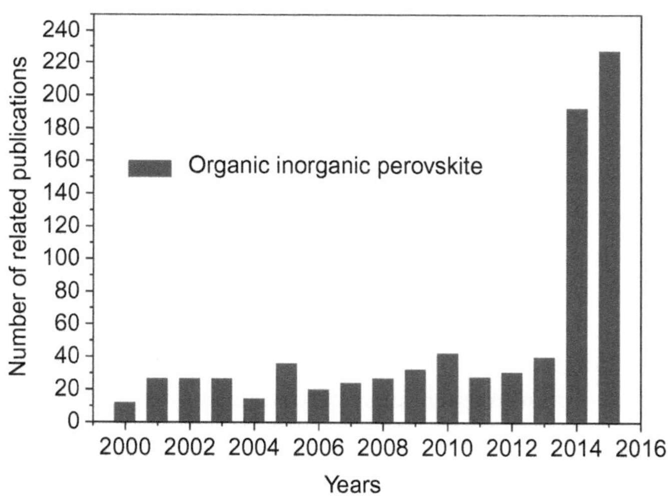

Figure 5.4 Publication statistics with respect to perovskites ABX_3. *Published with permission from J. Chen, S. Zhou, S. Jin, H. Li, T.Zhai, Crystal organometal halide perovskites with promising optoelectronic applications, J. Mater. Chem. C 4 (1) (2016) 11–27.*

like CH_3NH_3, $Pb(NO_3)_2$ and aqueous solution of Br_2 and stirred at 100°C. Later, mainly two methods were standardized for the convenient synthesis of organometallic hybrid halide perovskites. These two methods are solution-processed method and vapor-evaporation method. In between a two-step process for developing the same was developed that involves deposition of metal halide on certain substrate that include ash glass or quartz by vacuum evaporation or spin coating and subsequent dipping into organic ammonium iodide solution for a very short time at 300K.

5.3.1 Solution-Processed Method

From the very early days (1970), solution-processed method was proved to be one of the most effective in synthesizing crystalline organometal halide perovskites where the precipitation of both metal halide as well as organic halides take place by evaporation of corresponding solution through heating as well as spin coating where the last technique was found to be extremely effective in producing perovskite thin films since the discovery of perovskite solar cell [32].

There are solvents like γ-butyrolactone (GBL), *N,N* DMF and dimethylsulfoxide (DMSO) that are commonly used for the associated perovskite crystal. There are reports of producing $CH_3NH_3PbI_3$ or $CH_3NH_3PbBr_3$ at room temperature by dissolving the organic component as well as inorganic part that include CH_3NH_3Y and PbY_2 (Y stands for I_2 and Br_2 in) in appropriate solvents as mentioned before [33]. Also, mixed halide perovskite can also be synthesized by the same process simultaneously dissolving MaY and PbY_2 ($Y = I_2$, Cl_2) in proper ratio to synthesis $CH_3NH_3PbI_xCl_{3-x}$. Also, there are reports of producing extremely uniform $MAPb(I_{3-x}Br_x)_3$ layers done by Jeon et al. They followed a drop-casting method using toluene that was priorly treated by spin-coating techniques dissolving precursor in GBL and DMSO [34].

Though there is unprecedented improvement of solvent engineering technology for synthesizing this perovskite crystal of good quality. However, as discussed, the single-step solution process, suffers from the the inherent difficulties of producing Perovskite thin film of various morphology due to the uncontrolled synthesis process. Thus one should opt for a better ways of producing perovskite crystal of uniform morphology which is nothing but the two-stage single solution process.

There are plenty of reports regarding synthesis of perovskite crystal by double-step solution process that includes all the materials mentioned before as well as mixed cation perovskites like $(MA)_x(FA)_{1-x}PbI_3$, the later was synthesized by exposing PbI_2 infiltrated TiO_2 to an organic solution of FAI and MAI. Sometimes, thermal annealing after the normal synthesis was proven to be very effective in bringing better crystallinity into the system.

A systematic study in order to see the effect of temperature as well as precursor concentration on the phase of synthesized $MAPbI_3$ was done by Song et al. by systematic study of structural and optical properties of crystal perovskites that they synthesized by spin-coating MAI and PbI_2 precursor solutions with different mole fraction. However, the precise control over the morphology and phase of perovskite is still a matter of debate, and it has been seen that the presence small amount of Cl can dramatically improve the film morphology as well as the structures and proper reason is still under investigation. It has been shown that apart from synthesizing polycrystalline material, this solution-processed method has been proven to be very effective for producing single crystal perovskite material. There are reports of synthesizing different single crystalline perovskite material like $NH_2C(I)=[NH_2]_3MIs$ ($M=Pb/Sn$) or $(C_6H_5C_2H_4NH_3)_2PbCl_4$ by Mitzi et al. [35].

Dong et al. in 2015 reported the synthesis of the bulk single crystal perovskite with comparatively less defects, voids, grain boundary and thus with very high crystallinity by using a temperature gradient top-seeded-solution-growth method [36]. The schematic of the setup used by them has been shown in Fig. 5.5.

There is report of synthesizing $CH_3NH_3PbX_3$ single crystal as large as 100 mm^3 by Shi et al. [37] who used an antisolvent vapor-assisted crystallization approach. In the method, an antisolvent is made to diffuse into a solution containing the perovskite precursors that started the formation of the perovskite material with less defects and voids having a length of several micrometers. Another fascinating approach of synthesizing single-crystalline $CH_3NH_3PbBr_3$ is made by Liao et al., where they have used modified solvent/antisolvent vapor-assisted crystallization approach through evaporation of the solution for 24 hours [38]. It was shown that the as-synthesized samples consisted of sharpness; however, overall the samples were smooth. Another disadvantage of this solution process mediated growth of single crystal perovskites suffers is that it is time consuming. Effort was put by Saidaminov et al. to get rid of this difficulty in synthesizing single crystalline $CH_3NH_3PbX_3$, and they invented an inverse temperature crystallization technique which could synthesis the bulk single crystal of very high quality within time as low as few minutes and thus opens a new road map for synthesizing the material mentioned within a very short time [39].

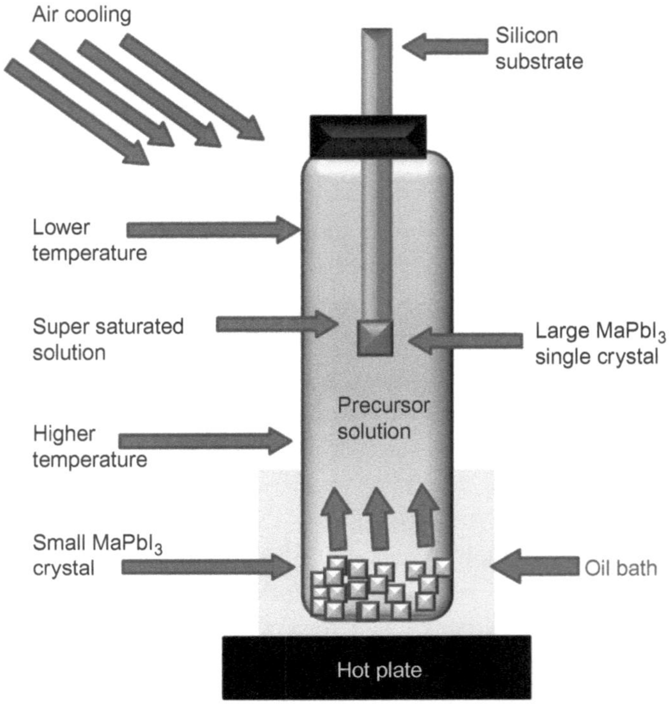

Figure 5.5 Schematic diagram of the setup for the growth of large MAPbI single crystal. *Reproduced from Q. Dong, Y. Fang, Y. Shao, P. Mulligan, J. Qiu, L. Cao, et al., Electron-hole diffusion lengths >175 μm in solution-grown $CH_3NH_3PbI_3$ single crystals, Science 347 (6225) (2015) 967–970.*

It is a point of note that the efficient performance of the bulk perovskites in the fields of optoelectronics has given the material a great importance, and with the advent of nanotechnology, researchers got propelled to employ the same material of lower dimensions and thus of different particular morphology. From this urge, Zhang et al. reported the synthesis of $CH_3NH_3PbX_3$ (X = Br, I, Cl) by their own developed ligand-assisted reprecipitation method [40]. The as-synthesized materials were highly luminescence color-tunable zero-dimensional quantum dot. The dots showed a quantum efficiency of as high as 70% at normal 300K and low excitation fluencies. In another report, Zhou et al. reported a room temperature template assisted growth of one-dimensional $CH_3NH_3PbBr_3$ nanowires [41]. Apart from these, there are tremendous efforts of combing fascinating features of single crystalline perovskite and low-dimensional uniqueness of the material worldwide for the development of different optoelectronic device. Apart from the work mentioned before, there are other work also, for instance, reported by Zhu et al. who developed a surface initiated growth process for fabricating high-quality single-crystal nanowires [42]. As synthesized, single-crystal $CH_3NH_3PbX_3$ one-dimensional structure has length and diameter around 20 μm and 200 nm, respectively, and thus having very high aspect ratio. Apart from these 1-D structures, there are reports of synthesizing single crystalline plate like two-dimensional $MAPbBr_{3-x}Cl_x$ and $MAPbBr_{3-x}I_x$ by simple halide exchange reaction in isopropyl alcohol. The schematic process of the reversible halide exchange reaction is shown in Fig. 5.6.

Another report of synthesizing 2D $(C_4H_9NH_3)_2PbBr_4$ nanocrystal via a ternary cosolvent solution-phase growth technique has been given by Dou et al. [43]. They could be able to synthesis 2D crystals with thickness varied between few and tens of nanometers. They also have reached the conclusions that mechanical-and solvent-exfoliation method is not as much effective as solution-growth process for synthesizing two-dimensional perovskite, and thus they open up a new aspect of synthesizing 2D perovskites and employed it in developing different optoelectronic devices.

Based on the several references mentioned above, it is seen that for the synthesis of $MAPbX_3$ and related perovskite crystal, the solution process is highly applicable and rather efficient. The method also has some discrepancy therefore hindering their proper use in industry scale. The

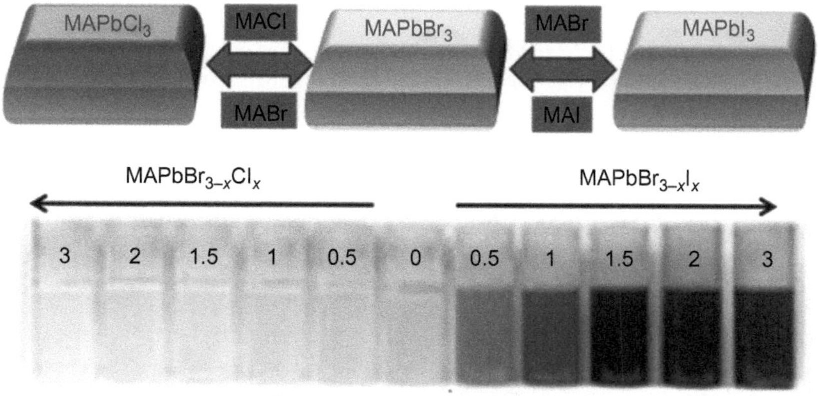

Figure 5.6 Schematic of anion exchange reaction and corresponding color change with relative abundance of halides. *Reproduced from D. Shi, V. Adinolfi, R. Comin, M. Yuan, E. Alarousu, A. Buin, et al., Low trap-state density and long carrier diffusion in organolead trihalide perovskite single crystals, Science 347 (6221) (2015) 519–522.*

disadvantage includes uncontrolled growth and thus lack of reproducibility, which is extremely important for the patentable synthesis. Another limitation suffered by the method is the pinhole. All these issues raise a necessity of developing another effective method for controllable synthesis of high-quality $MAPbX_3$ crystal.

5.3.2 Vapor-Evaporation Methods

Apart from solution process, another effective ways of synthesizing hybrid halide perovskite came out to be vapor-evaporation method. It has the advantage over the other in the sense that it always produces sample with higher crystallinity compared to the solution process. However, the method has limitation in terms of the yield and thus is not very successfully employed in industry scale.

It was found that the precursor flux has markable effects on the composition of the sample as obtained in this regard one should mention that unlike the tetragonal phase shown by $CH_3NH_3PbI_3$ at room temperature the same material synthesized by vapor-evaporation method show a cubic phase. Also, in situ XRD confirms that there are gaps between different phases. Following the other methods, people also employed single and sequential steps in evaporation to crystallinity to the as-synthesized perovskite.

The thermal evaporation both single source or coevaporation are being used since early 1990 for synthesizing hybrid perovskite with good crystallinity. Both approaches have their respective challenges. When for the single-source evaporation one needs a dedicated instrument, for the other one it needs to have a precise control on the flow rate, i.e., flux of either precursor, which is itself a challenge as the vapor pressure of the organic component is very much higher compared to the inorganic one. There are separate reports of synthesizing mixed halide perovskite with smooth surface, small grain size and large surface coverage by low-tempearature vapor-assisted solution process reported by Chen et al. and Liu et al. [44,45]. Further noteworthy is that Liu et al. was able to synthesize $CH_3NH_3PbI_{3-x}Cl_x$ with good crystallinity by employing dual-source vapor-deposition method.

Chemical vapor deposition (CVD) has been reported to be another effective ways for fabricating low-dimensional perovskites as reported by Xiong's group who synthesized free standing $CH_3NH_3PbI_3$ nanowires or platelets or $CH_3NH_3PbI_{3-a}X_a$ (where X = I, Cl, Br) nanoplatelets [46].

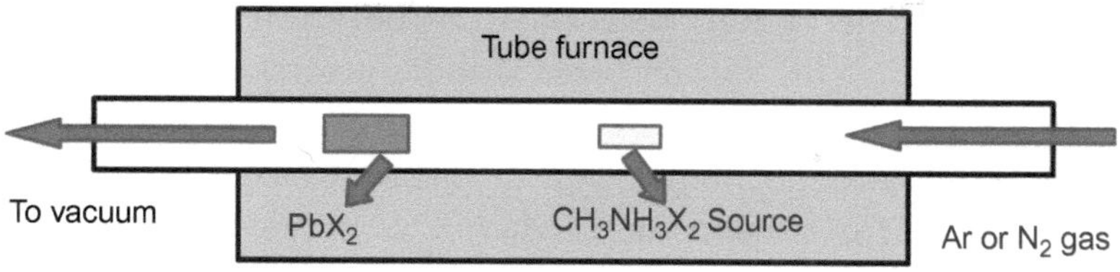

Figure 5.7 Schematic diagram of the setup for the growth of large MAPbI single crystal by Chemical vapor deposition.

Fig. 5.7 has shown the schematic of the experimental setup that was used by them. They have reported synthesizing of $CH_3NH_3PbI_3$ nanowires with length as large as 20 μm and diameter up to 500 nm. The final products of $CH_3NH_3PbI_3$ were achieved by intercalation CH_3NH_3I with previously synthesized PbI_2 platelets synthesized by CVD at 380°C on Si substrate. It is seen that the final product preserves the structure of PbI_2 with the thickness differing by a factor of 1.81.

5.4 OPTOELECTRONIC PROPERTIES

The exceptionally remarkable performances of hybrid perovskites result from their unique characteristic properties. Hybrid perovskites exhibit a significant optical absorption, an modifiable bandgap, long diffusion lengths, ambipolar charge transport, high carrier mobility, and a high sufferance for defects [33,45,47]. As discussed in the previous sections, the easily tunable electronic and optical properties of hybrid perovskite make it a major source of attraction. However, for large-scale performance, there are still various factors that cannot be completely perceived and immense obstacles in terms of stability and toxicity are required to be overcome.

5.4.1 Absorption

A typical solar cell active layer generally requires a film thickness of 2 μm or more to effectively facilitate the collection of charge carriers. However, hybrid perovskites permits a much lower film thickness of only 500 nm to achieve appreciable absorption across the entire visible region due to its significant optical absorbance [33]. $MAPbI_3$ and $MASnI_3$ show sharp absorption onset, which indicates the possibility of direct bandgaps. $MAPbI_3$-based devices are capable of achieving absorption over the entire visible spectrum as far as the tail end of the red region, i.e., approximately up to 800 nm. The optical absorbance can be further extended up to 1000 nm of the NIR by Sn-based halide perovskites devices [48]. Moreover, for a 0.3−1.0 increase in x in case of mixed $MASn_xPb_{1-x}I_3$ devices, there is a corresponding shift in valence band edge from −5.12 to −4.73 eV and a conduction band shift from −3.18 to −3.63 eV due to which the absorption edge can be again increased as far as 1300 nm [48].

The larger shift in valence band (VB) as compared to conduction band (CB) results in the reduction of bandgap. The absorption onsets are situated around 800 nm for both $MAPbI_3$ and $MAPbI_{3-x}Cl_x$, near 700 nm for $MAPbI_{3-x}Br_x$ and near 850 nm for $FAPbI_3$-based devices. $FAPbI_3$ devices have a reduced bandgap of 1.48 eV near the specified absorption edge. This bandgap can be successfully tuned and increased up to a value of 2.23 eV by introducing Br in $FAPbI_{3-x}Br_x$ for $x = 0$ (100% Br content).

$MAPbI_3$ has a reported Urbach energy of 15 meV [49]. There is an exponential increase in the optical absorption higher than four orders of magnitude (decades) in the below bandgap region, with absence of any optically detected deep states below photon energies of 1.5 eV. This property of the perovskites makes it a potential candidate for tandem devices.

5.4.2 Carrier Diffusion Length

PL measurements reveal the carrier diffusion lengths (L_D) for both electrons and holes to be up to 100 nm for $MAPbI_3$ and exceeding 1 μm for the mixed halide $MAPbI_{3-x}Cl_x$. Mesoporous ETM is necessary for $MAPbI_3$-based devices since there are reports regarding the more effective extraction of holes than electrons [50]. On the other hand, there is absolutely no necessity of a mesoporous ETM in $MAPbI_{3-x}Cl_x$-based devices since diffusion lengths for both electrons and holes exceeds 1 μm. Very recently, diffusion lengths over 175 μm under 1 sun illumination in a monocrystalline $MAPbI_3$ film were exhibited by Dong et al., which resulted in enhancement of the carrier mobility, lifetime, and reduced number of defects [36]. FA-based devices have diffusion lengths which lie in between to those of $MAPbI_3$ and $MAPbI_{3-x}Cl_x$.

5.4.3 Carrier Mobility

$MAPbI_3$ exhibits a near instant charge generation, which within 2 ps dissociates into essentially balanced free charge carriers of high mobility of about 25 cm^2/V s. Electron is injected into a mesoporous TiO_2 electron transport layer (ETL) in less than 1 ps, but unfortunately, the languid mobility of electrons in the intrinsic TiO_2 hinders mobility which subsequently leads to unbalanced charge transport. Ponseca calculated the mobilities of holes and electrons to be 7.5 and 12.5 cm^2/V s, respectively, at THz frequencies, and the corresponding values in the microseconds range are 2 and 1 cm^2/V s, respectively. The electron-to-hole mobility ratio generated from this is about 2.0, which match exactly with the calculated values. Heo et al. reported the fabrication of a perovskite-based thin film transistor which showed ambipolar semiconductor behavior, and the hole-transport mobility in the device was measured to be 10^{-5} cm^2/V s [51].

5.4.3.1 Electronic Structure

There have been a number of studies revealing the electronic band structures of 2D and 3D hybrid perovskite materials. The valence band maximum (VBM) is comprised of an antibonding hybrid state between the B-s and X-p orbitals, whereas the conduction band minimum (CBM) consists of a nonbonding hybrid state between the B-p and X-p orbitals [15].

A detailed insight into the electronic properties of $MAPbX_3$- and $CsPbX_3$-based devices were demonstrated through first-principles pseudopotential studies conducted initially by Park and Chang in 2004 [15]. This study elucidated that MA has minimum contribution toward electronic properties due to the fact that its electronic levels lie deep within the valence and conduction bands. The study also revealed the fact that BX_6 octahedra are the primary source of the resulting band edges. Hence from the above facts it is evident that a change from organic (MA) to inorganic (Cs) does not affect the band structures significantly as a result of size effects.

On the other hand, substitution of halide components in the order Cl → Br → I significantly affects the corresponding VB transition from 3p → 4p → 5p, which results in the subsequent lowering of the ionization potential (binding energy) [52]. It is also reported that a redshift in the electronic energy bandgap of hybrid perovskites occurs with an increase in the pressure, which is due to the pressure-sensitive antibonding and nonbonding natures of the VBM and CBM, respectively. This extraordinary property of hybrid perovskites devices is in stark contrast to most typical semiconductor materials [52].

A precise spin−orbit coupling-GW (SOC-GW) model was recently developed by Umari et al. in order to make the density functional theory (DFT) of electronic band structure models to be more consistent and valid [53]. Variations observed in electronic properties of $MAPbI_3$ are caused by relativistic effects, which have a dual effect on the perovskite. It improves the stability of $MAPbI_3$ toward oxidation, but on the other hand, it also causes an unwanted increase in the bandgap. The bandgap values obtained from the first principle studies by Umari matches approximately with the experimental values along with an uncertainty factor of only ± 0.1 eV.

The maximum short circuit current (J_{sc}) was evaluated to be 25 mA/cm^2 as reported by Umari et al. which matches more or less with the experimentally determined value of 28 mA/cm^2. Due to the decreased absorption efficiency in the 600−800-nm range for $MAPbI_3$, short circuit current J_{sc} reaches the highest value of 21 mA/cm^2. It was also contemplated that differences in structures depending on the degree of BX_6 octahedra tilting results in the approximate 0.2 eV difference between $MASnI_3$ and $MAPbI_3$.

5.4.3.2 Density of States

Umari et al. performed a detailed density of states analysis to have a proper insight into the bandgap variations. The report delineates the existence of certain additional peaks for states within $MASnI_3$ extending from the main valence band peak which can be significantly attributed to the I-p states along with the considerable antibonding interactions of Sn-5s and I-5p orbitals. These peaks are absent in case of $MAPbI_3$. The interaction of Pb-6s orbitals with I-5p orbitals is preferably less since Pb-6s orbitals occupy a lower energy level. This results in the rapid increase of the valence band levels of $MAPbI_3$ as compared to $MASnI_3$. It is less probable for Pb-6s orbital to react or share its electrons since the 5 s shells in Sn or 6p shells in Pb stabilizes the Pb-6s orbitals. Hence spin−orbit coupling is an important factor that is to be taken into considerations in both theoretical and experimental analysis of electronic band structure analysis [52,53]. In case of $MAPbI_3$, the valence band shift is approximately 0.5 eV greater than the conduction band shift, which results in the increment of the bandgap as compared to that of $MASnI_3$.

5.4.3.3 Effective Mass

The electronic properties of hybrid perovskite halides like $MAPbX_3$ ($X = I$, Cl, Br) were initially evaluated by Park and Chang. The effective masses in $MAPbBr_3$ for electrons and holes at the CBMs and the VBMs were evaluated to be $m_e^* = 0.25m_0$ and $m_h^* = 0.12m_0$ (where m_0 is the mass of an electron), respectively [15], whereas these values as observed in other common semiconductors like GaAs are $m_e^* = 0.066m_0$ at CBMs and $m_h^* = 0.05m_0$ for heavy holes and $0.08\mathrm{m}_0$ for light holes at VBMs. It can be clearly seen from the abovementioned values of effective masses that a hole at the VBM has a lower value than that of an electron at the CBM. Moreover, an organic cation can increase the effective masses considerably due to the larger lattice constants as compared to an inorganic cation (Cs). Owing to this, it was found that for $MAPbBr_3$, the value of m^* for holes at the VBM was $0.15m_0$, whereas for $MASnBr_3$, the value was $0.12m_0$. Carrier confinement effect resulting from large number of stacked layers can significantly lower the bandgap of 2D perovskite structures.

Maximum values of effective masses for pseudocubic $MAPbI_3$ can be achieved by introducing SOC effects and the corresponding values are found to be of $m_e^* = 0.23m_0$ and $m_h^* = 0.29m_0$ [54]. The first principle simulation results give a proper insight into the ambipolar carrier transport characteristics in perovskite-based solar cell devices. The calculation of these effective masses is performed considering an ideal situation by deliberately neglecting phonon-scattering effects, defects or any impurities. Hence the calculated effective masses of hybrid perovskites matches approximately with the values as observed in case of Silicon ($m_e^* = 0.19m_0$, $m_h^* = 0.53m_0$ for heavy holes and $0.16m_0$ for light holes). SOC-GW method as recently employed by Umari et al. has significantly enhanced the accuracy in theoretical calculations to make them more relevant to experimental values [53]. The evaluation of effective masses as observed for different symmetries across the band structure reveals the fact that relativistic effects can have a considerable impact on band dispersion, which is conspicuously observed in the case of the valence band of $MAPbI_3$ as compared to $MASnI_3$. From the above discussions, it is evident that $MASnI_3$ can demonstrate improved hole transport along with reduced electron transport as compared to $MAPbI_3$.

5.4.3.4 Defects and Band Structure

Some of the distinctive properties of hybrid perovskite materials are their unusually long carrier lifetimes and carrier diffusion lengths along with effective PL quantum yield and optically pumped lasing effects. The nonradiative recombination is hindered to a great extent which subsequently results in impressive emissive quantum yields accompanied by long carrier lifetimes. The defects of perovskites were studied in great details by Yin et al. through DFT calculations [55]. Positively charged MA^+, lead (Pb^{2+}), and negatively charged iodide (I^-) are the three primary ions that constitute the Hybrid perovskites materials. It is to be noted that the interstitial MA^+ along with the Pb^{2+} vacancies can form shallow charge-carrier trap states near the band edges of the perovskite material. Both of them also generally create defects with the lowest formation energies.

Energy levels of deep lying defects are situated near the mid bandgap region, whereas the energy levels in case of shallow defects lie near the band edges. Nonradiative recombination cannot occur in these shallow defect regions and this has been experimentally observed and reported [56]. Again there are reports supported by simulation demonstrating the fact that films which are grown under iodine affluent conditions are susceptible to generate a large number of deep electronic trap levels which acts as recombination centers [57]. Also, films grown in the abovementioned condition exhibits higher charge diffusion lengths. The extended carrier lifetime of the perovskite materials can be attributed to the fact that there is absence of any deep trap states located within the bandgap region of the material. Defects created by Pb vacancies and interstitial MA^+ ions are extremely sensitive to synthesis conditions, and they, respectively, function as an electron acceptor and electron donor. Synthesis techniques also shed light on the fact that whether a perovskite material is p- or n-type. First principle theoretical studies reveal that deeply located defect states are not initiated by grain boundaries. It also unveils the fact that the defect states located at the interface between the perovskite and hole transport layer (HTL)/ETL can act as interfacial recombination sites thereby hampering the device performances. Chen et al. and others have reported the possibility of passivizing such surface states [58–60]. Interaction with the mesoporous scaffold can influence the Fermi level which may correspond to the trap sites located at the subgap region of the electronic bandgap. The Fermi level can be raised up to a level of 50 meV just below the conduction band by mesoporous TiO_2 and Al_2O_3 scaffolds, whereas the Fermi level is situated in the mid-region of the bandgap for a solid perovskite film deposited on fluorine doped tin oxide (FTO) glass. There are reports stating that by simply varying MAI to PbI_2 proportion, one can generate either n- or p-doped $MAPbI_3$ [61]. Manser et al. have reported that accumulation of charge within perovskite films results in an increase in the intrinsic bandgap [62]. This phenomenon follows the Burstein–Moss band filling model.

5.4.3.5 Carrier Dynamics

It has been previously stated that the long and unbiased charge carrier diffusion length of hybrid perovskite semiconductor makes it a promising candidate for various potential applications. The THz photoinduced absorption transient spectroscopy approximates a charge carrier mobility of $\sim 10\ cm^2/V\ s$ as the lower limit of MA lead trihalide [63], which is in good accordance with results obtained from time resolved microwave conductivity [64] and excitation density perturbation analysis [65]. The Langevin theory overestimates the ratio of biomolecular rate of recombination to charge mobility. The ratio is calculated to be four times lower than the predicted value. The origin of this distinctive property is from the spatial dissociation of electrons and holes either inside the metal-halide hybrid structure or across a crystalline domain. Till now only the perovskite film has been taken into consideration while calculating all the values of the respective mobilities. However, while considering the actual scenario by taking into account the interfacial effects, it has to be noted that the values of the effective mobility in the hybrid devices significantly decrease as compared to the reported values which is why it is advisable to measure the effective charge carrier mobility directly within full devices.

There are tremendous efforts that have been put to have a thorough understanding regarding the detailed photovoltaic process across entire perovskite solar cell. This effort includes locating exact location carrier generation charge dissociation of electron hole pair generation and dissociation within couple of picoseconds creating mobile charges. A simple equation was developed for describing the photocarrier dynamics proving a free carrier model better describe the system compared to the exciton mode. It is also separately confirmed by impedance spectroscopy as well as by PL spectroscopy that charge accumulation takes place within the material and that in turn makes influence on the band-edge shift and hence the capacitor behavior in of photovoltaic device.

5.5 APPLICATIONS

There are numbers of properties that have made the hybrid perovskites suitable for applying in various optoelectronic devices. The properties that have been mentioned in previous sections include strong PL properties, low nonradiative recombination rates and long carrier lifetime, and color-tunable absorption and emission properties. There are numbers of different optoelectronic devices have been successfully developed by using the single or polycrystalline hybrid perovskites. Thus a new branch of research centering the development of new hybrid perovskite-based optoelectronic devices has been opened worldwide. In the coming section, few application would be discussed that will include solar cell, light-emitting electrochemical cells (LECs), lasing, field-effect transistor (FET), and photodetectors.

5.5.1 Perovskite Solar Cells

One of the main applications of the hybrid perovskite material is its use in different solar cells. There are many reports regarding development of organometal halide perovskite solar cells. These reports also cover the basic science and fundamentals of organometal halide perovskite solar cells for the better understanding of the basic understanding of the corresponding mechanism. In 2009, there was first report of synthesizing dye synthesized solar cell giving an efficiency of 3.8% and the particular material used was $CH_3NH_3PbI_3$. However, this early solar cell was not of much practical use as the used perovskite generally gets dissolved in organic liquids. Thus a breakthrough was needed which was achieved in 2012 by Kim et al. who were able to bring an efficiency as high as 9.7 by replacing the organic liquid electrolyte by solid hole conductor spiro-MeOTAD (2,2′,7,7′-tetrakis-*x*-2′,7,7′-tetrakis-9,9′-spirobifluorene) [66]. In the next year, Heo et al. further enhanced the PCE to 12% by using polytriarylamine in perovskite solar cells [51]. A PCE as high as 15% was achieved by Burschka et al. who developed perovskites within the porous metal oxide thin films by following sequential deposition method. Another advantage of their approach was very good reproducibility which has made it usable to real industry scale [67].

Also, the as mentioned sequential process for the fabrication of perovskite solar cell has other advantages over the previously existing single deposition technique regarding high efficiency and stability. There are reports of developing 2D perovskite solar cells at room temperature, by simple combination of interface engineering and metal codoped spiro-OMeTAD process giving a very high PCE of 19.3%

The solar cell which has been modified with indium tin oxide coated poly-ethyleneimineethoxylated has work function much lesser compared to common perovskite solar cells and thus better response. When Y-TiO_2 (Yttrium-doped TiO_2) has been used, there are much better matching of energy level between CBMs and highest occupied molecular orbital spiro-OMeTAD with that of the perovskites giving rise to more numbers of electrons that get generated due to the excitation with photon within the perovskite that ultimately gives a PCE of 19.3% under AM 1.5 illumination and that to without exploiting any kind of antireflection coating. Also, there is report of a giant PCE of 20.1% in 2015 by Jeone et al. who employed an intramolecular exchange process to replace CH_3NH_3 in $CH_3NH_3PbI_3$ by FA analog [68]. The next milestone regarding the betterment of perovskite-based solar cell lays in the increase of grain size and crystallinity thus reducing the defects which is one of the major challenges for the development of the perovskite-based optoelectronic device.

5.5.2 Light-Emitting Electrochemical Cells

In 1995, Pei et al. reported the synthesis of LEC which is basically a device that helps converting light energy from electric energy by an appropriate addition of inorganic salt to a mixture of ionic conductive and conjugated luminescent polymer [69]. The basic construction of the LEC is nothing but a semiconductor material that is kept within two metal electrodes thus forming a sandwich structure. The structure is rather simple, consisting of easy and environment-resistive construction, which made LEC attractive for using as substitute of organic light-emitting diode (LED) [70].

Reports are there where LEC with high stability of 1–2 cd/m^2 at low input current has been reported by Ayguler et al. in 2015 who replaced $CH_3NH_3^+$ by FA^+ to synthesis $FAPbBr_3$ that was found to be effective for using in LEC device as mentioned before [71]. They have seen that both the FA- and MA-based materials give a band-edge emission of very narrow full-width-at-half-maxima (FWHMs) of 19 nm thus showing a high degree of crystallinity into the sample, i.e., as prepared nanoparticle. Also it is seen that when MA gets replaced by FA, a redshift in the emission spectrum from 532 to 545 nm takes place. This report thus opened a new horizon for the FAPbBr-based perovskite for being used in new LEC-based devices.

5.5.3 Light-Emitting Diodes

LED is a semiconductor device that emits visible light when an electric current passes through it. The output from a LED can range from red (at a wavelength of approximately 700 nm) to blue-violet (about 400 nm).

Some LEDs emit infrared (IR) energy (830 nm or longer); such a device is known as an *infrared-emitting diode* (IRED). An LED or IRED consists of two elements of processed material called *P-type semiconductors* and *N-type semiconductors*. These two elements are placed in direct contact, forming a region called the *P—N junction*. In this respect, the LED or IRED resembles most other diode types, but there are important differences. The LED or IRED has a transparent package, allowing visible or IR energy to pass through. Also, the LED or IRED has a large P—N-junction area whose shape is tailored to the application. Nowadays, the discovery of white LED has made it a material of extreme importance due its very low power consumption. However, still challenges remain for marketization of white light LED that include highly efficient light emission spanning the visible spectrum and exceptional color quality specially at low cost. However, the exceptional optical properties of organometallic hybrid perovskites are proven to be capable of facing the challenges with success. The perovskite-based LED was first reported by Mitzi el al. in 2000 with two-dimensional organometal halide perovskite.

In 2014, Tan first reported organometallic halide perovskite-based LED where the material was synthesized by solution-processed method. They have tuned color of the electroluminescence spectra in the near IR green and red region by varying the percentage of halides in their as-synthesized $CH_3NH_3PbI_{3-x}Cl_x$, $CH_3NH_3PbBr_x$, or $CH_3NH_3PbI_{3-x}Br_x$ at room temperature. They have shown that the device based on $CH_3NH_3PbBr_3$ and $CH_3NH_3PbBr_2I$ exhibits stable green and red emission, respectively. The pure bromide, mixed bromide—iodide, and mixed iodide—chloride perovskites emit spectra in green region with luminescence of 364 cd/m^2 and red region with an IR radiance of 13.2 W/sr/m^2. Also, it should be noted that the electroluminescence (EL) quantum efficiency of the pure bromide-based perovskite has been found to be enhanced with increasing input current density suggesting a need for high charge density in order to achieve efficient radiative recombination.

A fluorine-free perovskite LED was developed by Hoye et al. by spatial atmospheric atomic layer deposition technique which made them able to deposit ZnO directly onto lead- and bromide-based perovskite and thus achieved lower turn on voltage and higher luminescence much better compared to previous F8-based reports [72]. The elimination of fluorine also lowered the cost and unstability of the device and increased conductivity. In this regard, the interface engineering techniques was proven to be an important and effective technique for controlling the carrier transport [73]. There are reports of chorine-, bromine-, iodine-based mixed perovskites ($CH_3NH_3PbCl_xBr_yI_{3-x-y}$) with very high luminescence. The mixed halide-based material facilitate the development of large area flat panel display with high efficiency and flexibility and also the solid state lighting device with very high color purity [74].

So far, the main challenge regarding these perovskite-based LEDs that get operated at high current densities is lying on the fact that a proper contrast electrode with perfect matching band alignment should be found in order to get high charge injection along with the minimizing of the associated series resistance.

Also now there are tremendous thrusts regarding the low-dimensional perovskite-based LEDs that can be tuned apart from the optical response to a proper morphology.

In this regard, the work of Wong et al. in 2015 is worth mentioning in which they have transformed a bromide-based perovskite to iodide-based perovskite nanorods with definite morphology by a simple low-temperature ion-exchange process [75]. The structure showed good EL spectra at room temperature and thus further used in associated LED devices. Thus, these low-dimensional perovskite-based LEDs with high color purity and easy cost-effective synthesis techniques like their bulk analog have further accelerated the work on the perovskite-based LEDs.

5.5.4 Lasing

Laser, i.e., light amplification by stimulated emission of radiation, is associated with generation of very high directional, monochromatic, and coherent light; it has found tremendous applications in different fields of applications like lighting, telecommunications, displays, optical fibers, and many others. Now it should be noted that there exists a huge mismatch between the lattice parameter of the silicon and high performance III–V semiconductors which is one of the greatest problem for developing monolithic integration of laser onto silicon substrates and also the different growth temperature further hindered the same. However, these are very crucial for developing integrated optical connection within silicon chips which now serves as the fundamentals for the inter- and intrachip connection. These difficulties further create problems in searching or developing cost-effective, large-scale and steady-flow synthesis of on chip optical device.

In this context, it should be noted that the perovskite-based laser at low temperature (40K) was reported long before at the early 1990s where they have used layered structures $(C_6H_{13}NH_3)_2PbI_4$ [76]. Since then, efforts were made with great thrust to develop hybrid organometallic perovskite-based lasing material at room temperature. There is a report of developing an ultrastable amplified spontaneous emission (ASE) at strikingly low thresholds of 12 ± 2 μJ cm due to the observed ASE in the solution-processed $CH_3NH_3PbX_3$ perovskites.

Xing et al. reported the room temperature lasing action of $CH_3NH_3PbI_3$ single crystals with wide color tenability in the visible region. The color tenability was achieved by proper substitution of halides in mixed halide materials $CH_3NH_3PbI_xCl_yBr_{3-x-y}$ successfully. In this case, emission intensity was found to be varying linearly above the threshold fluence [77].

Also the performance of the system was found to be exceptionally stable. The ultrastable performance of the system has been found to be existing due to the very low bulk defect density, large absorption coefficient, and slow Auger recombination rate in single crystal $CH_3NH_3PbI_3$. Furthermore, optically pumped lasing has also been demonstrated at room temperature in $CH_3NH_3PbI_3$-based polygonal spherical resonators.

It is seen that better performance regarding the lasing action can be achieved for low-dimensional, i.e., nanostructured perovskite compared to their bulk form [78].

A lasing action with very high Q factor (Q ~ 3600) and low threshold (220 nJ cm^2) and most importantly with near unity quantum yield was reported by Zhu et al. for their single crystalline $CH_3NH_3PbX_3$(X = CI, Br, Cl) nanowire at room temperature [42]. The extremely good performance of their as-synthesized material was reported to be due to the long carrier lifetimes and low nonradiative recombination rates. Also, reports related to visible and near-IR lasing action of free standing pure or mixed halides perovskite was published by Xing et al. [79]. Another reports of room temperature lasing action of mixed halide perovskite nanoplatelets $CH_3NH_3PbI_{3-a}X_a$ (X = I, Br, Cl) optically pumped lasing in polygonal nanocavities was reported by Zhang et al. [80]. A single mode lasing with was achieved with $CH_3NH_3PbBr_3$ by Liao et al. with a threshold of 3.6 ± 0.5 μJ cm and a tunability of color between 525 and 557 nm, i.e., in the green range [38]. It should be noted that the hybrid organometallic perovskites-based material showed better lasing performance due to mainly its low defect density and large exciton binding energy.

The efficient performance of single crystal perovskite-based nanowire in lasing action has emerged a new field of low-dimensional perovskite-based laser material. The perovskite material-based laser due to its facile solution-processed growth technique and excellent color tenability within the visible range have made it an excellent candidate for being used in different fields of application that include nanophotonics, optical computing and chemical/biological sensing, and others.

5.5.5 Field-Effect Transistors

Organometallic hybrid perovskites are considered to be one of the main materials for using it in FET. This is due to the fact that generally FET is a semiconducting channel layer (p- or n- type). Here the conductivity gets controlled by applying appropriate biasing voltage w.r.t. a associated isolated conducting gate. Semiconductor materials that are of low cost and having mobility close to the amorphous silicon can be very effectively used in FET materials and thus can be applied in display as well as storage technology. As mentioned before, hybrid perovskite satisfies all these criteria and thus become one major part of these applications.

In this field, one cannot avoid the extensive work of the group led by Mitzi et al. mainly in the beginning of the 21st century. His work mainly associated with tin-based perovskite FET where they were able to achieve very high mobility and I_{on}/I_{off} ratio [81]. In their spin-coated synthesized $(C_6H_5C_2H_4NH_3)_2SnI_4$, they reported to have a field-effect mobility up to $0.62\ cm^2/V/s$ and on/off ratio of the order of 10^4. They further improved the performance of their device by substituting organic cation and finally massive results came from their group where they reported saturation and linear mobility of 2.6 and $1.7\ cm^2/V/s$ and an on/off ratio value of 10^6 [82].

Another noteworthy report came in the couple of years back by Chin et al. who has shown that the screening effect associated to the ionic transport can be lowered by lowering the operating temperature in their $CH_3NH_3PbI_3$-based hybrid perovskite FETs.

In the work by Mitzi mentioned in the previous section, the group has shown that electroluminescence spectra of $CH_3NH_3PbI_3$-based light-emitting FET consist of three peaks at 750, 780, and 800 nm. They also have confirmed that how lowering the operating temperature is an effective way to reduce the hysteresis effect.

These systematic works have established hybrid halide organometallic perovskite as one of the best candidates for use in FET with superiority regarding the cost effectiveness and performance over the other conventional semiconductors.

Again the lead- and tin-based perovskites can be optimized in terms of intrinsic transport parameters that will help in developing different optoelectronic devices like solar cells, LEDs with optimized performance, or laser operating at room temperatures.

5.5.6 Photodetector

Photodetectors are sensors that can convert the photon energy of light into electrical signal. They are absolutely necessary for various scientific implementations like fiber optic communication systems, process control, environmental sensing, safety and security, and also in defense-related applications. Impressive performance of photodetectors can be attained by successfully utilizing semiconductor materials having significant absorption extinction coefficient along with high mobility of charge carriers. These properties ensure adequate absorption of light and large photocurrent. $CH_3NH_3PbI_3$ perovskite exhibits significant light-absorption property and impressive activity of long-lived photogenerated carriers, which results in high quantum yield conversion along with high gain in photoconductivity as required for photodetector activities. The very first high performance flexible organolead halide perovskite $MAPbI_3$-based photodetector was developed by Xie et al. who reported a wide range of photoresponse from 780 to 310 nm. About 3.49 A/W photoresponse and external quantum efficiency (EQE) of $1.19 \times 10^3\%$ at 365 nm was possible to be achieved on the flexible $MAPbI_3$ film photodetectors, thereby displaying high sensitivity, quick speed of response, and outstanding stability [83]. Following this, a photodetector based on hybrid perovskite $CH_3NH_3PbI_{3-x}Cl_x$ was reported by Dou et al. who displayed an inverted device structure.

The three types of devices which were reported to be fabricated are PD1 without buffer, PD2 with 2,9-dimethyl-4,7-diphenyl-1,10-phenanthroline (BCP) (2,9-dimethyl-4,7-diphenyl-1,10-phenanthroline), and PD3 with poly[(9,9-bis(3'-(N,N-dimethylamino)propyl)-2,7-fluorene)-alt-2,7-(9,9-dioctylfluorene) (PFN) (poly(9,9-bis(3′-(*N*,*N*-dimethylamino) propyl)-2,7-fluorene)-alt-2,7-(9,9-dioctylfluorene))—both as cathode buffer layers. It was shown that PD3 exhibits a remarkably decreased dark current J_d, an impressive superior rectification ratio of $\sim 10^5$ (± 1 V), and a superior detection limit of 8×10^{13} Jones at -100 mV. The generation of a dipole layer at the interface of [6,6]- phenyl C61-butyric acid methyl ester (PCBM) and Al resulted in a distinguished increase in the forward bias injection of PD3. The dipole layer contributes an additional electric field to impede the hole injection, whereas on the other hand it improves electron injection.

The overwhelming advancements of photodetectors based on organometal halide perovskite films led to the increasing attraction toward low-dimensional perovskites consisting of many active sites. Porous $CH_3NH_3PbBr_3$ nanowires exhibiting visible-light photodetection ability having a photoresponse time as small as 0.12 s and a decay time of approximately 0.086 s was reported by Zhuo et al. [41]. The exceptionally porous one-dimensional structure accompanied by multiple active sites and significant light absorption properties resulted in the strong sensitivity and high stability of $CH_3NH_3PbBr_3$ nanowires. An intense association between the photocurrent and the PL decay time was detected in plate type NCs $MAPbBr_{3-x}Cl_x$ and $MAPbBr_{3-x}I_x$ photodetectors as developed and fabricated by Jang et al. [84]. Iodine-rich halide perovskite $MAPbBrI_2$ has tetragonal phase character which attributes to its highest observed photocurrent along with longest PL decay time. The fact that the photoconversion efficacy is dependent on the composition of the perovskite was confirmed by first-principles density-functional theory-related calculations.

Pristine perovskites exhibited fast recombination of photogenerated charge carriers. In order to prevent this, hybrid perovskite-based photodetectors were developed, which rapidly gained huge importance. A hybrid photodetector based on MA lead halide/single-layer graphene was developed by Lee et al. which showed a photoresponse of 180 A/W and a quantum efficiency of 5×10^4% at a comparatively high power of illumination of 1 μW [85].

An elevation in the optical absorption accompanied with the photogating mechanisms of the hybrid perovskites results in the wide range of spectral photoresponsivity between 800 and 400 nm. The time required for the rise and fall of the perovskite-graphene based hybrid photodetector were calculated to be 87 and 540 ms, respectively. This implies an impressive on−off switching behavior of the hybrid device.

The perovskite/graphene based hybrid device can sustain its on−off switching performance for multiple cycles which indicates the devices' robustness and reproducibility. Following this, a $MAPbI_3$ nanowire/graphene based hybrid photodetector was fabricated by Spina et al. [86]. This device acquired a significantly high photoresponsivity of 2.6×10^6 A/W. The unique nanowire-like morphology resulted in the high performance of the hybrid device toward photodetection. The development of graphene/perovskite-based hybrid photodetector devices would act as a motivation toward the future exploration of the devices' performances. Recently, there has been a quantum leap in the intrinsically narrowband photodetection accompanied with a broad range of tunable photoresponse utilizing single crystal-based hybrid perovskite $MAPbBr_{3-x}Cl_x$ and $MAPbI_{3-x}Br_x$.

An individual narrow peak was observed near the absorption edge for each kind of the various halide single crystals. It could also be observed that the FWHM of the respective peaks is less than 20 nm which validate photodetection in narrowband region. For shorter wavelength excitation, collection of charge carriers gets restrained due to surface charge recombination which attributes to the narrowband photodetection.

Moreover, the composition of the halides in the halide-based single crystal perovskite materials can be tuned in order to incessantly regulate the response spectrum of photodetector across the complete visible range from blue to red. The temporal response of the hybrid device is recorded with 6 $\mu W/cm^2$ intensity of light under an LED illumination of 570 nm with a bias of -4 V and a modulation frequency of 150 Hz.

The single crystals of both the single halide and mixed halide perovskite materials exhibit persistent nature in the corresponding EQE spectra even after extended exposure to air. This indicates their superior stability as compared to their complementary polycrystalline perovskite thin films. Hence an innovative economically viable design model following a simple fabrication procedure for narrowband photodetection has been reported. One of the drawbacks that are required to be prevented is the background noise emission. In contrast to the conventional inorganic semiconductors, organometal halide perovskites can be fabricated on an extensive range of substrates including plastics owing to the low-temperature solution process. This makes halide perovskite a promising candidate for flexible electronic and potential photonic applications. Moreover, the extraordinary optoelectronic properties of perovskites results in impressive detection capability along with rapid response which are absolutely necessary features required for perovskite-based photodetectors. The rapid burgeoning of hybrid perovskite-based photodetectors in recent times imparts endless application possibilities in developing unique device configurations. This can facilitate the fabrication of imaging sensors which are appropriate for low-light photography, UV detectors, and smart skin sensors.

The astounding optical and photophysical properties of hybrid organometal halide perovskites have enabled its application to spread beyond solar cells into multiple other optoelectronic applications. Optoelectronic devices based on perovskites have gained the attraction of the scientific community at large and will continue to do so in near future. Single crystal and low-dimensional perovskites are the new emergent classes of devices demonstrating highly appreciable optoelectronic-based performances. There are still a plethora of opportunities those are required to be ventured in order to develop single crystal and low-dimensional perovskites-based devices for light-emitting and optoelectronic applications.

5.6 CONCLUSION

The chapter deals with a detailed discussion on the optoelectronic application of the hybrid organic electronic perovskites. In this consequence, the detailed structure—property as well as symmetry—phase relationship has been discussed. Basic synthesis procedure that has been given with associated detailed literature review has been included. Different properties which are directly related to the optoelectronic application have been given, and in this context, the current existing challenges are mentioned precisely. Little more emphasis was put into the halide-based perovskites. Different optoelectronic applications have been discussed in detail with their corresponding challenges. Thus this article can act as bridge between organic and inorganic material-based perovskites for optoelectronic applications. Also, it gives an idea about the need of coupling between experimental and theory-based investigation to get an optimum performance/efficiencies from the associated devices.

ACKNOWLEDGMENTS

The author expresses sincere thanks to Miss Dimitra Das of Jadavpur University, Miss Debaleena Nawn of IIT Kharagpur, and Mr. Pritam Kishore Chakrabarty of Indian Institute of Engineering Science and Technology for helping out with this chapter.

REFERENCES

[1] J. Chen, S. Zhou, S. Jin, H. Li, T. Zhai, Crystal organometal halide perovskites with promising optoelectronic applications, J. Mater. Chem. C 4 (1) (2016) 11–27.

[2] R.S. Sanchez, V. Gonzalez-Pedro, J.W. Lee, N.G. Park, Y.S. Kang, I. Mora-Sero, et al., Slow dynamic processes in lead halide perovskite solar cells. Characteristic times and hysteresis, J. Phys. Lett. 5 (13) (2014) 2357–2363.

[3] V.W. Bergmann, Y. Guo, H. Tanaka, I.M. Hermes, D. Li, A. Klasen, et al., Local time-dependent charging in a perovskite solar cell, ACS Appl. Mater. Interfaces 8 (30) (2016) 19402–19409.

[4] T.M. Brenner, D.A. Egger, L. Kronik, G. Hodes, D. Cahen, Hybrid organic–inorganic perovskites: low-cost semiconductors with intriguing charge-transport properties, Nat. Rev. Mater. 1 (2016) 15007.

[5] E. Edri, S. Kirmayer, S. Mukhopadhyay, K. Gartsman, G. Hodes, D. Cahen, Elucidating the charge carrier separation and working mechanism of $CH_3NH_3PbI_{3-x}Cl_x$ perovskite solar cells, Nat. Commun. 5 (2014) 3461.

[6] H. Zhu, K. Miyata, Y. Fu, J. Wang, P.P. Joshi, D. Niesner, et al., Screening in crystalline liquids protects energetic carriers in hybrid perovskites, Science 353 (6306) (2016) 1409–1413.

[7] H. Zhu, M.T. Trinh, J. Wang, Y. Fu, P.P. Joshi, K. Miyata, et al., Organic cations might not be essential to the remarkable properties of band edge carriers in lead halide perovskites, Adv. Mater. 29 (1) (2017) 1–6.

[8] W. Zhang, V.M. Burlakov, D.J. Graham, T. Leijtens, A. Osherov, V. Bulović, et al., Photo-induced halide redistribution in organic–inorganic perovskite films, Nat. Commun. 7 (2016) 11683. ncomms 11683.

[9] E. Mosconi, D. Meggiolaro, H.J. Snaith, S.D. Stranks, F. De Angelis, Light-induced annihilation of Frenkel defects in organo-lead halide perovskites, Energy Environ. Sci. 9 (10) (2016) 3180–3187.

[10] O. Almora, I. Zarazua, E. Mas-Marza, I. Mora-Sero, J. Bisquert, G. Garcia-Belmonte, Capacitive dark currents, hysteresis, and electrode polarization in lead halide perovskite solar cells, J. Phys. Chem. Lett. 6 (9) (2015) 1645–1652.

[11] G. Garcia-Belmonte, J. Bisquert, Distinction between capacitive and noncapacitive hysteretic currents in operation and degradation of perovskite solar cells, ACS Energy Lett. 1 (4) (2016) 683–688.

[12] V.M. Goldschmidt, Ber. Dtsch. Chem. Ges. 60 (1927) 1263–1268.

[13] M. Johnsson, P. Lemmens, Handbook of Magnetism and Advanced Magnetic Materials, John Wiley & Sons, Ltd., New York, USA, 2007.

[14] C. Li, X. Lu, W. Ding, L. Feng, Y. Gao, Z. Guo, Formability of ABX_3 (X = F, Cl, Br, I) Halide Perovskites, Acta Crystallogr., Sect. B: Struct. Sci 64 (6) (2008) 702–707.

[15] Y.H. Chang, C.H. Park, K. Matsuishi, First-principles study of the structural and the electronic properties of the lead-halide-based inorganic–organic perovskites $(CH_{\sim3}NH_{\sim3})$ $PbX_{\sim3}$ and $CsPbX_{\sim3}$ (X = Cl, Br, I), J. Korean Phys. Soc. 44 (2004) 889–893.

[16] Q. Chen, N. De Marco, Y.M. Yang, T.B. Song, C.C. Chen, H. Zhao, et al., Under the spotlight: the organic–inorganic hybrid halide perovskite for optoelectronic applications, Nano Today 10 (3) (2015) 355–396.

[17] F. Brivio, A.B. Walker, A. Walsh, Structural and electronic properties of hybrid perovskites for high-efficiency thin-film photovoltaics from first-principles, APL Mater. 1 (4) (2013) 042111.

[18] T. Baikie, Y. Fang, J.M. Kadro, M. Schreyer, F. Wei, S.G. Mhaisalkar, et al., Synthesis and crystal chemistry of the hybrid perovskite (CH_3NH_3) PbI_3 for solid-state sensitised solar cell applications, J. Mater. Chem. A 1 (18) (2013) 5628–5641.

[19] E. Mosconi, A. Amat, M.K. Nazeeruddin, M. Grätzel, F. De Angelis, First-principles modeling of mixed halide organometal perovskites for photovoltaic applications, J. Phys Chem C 117 (27) (2013) 13902–13913.
[20] A. Poglitsch, D. Weber, Dynamic disorder in methylammoniumtrihalogenoplumbates (II) observed by millimeter-wave spectroscopy, J. Chem. Phys. 87 (11) (1987) 6373–6378.
[21] C.C. Stoumpos, C.D. Malliakas, M.G. Kanatzidis, Semiconducting tin and lead iodide perovskites with organic cations: phase transitions, high mobilities, and near-infrared photoluminescent properties, Inorg. Chem. 52 (15) (2013) 9019–9038.
[22] M.I. Dar, N. Arora, P. Gao, S. Ahmad, M. Grätzel, M.K. Nazeeruddin, Investigation regarding the role of chloride in organic–inorganic halide perovskites obtained from chloride containing precursors, Nano Lett. 14 (12) (2014) 6991–6996.
[23] G.E. Eperon, S.D. Stranks, C. Menelaou, M.B. Johnston, L.M. Herz, H.J. Snaith, Formamidinium lead trihalide: a broadly tunable perovskite for efficient planar heterojunction solar cells, Energy Environ. Sci. 7 (3) (2014) 982–988.
[24] N. Pellet, P. Gao, G. Gregori, T.Y. Yang, M.K. Nazeeruddin, J. Maier, et al., Mixed-organic-cation perovskite photovoltaics for enhanced solar-light harvesting, Angew. Chem. Int. Ed. 53 (12) (2014) 3151–3157.
[25] S. Pang, H. Hu, J. Zhang, S. Lv, Y. Yu, F. Wei, et al., NH2CH = NH2PbI3: an alternative organolead iodide perovskite sensitizer for mesoscopic solar cells, Chem. Mater. 26 (3) (2014) 1485–1491.
[26] D.B. Mitzi, C.A. Feild, Z. Schlesinger, R.B. Laibowitz, Transport, optical, and magnetic properties of the conducting halide perovskite CH3NH3SnI3, J. Solid State Chem. 114 (1) (1995) 159–163.
[27] A. Mei, X. Li, L. Liu, Z. Ku, T. Liu, Y. Rong, et al., A hole-conductor–free, fully printable mesoscopic perovskite solar cell with high stability, Science 345 (6194) (2014) 295–298.
[28] C. Bernal, K. Yang, First-principles hybrid functional study of the organic–inorganic perovskites $CH_3NH_3SnBr_3$ and $CH_3NH_3SnI_3$, J. Phys Chem C 118 (42) (2014) 24383–24388.
[29] N.K. Noel, S.D. Stranks, A. Abate, C. Wehrenfennig, S. Guarnera, A.A. Haghighirad, et al., Lead-free organic–inorganic tin halide perovskites for photovoltaic applications, Energy Environ. Sci. 7 (9) (2014) 3061–3068.
[30] E.L. Unger, A.R. Bowring, C.J. Tassone, V.L. Pool, A. Gold-Parker, R. Cheacharoen, et al., Chloride in lead chloride-derived organo-metal halides for perovskite-absorber solar cells, Chem. Mater. 26 (24) (2014) 7158–7165.
[31] A. Dualeh, N. Tétreault, T. Moehl, P. Gao, M.K. Nazeeruddin, M. Grätzel, Effect of annealing temperature on film morphology of organic–inorganic hybrid perovskite solid-state solar cells, Adv. Funct. Mater. 24 (21) (2014) 3250–3258.
[32] M.M. Lee, J. Teuscher, T. Miyasaka, T.N. Murakami, H.J. Snaith, Efficient hybrid solar cells based on meso-superstructured organometal halide perovskites, Science 338 (6107) (2012) 643–647.
[33] A. Kojima, K. Teshima, Y. Shirai, T. Miyasaka, Organometal halide perovskites as visible-light sensitizers for photovoltaic cells, J. Am. Chem. Soc. 131 (17) (2009) 6050–6051.
[34] N.J. Jeon, J.H. Noh, Y.C. Kim, W.S. Yang, S. Ryu, S.I. Seok, Solvent engineering for high-performance inorganic–organic hybrid perovskite solar cells, Nat. Mater. 13 (9) (2014) 897–903.
[35] D.B. Mitzi, A layered solution crystal growth technique and the crystal structure of $(C_6H_5C_2H_4NH_3)_2$ $PbCl_4$, J. Solid State Chem. 145 (2) (1999) 694–704.

[36] Q. Dong, Y. Fang, Y. Shao, P. Mulligan, J. Qiu, L. Cao, et al., Electron-hole diffusion lengths >175 μm in solution-grown $CH_3NH_3PbI_3$ single crystals, Science 347 (6225) (2015) 967−970.
[37] D. Shi, V. Adinolfi, R. Comin, M. Yuan, E. Alarousu, A. Buin, et al., Low trap-state density and long carrier diffusion in organolead trihalide perovskite single crystals, Science 347 (6221) (2015) 519−522.
[38] Q. Liao, K. Hu, H. Zhang, X. Wang, J. Yao, H. Fu, Perovskite Microdisk Microlasers Self-Assembled from Solution, Adv. Mater. 27 (22) (2015) 3405−3410.
[39] M.I. Saidaminov, A.L. Abdelhady, B. Murali, E. Alarousu, V.M. Burlakov, W. Peng, et al., High-quality bulk hybrid perovskite single crystals within minutes by inverse temperature crystallization, Nat. Commun. 6 (2015) 7586.
[40] F. Zhang, H. Zhong, C. Chen, X.G. Wu, X. Hu, H. Huang, et al., Brightly luminescent and color-tunable colloidal $CH_3NH_3PbX_3$ (X = Br, I, Cl) quantum dots: potential alternatives for display technology, ACS Nano 9 (4) (2015) 4533−4542.
[41] S. Zhuo, J. Zhang, Y. Shi, Y. Huang, B. Zhang, Self-template-directed synthesis of porous perovskite nanowires at room temperature for high-performance visible-light photodetectors, Angew. Chem. Int. Ed. 54 (19) (2015) 5693−5696.
[42] H. Zhu, Y. Fu, F. Meng, X. Wu, Z. Gong, Q. Ding, et al., Lead halide perovskite nanowire lasers with low lasing thresholds and high quality factors, Nat. Mater. 14 (6) (2015) 636.
[43] L. Dou, A.B. Wong, Y. Yu, M. Lai, N. Kornienko, S.W. Eaton, et al., Atomically thin two-dimensional organic−inorganic hybrid perovskites, Science 349 (6255) (2015) 1518−1521.
[44] Q. Chen, H. Zhou, Z. Hong, S. Luo, H.S. Duan, H.H. Wang, et al., Planar heterojunction perovskite solar cells via vapor-assisted solution process, J. Am. Chem. Soc. 136 (2) (2014) 622−625.
[45] M. Liu, M.B. Johnston, H.J. Snaith, Efficient planar heterojunction perovskite solar cells by vapour deposition, Nature 501 (7467) (2013) 395−398.
[46] S.T. Ha, X. Liu, Q. Zhang, D. Giovanni, T.C. Sum, Q. Xiong, Synthesis of organic−inorganic lead halide perovskite nanoplatelets: towards high-performance perovskite solar cells and optoelectronic devices, Adv. Opt. Mater. 2 (9) (2014) 838−844.
[47] J. Burschka, N. Pellet, S.J. Moon, R. Humphry-Baker, P. Gao, M.K. Nazeeruddin, et al., Nature 499 (2013) 316.
[48] Y. Ogomi, A. Morita, S. Tsukamoto, T. Saitho, N. Fujikawa, Q. Shen, et al., $CH_3NH_3Sn_x\ Pb_{(1-x)}I_3$ perovskite solar cells covering up to 1060 nm, J. Phys. Lett. 5 (6) (2014) 1004−1011.
[49] S. De Wolf, J. Holovsky, S.-J. Moon, P. Löper, B. Niesen, M. Ledinsky, et al., J. Phys. Chem. Lett. 5 (2014) 1035−1039.
[50] E. Edri, S. Kirmayer, A. Henning, S. Mukhopadhyay, K. Gartsman, Y. Rosenwaks, et al., Why lead methylammonium tri-IODIDE perovskite-based solar cells requires a mesoporous electron transporting scaffold (but not necessarily a hole conductor), Nano Lett. 14 (2) (2014) 1000−1004.
[51] J.H. Heo, S.H. Im, J.H. Noh, T.N. Mandal, C.S. Lim, J.A. Chang, et al., Efficient inorganic−organic hybrid heterojunction solar cells containing perovskite compound and polymeric hole conductors, Nat. Photonics 7 (6) (2013) 486−491.
[52] J.M. Frost, K.T. Butler, F. Brivio, C.H. Hendon, M. Van Schilfgaarde, A. Walsh, Atomistic origins of high-performance in hybrid halide perovskite solar cells, Nano Lett. 14 (5) (2014) 2584.
[53] P. Umari, E. Mosconi, F. De Angelis, Relativistic GW calculations on $CH_3NH_3PbI_3$ and $CH_3NH_3SnI_3$ perovskites for solar cell applications, Sci. Rep. 4 (2014) 4467.

[54] G. Giorgi, J.I. Fujisawa, H. Segawa, K. Yamashita, Small photocarrier effective masses featuring ambipolar transport in methylammonium lead iodide perovskite: a density functional analysis, J. Phys. Lett. 4 (24) (2013) 4213–4216.

[55] W.J. Yin, T. Shi, Y. Yan, Unique properties of halide perovskites as possible origins of the superior solar cell performance, Adv. Mater. 26 (27) (2014) 4653–4658.

[56] H.S. Duan, H. Zhou, Q. Chen, P. Sun, S. Luo, T.B. Song, et al., The identification and characterization of defect states in hybrid organic–inorganic perovskite photovoltaics, Phys. Chem. Chem. Phys. 17 (1) (2015) 112–116.

[57] A. Buin, P. Pietsch, J. Xu, O. Voznyy, A.H. Ip, R. Comin, et al., Materials processing routes to trap-free halide perovskites, Nano Lett. 14 (11) (2014) 6281–6286.

[58] Q. Chen, H. Zhou, T.B. Song, S. Luo, Z. Hong, H.S. Duan, et al., Controllable self-induced passivation of hybrid lead iodide perovskites toward high performance solar cells, Nano Lett. 14 (7) (2014) 4158–4163.

[59] T. Supasai, N. Rujisamphan, K. Ullrich, A. Chemseddine, T. Dittrich, Formation of a passivating $CH_3NH_3PbI_3/PbI_2$ interface during moderate heating of $CH_3NH_3PbI_3$ layers, Appl. Phys. Lett. 103 (18) (2013) 183906.

[60] S. Lv, L. Han, J. Xiao, L. Zhu, J. Shi, H. Wei, et al., Mesoscopic $TiO_2/CH_3NH_3PbI_3$ perovskite solar cells with new hole-transporting materials containing butadiene derivatives, Chem. Commun. 50 (52) (2014) 6931–6934.

[61] J. Wang, S. Wang, X. Li, L. Zhu, Q. Meng, Y. Xiao, et al., Novel hole transporting materials with a linear π-conjugated structure for highly efficient perovskite solar cells, Chem. Commun. 50 (44) (2014) 5829–5832.

[62] J.S. Manser, P.V. Kamat, Band filling with free charge carriers in organometal halide perovskites, Nat. Photonics 8 (9) (2014) 737–743.

[63] C. Wehrenfennig, G.E. Eperon, M.B. Johnston, H.J. Snaith, L.M. Herz, High charge carrier mobilities and lifetimes in organolead trihalide perovskites, Adv. Mater. 26 (10) (2014) 1584–1589.

[64] H. Oga, A. Saeki, Y. Ogomi, S. Hayase, S. Seki, Improved understanding of the electronic and energetic landscapes of perovskite solar cells: high local charge carrier mobility, reduced recombination, and extremely shallow traps, J. Am. Chem. Soc. 136 (39) (2014) 13818–13825.

[65] T. Leijtens, S.D. Stranks, G.E. Eperon, R. Lindblad, E.M. Johansson, I.J. McPherson, et al., Electronic properties of meso-superstructured and planar organometal halide perovskite films: charge trapping, photodoping, and carrier mobility, ACS Nano 8 (7) (2014) 7147–7155.

[66] H.S. Kim, C.R. Lee, J.H. Im, K.B. Lee, T. Moehl, A. Marchioro, et al., Lead iodide perovskite sensitized all-solid-state submicron thin film mesoscopic solar cell with efficiency exceeding 9%, Sci. Rep. 2 (2012) 1–7.

[67] J. Burschka, N. Pellet, S.J. Moon, R. Humphry-Baker, P. Gao, M.K. Nazeeruddin, et al., Sequential deposition as a route to high-performance perovskite-sensitized solar cells, Nature (2013).

[68] N.J. Jeon, J.H. Noh, W.S. Yang, Y.C. Kim, S. Ryu, J. Seo, et al., Compositional engineering of perovskite materials for high-performance solar cells, Nature 517 (7535) (2015) 476–480.

[69] Q. Pei, G. Yu, C. Zhang, Y. Yang, A.J. Heeger, Polymer light-emitting electrochemical cells, Science 269 (5227) (1995) 1086.

[70] R.D. Costa, E. Ortí, H.J. Bolink, F. Monti, G. Accorsi, N. Armaroli, Luminescent ionic transition-metal complexes for light-emitting electrochemical cells, Angew. Chem. Int. Ed. 51 (33) (2012) 8178–8211.

[71] M.F. Aygüler, M.D. Weber, B.M. Puscher, D.D. Medina, P. Docampo, R.D. Costa, Light-emitting electrochemical cells based on hybrid lead halide perovskite nanoparticles, J. Phys Chem C 119 (21) (2015) 12047–12054.

[72] R.L. Hoye, M.R. Chua, K.P. Musselman, G. Li, M.L. Lai, Z.K. Tan, et al., Enhanced performance in fluorene-free organometal halide perovskite light-emitting diodes using tunable, low electron affinity oxide electron injectors, Adv. Mater. 27 (8) (2015) 1414–1419.
[73] L. Gil-Escrig, G. Longo, A. Pertegás, C. Roldán-Carmona, A. Soriano, M. Sessolo, et al., Efficient photovoltaic and electroluminescent perovskite devices, Chem. Commun. 51 (3) (2015) 569–571.
[74] Y.H. Kim, H. Cho, J.H. Heo, T.S. Kim, N. Myoung, C.L. Lee, et al., Multicolored organic/inorganic hybrid perovskite light-emitting diodes, Adv. Mater. 27 (7) (2015) 1248–1254.
[75] A.B. Wong, M. Lai, S.W. Eaton, Y. Yu, E. Lin, L. Dou, et al., Growth and anion exchange conversion of $CH_3NH_3PbX_3$ nanorod arrays for light-emitting diodes, Nano Lett. 15 (8) (2015) 5519–5524.
[76] T. Kondo, T. Azuma, T. Yuasa, R. Ito, Biexciton lasing in the layered perovskite-type material $(C_6Hr_3NH_3)_2Pb_{I4}$, Solid State Commun. 105 (4) (1998) 253–255.
[77] G. Xing, N. Mathews, S.S. Lim, N. Yantara, X. Liu, D. Sabba, et al., Low-temperature solution-processed wavelength-tunable perovskites for lasing, Nat. Mater. 13 (5) (2014) 476–480.
[78] A. Fu, P. Yang, Organic–inorganic perovskites: lower threshold for nanowire lasers, Nat. Mater. 14 (6) (2015) 557–558.
[79] J. Xing, X.F. Liu, Q. Zhang, S.T. Ha, Y.W. Yuan, C. Shen, et al., Vapor phase synthesis of organometal halide perovskite nanowires for tunable room-temperature nanolasers, Nano Lett. 15 (7) (2015) 4571–4577.
[80] Q. Zhang, S.T. Ha, X. Liu, T.C. Sum, Q. Xiong, Room-temperature near-infrared high-Q perovskite whispering-gallery planar nanolasers, Nano Lett. 14 (10) (2014) 5995–6001.
[81] C.R. Kagan, D.B. Mitzi, C.D. Dimitrakopoulos, Organic–inorganic hybrid materials as semiconducting channels in thin-film field-effect transistors, Science 286 (5441) (1999) 945–947.
[82] D.B. Mitzi, C.D. Dimitrakopoulos, J. Rosner, D.R. Medeiros, Z. Xu, C. Noyan, Hybrid field-effect transistor based on a low-temperature melt-processed channel layer, Adv. Mater. 14 (23) (2002) 1772–1776.
[83] X. Hu, X. Zhang, L. Liang, J. Bao, S. Li, W. Yang, et al., High-performance flexible broadband photodetector based on organolead halide perovskite, Adv. Funct. Mater. 24 (46) (2014) 7373–7380.
[84] D.M. Jang, K. Park, D.H. Kim, J. Park, F. Shojaei, H.S. Kang, et al., Reversible halide exchange reaction of organometal trihalide perovskite colloidal nanocrystals for full-range band gap tuning, Nano Lett. 15 (8) (2015) 5191–5199.
[85] Y. Lee, J. Kwon, E. Hwang, C.H. Ra, W.J. Yoo, J.H. Ahn, et al., High-performance perovskite–graphene hybrid photodetector, Adv. Mater. 27 (1) (2015) 41–46.
[86] M. Spina, M. Lehmann, B. Náfrádi, L. Bernard, E. Bonvin, R. Gaál, et al., Microengineered $CH_3NH_3PbI_3$ nanowire/graphene phototransistor for low-intensity light detection at room temperature, Small 11 (37) (2015) 4824–4828.

CHAPTER 6

Ion Migration in Hybrid Perovskites: Evolving Understanding of a Dynamic Phenomenon

Christie L.C. Ellis, Emily Smith, Hamza Javaid, Gabrielle Berns and Dhandapani Venkataraman
University of Massachusetts Amherst, Amherst, MA, United States

6.1 INTRODUCTION

Hybrid organic–inorganic perovskites (HOIPs) have emerged as promising new materials for fundamental studies and practical applications, with unique and desirable properties arising from their combination of organic and inorganic components [1]. Within a short time span, HOIPs have been demonstrated in diverse applications such as thermoelectrics, piezoelectrics, field effect transistors, lasing, light-emitting diodes, memory storage, computing, sensing, and, most prominently, photovoltaics [2–10]. In spite of rapid advances in device properties, the underlying origins of some of these properties remain a matter of debate. Ion transport has been implicated as the cause of many important properties associated with HOIPs (see Fig. 6.1), yet the chemical identity of the migrating ion or ions and the mechanism of transport remain unclear. The lack of understanding of ion transport in these systems is a stumbling block for designing new HOIPs with desired and improved properties. In this chapter, we discuss the key results in the area of ion migration in HOIPs and our current understanding on nature of ion migration in these materials.

HOIPs, like traditional oxide-based perovskites, exhibit several experimental hallmarks of ion transport. Nonetheless, studying ion transport in these materials has not been straightforward due to several factors that complicate data analysis and interpretation [11,12]. First, there are multiple possible mobile ions in HOIPs, and it is possible that all of them are

Perovskite Photovoltaics
DOI: https://doi.org/10.1016/B978-0-12-812915-9.00006-X

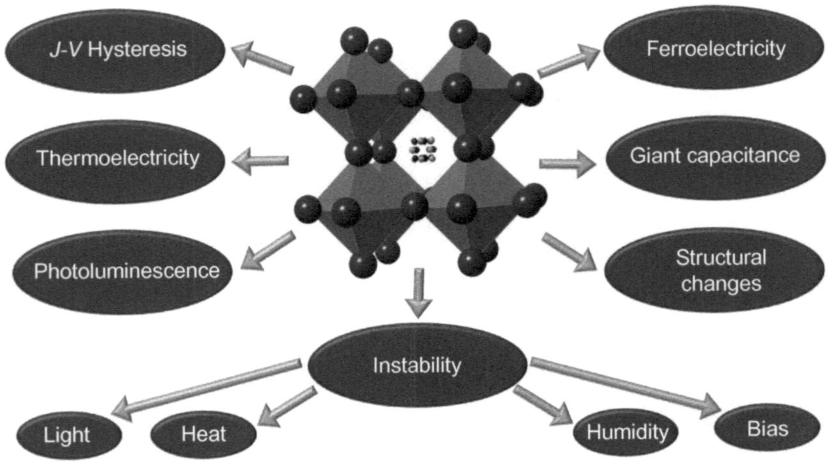

Figure 6.1 Observed phenomena in hybrid organic–inorganic perovskites with possible relationship to ion transport.

responsible for the experimental observations of ion transport [13–15]. Second, HOIPs are both semiconductors and ionic conductors; thus, there is a possibility of coupled ionic–electronic transport. Third, HOIPs are relatively unstable materials; they have a high and variable defect density and degrade quickly under many device-relevant conditions such as light, heat, humidity, and electrical bias [16]. Fourth, the nonspherical organic counterions in HOIPs interact with the lattice framework through a diverse array of supramolecular interactions such as van der Waals, dipole-dipole, and hydrogen bond interactions. These interactions, coupled with lattice distortions, phase transitions, grain boundaries, and interfaces, complicate the interpretation of the experimental data [17–19]. With several interconnected and synergistic factors influencing ion transport coupled with a flood of publications (and data) in a short time span, there is diversity in opinion on the nature of ion transport in HOIPs.

6.2 BACKGROUND ON HOIPS

HOIPs have AMX_3 stoichiometry, where A is the organic cation, M is the metal ion, and X is the halide (see Fig. 6.2) [10]. Methylammonium lead iodide, $(CH_3NH_3)PbI_3$, is the most commonly studied HOIP. Other HOIPs containing formamidinium (FA^+) cations instead of methylammonium (MA^+), Br^- or Cl^- instead of I^-, and Sn^{2+} instead of Pb^{2+}, as well as their mixtures have also been explored [20–23]. HOIPs exhibit a

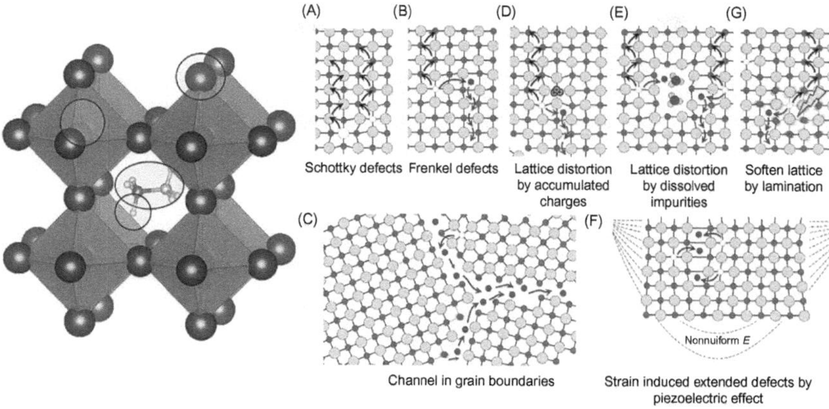

Figure 6.2 Left: Possible mobile ions in $MAPbI_3$: I^-, Pb^{2+}, MA^+, and $H^{+/0/-}$ [13]. Right: Selected possible pathways for ion migration [12]. *Left: Adapted with permission from C. Eames, J.M. Frost, P.R.F. Barnes, B.C. O'Regan, A. Walsh, S.M. Islam, Ionic transport in hybrid lead iodide perovskite solar cells, Nat. Commun. 6 (2015). Right: Reprinted with permission from Y. Yuan, J. Huang, Ion migration in organometal trihalide perovskite and its impact on photovoltaic efficiency and stability, Acc. Chem. Res. 49 (2) (2016) 286–293, doi:10.1021/acs.accounts.5b00420. Copyright 2016 American Chemical Society.*

rich structural diversity. $MAPbI_3$, for example, is orthorhombic (space group = *Pnma*) below 165K when MA^+ dipoles are all aligned, undergoes a sharp first-order transition to the tetragonal phase (*I4/mcm*) at 165K when MA^+ gains some rotational freedom, and undergoes a continuous second-order transition to the high-symmetry cubic phase (*Pm3m*) at 327K when MA^+ dipoles are fully disordered [24]. The PbI_6 octahedra can significantly distort even in the cubic phase, which gives rise to bulk flexibility in these materials [17,25].

All-inorganic alkali-metal lead and tin halide perovskites were first synthesized in 1893, and their crystallographic structures were determined in 1958 by Christian Møller, who noted that these materials exhibited photoconductivity [26,27]. The first HOIP, methylammonium lead iodide ($MAPbI_3$), was fabricated by Dieter Weber in 1978, and the optoelectronic properties of these materials were investigated in the early 1990s by Mitzi et al. [28,29]. HOIPs first emerged in the photovoltaic community not as a primary active material but as a light-absorbing dye in dye-sensitized solar cells (DSSCs). The first liquid hybrid perovskite DSSC was fabricated in 2009 and had a power conversion efficiency (PCE) of 3.8% [30]. In 2012, Lee et al. and Kim et al. fabricated solid-state HOIP active material photovoltaics using $MAPbI_3$ and

$MAPbI_xCl_{1-x}$, respectively, with PCEs of 8%–10%, surpassing the best known solid-state DSSCs and many of the organic photovoltaic materials at the time [31,32]. The development of high efficiency HOIP photovoltaics steadily progressed from this point to reach the current device efficiency record of 22.1%, rivaling the most efficient photovoltaics available today such as CIGS and CdTe (~20% max PCE) and crystalline silicon (~25% max PCE) [33–35].

The potential of HOIP materials for photovoltaic applications is inarguably large. In the Shockley–Queisser limit, single-junction HOIPs have a theoretical potential PCE of nearly 30% [36,37]. HOIP-based devices have the potential to exceed this limit using tandem cell architectures. Catchpole et al. have recently reported PCE values of 26.4% in a tandem cell containing HOIP and silicon as active layers [38,39]. High efficiency is one of many indicators of the anomalously low electron-hole recombination rates in HOIPs, which persist even in low temperature solution-processed devices with high bulk defect densities. The band gap of $MAPbI_3$ is close to ideal for photovoltaic applications at 1.55 eV and can be tuned by doping or substitution with different organic cations or halides [40–43]. Binding energies of excitons in HOIPs are low, 0.03 eV for $MAPbI_3$. Therefore, excitons dissociate into free carriers at room temperature. HOIPs are also ambipolar charge transporters with high electron and hole mobilities on the order of $\sim 10^1$ $cm^2/V/s$ for $MAPbI_3$, and long carrier diffusion lengths of 100–1000 nm—much longer than the film thickness needed for full optical absorbance [1].

6.3 LONG TIMESCALE PHENOMENA IN HOIPS

A curious feature of HOIP materials is that they exhibit processes with longer timescales—ranging from milliseconds to minutes or even days—which are not typically present in electronically semiconducting materials. These processes ultimately led to the investigation of a variety of potential material properties, including ion transport.

One of the most widely observed hallmarks of a long timescale process in HOIPs is the current density-voltage ($J-V$) hysteresis in photovoltaic device measurements [44–46]. In nearly all known photovoltaic materials, the current density (J) vs voltage (V) curve obtained by scanning from positive to negative bias is similar to the curve obtained by scanning from negative to positive bias [47]. However, many HOIP photovoltaic cells exhibit severe hysteresis in this measurement, giving drastically different

J–*V* curves and PCEs depending on factors such as scan direction, scan rate, and poling [46,47]. Additionally complicating the matter, different device architectures and fabrication methods produce widely varying degrees of hysteresis across devices. Observed *J*–*V* hysteresis clearly points to a dynamic process occurring within the perovskite active layer in the second—minute long timescale of these measurements, under controlled device conditions of illumination and applied bias. In addition to hysteresis in *J*–*V* measurements, specific electrical phenomena operating on these timescales have also been observed. They include large capacitance values, large photoinduced dielectric constants, and ferroelectric polarization [48–52]. Structural changes, such as the degradation of these materials over periods of hours to months, also indicate that there is a time-dependent factor involved in material stability [53].

While many researchers have observed similar phenomena, their explanations for these observations vary. Some attribute these phenomena to accumulation of electronic charge carriers in deep trap states near interfaces or distortion of structure by local concentrations of charge carriers, leading to high capacitance values across the film [54–56]. Others have argued that polarization through structural distortion occurs first, causing the observed ferroelectricity, and then contributes to changes in electronic charge carrier behavior [57].

Ion transport through HOIPs has surfaced more recently as a possible explanation for these long timescale phenomena [11,45]. Ions can respond to an electric field at a slower timescale than electronic charge carriers, and their accumulation at interfaces can cause similar behaviors to those attributed to electronic charge carrier accumulation. Although this explanation was initially received with skepticism, over the past few years, substantial evidence has been gathered which supports the hypothesis that HOIPs are in fact ion-transporting materials. Computational evidence has provided a strong foundation for both the feasibility of ion movement and a focus for experimental investigations of it [15,58]. Movement of ions or related phenomena have been observed qualitatively in many different HOIP systems, and in some cases, relevant values such as activation energies or diffusion coefficients of mobile ions have been quantified [13,59–61]. While the evidence gathered to reach this consensus represents progress in our understanding of these multifaceted and unique materials, many questions remain unanswered. The HOIP research community must address these questions to fully realize the potential of these materials.

6.4 QUALITATIVE EVIDENCE OF ION TRANSPORT

Structural and electrochemical studies on HOIPs have been a rich source of evidence in support of ion transport. Gr tzel et al. performed a classic solid-state electrochemical experiment under dark conditions on a cell with architecture (cathode to anode) of $Pb/MAPbI_3/AgI/Ag$. Under electrical bias, the formation of PbI_2 or MAI at the interface of $MAPbI_3$ with either Pb or AgI would indicate that certain ions were able to move to these reactive interfaces [62]. Based on energy-dispersive X-ray spectroscopy (EDX) and X-ray diffraction (XRD) data, the authors concluded that PbI_2 formed at the $Pb/MAPbI_3$, showing that I^- is the fastest moving ion. However, they did not examine the $MAPbI_3/AgI$ interface for MAI formation since they did not observe any change at this interface in scanning electron microscopy (SEM) images. This evidence does not conclusively rule out the possibility of mobile MA^+ ions. Huettner et al. observed uneven distribution of iodine and lead using X-ray photoelectron spectroscopy (XPS) in a micron-sized $MAPbI_3$ crystal after the application of electric bias. They observed a range around the ideal I/Pb ratio of 3 with ratios as low as 1 in the center of the crystal and as high as 6 at the cathode [63]. Additionally, Huang et al. have reported several key results in this area; they fabricated a symmetrical horizontal micron-length $MAPbI_3$ device and observed the movement of a PbI_2 "thread" through the film under small applied bias using optical microscopy, EDX, SEM, and XRD [60]. They reasoned that, for this to occur, both MA^+ and I^- had to be mobile and undergo forward and backward reactions in the solid state at comparable rates. They observed the movement of MA^+ ions through an $MAPbI_3$ film by photothermal-induced resonance (PTIR) under applied bias on a timescale of hundreds of seconds [59]. They also used XPS and ultraviolet photoelectron spectroscopy to demonstrate that changing the stoichiometry of the two $MAPbI_3$ precursors, MAI and PbI_2, as well as thermal annealing, could drastically vary the energy levels and p- or n-doped character of the films, as shown in Fig. 6.3 [64]. While this last result does not directly indicate ion movement, it does demonstrate that presence of excess ion species/defects can alter the HOIP energy levels, supporting the hypothesis that ion movement and accumulation is related to $J-V$ hysteresis.

Techniques traditionally used to study simple electrically conducting systems have also proven useful for examining ion movement. Huang et al. discovered that the photocurrent direction in ITO/PEDOT:PSS/

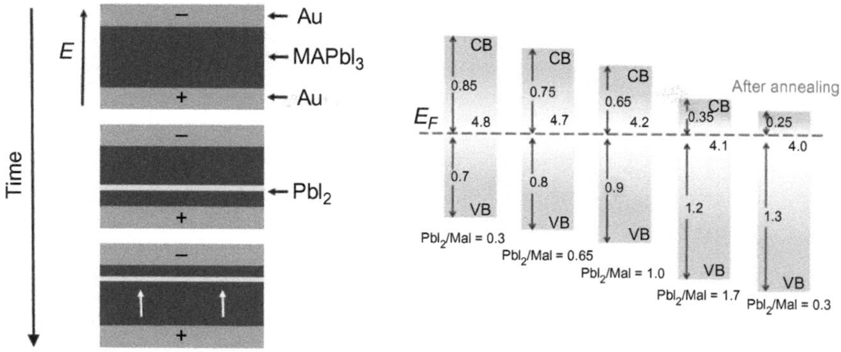

Figure 6.3 Left: Illustration of PbI_2 thread moving through $MAPbI_3$ film under applied bias [60]. Right: Variation of $MAPbI_3$ energy levels by stoichiometric ratio and annealing [64]. *Left: Illustrated using data from Y. Yuan, Q. Wang, Y. Shao, H. Lu, T. Li, A. Gruverman, et al., Electric-field-driven reversible conversion between methylammonium lead triiodide perovskites and lead iodide at elevated temperatures, Adv. Energy Mater. 6 (2) (2016), doi:10.1002/aenm.201501803. Copyright 2016 John Wiley and Sons. Right: Reprinted with permission from Q. Wang, Y. Shao, H. Xie, L. Lyu, X. Liu, Y. Gao, et al., Qualifying composition dependent p and n self-doping in $CH_3NH_3PbI_3$, Appl. Phys. Lett. 105 (16) (2014). Copyright 2014 AIP Publishing LLC.*

$MAPbI_3$/Au devices was switchable with applied bias directions by reversibly doping to either n–i–p or p–i–n structure as measured by Kelvin probe force microscopy [48]. They attributed this effect to ion movement—speculating that I^-, MA^+, and Pb^{2+} might be involved—as polarization information could be retained in the film for many months after initial poling, poling was frozen at low temperatures, and reversible color change was observed near cathode and anode after poling. Bisquert et al. fabricated an $MAPbI_3$ photovoltaic (PV) device that they poled in the dark for 30 seconds with 0, −1, or −2 V, short circuited for 1 minute to eliminate electronic charge accumulation, then switched quickly from V_{oc} to J_{sc} conditions and measured time-dependent photocurrent density, shown in Fig. 6.4 [46]. They observed an initial nonsteady-state photocurrent process with a fast decay time in all samples when the bias was switched from V_{oc} to 0 V. Additionally, they observed a second nonsteady-state process, where the photocurrent density was initially much lower and recovered to the 0-V value after several minutes in the −1-V poled cell and more so in the −2-V poled cell. They reasoned that this process was too slow to be accounted for by electronic carrier phenomena and instead fit well with reduced band bending due to ion redistribution; additionally, they note that this hysteretic effect cannot be accounted for

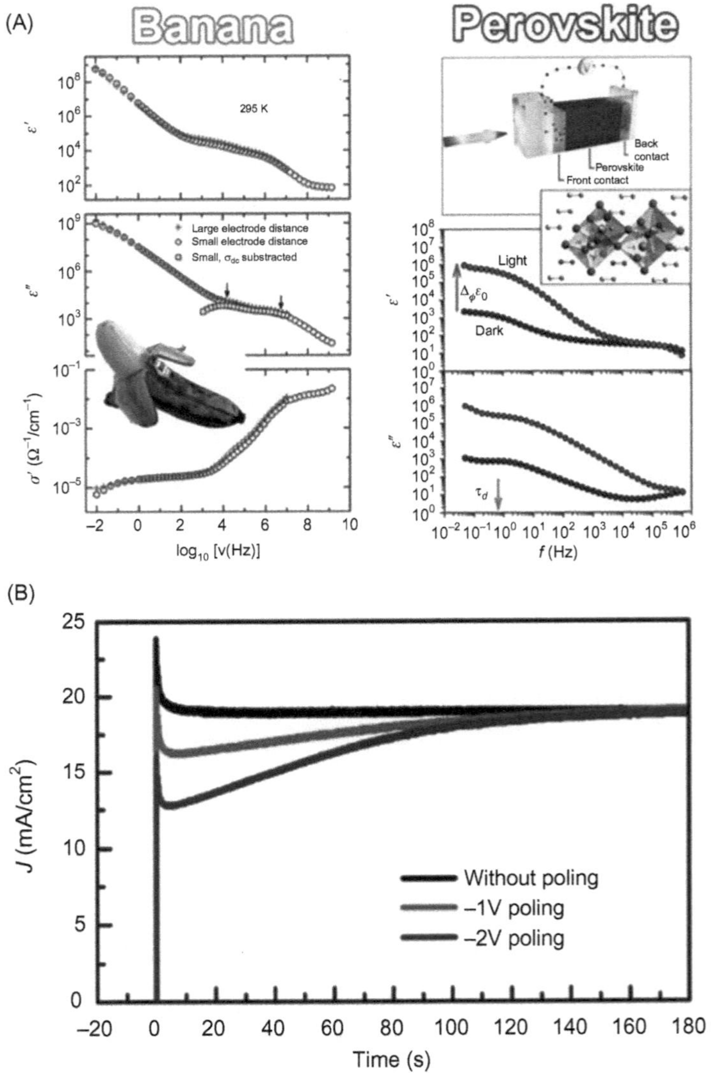

Figure 6.4 (A) Comparison of real and imaginary dielectric permittivity of banana skin and $MAPbI_3$ show similar frequency dependence [11,51,65]. (B) Minute-long recovery of steady state photocurrent [46]. *(A) Reprinted with permission from J.M. Frost, A. Walsh, What is moving in hybrid halide perovskite solar cells?, Acc. Chem. Res. (2016), doi:10.1021/acs.accounts.5b00431. Copyright 2016 American Chemical Society. (B) Reprinted with permission from B. Chen, M. Yang, X. Zheng, C. Wu, W. Li, Y. Yan, et al., Impact of capacitive effect and ion migration on the hysteretic behavior of perovskite solar cells, J. Phys. Lett. 6 (23) (2015) 4693–4700. Copyright 2015 American Chemical Society.*

by ferroelectric band bending, as a voltage $>V_{oc}$ would need to be applied to recover unpoled steady state current in this case. Frost and Walsh further addressed potential causes of observed slow electrical phenomena by demonstrating the similarity of dielectric permittivity measurements of $MAPbI_3$ made by Bisquert et al. to a 2008 paper by Lunkenheimer et al. [11,51,65]. Lunkenheimer used banana skin—a known paraelectric ion conductor—to demonstrate that simple dielectric permittivity measurements can differentiate between polarization caused by ion conduction and ferroelectricity; Bisquert's permittivity measurements, as shown besides those of the banana in Fig. 6.4, have a strikingly similar frequency dependence, with greatly increased permittivity at low frequency corresponding to ion conduction [65].

6.5 COMPUTATIONAL STUDIES OF ION TRANSPORT

Computational modeling using density functional theory provides a useful tool to independently model multiple phenomena in HOIPs and thus has been a critical tool in gaining insight into ion transport. Modeling has been used to predict quantitative values such as formation energies of ion-related defects or activation energies of defect/ion migration that could be validated by experiments.

Yan et al. calculated formation energies of various defects such as vacancies, interstitials, cation substitutions, and antisite substitutions for I^-, Pb^{2+}, and MA^+ in $MAPbI_3$ [36]. They found that these formation energies were sufficiently low—many <1 eV and as low as 0.2 eV—and predicted that many of these defects are likely present in this material. In addition, they observed that those ionic defects with low formation energies lead to shallow electronic traps. This prediction is consistent with the low nonradiative recombination rates and inconsistent with long-timescale (de)trapping processes. Following this study, several groups calculated activation energies (E_a) for migration of several ionic defects using transition state modeling (see Table 6.1).

Islam et al. calculated an E_a of 0.58 eV for I^- vacancies, 0.84 eV for MA^+ vacancies, and 2.31 eV for Pb^{2+} vacancies [13]. They concluded from these numbers, as well as the knowledge that many perovskites are known oxide anion conductors [66,67], that iodide is the most likely mobile species while methylammonium and lead are unlikely to be mobile. They also concluded that collection of iodide ions and vacancies at device interfaces could influence the band energies and create

Table 6.1 Computationally determined activation energy values of possible mobile ions by migration mechanism

Ref.	I^- (vacancy) (eV)	I^- (interstitial) (eV)	MA^+ (vacancy) (eV)	Pb^{2+} (vacancy) (eV)
Islam [13]	0.58	–	0.84	2.31
Tateyama [14]	0.33	–	0.55	–
De Angelis [15]	0.16	0.16	0.46	0.80
Mattoni [68]	0.10	0.24	–	–
Du [81]	0.26	0.19	0.62	–

hysteresis. Mattoni et al. modeled I^- movement through two possible mechanisms—vacancy diffusion or interstitial diffusion. They found a difference in E_a for the two different mechanisms as well as much lower overall E_a values—0.10 eV for vacancy diffusion and 0.24 eV for interstitial diffusion—indicating that the vacancy diffusion is the preferred ion diffusion pathway [68] (Fig 6.5). Tateyama et al. calculated E_a in both $MAPbI_3$ and $FAPbI_3$ for I^- vacancies and MA^+/FA^+ vacancies along different crystallographic axes [14]. They calculated E_a values for both I^- diffusion (0.33 eV) and MA^+ diffusion (0.55 eV) that were lower than those calculated by Islam et al. They also found that E_a for formamidinium (FA^+) diffusion was comparable to MA^+ diffusion (0.57 eV). They criticized the comparison of HOIPs to oxide perovskites, as oxide conducting perovskites have stronger ionic bonding character. However, both Tateyama and Mattoni note that the E_a values they obtained are well within the ranges of those in known ion-transporting materials, suggesting the likelihood that both the halide anion and the organic cation could be mobile. De Angelis et al. found lower, identical I^- vacancy and I^- interstitial E_a values (0.16 eV), a MA^+ vacancy E_a value comparable to Tateyama et al. (0.46 eV), and a significantly lower Pb^{2+} vacancy E_a than Islam et al. (0.80 eV) [15]. They reasoned that I^- was moving on the scale of ns, too quick to account for the ms/s timescale hysteresis effect, and that mobile MA^+ or possibly Pb^{2+} was more likely responsible. Additionally, De Angelis et al. performed a density of states calculation for an $MAPbI_3/TiO_2$ ETL planar interface and found that accumulation of oppositely charged ions/defects at the interface could cause band bending which could either impede or assist charge extraction, depending on which direction the applied electric field caused ions to migrate. Du et al. calculated E_a values for both interstitial and vacancy movement of I^- and

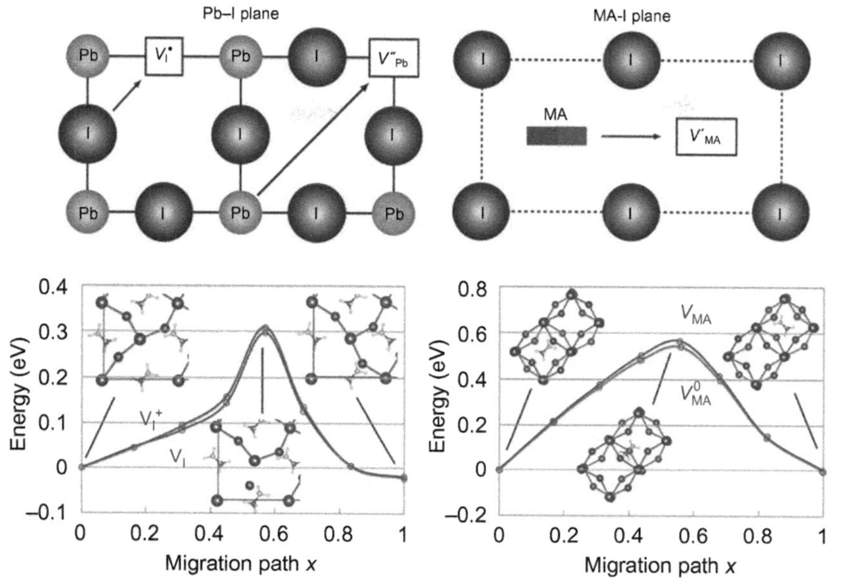

Figure 6.5 Top: Illustration of vacancy-mediated diffusion paths of I^-, Pb^{2+}, and MA^+ [13]. Bottom: Transition state calculations of I^- and MA^+ vacancy-mediated diffusion to determine activation energy of ion transport by this mechanism [14]. *Top: Reprinted with permission from C. Eames, J.M. Frost, P.R.F. Barnes, B.C. O'Regan, A. Walsh, S.M. Islam, Ionic transport in hybrid lead iodide perovskite solar cells, Nat. Commun. 6 (2015). Copyright 2015 Nature Publishing Group. Bottom: Reprinted with permission from J. Haruyama, K. Sodeyama, L. Han, Y. Tateyama, First-principles study of ion diffusion in perovskite solar cell sensitizers. J. Am. Chem. Soc. 137 (32) (2015) 10048–10051. Copyright 2015 American Chemical Society.*

MA^+ in $MAPbI_3$. They found that, while these values were low enough to potentially be occurring, interstitial movement was favored for both. For I^-, they calculated a minimum E_a of 0.26 eV for a vacancy pathway and a minimum E_a of 0.19 eV for an interstitial pathway, and for MA^+, a minimum E_a of 0.62 eV for a vacancy pathway and a minimum E_a of 0.38 eV for an interstitial pathway.

Rappe et al. calculated low-activation energies for hydrogen movement by the Bourgoin–Corbett mechanism, where hydrogen is able to capture or release free carriers and migrate along the I–Pb–I bond axis as $H^{+/0/-}$ as it is selectively stabilized by local anionic, neutral, or cationic character [69]. Hydrogen migration is not treated extensively in the literature as hydrogen defects are considered to be minor in HOIPs with a predicted concentration of $\sim 10^{11}\ cm^{-3}$. In comparison, the concentrations of Pb^{2+}, I^-, and MA^+ defects are predicted to be $\sim 10^{17}–10^{20}\ cm^{-3}$.

Nonetheless, it is acknowledged that water infiltration into HOIP films may affect these defect concentrations [11].

Although computation provides insights into possible mechanistic pathways and activation energy barriers, the prediction is to be computationally verified. In the next section, we describe key techniques to measure ion diffusion in solids. We first look at the direct techniques for studying ion diffusion and the limitations of these techniques to probe ion diffusion in HOIPs.

6.6 DIRECT TECHNIQUES FOR STUDYING ION DIFFUSION

Researchers have employed various direct methods to probe the long-range diffusion in solids. One standard technique is the radiotracer method, which uses the suitable/stable radioisotopes to measure the diffusion within a solid. This technique has superiority over other techniques in terms of sensitivity and element selectivity. Combining this tracer method with both mechanical and sputter sectioning techniques like secondary ion mass spectroscopy (SIMS) has been used to obtain a depth profile of solid-state materials [70]. Researchers have used these techniques on HOIPs to extract the diffusion coefficients of the mobile ion/ions directly, thereby revealing their identity [71].

6.7 TRACER EXPERIMENTS

A typical tracer diffusion experiment uses radioactive or stable isotopes to examine the self-diffusion of a target atom (i particle) within a solid through the use of chemically identical but isotropically labeled species i^*. Since i and i^* are distinguishable from one another, the movement of identical i particles in condensed matter can be traced with the help of i^* tracers [72]. The diffusion process of the tracer is entropically driven by its chemical potential, while the chemical potential of other species is constant throughout the system. The relationship between the measured diffusion coefficient of a tracer and the self-diffusion coefficient of the target atom is directly related by the equation

$$D_{i*} = f * D_i$$

where D_i^* is the tracer diffusion coefficient, D_i is the self-diffusion coefficient, and f^* is the tracer correlation factor; a metric of the migration mechanism and geometry of the sublattice [72].

There are four steps to the tracer method: deposition of the tracer, diffusion annealing, serial sectioning, and the analysis of the depth profile [70]. The radiotracer can either be electrodeposited or vapor deposited onto a sample without much concern on homogeneity as long as the sample area is kept constant and the variation in the tracer layer thickness is small in comparison to the diffusion length. Following the tracer deposition, isothermal annealing is performed at a constant temperature for a desired time. Next, the concentration-depth profile of the tracer is studied to determine the diffusion coefficient. This is done by comparing it with the Gaussian solution of the Fick's second law [72]. The best way to conduct this evaluation is by serial sectioning of the sample layer by layer and determining the concentration of the tracer at different depths. This could be pictured in terms of isoconcentrated contours that are parallel to the surface for lattice diffusion but are perpendicular to the diffusion direction. For sectioning, it is salient that the sections should be parallel to the isoconcentrated contours [70]. The specific activity per section of a tracer, $A(x)$, is proportional to the tracer concentration:

$$A(x) = kC(x)$$

where k is a constant that depends on the nature and energy of the nuclear radiation and on the efficiency of the counting device, $A(x)$ is the specific activity per section of a tracer, and $C(x)$ is the tracer concentration. In the diffusion-depth profile, the specific activities per section are plotted against the position of the distance of center of each section from the surface [70]. A frequently used technique for the serial sectioning is SIMS.

Several research groups have employed tracer methods to deduce the diffusion coefficient in conventional metal oxide perovskites. De Souza has extensively used tracer methods to calculate the oxygen diffusion coefficient in $SrTiO_3$. Based on the results, he was able to estimate a fundamental limit for the oxygen-vacancy diffusivity in ABO_3 perovskite [72]. Recently, Maier et al. have used the tracer experiments to deduce the mobile species in HOIPs. They took two radiolabeled $MAPbI_3$ samples, one with ^{13}C and the other rich with ^{15}N, and placed them in contact, allowing them to diffuse for a period of time. The depth profile evaluation with SIMS showed that the MA^+ did not diffuse significantly under dark. They provided an upper limit for the MA^+ tracer diffusion coefficient of 9×10^{-14} cm^2/s at 378K, as well as concluded that the MA^+ ion is likely not mobile under their experimental conditions [71].

6.8 SECONDARY ION MASS SPECTROMETRY

SIMS is a technique used to analyze the surface composition of solid surfaces by sputtering the surfaces with primary ions and collecting ejected secondary ions. During the primary ion collision with the surface atoms, primary ions lose their kinetic energy while the target surface becomes charged. Once the target surface reached a threshold energy, surface atoms are ejected from the sample and analyzed by mass spectrometry that filters out ions with a chosen mass/charge ratio. The sputtering yield of the atomic and molecular species from the target surface is between 0.1 and 10 atoms per primary ion [70]. Time-of-flight SIMS (TOF-SIMS) is an efficient depth-profiling instrument because of its high-mass resolution capability, ability to detect and provide chemical information of isotopes, and detection limits in the range of ppm [70].

Of importance to the study of ion transport in HOIPs is diffusion profile measurements. In these measurements, the sample is slowly etched away from the surface, allowing the elemental composition to be profiled as a function of depth. In typical SIMS, the profile of the sample composition as a function of depth is obtained by eroding the sample surface with the ion beam and recording the subsequent SIMS spectra. A diffusion profile for the sample is then generated by plotting the tracer and host atom ratio against the sputtering time. The diffusion distance is measured from the total crater depth based on the assumption that the surface of the material is uniformly removed with time [73].

SIMS enables the measurement of small diffusion coefficients in solutes with high resolution and sensitivity in the 10-nm range and at the ppm level. However, the depth resolution decreases over time due to surface roughening and atomic mixing. Depth profile SIMS measurements have historically been used to study diffusion of foreign atoms into a sample, but it can also separate different isotopes of the same element. Moreover, recent research employs SIMS for self-diffusion processes in oxide-based perovskites. Steele et al. used SIMS to determine the proton concentration in oxide perovskites directly [74]. Moreover, they studied the $La_{1-x}Sr_xMn_{1-y}Co_yO_{3\pm\delta}$ perovskite oxide system using SIMS and showed that the oxygen tracer diffusivity of these materials increases significantly with strontium site fraction, x, and cobalt site fraction, y [75]. In HOIPs, this technique has also been used; Lin et al. constructed 3D images of HOIPs under controlled humidity to study the degradation process that these materials undergo. They used deuterium oxide instead of water as the moisture source and monitored its penetration into the sample through the H/D exchange of the organic MA^+ ion [76]. Yang et al. carried out TOF-SIMS-based studies on HOIP solar cells and showed that ion migration caused the degradation of these cells at the device level—not only that the mobile I^- ion was responsible for this degradation but also that the Ag atoms/ions from the metal electrodes moved into the perovskite films [77]. Di Carlo et al. employed SIMS to measure the conversion efficiency of various HOIP precursors into the perovskite layer by looking into the characteristic molecular fragments in the HOIPs and the precursors [78].

6.9 NUCLEAR MAGNETIC RESONANCE

Nuclear magnetic resonance (NMR) [79,80] is a technique that has been utilized to quantitatively study dynamic processes in HOIPs. Nuclei containing an angular spin $I \geq 1/2$ experience reorientation in the presence of a net magnetic field. An NMR experiment consists of applying an external magnetic field to a sample containing NMR active nuclei, aligning them either with (α) or against (β) the applied field, and introducing an energy pulse in the form of electromagnetic radiation (radio waves) to saturate the population of nuclei in the β state. When the radiofrequency pulse is complete, the excited nuclei can relax into their original state, resulting in a fluctuation of the magnetic field that is measured.

NMR can give both qualitative and quantitative information about a sample. The chemical shift of a peak in the NMR spectrum provides information on the chemical environment of that nuclei. Additionally, peak splitting is often observed, which is an indication of a single nucleus coupling to the magnetic moments of neighboring active nuclei, resulting in a net increase or decrease of its signal.

Of greater importance with regard to the study of HOIPs is the observance of nuclei relaxation, a characteristic that can be studied by manipulating the timing of applied radiofrequency pulses. In these studies, peak broadening or narrowing is observed and correlates to nuclei relaxation time. Relaxation can occur by two mechanisms: spin–lattice relaxation (T_1) and spin–spin relaxation (T_2). T_1 relaxation time corresponds to longitudinal relaxation along the z direction of the magnetic field and is indicative of nuclei returning to their normal Gaussian distribution of states after a pulse is applied. T_2 relaxation occurs due to a loss of phase coherence among the nuclei and governs the width of each peak. T_2

relaxation is heavily impacted by molecular reorientation; as reorientation rates slow down, T_2 relaxation becomes shorter. By studying T_2 relaxation rates and subsequent line broadening, information is available about dynamic processes occurring at the molecular level.

NMR is traditionally conducted in a liquid state, where molecules tend to have low energetic barriers to reorientation and are therefore inherently isotropic in their interactions. However, when considering solid-state samples such as HOIPS, NMR becomes more difficult to use. First, solid-state samples often introduce anisotropy into their nuclear spin interactions that can result in significant peak overlap and broad featureless peak shapes. Additionally, many active nuclei of interest in HOIP samples have a low natural abundance (i.e., ^{13}C and ^{15}N), which results in a low signal to noise ratio and increased experiment times. One way to increase the signal to noise ratio in solid-state NMR is through magic angle spinning NMR (MAS NMR). In this technique, the sample is mechanically rotated at a desired frequency at some angle, β_R, to the magnetic field. The desired effect is to simulate the "tumbling" of molecules in solution and restore completely isotropic interactions in the sample. In fact, when $\beta_R = 54.7^\circ$ (deemed the "magic angle") with respect to the net magnetic field, the anisotropic portion of the chemical shift Hamiltonian averages to zero, retaining only the isotropic portion of the chemical shift. When thinking about HOIPs, if one considers the organic cation as a component tumbling within a molecular frame made up of the PbI_6 octahedra, one can experimentally probe the change in NMR linewidths corresponding to a dynamic cation reorientation process occurring. In fact, Emsley et al. were able to utilize this concept in the ^{14}N MAS NMR study of mixed cation $(MA)_x(FA)_{1-x}PbI_3$ HOIPs to determine cation reorientation correlation times for both FA^+ and MA^+ [81]. Other groups have also used temperature-dependent ^1H and ^{13}C MAS NMR to study dynamic reorientation of ions in HOIPs over the course of phase changes [82,83].

Outside of short range dynamics such as cation reorientation, Maier et al. were able to utilize NMR as a tool to study long-range dynamics (or lack thereof) of the MA^+ and Pb^{2+} ions in $MAPbI_3$. With temperature-dependent ^{14}N NMR measurements conducted in tetragonal phase $MAPbI_3$, they observed quadrupolar splitting of the MA^+ cation peak and rationalized that this could only occur in the absence of isotropic motion of MA^+. Above the phase transition into the cubic phase, only a single isotropic motion is expected, so they performed temperature-dependent ^{1}H NMR and observed a negligible decrease in signal linewidth from 193 to 503K, which would be expected to decrease 1000 fold if there was a fast and isotropic solid-state diffusion of MA^{+} occurring. Lastly, they studied the dynamics of the Pb^{2+} ion with ^{207}Pb NMR and observed a chemical shift indicative of Pb^{2+} coordination to I^{-}, downfield shifted from that of PbI_2. Furthermore, there was no appreciable anisotropy, nor variations of linewidth with temperature observed in the Pb^{2+} peak, which is indicative of a strong and symmetrical coordination of Pb^{2+} to I^{-} as well as negligible dynamic processes occurring on either a short or long-range scale. From this, they rationalize that, under NMR relevant conditions, Pb^{2+} is not mobile and MA^{+} is mobile only in terms of a reorientation of the cation. Based on these observations, and the observed magnitude of I^{-} short range dynamics by ^{127}I NMR, along with other complementary techniques, they come to the conclusion that I^{-} is likely the only mobile ion in HOIPs [71].

6.10 COMPLICATING FACTORS IN MEASUREMENT OF ION DIFFUSION IN HOIPS

Although the above techniques have been widely used for studying ion transport in solids, they are difficult to deploy for *in-operando* studies of HOIP-based devices. The fact remains that not all are well suited for the task. For example, it has been shown that light may impact ion transport in HOIPs; many researchers have observed an enhancement of the ionic transport as a factor of illumination [12,84–86]. However, performing NMR and SIMS studies *in-operando* are nontrivial. The discrepancy in the ionic conductivity observed under dark and light has led some groups to question whether the nature of the mobile ion is the same under both conditions. Therefore, comprehensive studies need to be conducted in both light and dark in order to gain a full perspective of ionic mobility. Unfortunately, introducing light to an experiment presents a logistical barrier for many analytical techniques and, as a result, many studies have been conducted only under dark. While measurements taken in the dark are undoubtedly useful to gain a basis of knowledge on the matter, it is important that analogous studies are conducted under illumination.

Stability of HOIPs also needs to be considered when thinking about techniques to probe ion transport. Device stability has long been an issue for HOIPs, and device degradation under certain conditions has been well documented in the literature. For instance, $MAPbI_3$ photovoltaic devices stored under dark in an N_2 environment have been reported to lose 20% of their original PCE over 24 hours. Conversely, devices stored in ambient air under dark lose nearly 80% of their efficiency over 24 hours [87]. When illuminated, devices degrade even more rapidly, losing 40% of their efficiency in an N_2 environment over 20 hours [84]. Strikingly, devices under illumination in dry air can degrade exceptionally fast, showing PCE depletion within minutes of exposure and rapidly declining to around 5% of their original PCE within 6 hours [88]. Though overall device stability relies on a multitude of factors including fabrication technique, device architecture, grain size, chemical composition, and temperature, HOIP devices are notoriously unstable to many device-relevant conditions. The rapid degradation of HOIPs impacts the ability of researchers to conduct experiments that need to be taken over more than a few hours, especially in ambient air.

Time sensitivity is not the only stability-related difficulty when taking measurements; many techniques themselves have been shown to degrade devices more rapidly. For example, XRD, which has both been used to indirectly probe ion transport and is commonly used to characterize HOIPs, has been shown to be destructive to the material [89]. Beam damage can begin to occur just 1 hour after exposure, and layers are visibly degraded within 24 hours of exposure. Beam damage is not limited to XRD; EDX, XPS, Auger electron spectroscopy, X-ray fluorescence, electron energy loss spectroscopy, and Rutherford backscattering are all capable of inducing beam damage to HOIPs [89]. The instability of HOIPs to beam damage limits the reliable data that can be obtained by using these techniques and must be considered when designing experiments. Degradation of the material during measurement can leave artifacts in the data and ultimately lead to inaccurate analysis. When selecting methods to study ion transport, therefore, it is imperative that a nondestructive technique that allows for illumination is used, preferably in a dry or anaerobic environment.

6.11 IMPEDANCE SPECTROSCOPY: A PROMISING INDIRECT METHOD TO PROBE ION TRANSPORT

Electrochemical impedance spectroscopy (EIS) has recently emerged as a tool in HOIP literature to examine processes that occur at different timescales. The concept of "impedance" is analogous to resistance but allows for description of complex circuits which have nonlinear current–voltage relationships, such as those which exhibit capacitance, inductance, or mass diffusion. EIS takes advantage of this concept by relating theoretical circuit elements to actual electrochemical processes occurring in a material and allowing fitting of current, voltage, and frequency data to equivalent circuit models. EIS data is generated by applying an AC potential to an electronic device, measuring the AC current response, and recording phase shift and amplitude changes over a range of applied frequencies. Examining the current response over a range of frequencies allows separation of processes which occur on different timescales, making it ideal for separating electronic and ionic processes in mixed conductors. This is a key advantage of EIS making it an apt tool to study ion diffusion in HOIPs and resolve the chemical identity of the ion.

6.12 EIS EQUIVALENT CIRCUIT MODELS

EIS results are plotted by convention as a Nyquist plot, where the AC current response signal is separated into the real and imaginary components of impedance and plotted in complex space. The data closest to the origin is associated with the higher frequency spectrum, whereas the data farther from the origin is associated with the lower frequency spectrum. The analysis of data from impedance spectroscopy is not straightforward. As a first step, Nyquist plots of EIS data, with the real component of the impedance (Z') on the horizontal axis, and the imaginary component (Z'') on the vertical axis, are fit with equivalent circuit models to quantitatively model device properties, as shown in Fig. 6.6. The initial guesses for these models are based on knowledge of underlying physical processes in the system. The shape of the Nyquist plots also provides clues to the physical processes that may be involved. For example, if the Nyquist plot is a line parallel to the y-axis, then the equivalent circuit is composed of a capacitor in series with a resistor. The value of the resistance is the intercept on the x-axis. If the Nyquist plot is a semicircle and has an intercept at the origin, the equivalent circuit is composed of a capacitor in parallel with the resistor. If there is a resistor in series to this circuit, the semicircle is displaced on the x-axis and series resistance is the value of the displacement. Thus, visual examination of features of the Nyquist plot combined with knowledge or supposition of inner workings of the device allows for creation of an equivalent circuit model, where each circuit element represents a specific process. For example, a simple photovoltaic device which is only an electronic conductor can be modeled with just two elements, a

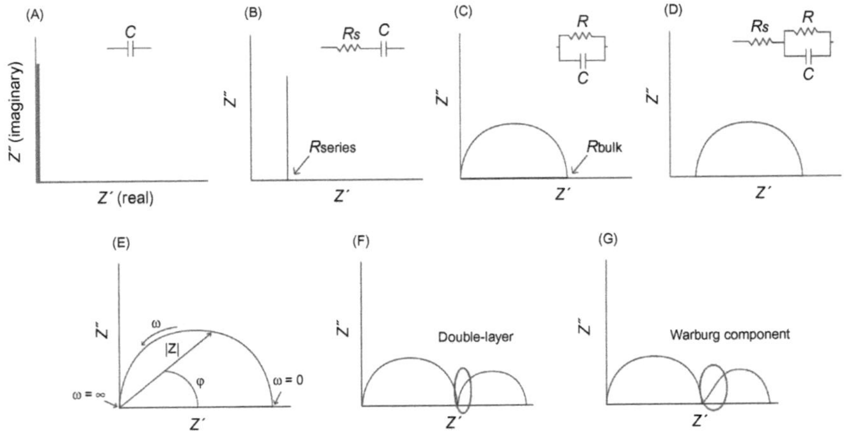

Figure 6.6 Top: Illustration of the form that basic circuit elements take on Nyquist plots [90]. Bottom Left: Basic schematic for reading Nyquist plots [90]. Bottom Center and Right: Nyquist plots of a pure double-layer capacitor vs a circuit which includes a linear Warburg diffusion element [84,91].

resistor and a capacitor, where the resistor represents the resistance experienced by electronic charge carriers when moving through the active material, and the capacitor represents the capacitance generated by differential separation of electrons and holes. This simple RC circuit will appear as a semicircle in the high-frequency region of the Nyquist plot (closer to the origin). Values for different circuit elements obtained upon fitting can give both qualitative ideas of how separate processes in a device change with varying conditions as well as quantitative kinetic data such as time constants or activation energies for each separate process.

As mentioned before, HOIPs are semiconductors and ionic conductors. Thus, impedance plots contain information about both electronic and ionic processes. Often two distinct features are seen in the Nyquist plots of HOIP-based devices: a high-frequency semicircle and a low-frequency semicircle or linear feature. At high frequencies, electronic process typically dominates, and devices can be modeled with a simple electronic RC circuit [92]. At relatively low frequencies, ionic transport dominates and the equivalent circuit model mirrors a Randles circuit model. Ionic transport does indeed dominate at low frequency and electronic dominates at higher frequency, but there are comparable ionic/electronic influences in the middle frequency ranges. However, these features change based on the device architecture and components. Hence, there is considerable debate about the physical process involved within a HOIP device.

Fig. 6.7 shows equivalent circuit models tested for HOIP-based devices in the literature [93]. Model A in this figure is the simplest circuit model, while Models B–D are slightly more complex. At high frequencies, Model B is a simple electronic *RC* circuit because the double-layer capacitance element is effectively shortened. At low frequencies, however, the circuit model mirrors a Randles circuit model because ionic transport

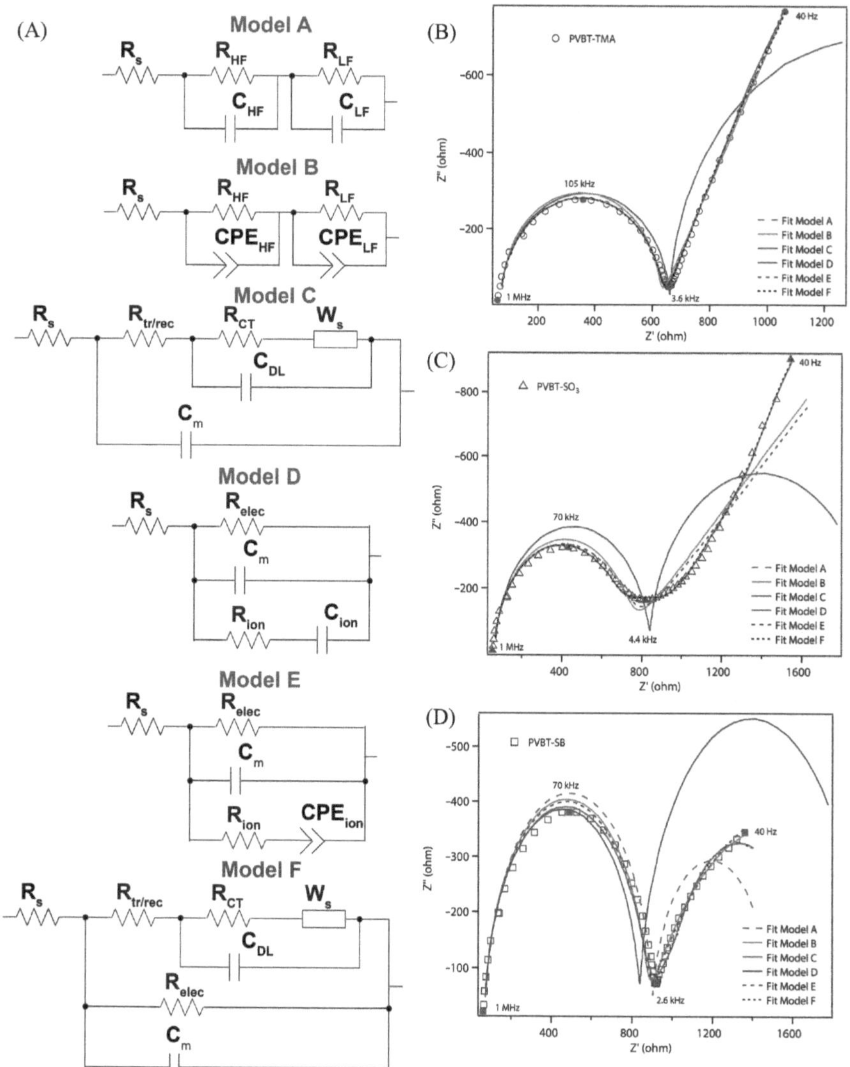

Figure 6.7 Examples of equivalent circuit models tested for HOIP-based devices in literature [93].

dominated at low frequencies. Model C captures the coupled electronic/ionic transport, but the parallel resistor, R_{elec}, is not coupled with ionic transport. Model D contains parallel electronic and ionic transport resistances, unlike the previous models. Model E replaces C_{ion} with a constant phase element (CPE), CPE_{ion}, and shows an improved fit from Model C. In Models D and E, R_{ion} captures the resistance associated with lower frequency. Model E does couple electronic/ionic transport, unlike the resistor in Model C. Overall, Model F is used to create a robust equivalent circuit to fit a greater number of HOIP-containing samples. In Model F—developed by Venkataraman et al. and used successfully by Fan et al.—electronic transport separated from ionic transport is modeled as a resistor (R_{elec}), and the ion diffusion is modeled with a special CPE known as a Warburg (W_s) [84,94]. The Warburg element captures mass transfer due to ion diffusion and appears as a diagonal line with the slope of 45° in the low-frequency part of the Nyquist plots [95,96]. Mathematically, the impedance of a CPE is defined by the equation

$$Z(j\omega) = \frac{1}{Q(j\omega)^{\alpha}}$$

where Q is the CPE coefficient, $j = \sqrt{-1}$, ω is the angular frequency, and α is an exponent equal to 1 for an ideal capacitor. An ideal Warburg element has impedance

$$Z_W(j\omega) = \frac{A_W}{\omega^{\alpha}} - \frac{jA_W}{\omega^{\alpha}}$$

where A_W is Warburg resistance, and α is equal to 0.5 [96]. The Warburg element gives insight to diffusion-based processes at high and low-frequency ranges and is therefore an ideal element to consider when analyzing ion transport in the EIS spectra of HOIPs.

In order to accurately model EIS data for a device, each circuit element in the proposed model must correspond to a physical process. Fig. 6.8 shows a schematic of the corresponding physical processes in an HOIP PV device for the circuit model developed by Venkataraman et al., including the accumulation region and the region of ion diffusion that extends into the bulk HOIP film. The equivalent circuit models that represent these EIS spectra combine both slow and fast charge transport processes [84,92,97–99]. In this model, R_{elec} is not coupled with ionic transport. In the model developed by Venkataraman et al., high-frequency impedance data is modeled with bulk electronic charge transport/recombination

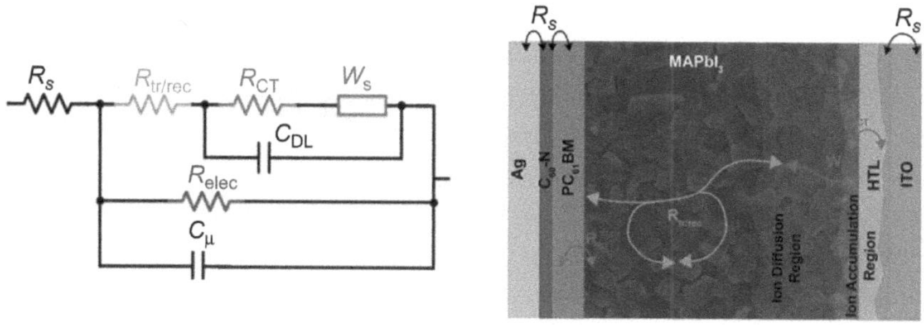

Figure 6.8 Schematic representation of the equivalent circuit model used in studies by Venkataraman et al. [84] and Fan et al. [100]. Additional resistance (R_{elec}) is added in parallel for electronic transport which is not coupled with ionic transport.

resistance ($R_{tr/rec}$), and charge stored in the bulk of the HOIP active layer is modeled with a chemical capacitance element (C_μ) [84,95,100]. The low-frequency impedance data is modeled with a charge-transfer resistance (R_{CT}) and a double-layer capacitance (C_{DL}) to quantify changes in resistance and capacitance at the active layer/electrode barriers caused by ionic/electronic charge accumulation, as well as the Warburg element (W_s) to quantify ion diffusion through the HOIP active layer.

Low-frequency features in the EIS spectra of HOIP photovoltaics—corresponding to longer timescale processes—have been interpreted in many different ways in regard to ion transport. Initial studies of HOIPs which included electrochemical impedance analysis, such as by Gr tzel et al. and Cameron et al., observed irregularities in the low-frequency regime of their EIS spectra but did not further analyze these features [101,102]. Gr tzel et al. later attempted to adapt an EIS circuit model developed for DSSCs to a solid-state HOIP photovoltaic device [103]. They noticed a feature in the low-frequency regime that could appear alternately as a negative capacitance (inductance) or as an additional capacitive semicircle and noted that it might be related to the reorganization of mobile ions at the interfaces. Interestingly, in moving from the DSSC circuit model to the solid-state HOIP model, they removed a circuit element known as the Warburg diffusion element—which is typically used for modeling of ion diffusion in EIS experiments. They instead used a simple RC circuit element that models ion accumulation at the ETL and HTL interfaces but does not account for mass transport through the film. Cahen et al. also observed a negative capacitance feature in the low-frequency regime of their EIS spectra but rejected the hypothesis that it was related to ion/vacancy accumulation. They stated that this would only be the case if the ions/vacancies were able to affect the barrier for charge injection into the ETL/HTL and reasoned that this is not true for HOIPs [104]. However, prior evidence has demonstrated that ion and vacancy accumulation is likely able to cause band bending at the ETL/HTL interfaces which would affect the barrier for charge injection; additionally, they noted that this low-frequency feature disappears in samples measured in ambient air, which were speculated to be heavily n-doped. Cheng et al. did a seminal study by performing EIS analysis on HOIP cells with varying architecture and doping levels. In their paper, they observe the signature of a Warburg diffusion element in their EIS spectra, mentioning, "the revealing 45° transport feature is observed in some cell architectures but not others." [50]. None of their analysis, however, includes Warburg diffusion or attributes this low-frequency response to ion transport. Instead, they ascribe the low-frequency features to dielectric relaxation and view the relative doping only in terms of electronic charge carriers, despite prior evidence that ionic species can strongly influence doping levels.

While EIS has clearly uncovered a low-frequency responsive feature in HOIPs which corresponds to a slow timescale process, analysis of this data and the circuit model which best describes it remains controversial. It is clear from the literature that the Warburg diffusion element recurs as a possibility without ever being seriously integrated into EIS analysis, despite mounting certainty that HOIPs are indeed capable of ion transport through the film. Venkataraman et al. incorporated the Warburg diffusion element in their analysis of the EIS spectra of $MAPbI_3$, $MA_xFA_{1-x}PbI_3$, and $FAPbI_3$ [84]. They were able to use EIS to qualitatively identify that it corresponded to mass transport of the organic cation, providing evidence for its mobility, and quantify values for this ion such as diffusion coefficient and activation energy barriers. Fan et al. also observed the Warburg element in their EIS data and found evidence supporting the hypothesis that it arises from movement of the organic cation.

6.13 MEASURED DIFFUSION CONSTANTS IN HOIPS

Researchers have also applied a wide variety of techniques aimed at quantifying ion transport. Though concrete numerical measures have the potential to make a strong case for or against ion transport, the evidence so far gathered shows few signs of consensus.

Activation energy (E_a) measurements have been pursued widely as a way of quantifying ion movement, but the measured values and the assigned ion identities vary. In addition to computational modeling of ion transport, Islam et al. employed chronophotoamperometry to determine the lowest activation energy for ion mobility in $MAPbI_3$ [13], and found this E_a to be 0.60–0.68 eV. They concluded that I^- was the mobile ion since their experimental E_a was in close agreement with their calculated E_a (I^-) of 0.58 eV, and the mobile ion was likely responsible for $J-V$ hysteresis. Huang et al. observed two mobile species, termed "fast ions" and "slow ions," from their PTIR experiments, and calculated an E_a of 0.36 eV for the fast ions [59]. They hypothesized that both these mobile species could be MA^+, perhaps moving through different pathways. In another study, Huang et al. investigated $MAPbI_3$ using temperature-dependent conductivity and found that E_a values obtained under dark (0.08 eV) differ from values obtained under illumination (0.27 eV) in a variety of samples. They concluded that I^- was the mobile ion; additionally, they noted that ion migration was not observed in single crystals over the course of several days [85]. Kingon et al. observed two separate thermally activated E_a values, 0.10 and 0.41 eV, by temperature-dependent pulsed voltage–current measurements designed to separate electronic and ionic conductivity, which they assigned to I^- and MA^+ migration, respectively [105]. Venkataraman et al. used EIS to determine activation energies using the Warburg element feature present in the spectra of $MAPbI_3$, $MA_xFA_{1-x}PbI_3$, and $FAPbI_3$ photovoltaic devices [84]. They found E_a values of 0.58 eV below 320K and 0.23 eV above 320K for $MAPbI_3$, 0.63 eV for mixed $MA_xFA_{1-x}PbI_3$, and 0.22 eV below 330K and 0.79 eV above 330K for $FAPbI_3$. The activation energies measured from the Warburg element were attributed to the organic cation transport as there was a change in E_a upon changing organic cation from MA^+ to FA^+. Also, the measured values were similar to the computed vales for MA^+ transport, and the observed discontinuity in E_a for $MAPbI_3$ matched with the phase transition temperature from tetragonal to cubic. Fan et al. also used EIS to determine $MAPbI_3$ E_a, verifying the model used and Warburg element observed by Venkataraman et al. [94]. They found a discontinuity in the E_a at ~320K as well, but with larger values of 0.7 eV at low temperatures and 0.5 eV at high temperatures. While these studies agree that ion movement occurs in HOIPs, they disagree on the nature of the mobile ion and the mechanism of ion movement.

6.14 UNANSWERED QUESTIONS ON THE NATURE OF ION TRANSPORT IN HOIPS

In the past few years, enough evidence has been compiled such that the majority of the HOIP research community supports the idea that these materials are capable of ion transport, and most seem to agree that the halide anion is the only or primary mobile ion. However, this only scratches the surface in understanding ion transport as a phenomenon in HOIP materials, as conflicting evidence, roadblocks in experimental design, and widespread assumptions still leave many questions unanswered.

Many studies assume that only one ion is mobile—typically the halide anion. These studies either explicitly state that there is one mobile ion or consider the effects from other species to be minimal. While there is ample evidence that the halide ion is mobile and that it is likely to be the fastest moving species, there is far too much evidence pointing to the possibility of other mobile ions to discount their potential existence and impact. Both computational and experimental studies detailed above have shown evidence of MA^+ movement with activation energies similar to those of halide movement. MA^+ defects can also be just as prevalent as halide defects. Many researchers have concluded that, though the halide is indeed mobile, it is moving too fast to account for much of the observed phenomena that point to ion transport in the first place. In addition to the organic counterion, there is also the possibility of Pb^{2+} or $H^{+/0/-}$ movement [15,69,82,106]; currently, there is little evidence either supporting or discounting these as mobile species, but this possibility is far from being conclusively decided and thus should not be ruled out without further investigation. Identifying all the mobile species will be critical to designing materials with increased stability and fully understanding the properties of HOIPs.

Another common assumption that may be a stunting progress in this area is the mechanism of ion movement; many studies assume movement through vacancies, but there is no unequivocal evidence for this mechanism. Some computational evidence suggests that interstitial movement could have very similar activation energies to vacancy movement for some ions [15,47]. Additionally, the role of grain boundaries in ion transport is still poorly understood, as this type of transport could obviously have very different activation energies and occur in parallel with ion transport in the bulk of the material. With the possibility of multiple ion movement mechanisms and multiple mobile ions, it is evident that this is a multifaceted problem and will require careful experimental design to isolate the multitude of possible phenomena. Clearly delineating the path(s) these ions are taking will be critical to determining suitable fabrication methods—balancing the presence of particular defects and grain boundaries with ease and lower cost of fabrication—for devices using these materials.

Fabrication of active materials and devices is perhaps the largest contributing factor to the widespread, seemingly contradictory evidence throughout the HOIP field. There are many HOIP active layer fabrication methods, and each creates different types and densities of defects, degrees of crystallinity, preferred crystal growth orientations, and more. Moreover, due to the high sensitivity of HOIP materials to stoichiometric variations and environmental conditions such as heat, humidity, and light, even the same fabrication method can produce inconsistent results. This is hugely compounded when one considers that these active layers are usually not constructed on their own, but rather as part of devices, introducing new effects from the myriad hole and electron transporting and electrode materials. The energy levels of these materials, ion absorbing or blocking character, roughness, geometry, crystallinity, and reactivity can dramatically shape measurements of electrical and structural characteristics of HOIP devices, making comparison of results between different publications and research groups extremely challenging. However, with careful experimental design, a systematic exploration of these variances could be a powerful tool in understanding ion transport in HOIPs.

One of the most intriguing yet difficult roadblocks in making determinations on whether or not ion transport is occurring and/or responsible for certain phenomena is the necessary entwinement of electronic and ionic transport. Capacitance due to deep trap states is often cited as an alternative explanation for observed phenomena that others attribute solely to ion transport [47,103]. However, there is a distinct possibility, which has been explored by some, that these two processes are connected: that the accumulation of mobile ions at interfaces creates these deep trap states [107,108]. There are a multitude of ways in which ionic and electronic charge carrier transport could be coupled and, while this makes theoretical sense, simultaneous measurement of ionic and electronic conductivities in mixed ionic–electronic conductors is a challenge.

The potential role of light in ion transport also presents an interesting research question. Ion transport has been observed both in the dark and under illumination, as well as under specific sources with different UV, visible, or IR components [84]. However, the activation energy barriers and behaviors are different under these conditions, and it is unclear whether the same ion is mobile under light and dark. Thus, the effect of light on ion transport in HOIPs needs to be ascertained.

The ultimate goal of studies in ion transport is to improve stability, lifetime, and reproducibility of perovskite devices. A common expectation is that halting ion transport will improve device characteristics [109]. However, it is unclear whether what effect halting ion transport would have on the various interesting properties of HOIPs [110]. Additionally, an intriguing question that has not been addressed extensively is whether the ion-transporting properties of these materials could be exploited for applications such as memristor fabrication, neuromorphic computing, or bioelectronics devices.

REFERENCES

[1] M. Grätzel, The light and shade of perovskite solar cells, Nat. Mater. 13 (9) (2014) 838–842.
[2] Y. He, G. Galli, Perovskites for solar thermoelectric applications: a first principle study of $CH_3NH_3AI_3$(A = Pb and Sn), Chem. Mater. 26 (18) (2014) 53945400.
[3] Y.-J. Kim, T.-V. Dang, H.-J. Choi, B.-J. Park, J.-H. Eom, H.-A. Song, et al., Piezoelectric properties of $CH_3NH_3PbI_3$ perovskite thin films and their applications in piezoelectric generators, J. Mater. Chem. A 4 (3) (2015) 756–763.
[4] X.Y. Chin, D. Cortecchia, J. Yin, A. Bruno, C. Soci, Lead iodide perovskite light-emitting field-effect transistor, Nat. Commun. 6 (2015) 7383-1–7383-9.
[5] G. Xing, N. Mathews, S. Lim, N. Yantara, X. Liu, D. Sabba, et al., Low-temperature solution-processed wavelength-tunable perovskites for lasing, Nat. Mater. 13 (5) (2014) 476–480.
[6] Z.-K. Tan, R. Moghaddam, M. Lai, P. Docampo, R. Higler, F. Deschler, et al., Bright light-emitting diodes based on organometal halide perovskite, Nat. Nanotechnol. 9 (9) (2014) 687–692.
[7] E.J. Yoo, M. Lyu, J.H. Yun, C.J. Kang, Y.J. Choi, Resistive switching behavior in organic–inorganic hybrid $CH_3NH_3PbI_{3-x}Cl_x$ perovskite for resistive random access memory devices, Adv. Mater. 27 (2015) 6170–6175.
[8] G. Lin, Y. Lin, R. Cui, H. Huang, X. Guo, C. Li, et al., An organic–inorganic hybrid perovskite logic gate for better computing, J. Mater. Chem. C 3 (41) (2015) 10793–10798.
[9] X. Hu, X. Zhang, L. Liang, J. Bao, S. Li, W. Yang, et al., High-performance flexible broadband photodetector based on organolead halide perovskite, Adv. Funct. Mater. 24 (46) (2014) 7373–7380.
[10] M.A. Green, A. Ho-Baillie, H.J. Snaith, The emergence of perovskite solar cells, Nat. Photonics 8 (7) (2014) 506–514.
[11] J.M. Frost, A. Walsh, What is moving in hybrid halide perovskite solar cells? Acc. Chem. Res. 3 (2016) 528–535.
[12] Y. Yuan, J. Huang, Ion migration in organometal trihalide perovskite and its impact on photovoltaic efficiency and stability, Acc. Chem. Res. 49 (2) (2016) 286–293.

[13] C. Eames, J.M. Frost, P.R.F. Barnes, B.C. O'Regan, A. Walsh, S.M. Islam, Ionic transport in hybrid lead iodide perovskite solar cells, Nat. Commun. 6 (2015) 7497–7505.
[14] J. Haruyama, K. Sodeyama, L. Han, Y. Tateyama, First-principles study of ion diffusion in perovskite solar cell sensitizers, J. Am. Chem. Soc. 137 (32) (2015) 10048–10051.
[15] J.M. Azpiroz, E. Mosconi, J. Bisquert, F. Angelis, Defect migration in methylammonium lead iodide and its role in perovskite solar cell operation, Energy Environ. Sci. 8 (7) (2015) 2118–2127.
[16] J.S. Manser, M.I. Saidaminov, J.A. Christians, O.M. Bakr, P.V. Kamat, Making and breaking of lead halide perovskites, Acc. Chem. Res. 49 (2) (2016) 330–338.
[17] D.A. Egger, A.M. Rappe, L. Kronik, Hybrid organic–inorganic perovskites on the move, Acc. Chem. Res. 3 (2016) 573–581.
[18] M. Kepenekian, R. Robles, C. Katan, D. Sapori, L. Pedesseau, J. Even, Rashba and dresselhaus effects in hybrid organic–inorganic perovskites: from basics to devices, ACS Nano 9 (12) (2015) 11557–11567.
[19] X.Y. Zhu, V. Podzorov, Charge carriers in hybrid organic–inorganic lead halide perovskites might be protected as large polarons, J. Phys. Lett. 6 (23) (2015) 4758–4761.
[20] S.P. Pang, H. Hu, J.L. Zhang, S.L. Lv, Y.M. Yu, F. Wei, et al., $NH_2CH = NH_2PbI_3$: an alternative organolead iodide perovskite sensitizer for mesoscopic solar cells, Chem. Mater. 26 (3) (2014) 1485–1491.
[21] S.J. Yoon, K.G. Stamplecoskie, P.V. Kamat, How lead halide complex chemistry dictates the composition of mixed halide perovskites, J. Phys. Chem. Lett. 7 (7) (2016) 1368–1373.
[22] I.L. Braly, H.W. Hillhouse, Optoelectronic quality and stability of hybrid perovskites from MAPbI(3) to MAPbI(2)Br using composition spread libraries, J. Phys. Chem. C 120 (2) (2016) 893–902.
[23] F. Hao, C.C. Stoumpos, R.P.H. Chang, M.G. Kanatzidis, Anomalous band gap behavior in mixed Sn and Pb perovskites enables broadening of absorption spectrum in solar cells, J. Am. Chem. Soc. 136 (22) (2014) 8094–8099.
[24] M.T. Weller, O.J. Weber, P.F. Henry, A.M. Di Pumpo, T.C. Hansen, Complete structure and cation orientation in the perovskite photovoltaic methylammonium lead iodide between 100 and 352 K, Chem. Commun. 51 (20) (2015) 4180–4183.
[25] C.C. Stoumpos, M.G. Kanatzidis, The renaissance of halide perovskites and their evolution as emerging semiconductors, Acc. Chem. Res. 48 (10) (2015) 2791–2802.
[26] H.L. Wells, Über die Cäsium- und Kalium-Bleihalogenide, Zeitschrift für anorganische Chemie 3 (1) (1893) 195–210.
[27] C.K. Moller, Crystal structure and photoconductivity of caesium plumbohalides, Nature 182 (4647) (1958) 1436.
[28] D. Weber, $CH_3NH_3PbX_3$, ein Pb(II)-System mit kubischer Perowskitstruktur/ $CH_3NH_3PbX_3$, a Pb(II)-System with Cubic Perovskite Structure, Z. Naturforsch. B 33 (1978) 1443–1445.
[29] D.B. Mitzi, Synthesis, structure, and properties of organic–inorganic perovskites and related materials, Progress in Inorganic Chemistry, John Wiley & Sons, Inc., 2007, pp. 1–121.
[30] A. Kojima, K. Teshima, Y. Shirai, T. Miyasaka, Organometal halide perovskites as visible-light sensitizers for photovoltaic cells, J. Am. Chem. Soc. 131 (17) (2009) 6050–6051.
[31] M.M. Lee, J. Teuscher, T. Miyasaka, T.N. Murakami, H.J. Snaith, Efficient hybrid solar cells based on meso-superstructured organometal halide perovskites, Science (New York, N.Y.) 338 (6107) (2012) 643–647.

[32] H.-S. Kim, C.-R. Lee, J.-H. Im, K.-B. Lee, T. Moehl, A. Marchioro, et al., Lead iodide perovskite sensitized all-solid-state submicron thin film mesoscopic solar cell with efficiency exceeding 9%, Sci. Rep. 2 (2012) 591.
[33] H.J. Snaith, Perovskites: the emergence of a new era for low-cost, high-efficiency solar cells, J. Phys. Lett. 4 (21) (2013) 3623–3630.
[34] Laboratory, N. R. E., Best Research-Cell Efficiencies. **2017**.
[35] A. Polman, M. Knight, E.C. Garnett, B. Ehrler, W.C. Sinke, Photovoltaic materials: present efficiencies and future challenges, Science 352 (6283) (2016) aad4424-1 – aad4424-10.
[36] W.-J. Yin, T. Shi, Y. Yan, Unusual defect physics in $CH_3NH_3PbI_3$ perovskite solar cell absorber, Appl. Phys. Lett. 104 (6) (2014) 063903-1–063903-4.
[37] W. Shockley, H.J. Queisser, Detailed balance limit of efficiency of p–n junction solar cells, J. Appl. Phys. 32 (3) (1961) 510.
[38] K.A. Bush, A.F. Palmstrom, Z.J. Yu, M. Boccard, R. Cheacharoen, J.P. Mailoa, et al., 23.6%-efficient monolithic perovskite/silicon tandem solar cells with improved stability, Nat. Energy 2 (4) (2017) 17009.
[39] T. Duong, Y. Wu, H. Shen, J. Peng, X. Fu, D. Jacobs, et al., Rubidium multication perovskite with optimized bandgap for perovskite-silicon tandem with over 26% efficiency, Adv. Energy Mater. 7 (2017) 1700228-1–1700228-11.
[40] H. Choi, J. Jeong, H.-B. Kim, S. Kim, B. Walker, G.-H. Kim, et al., Cesium-doped methylammonium lead iodide perovskite light absorber for hybrid solar cells, Nano Energy 7 (2014) 80–85.
[41] A. Amat, E. Mosconi, E. Ronca, C. Quarti, P. Umari, K. Nazeeruddin, et al., Cation-induced band-gap tuning in organohalide perovskites: interplay of spin–orbit coupling and octahedra tilting, Nano Lett. 14 (6) (2014) 3608–3616.
[42] M.R. Filip, G.E. Eperon, H.J. Snaith, F. Giustino, Steric engineering of metal-halide perovskites with tunable optical band gaps, Nat. Commun. 5 (2014) 5757.
[43] J.H. Noh, S.H. Im, J.H. Heo, T.N. Mandal, S.I. Seok, Chemical management for colorful, efficient, and stable inorganic–organic hybrid nanostructured solar cells, Nano Lett. 13 (4) (2013) 1764–1769.
[44] R.S. Sanchez, V. Gonzalez-Pedro, J.-W. Lee, N.-G. Park, Y. Kang, I. Mora-Sero, et al., Slow dynamic processes in lead halide perovskite solar cells. Characteristic times and hysteresis, J. Phys. Lett. 5 (13) (2014) 2357–2363.
[45] G. Richardson, S. O'Kane, R. Niemann, T. Peltola, J.M. Foster, P. Cameron, et al., Can slow-moving ions explain hysteresis in the current-voltage curves of perovskite solar cells?, Energy Environ. Sci. 9 (2016) 1476–1485.
[46] B. Chen, M. Yang, X. Zheng, C. Wu, W. Li, Y. Yan, et al., Impact of capacitive effect and ion migration on the hysteretic behavior of perovskite solar cells, J. Phys. Lett. 6 (23) (2015) 4693–4700.
[47] H.J. Snaith, A. Abate, J.M. Ball, G.E. Eperon, T. Leijtens, N.K. Noel, et al., Anomalous hysteresis in perovskite solar cells, J. Phys. Lett. 5 (9) (2014) 1511–1515.
[48] Z. Xiao, Y. Yuan, Y. Shao, Q. Wang, Q. Dong, C. Bi, et al., Giant switchable photovoltaic effect in organometal trihalide perovskite devices, Nat. Mater. 14 (2) (2015) 193–198.
[49] D.P. Almond, C.R. Bowen, An explanation of the photoinduced giant dielectric constant of lead halide perovskite solar cells, J. Phys. Lett. 6 (9) (2015) 1736–1740.
[50] A.R. Pascoe, N.W. Duffy, A.D. Scully, F. Huang, Y.-B. Cheng, Insights into planar $CH_3NH_3PbI_3$ perovskite solar cells using impedance spectroscopy, J. Phys. Chem. C 119 (9) (2015) 4444–4453.
[51] E.J. Juarez-Perez, R.S. Sanchez, L. Badia, G. Garcia-Belmonte, Y. Kang, I. Mora-Sero, et al., Photoinduced giant dielectric constant in lead halide perovskite solar cells, J. Phys. Lett. 5 (13) (2014) 2390–2394.

[52] H.-S. Kim, S. Kim, B. Kim, K.-S. Shin, M. Gupta, H. Jung, et al., Ferroelectric polarization in $CH_3NH_3PbI_3$ perovskite, J. Phys. Lett. 6 (9) (2015) 1729–1735.
[53] T. Berhe, W.-N. Su, C.-H. Chen, C.-J. Pan, J. Cheng, H.-M. Chen, et al., Organometal halide perovskite solar cells: degradation and stability, Energy Environ. Sci. 9 (2015) 323–356.
[54] A. Baumann, S. Väth, P. Rieder, M.C. Heiber, K. Tvingstedt, V. Dyakonov, Identification of trap states in perovskite solar cells, J. Phys. Lett. 6 (12) (2015) 2350–2354.
[55] W. Nie, J.C. Blancon, A.J. Neukirch, K. Appavoo, H. Tsai, M. Chhowalla, et al., Light-activated photocurrent degradation and self-healing in perovskite solar cells, Nat. Commun. 7 (2016) 11574-1–11574-9.
[56] W. Tress, N. Marinova, T. Moehl, S.M. Zakeeruddin, M.K. Nazeeruddin, M. Gratzel, Understanding the rate-dependent $J-V$ hysteresis, slow time component, and aging in $CH_3NH_3PbI_3$ perovskite solar cells: the role of a compensated electric field, Energy Environ. Sci. 8 (3) (2015) 995–1004.
[57] S. Meloni, T. Moehl, W. Tress, M. Franckevičius, M. Saliba, Y.H. Lee, et al., Ionic polarization-induced current–voltage hysteresis in $CH_3NH_3PbX_3$ perovskite solar cells, Nat. Commun. 7 (2016) 10334.
[58] L.D. Whalley, J.M. Frost, Y.-K. Jung, A. Walsh, Perspective: theory and simulation of hybrid halide perovskites, arXiv 146 (2017) 220901-1–220901-11.
[59] Y. Yuan, J. Chae, Y. Shao, Q. Wang, Z. Xiao, A. Centrone, et al., Photovoltaic switching mechanism in lateral structure hybrid perovskite solar cells, Adv. Energy Mater. 5 (15) (2015) 1500615-1–1500615-7.
[60] Y. Yuan, Q. Wang, Y. Shao, H. Lu, T. Li, A. Gruverman, et al., Electric-field-driven reversible conversion between methylammonium lead triiodide perovskites and lead iodide at elevated temperatures, Adv. Energy Mater. 6 (2) (2016) 1501803-1–1501803-7.
[61] J. Peng, L. Lu, H. Yang, T. Ma, Validation of the Sandia model with indoor and outdoor measurements for semi-transparent amorphous silicon PV modules, Renewable Energy 80 (2015) 316–323.
[62] T.-Y.Y. Yang, G. Gregori, N. Pellet, M. Grätzel, J. Maier, The significance of ion conduction in a hybrid organic–inorganic lead-iodide-based perovskite photosensitizer, Angew. Chem. Int. Ed. 54 (27) (2015) 7905–7910.
[63] C. Li, S. Tscheuschner, F. Paulus, P.E. Hopkinson, J. Kießling, A. Köhler, et al., Iodine migration and its effect on hysteresis in perovskite solar cells, Adv. Mater. 28 (2016) 2446–2454.
[64] Q. Wang, Y. Shao, H. Xie, L. Lyu, X. Liu, Y. Gao, et al., Qualifying composition dependent p and n self-doping in $CH_3NH_3PbI_3$, Appl. Phys. Lett. 105 (16) (2014) 163508-1–163508-5.
[65] A. Loidl, S. Krohns, J. Hemberger, P. Lunkenheimer, Bananas go paraelectric, J. Phys.: Condens. Matter 20 (19) (2008) 191001.
[66] J.B. Goodenough, Electronic and ionic transport properties and other physical aspects of perovskites, Rep. Prog. Phys. 67 (11) (2004) 1915.
[67] J. Mizusaki, K. Arai, K. Fueki, Ionic conduction of the perovskite-type halides, Solid State Ionics 11 (3) (1983) 203–211.
[68] P. Delugas, C. Caddeo, A. Filippetti, A. Mattoni, Thermally activated point defect diffusion in methylammonium lead trihalide: anisotropic and ultrahigh mobility of iodine, J. Phys. Lett. 7 (13) (2016) 2356–2361.
[69] D.A. Egger, L. Kronik, A.M. Rappe, Theory of hydrogen migration in organic–inorganic halide perovskites, Angew. Chem. Int. Ed. 54 (42) (2015) 12437–12441.

[70] H. Mehrer, Diffusion in Solids: Fundamentals, Methods, Materials, Siffusion-Controlled Processes, Vol. 155, Springer Science & Business Media, Berlin, Germany, 2007.
[71] A. Senocrate, I. Moudrakovski, G.Y. Kim, T.-Y. Yang, G. Gregori, M. Grätzel, et al., The nature of ion conduction in methylammonium lead iodide: a multi-method approach, Angew. Chem. Int. Ed. 56 (27) (2017) 7755–7759.
[72] R.A. De Souza, Oxygen diffusion in $SrTiO_3$ and related perovskite oxides, Adv. Funct. Mater. 25 (40) (2015) 6326–6342.
[73] N.S. Kulkarni, R.J. Bruce Warmack, B. Radhakrishnan, J.L. Hunter, Y. Sohn, K.R. Coffey, et al., Overview of SIMS-based experimental studies of tracer diffusion in solids and application to Mg self-diffusion, J. Phase Equilib. Diffus. 35 (6) (2014) 762–778.
[74] R.A. De Souza, J.A. Kilner, B.C.H. Steele, A SIMS study of hydrogen in acceptor-doped perovskite oxides, Solid State Ionics 77 (1995) 180–184.
[75] R.A. De Souza, J.A. Kilner, Oxygen transport in $La_{1-x}Sr_xMn_{1-y}Co_yO_{3\pm\delta}$ perovskites, Solid State Ionics 126 (1) (1999) 153–161.
[76] W.-C. Lin, H.-Y. Chang, K. Abbasi, J.-J. Shyue, C. Burda, 3D in situ ToF-SIMS imaging of perovskite films under controlled humidity environmental conditions, Adv. Mater. Interfaces 4 (2) (2017) 1600673.
[77] T. Zhang, X. Meng, Y. Bai, S. Xiao, C. Hu, Y. Yang, et al., Profiling the organic cation-dependent degradation of organolead halide perovskite solar cells, J. Mater. Chem. A 5 (3) (2017) 1103–1111.
[78] F. Matteocci, Y. Busby, J.-J. Pireaux, G. Divitini, S. Cacovich, C. Ducati, et al., Interface and composition analysis on perovskite solar cells, ACS Appl. Mater. Interfaces 7 (47) (2015).
[79] K. Schmidt-Rohr, H.W. Spiess, Multidimensional Solid-State NMR and Polymers, Elsevier Science, Burlington, 2012.
[80] C. Dybowski, R.L. Lichter, NMR Spectroscopy Techniques, Dekker, New York, NY, 1989.
[81] D. Kubicki, D. Prochowicz, A. Hofstetter, P. Pechy, S.M. Zakeeruddin, M. Grätzel, et al., Cation dynamics in mixed-cation $(MA)_x(FA)_{1-x}PbI_3$ hybrid perovskites from solid-state NMR, J. Am. Chem. Soc. 139 (2017) 10055–10061.
[82] T. Baikie, N.S. Barrow, Y. Fang, P.J. Keenan, P.R. Slater, R.O. Piltz, et al., A combined single crystal neutron/X-ray diffraction and solid-state nuclear magnetic resonance study of the hybrid perovskites $CH_3NH_3PbX_3$ (X = I, Br and Cl), J. Mater. Chem. A 3 (17) (2015) 9298–9307.
[83] Y. Chen, M. He, J. Peng, Y. Sun, Z. Liang, Structure and growth control of organic–inorganic halide perovskites for optoelectronics: from polycrystalline films to single crystals, Adv. Sci. 3 (4) (2016) 1500392.
[84] M. Bag, L.A. Renna, R.Y. Adhikari, S. Karak, F. Liu, P.M. Lahti, et al., Kinetics of ion transport in perovskite active layers and its implications for active layer stability, J. Am. Chem. Soc. 137 (40) (2015) 13130–13137.
[85] J. Xing, Q. Wang, Q. Dong, Y. Yuan, Y. Fang, J. Huang, Ultrafast ion migration in hybrid perovskite polycrystalline thin films under light and suppression in single crystals, Phys. Chem. Chem. Phys. 18 (44) (2016) 30484–30490.
[86] Y.-C. Zhao, W.-K. Zhou, X. Zhou, K.-H. Liu, D.-P. Yu, Q. Zhao, Quantification of light-enhanced ionic transport in lead iodide perovskite thin films and its solar cell applications, Light: Sci. Appl. 6 (2017) e16243.
[87] H. Zhou, Q. Chen, G. Li, S. Luo, T.-B. Song, H.-S. Duan, et al., Interface engineering of highly efficient perovskite solar cells, Science 345 (6196) (2014) 542–546.

[88] D. Bryant, N. Aristidou, S. Pont, I. Sanchez-Molina, T. Chotchunangatchaval, S. Wheeler, et al., Light and oxygen induced degradation limits the operational stability of methylammonium lead triiodide perovskite solar cells, Energy Environ. Sci. 9 (5) (2016) 1655–1660.
[89] R.L.Z. Hoye, P. Schulz, L.T. Schelhas, A.M. Holder, K.H. Stone, J.D. Perkins, et al., Perovskite-inspired photovoltaic materials: toward best practices in materials characterization and calculations, Chem. Mater. 29 (5) (2017) 1964–1988.
[90] J.F. Rubinson, Y.P. Kayinamura, Charge transport in conducting polymers: insights from impedance spectroscopy, Chem. Soc. Rev. 38 (12) (2009) 3339–3347.
[91] V.F. Lvovich, Fundamentals of electrochemical impedance spectroscopy, Impedance Spectroscopy, John Wiley & Sons, Inc., 2012, pp. 1–21.
[92] M. Bag, L.A. Renna, S.P. Jeong, X. Han, C.L. Cutting, D. Maroudas, et al., Evidence for reduced charge recombination in carbon nanotube/perovskite-based active layers, Chem. Phys. Lett. 662 (2016) 35–41.
[93] Yao Liu, L.A. Renna, Hilary B. Thompson, ZachariahA. Page, Todd Emrick, MichaelD. Barnes, et al., Role of ionic functional groups on ion transport at perovskite interfaces, Adv. Energy Mater. 7 (2017) 1701235-1–1701235-13.
[94] M.N.F. Hoque, N. Islam, Z. Li, G.F. Ren, K. Zhu, Z.Y. Fan, Ionic and optical properties of methylammonium lead iodide perovskite across the tetragonal-cubic structural phase transition, ChemSusChem 9 (18) (2016) 2692–2698.
[95] I. Zarazua, G. Han, P.P. Boix, S. Mhaisalkar, F. Fabregat-Santiago, I. Mora-Seró, et al., Surface recombination and collection efficiency in perovskite solar cells from impedance analysis, J. Phys. Lett. 7 (24) (2016) 5105–5113.
[96] C. Ho, Application of A-C techniques to the study of lithium diffusion in tungsten trioxide thin films, J. Electrochem. Soc. 127 (2) (1980) 343–350.
[97] L. Contreras, J. Idigoras, A. Todinova, M. Salado, S. Kazim, S. Ahmad, et al., Specific cation interactions as the cause of slow dynamics and hysteresis in dye and perovskite solar cells: a small-perturbation study, Phys. Chem. Chem. Phys. 18 (45) (2016) 31033–31042.
[98] H. Zhang, X. Qiao, Y. Shen, T. Moehl, S.M. Zakeeruddin, M. Gratzel, et al., Photovoltaic behaviour of lead methylammonium triiodide perovskite solar cells down to 80 K, J. Mater. Chem. A 3 (22) (2015) 11762–11767.
[99] H. Zhang, L. Xue, J. Han, Y.Q. Fu, Y. Shen, Z. Zhang, et al., New generation perovskite solar cells with solution-processed amino-substituted perylene diimide derivative as electron-transport layer, J. Mater. Chem. A 4 (22) (2016) 8724–8733.
[100] M.N.F. Hoque, N. Islam, K. Zhu, Z. Fan, Hybrid perovskite phase transition and its ionic, electrical and optical properties, MRS Adv. 2 (2017) 1–7.
[101] N. Pellet, P. Gao, G. Gregori, T.Y. Yang, M.K. Nazeeruddin, J. Maier, et al., Mixed-organic-cation perovskite photovoltaics for enhanced solar-light harvesting, Angew. Chem. Int. Ed. 53 (12) (2014) 3151–3157.
[102] A. Pockett, G.E. Eperon, T. Peltola, H.J. Snaith, A. Walker, L.M. Peter, et al., Characterization of planar lead halide perovskite solar cells by impedance spectroscopy, open-circuit photovoltage decay, and intensity-modulated photovoltage/photocurrent spectroscopy, J. Phys. Chem. C 119 (7) (2015) 3456–3465.
[103] A. Dualeh, T. Moehl, N. Tétreault, J. Teuscher, P. Gao, M. Nazeeruddin, et al., Impedance spectroscopic analysis of lead iodide perovskite-sensitized solid-state solar cells, ACS nano 8 (1) (2014).
[104] A. Zohar, N. Kedem, I. Levine, D. Zohar, A. Vilan, D. Ehre, et al., Impedance spectroscopic indication for solid state electrochemical reaction in $(CH_3NH_3)PbI_3$ films, J. Phys. Lett. 7 (1) (2016) 191–197.

[105] O.S. Game, G.J. Buchsbaum, Y. Zhou, N.P. Padture, A.I. Kingon, Ions matter: description of the anomalous electronic behavior in methylammonium lead halide perovskite devices, Adv. Funct. Mater. (2017). 1606584-n/a.
[106] G.E. Eperon, D.S. Ginger, B-site metal cation exchange in halide perovskites, ACS Energy Lett. (2017) 1190–1196.
[107] Imran, H.; Butt, N.Z. Investigation of dominant hysteresis phenomenon in perovskite solar cell, in: 2016 IEEE 43rd Photovoltaic Specialists Conference (PVSC), 5–10 June 2016, 2016, pp. 0776–0780.
[108] S. van Reenen, M. Kemerink, H.J. Snaith, Modeling anomalous hysteresis in perovskite solar cells, J. Phys. Lett. 6 (19) (2015) 3808–3814.
[109] A. Merdasa, M. Bag, Y.X. Tian, E. Kallman, A. Dobrovolsky, I.G. Scheblykin, Super-resolution luminescence microspectroscopy reveals the mechanism of photoinduced degradation in $CH_3NH_3PbI_3$ perovskite nanocrystals, J. Phys. Chem. C 120 (19) (2016) 10711–10719.
[110] H. Zhu, K. Miyata, Y. Fu, J. Wang, P.P. Joshi, D. Niesner, et al., Screening in crystalline liquids protects energetic carriers in hybrid perovskites, Science 353 (6306) (2016) 1409–1413.

CHAPTER 7

Low-Dimensional Perovskites

P.C. Reshmi Varma
Indian Institute of Science Education and Research Thiruvananthapuram, Thiruvananthapuram, Kerala, India

7.1 INTRODUCTION

Perovskite materials have emerged as the most promising and efficient low-cost energy materials for various optoelectronic and photonic device applications. The discovery of calcium titanate ($CaTiO_3$) in 1839 by a Russian mineralogist Perovski was considered to be the origin of perovskite, and materials with the same type of crystal structure as that of $CaTiO_3$ were known as the perovskite materials (structure). The general chemical formula used to describe the perovskite materials is ABX_3, where A and B are cations with A larger than that of B and X is the anion usually oxides or halogens. The unique physical properties of perovskite materials such as high-absorption coefficient, long-range ambipolar charge transport, low exciton-binding energy, high dielectric constant, ferroelectric properties, etc. have gained a huge interest in these materials for optoelectronic and photovoltaic applications. Different classes of perovskite materials, such as chalcogenide perovskite (AMO_3) and halide perovskite (ABX_3) that is again classified as alkali halide and organometal halide, were widely studied. The superior ferroelectric and superconducting properties of oxide-based perovskites were extensively studied for various applications. In the case of metal halide perovskites, mainly cesium lead halide ($CsPbX_3$) and methylammonium lead halide ($CH_3NH_3PbX_3$) perovskites are the most popular ones. Halide perovskite structure was first observed in $CsPbX_3$ in 1958. Studies on methylammonium-based perovskites were started in the 1970s, and these materials attracted huge attention due to their improved optical and electrical properties. Owing to the strong excitonic properties of layered perovskite materials (organic–inorganic lead halides), these materials were studied for thin-film light-emitting diodes (LEDs) and field effect transistor (FET) applications. The first breakthrough in methylammonium halide (MAX_3) perovskites occurred with their use as light-absorbing

Perovskite Photovoltaics
DOI: https://doi.org/10.1016/B978-0-12-812915-9.00007-1

materials in the photovoltaic device by Miyasaka et al. [1]. The high-absorption coefficient, long diffusion length, superior charge-transport properties, low nonradiative emission, and solution processability make them suitable materials for photovoltaic applications [2–4]. Perovskite materials are also widely used in LEDs [5,6], photodetectors, nanolasers, and waveguides.

In perovskite thin film based optoelectronic and photovoltaic devices, the quality of the perovskite film (morphology, grain size, uniformity, coverage, etc.) plays a significant role in determining the device performance. The research in perovskite thin films has attained rapid progress that the efficiency of the photovoltaic devices based on perovskite materials has reached from the initial 3.8%–22.1% [2] within 6 years. These devices mainly comprised thin films of bulk perovskite materials, and different methods such as spin coating of perovskite solution, sequential deposition of the solutions, and thermal evaporation were used to deposit the perovskite films. Even though the devices showed enhanced photovoltaic performances, the fabrication procedure and characterization of the devices has to be carried out under the controlled (inert) atmosphere to avoid the degradation of the materials in ambient conditions. As the unavoidable formation of the defect states and grain boundaries in these thin films caused deterioration in the device performance, the purity of the materials and the uniformity of the thin films are crucial factors for improved performance of the devices. The attempts to enhance the efficiency and the environmental stability of the perovskite materials based devices were accompanied by the synthesis of new stable materials and designing better device architectures.

Recently increased interest has been grown in the development of the low-dimensional (LD) perovskite materials and in exploring their properties for the photonic and optoelectronic device applications. The materials with at least one dimension within the nanoscale (1–100 nm) range are referred to as LD materials and nanocrystals (NCs). Unique features of the LD perovskite materials mainly their tunable optical and electrical properties due to the quantum-sized effects and their mechanical flexibility are attaining increased attention in the semiconductor materials and optoelectronic devices. Different morphologies of LD perovskites developed are perovskite quantum dots (QDs; zero-dimensional), perovskite nanorods or nanowires (NWs; one dimensional), and perovskite nanoplatelets and nanosheets (two dimensional). The advantages of reducing the dimensionality of the bulk perovskite materials are prominent on the optoelectronic properties of LD perovskites. These properties of LD perovskites have already been used widely in LEDs, photovoltaics, photodetectors, and lasing applications. The initial results of these research showed the potential of these materials in photonics and optoelectronics. There is still room for further improvement in the efficiency of the device performance and the stability that needs to be attained with a thorough understanding of the properties of the materials.

Reducing the dimensionality of the three-dimensional bulk perovskites has led to the formation of another class of perovskite materials, known as layered perovskites. Layered perovskite structure comprises an infinite number of two-dimensional slabs with structure ABX_3 separated by an element. In the case of organic–inorganic perovskites, when the long organic cations are used, which do not fit into the octahedral structure, would act as a "barrier" and would "cap" the three-dimensional perovskite and results in the formation of layered structure. The general formula for the layered structure perovskites is $A_{n-1}B_nX_{(3n+1)}$. In this structural formula for the layered perovskites, n indicates the size of the two-dimensional slabs. For example, $n=1$ indicates one layer of BX_6 octahedron thick, and $n=2$ means that the structure is two BX_6 octahedra thick. The layered perovskites, depending on the number of layers possessed, show different physical and chemical properties compared to that of the three-dimensional perovskites. Layered perovskites can be considered as a new class of materials with confined structure and with huge potential for optoelectronic and electrical applications. Maruyama et al. reported the new novel layered material in 1986 for the first time [7]. The compound corresponds to the formula $(C_nH_{(2n+1)}NH_3)_2MX_4$. This structure is known as two-dimensional perovskites that form the two-dimensional structure by the self-assembly of the layers held together by van der Waals and intermolecular force. As mentioned above, the confined structure of layered perovskites makes them promising candidates for LED application, because the confined quantum well structure leads to high radiative recombination, which is advantageous for LED devices. Photovoltaic performance of layered perovskites was also studied, which exhibited lower performance compared to the three-dimensional counterpart but showed enhanced environmental stability.

This chapter on LD perovskite materials comprises the details of the fundamental properties of different categories of the LD perovskite materials. This chapter includes sections focusing on the different advance materials synthesis procedures as well as the application of the LD perovskite materials in various optoelectronic fields.

7.2 LOW-DIMENSIONAL MATERIAL SYNTHESIS

7.2.1 Zero-Dimensional Materials (Quantum Dots)

QDs are spherical particles in the dimension of nanometer range, formed by applying quantum confinement in all the three dimensions [8]. These materials show quantum confinement effects and exhibit unique electrical and optical properties, making them potentially suitable for nanoscale optoelectronic and photovoltaic devices. Zero-dimensional perovskite materials or perovskite QDs also show features similar to the semiconductor QDs. In the application level, the performance of the perovskite thin film based devices was mainly determined by the quality of the perovskite thin film. The primary challenge was to produce a defect-free, uniform, void-free perovskite film. Reducing the device dimension to nanoscale would be an alternative to overcome this issue. Therefore, several methods were used to synthesize perovskite QDs. Organometal halide perovskite nanoparticles were first demonstrated using a template method. Mesoporous metal oxides such as titania or alumina in which the arrays were confined in all the three dimensions to nanoscale were used to prepare zero-dimensional perovskite nanoparticle. Perovskite QD using porous alumina was first demonstrated by Kojima et al. in dye-sensitized solar cell (DSSC) [9]. In their report, they have used the porous Al_2O_3 as the medium, and the $CH_3NH_3PbBr_3$ NCs with high density and green photoluminescence were self-organized in the template. The size of the QDs was determined by the diameter of the Al_2O_3 pore, and it was difficult to control the crystal size. However, considering the large-area applications, development of a much versatile synthesis method was required. For the first time, Pérez-Prieto et al. reported the colloidal perovskite QDs using solution process technique. Mainly two methods, (1) ligand-assisted reprecipitate (LARP) and (2) hot injection method, have been used for the synthesis of perovskite QDs.

Reprecipitation is an efficient and versatile method for synthesizing the perovskite NCs through the simple mixing of solvents. Presence of ligands helps one to control the morphology and crystal growth. In LARP method, a good (polar) solvent that is capable of dissolving ammonium halide salts and lead with ligands is mixed into a nonpolar solvent and vigorously stirred to form nanoparticles. In 2015, Zhang et al. reported the $CH_3NH_3PbBr_3$ perovskite QDs with excellent luminescent properties and a diameter less than 5 nm using LARP method [10]. They have used *n*-octyl amine and oleic acid as the coligands. The precursor solution of the perovskite QD was prepared by mixing CH_3NH_3Br, $PbBr_2$ in *n*-octyl amine and oleic acid and dissolving it in the polar solvent *N*-methylformamide (DMF). The precursor solution further added in toluene (poor solvent) and treated to vigorous stirring, which results in the formation of $CH_3NH_3PbBr_3$ nanoparticles. The different steps involved in the synthesis of $CH_3NH_3PbBr_3$ using LARP method is shown in Fig. 7.1. The long-chain ligands in the LARP method ensure the control of the crystal size and morphology. The presence of *n*-octyl amine in the synthesis procedure prevents the fast crystallization, aggregation, and further precipitation of the particles from the solution. Oleic acid plays a role in the stability of the colloidal nanoparticles. The supersaturation induced by the solubility change with solvent mixing controls the crystallization process [10]. LARP strategy can also be used to form $CH_3NH_3PbI_3$, $CH_3NH_3PbCl_3$ perovskite nanoparticles by tuning the precursors and solvents (Fig. 7.1).

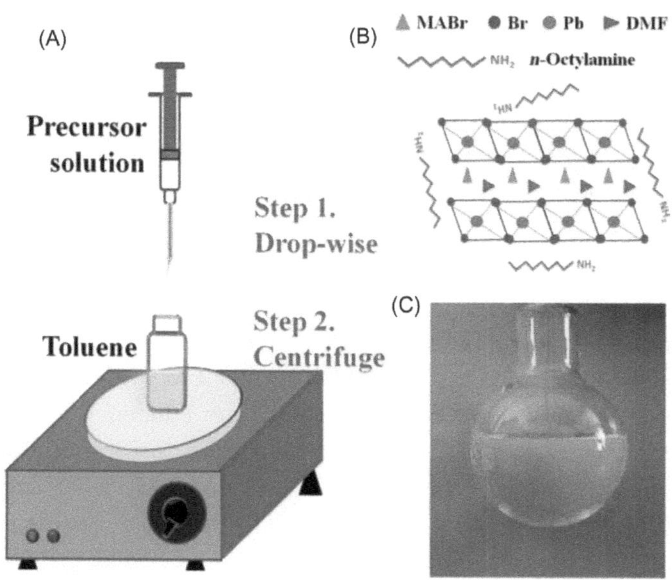

Figure 7.1 Steps involved in the synthesis of $CH_3NH_3PbBr_3$ QDs: (A) schematic illustration of the process of LARP technique and reaction system, (B) the schematic of the starting materials in precursor solution, (C) $CH_3NH_3PbBr_3$ colloidal solution [10]. *QDs*, quantum dots; *LARP*, ligand-assisted reprecipitate. *Reprinted with the permission from F. Zhang, H. Zhong, C. Chen, X.-G. Wu, X. Hu, H. Huang, et al., Brightly luminescent and color-tunable colloidal $CH_3NH_3PbX_3$ (X = Br, I, Cl) quantum dots: potential alternatives for display technology, ACS Nano 9 (2015) 4533, Copyright 2015, American Chemical Society.*

One of the fascinating features of the QDs is the tunability of their optical properties, which can be achieved by controlling the halide ion composition and the size of the NCs. To attain the tunability of the light emission of the QDs, Huang et al. [11] demonstrated that, in LARP method, the size of the $CH_3NH_3PbBr_3$ QDs could be controlled by varying the temperature at which the precipitation occurs.

For all inorganic perovskite QDs, $CsPbX_3$, which was much more stable and showed enhanced optical properties compared to that of hybrid perovskites, was found to be less soluble in common solvents, and a direct solution processing was challenging to achieve. An alternate method was used to prepare these perovskite QDs and is known as hot injection method. The synthesis of cesium lead halide perovskite ($CsPbX_3$, X = Cl, Br, I) QDs using hot injection method was reported by Kovalenko et al. for the first time in 2015. In this method, Cs-oliete reacted with Pb(II) halide in high boiling octadecene at 140–200°C. In octadecene, a mixture of oleylamine and oleic acid was added to solubilize PbX_2 and to stabilize the newly formed NCs. The size of the $CsPbX_3$ QDs was tuned to 4–15 nm by adjusting the reaction temperature from 140 to 200°C, and these NCs showed outstanding optical properties. They exhibited size-dependent bandgap energy tunability through the entire visible range 410–700 nm and exhibited a high photoluminescence quantum yield (PLQY) 50%–90%.

7.2.2 One-Dimensional Material

Perovskite NWs show application not only as light absorber but also as a charge transporting medium in optoelectronic and photovoltaic devices. NWs possess high crystallization, high quantum efficiency, and large diffusion length, which make them interesting entities for optoelectronic device application. In the case of perovskite thin film based devices, the film morphology plays a crucial role in determining the performance of the device. The perovskite film made using the spin-coating method usually shows large microwhiskers and void areas. The dendrite crystal growth shows that organolead halide perovskite exhibits the preferential growth, and this has become the inspiration to investigate on the synthesis of perovskite NWs.

One of the methods used to grow perovskite NWs is "one-step evaporation-induced self-assembly (EISA) method." In this method, the precursor solution, a mixture of methylammonium iodide (MAI) and

PbI_2 in DMF, was coated on a substrate, and the evaporation of the solvent led to the growth of perovskite supersaturated crystals. A subsequent annealing of the substrate at 80°C converted the NCs into NWs. Deng et al. [12] demonstrated the synthesis of $CH_3NH_3PbI_3$ perovskite NWs using the one-step method, and they have used the preheated substrates at different temperatures along with the room temperature (RT) one to deposit the NW precursor solution. For high-temperature substrates, the preferential growth seemed to be suppressed, and nanoparticles grow homogeneously. While in the case of RT substrates, the evaporation of solvents preferentially grew into NWs. The uniform distribution of NWs on the substrates was implemented by the UV-ozone treatment of the substrates. The surface treatment helped one to decrease the contact angle and in uniform spreading of the solution on the substrate. This eventually resulted in the uniform distribution of the NWs without any aggregation (Fig. 7.2).

Another solution-based method for the synthesis of NWs involves more than one step and is known as "dissolution recrystallization" method. Zhu et al. have demonstrated the conversion of the perovskite thin film into NWs using this method. In their study, they first formed the perovskite thin film by mixing $PbCl_2$ and CH_3NH_3I in DMF followed by annealing [13]. A mixed solvent of DMF and IPA is then spin coated on the perovskite film. Perovskite showing high solubility in DMF readily dissolves and the spinning of the substrate fastens the excess solvent evaporation. The quick evaporation of the solvent leads to the supersaturation of the solution and come to nucleation that eventually results in the crystallization of the NWs. Perovskite NW film shows enhanced conductivity compared to the perovskite films indicating better connectivity and increased charge carrier mobility.

Low-temperature solution-processed crystallization by slip coating was used to prepare the $MAPbI_3$ NWs [14]. This method involves the saturated solution of $MAPbI_3$ in DMF sandwiched between two glass slides and the sliding of the upper glass slide, keeping the bottom one fixed controls the distribution of NWs. The width of the $MAPbI_3$ NWs ranges from 50 to 200 nm and lengths up to 16 μm. Different parameters such as solvent concentration, temperature, sliding speed, fluid philicity/phobicity, etc. affect the kinetics of crystallization. Yang et al. performed anion exchange method to convert the $CH_3NH_3PbBr_3$ nanorod arrays to $CH_3NH_3PbI_3$ arrays. Growing the $CH_3NH_3PbBr_3$ nanorod arrays involved two steps [15]. First, the methanolic solution of lead acetate

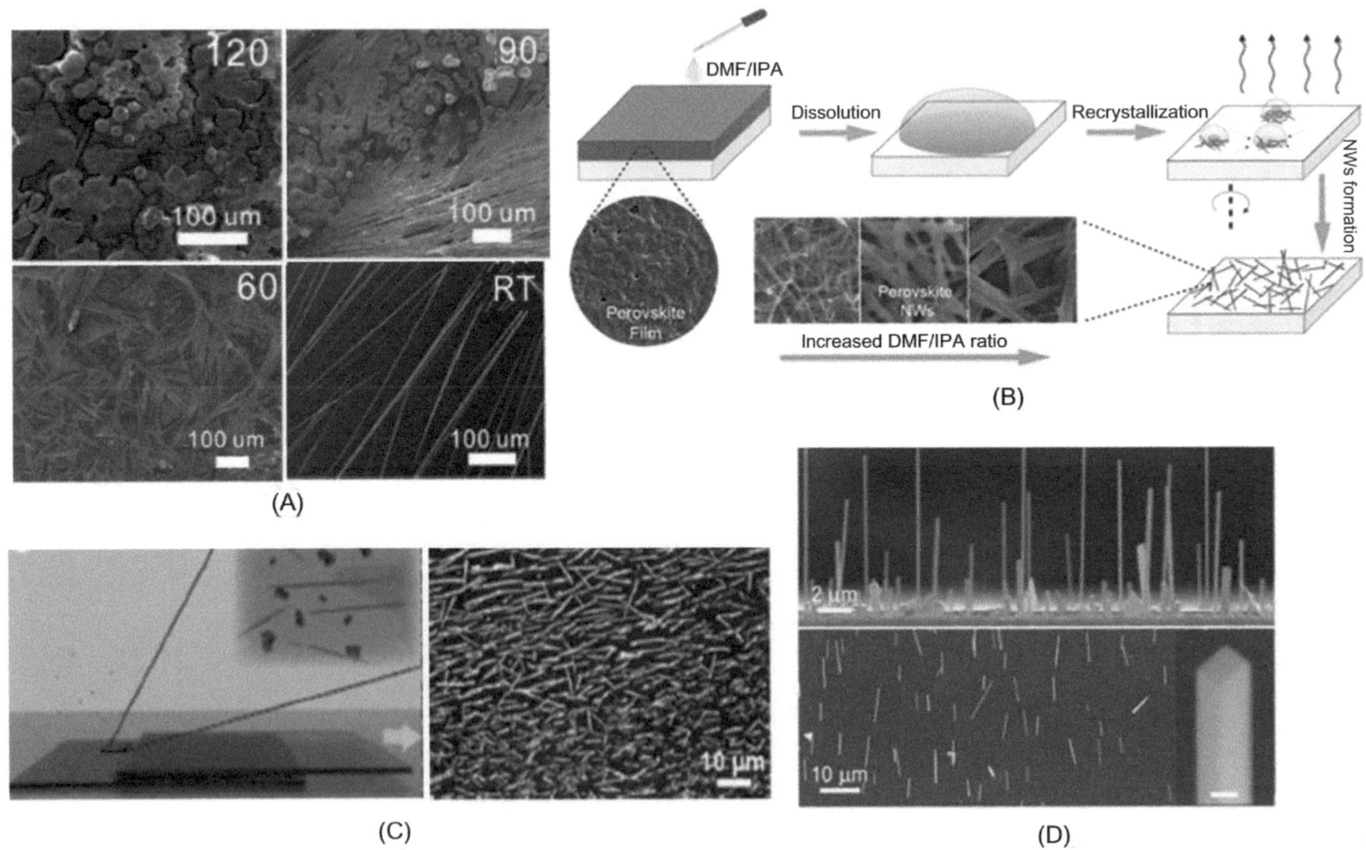
120
100 um
90
100 um
60
100 um
RT
100 um
(A)
DMF/IPA
Dissolution
Recrystallization
NWs formation
Perovskite Film
Perovskite NWs
Increased DMF/IPA ratio
(B)
10 μm
(C)
2 μm
10 μm
(D)

Figure 7.2 Different methods used in the synthesis of perovskite NWs: (A) SEM images of $CH_3NH_3PbI_3$ NWs at different temperatures using one-step EISA method; (B) schematic showing of the different steps involved in the dissolution recrystallization process; (C) (left) slip-coating fabrication of $MAPbI_3$ NWs, (right) their optical image; (D) SEM images of the vertically grown $CsPbBr_3$ NWs with rectangular cross section grown on a substrate. *NWs*, nanowires. *(A) Reprinted with the permission from H. Deng, D. Dong, K. Qiao, L. Bu, B. Li, D. Yang, et al., Growth, patterning and alignment of organolead iodide perovskite nanowires for optoelectronic devices, Nanoscale 7 (2015) 4163, Copyright 2015, Royal Society of Chemistry. (B) Reprinted with the permission from P. Zhu, S. Gu, X. Shen, N. Xu, Y. Tan, S. Zhuang, et al., Direct conversion of perovskite thin films into nanowires with kinetic control for flexible optoelectronic devices, Nano Lett. 16 (2016) 871, Copyright 2016, American Chemical Society. (C) Reprinted with the permission from E. Horváth, M. Spina, Z. Szekrényes, K. Kamarás, R. Gaal, D. Gachet, and et al., Nanowires of methylammonium lead iodide ($CH_3NH_3PbI_3$) prepared by low temperature solution-mediated crystallization, Nano Lett. 14 (2014) 6761, Copyright 2014, American Chemical Society. (D) Reprinted with the permission from K. Park, J.W. Lee, J.D. Kim, N.S. Han, D.M. Jang, S. Jeong, et al., Light—matter interactions in cesium lead halide perovskite nanowire lasers, J. Phys. Lett. 7 (2016) 3703, Copyright 2016, American Chemical Society.*

trihydrate was spin coated on the substrate. Then the substrate is immersed in the solution of CH_3NH_3Br in isopropanol to react with the lead acetate film uniformly. Several minutes of reaction resulted in the growth of orange color $CH_3NH_3PbBr_3$ nanorod arrays. SEM and TEM images of the nanorod arrays showed square cross section with a length ~1 μm and diameter about 100 nm. The anion exchange method to convert $CH_3NH_3PbBr_3$ nanorod arrays to $CH_3NH_3PbI_3$ involves annealing of $CH_3NH_3PbBr_3$ nanorod arrays in the presence of CH_3NH_3I vapor at 140−150°C. After several hours of reaction, anion exchange takes place to form $CH_3NH_3PbI_3$ nanorods retaining the initial morphology.

Keeping in mind the requirements of the large-scale application in optoelectronics, microwires (MWs) synthesis was done using doctor blade technique. Jie et al. demonstrated the synthesis of $CH_3NH_3PbI_3$ MW arrays using solution-processed doctor blade technique [16]. To prepare the MW arrays, the solution of $CH_3NH_3PbI_3$ in DMF is dragged using a blade on a heated substrate upon 100°C. The subsequent evaporation of the solvent yields highly oriented MW arrays. This method of synthesis has several advantages such as the possibility of one-step fabrication of highly crystalline MW arrays on large areas, solution processability helps its application on a flexible substrate, and the highly crystalline MWs with fewer grain boundaries make the perovskite material immune from degradation.

Similar to the case of semiconducting NWs, perovskite NWs were synthesized using vapor transport technique. Song et al. synthesized $CsPbX_3$ (X = Br, Cl, or I) perovskite NWs using chemical-vapor transport (CVT) method [17]. NWs grown were shown rectangular cross section with an average diagonal length of 300 nm and average NW length of 10 μm. The method is similar to chemical-vapor deposition (CVD) used for the semiconducting NWs. Powder sources of PbX_2 and CsX were kept inside the quartz tube reactor, and the substrate is kept 10 cm away from the source. Argon gas was used as the inert carrier gas maintaining a flow rate of 220 sccm. The argon gas flow was used to transport the source vapor on the Si substrate. In this process, the temperature of the PbX_2 and CxS were set at 470−600°C, and that of the substrate was set at 350−300°C.

NWs synthesis was also carried out using template method. This method promised more controlled growth and morphology. Porous $CH_3NH_3PbBr_3$ NWs were fabricated using self-template-directed synthesis. Perovskite porous NWs combine the advantages of perovskite properties and that of one-dimensional porous material [18]. Porous NWs shorten the charge carriers transport distance, improve the active areas on the surface and thus can find applications in advanced photoelectronic and electronic devices. To prepare the $CH_3NH_3PbBr_3$ porous NWs, Pb-containing precursor NWs were used as a template. The Pb-containing precursor NW, which also acts as Pb^{2+} ion source, reacts with CH_3NH_3Br and HBr in an organic solvent (2-propanol) at RT. The chemical transformation resulted in the conversion of Pb-containing NWs to $CH_3NH_3PbBr_3$ NWs with mesopores in the structure due to the release of the organic component into the solvent. Perovskite NWs were also synthesized using the anodized aluminum oxide (AAO) templates [19]. In this method, perovskite precursor solution was added to the template and is allowed to penetrate into the pores. The excess solvent is removed from the template surface by thermal annealing results in the formation of crystalline perovskite NWs. The advantage of using AAO method is the control on the NW dimensions by controlling the AAO-anodizing procedures.

7.2.3 Two-Dimensional Perovskites

Perovskite nanosheets, nanoplatelets, and nanodisks fall under the category of two-dimensional (2D) perovskite materials. These 2D materials showed high PLQY and found increased application in nanoelectronics, nanophotonics, and photovoltaics. Several methods of solution process and vapor deposition techniques have been developed for the synthesis of 2D-perovskite materials. Other than these techniques, exfoliation method similar to the one reported in graphene and other 2D materials was also implemented to produce the 2D-perovskite structures from the bulk [20]. Niu et al. synthesized 2D ultrathin flakes of organic–inorganic perovskite $(C_6H_9C_2H_4NH_3)_2PbI_4$ using micromechanical exfoliation method [21]. Organic–inorganic perovskite nanoplatelets were produced using the CVD. Compared to the solution-processing techniques, vapor deposition methods produce the high purity nanosheets but offer low yield. For the first time, CVD method was used to synthesis $CH_3NH_3PbI_3$ perovskite platelets by Xiong et al. [22]. The preparation of organic–inorganic perovskite nanoplatelets involves two steps. In the first step, the lead halide perovskite nanoplatelets on muscovite mica are grown in a CVD system. These platelets were converted into perovskites by gas–solid hetero-phase reaction with methylammonium halide molecules. The reaction of the halide nanoplatelets with the gas-phase methylammonium halide was carried out in a quartz tube under vacuum in the presence of an inert carrier gas nitrogen or argon, where the source methylammonium halide was kept in the center of the tube furnace with the temperature set at 120°C and the lead halide nanoplatelets was kept 5–6 cm away from the center in the downstream region inside the tube. In vapor-phase method, controlling the conversion process conditions, such as pressure, and temperature mixed halide perovskite can be produced, and this would be advantageous to achieve wavelength tunability to span a broad range of the spectrum.

Even though the vapor-phase method is giving high purity nanosheets and this more efficient to deliver monodispersed nanosheets compared to that of solution processing, it is not favoring the large-scale applications. The above-discussed vapor-phase deposition method delivered high production cost, and it is important to develop a scalable, inexpensive, high yield, and efficient technique to produce 2D perovskites for feasible applications. In this regard, solution-processing methods were employed to prepare the 2D-perovskite structures. Liao et al. [23] reported $CH_3NH_3PbBr_3$ perovskite microdisks using one-step solution self-assembly method. To prepare the microdisks, equal volumes of solutions of CH_3NH_3Br and $PbBr_2$ in *N,N*-DMF were mixed to form the $CH_3NH_3Br \cdot PbBr_2$ stock solution. The stock solution is dip coated on a glass substrate, which was immersed in the beaker containing dichloromethane of CH_2Cl_2. CH_2Cl_2 is a poor solvent for $CH_3NH_3PbBr_3$. The diffusion of CH_2Cl_2 vapor in the $CH_3NH_3Br \cdot PbBr_2$ stock solution coated on the glass substrate induces the nucleation and the growth of the microdisks. These were four side faces square microdisks with width ranging from 1 to 10 μm and a thickness of 1/10–1/4 of the width. For a further understanding of the properties of the 2D-perovskite materials, the size of the nanoplatelets or sheets should be confined to several unit cells. In this regard, atomically thin 2D hybrid perovskite $(C_4H_9NH_3)_2PbBr_4$ materials were produced from solution (Fig. 7.3).

Liu et al. developed 2D $CH_3NH_3PbX_3$ perovskite nanosheets using a combined method of solution processing, and vapor transport technique [24]. The two-step fabrication method involves the formation of the PbI_2 nanosheets by casting the saturated PbI_2 solution on the substrate followed by heating it at an elevated temperature (30–80°C). Since the number of nucleation centers is controlled by the temperature, the heating temperature plays an important role to control the nucleation and growth of 2D nanosheets. The PbI_2 nanosheets prepared on the SiO_2 substrates act as a template for the growth of 2D $CH_3NH_3PbI_3$ perovskite

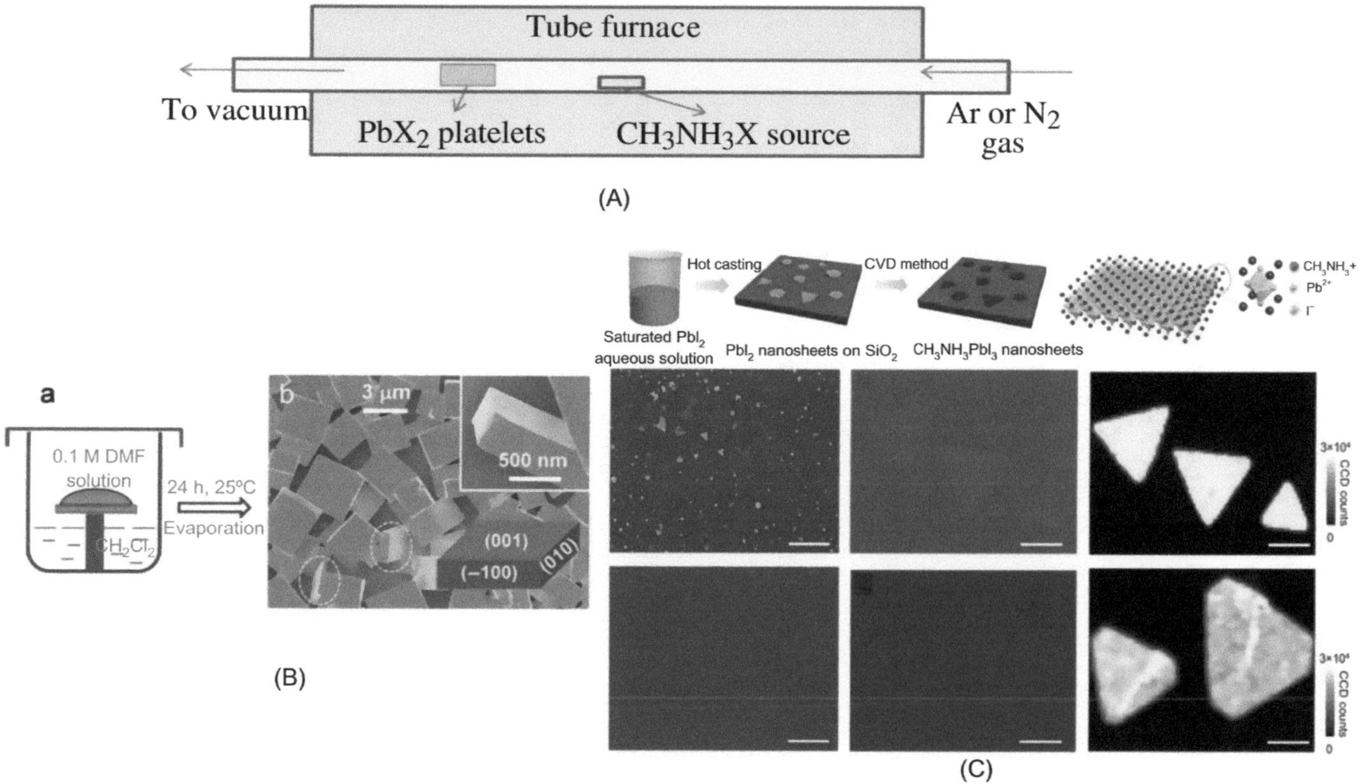

Figure 7.3 (A) Schematic of the synthesis setup of vapor transport system for the formation of $CH_3NH_3PbX_3$ (X = Cl, Br, I) nanoplatelets [22], (B) (left) preparation scheme of square microdisks using a one-step solution self-assembly method. Low- and high-magnification (top inset) SEM images of square microdisks of $CH_3NH_3PbBr_3$ perovskites. The right bottom inset shows a cartoon of a square microdisk [23]. (C) (Top) Schematic illustration of solution process to fabricate 2D PbI_2 nanosheets and vapor-phase conversion process to transfer PbI_2 into $CH_3NH_3PbI_3$ perovskite nanosheets. (Bottom) Crystal structure of single-unit-cell-thick 2D $CH_3NH_3PbI_3$ perovskite [24]. *Reprinted from J. Liu, Y. Xue, Z. Wang, Z.-Q. Xu, C. Zheng, B. Weber, et al., Two-dimensional $CH_3NH_3PbI_3$ perovskite: synthesis and optoelectronic application, ACS Nano 10 (2016) 3536–3542, Copyright 2016, American Chemical Society.*

nanosheets. In the second step that is the CVD method, the CH_3NH_3I molecules were intercalated into the cites of PbI_6 layers resulting in the formation of the 2D-perovskite nanosheets.

7.2.4 Structure and Synthesis of the Layered Perovskites

The layered perovskites are formed by choice of the different sizes of organic cations. A small-sized organic cation fits into the PbX_6 octahedral frame whereas bigger organic cation will not fit into the octahedral cage and leads to the formation of layered structure. This shows that the choice of a long organic cation caps the MAX_3 perovskite held together by van der Waals interactions and form the layered structure. Goldschmidt's tolerance factor formula enables the theoretical prediction of the structure. More precisely, it will give the information whether the organic cation will form either a two-dimensional or three-dimensional perovskite structures. Tolerance factor is given as $(R_A + R_X)/(\sqrt{2}(R_B + R_X))$, where R_A, R_B, R_X are the effective ionic radii of the corresponding ions. The calculated values of tolerance factor between 0 and 0.81 predict the formation of a three-dimensional structure, and tolerance factor values greater than 1 imply the formation of two-dimensional structure. In the structural formula for the two-dimensional layered perovskites, $n = 1$ represents one 2D-perovskite sheet and $n = \infty$ is the 3D-perovskite structure, and for all the values of n, $2 < n < \infty$ layered structure of perovskites are formed. The family of layered perovskites comprises different oriented layers such as $\langle 0\ 0\ 1 \rangle$, $\langle 0\ 1\ 1 \rangle$, and $\langle 1\ 1\ 1 \rangle$. Among these, $\langle 0\ 0\ 1 \rangle$ is the most popular oriented perovskite structure. The dimensionality of the perovskite or the number of perovskite sheets inside the organic layers is controlled by controlling the ratio of the organic cation, whereas the type of long organic cation chosen determines the crystallographic orientation of the inorganic framework.

The synthesis of two-dimensional perovskite involves the solution-processed technique. In this method, stoichiometric precursors are dissolved in solutions and are deposited on the substrate and assemble to form the 2D-perovskite structure. In the solution-processed technique, mainly there are two methods used—one-step method and two-step method. These methods were also used in the synthesis of three-dimensional perovskite materials prepared using the solution-processed technique. In the one-step method, all the precursors are dissolved in the solution and are deposited on the substrate by spin coating, or drop casting followed by the evaporation of the solvent results in the crystallization of the perovskite. For the preparation of the 3D organic–inorganic halide perovskites, metal halide and small organic cation were chosen in a ratio of 1:1, whereas the synthesis of two-dimensional perovskites required a solution of a mixture of metal halide and a long organic cation in a molar ratio of 1:2.

7.2.4.1 Optical Properties of Layered Perovskites

The attractive optical and electrical properties of the three-dimensional perovskite materials attained huge attention in the photovoltaics and optoelectronic device applications. The new class of materials layered perovskites exhibited different optical properties compared to that of the three-dimensional ones. LD hybrid perovskite layered materials possess high exciton-binding energy compared to the 3D analog mainly due to the variation in the electronic potential across the layers. The dielectric confinement effect, the difference in the dielectric constant of the organic cation and inorganic framework determines the exciton-binding energy. The high exciton-binding energy and large oscillation strength in LD-layered perovskite arise from its quantum confinement and the dielectric confinement effect. The bandgap of the perovskite materials increases with the decrease of the dimension of the materials that also affects its absorption spectra. The optical properties and the bandgap of the LD perovskites can be tubed by varying the number of layers separated by a long organic cation. The barrier organic cation in LD-layered perovskites makes the perovskite structure hydrophobic, or it provides more resistance to humidity and moisture. This means that 2D perovskites are more stable in the ambient air conditions compared to the 3D-perovskite structure.

7.3 APPLICATIONS

7.3.1 Photovoltaic Devices

Perovskite materials, mainly hybrid metal halide perovskites were extensively studied for photovoltaic applications. The major factors that determine the power conversion efficiency of the solar cells are its light absorption efficiency, efficient exciton dissociation, enhanced charge carrier transport, and collection at the electrodes. Perovskite materials that possess high-absorption coefficient, broad absorption spectrum extending from UV–vis to near-infrared (NIR) region, ambipolar charge carrier transport, and long charge carrier diffusion length were considered to be ideal materials for photovoltaic applications. Photovoltaic researches based on perovskite materials have gained huge advancement in a very short period. More research is undergoing extensively to optimize the high-efficiency devices and their environmental stability. Another significant development in the perovskite-based photovoltaic research is, exploring the photovoltaic properties of the LD perovskite materials, which, in turn, turned to be a hot topic of research recently. LD materials show unique optical and electrical properties due to the quantum confinement effects. Dimensionality of the material plays an important role in determining its physical properties of the material. In the case of perovskite materials, it was found that the recombination rate is lower in the LD perovskite materials compared to its thin-film counterpart, and these materials possess larger diffusion length. High photons to current conversion efficiency and unique optical properties have made them a strong potential candidate in photovoltaic research.

In photovoltaic studies, semiconductor QDs gained considerable attention mainly because of its size-dependent bandgap tunability, which helps one to tune the absorption spectrum of the active material from UV−vis to near-IR region, high extinction coefficient, and multiple exciton generations. Owing to the unique optical properties of QDs, perovskite QDs were used as sensitizers in solar cells. In QD sensitized solar cells, QDs are used as light absorbers, and the architecture of the solar cell is analogous to the DSSC where the dye is replaced by the QDs. $MAPbI_3$ perovskite QDs-sensitized solar cells showed power conversion efficiency of 6.5% [25]. These QDs of size 2−3 nm are prepared using perovskite precursor solution (equimolar CH_3NH_3I and PbI_2 in γ-butyrolactone solution) on nanocrystalline TiO_2. By increasing the concentration of the equimolecular mixture of CH_3NH_3I and PbI_2 solutions, the color is changed from yellow (10.05%) to black (40.26%) that was attributed to the quantum size effect or the increased density of the QDs. The color change of the $CH_3NH_3PbI_3$ QDs deposited on the TiO_2 film was explained by studying the UV−vis spectrum of the QDs, which shows a simultaneous increase in the absorption intensity or absorption onset wavelength as the concentration of perovskite precursor solution increases. $CH_3NH_3PbI_3$ QD sensitized solar cells used iodide/iodine-based redox electrolyte. These $CH_3NH_3PbI_3$ perovskite QD sensitized photovoltaic devices were found to be highly unstable under continuous irradiation, as the perovskite QDs gradually dissolved into the electrolyte. All-inorganic perovskite QD-based solar cells were more efficient and showed better stability compared to the solar cells based on organic/inorganic halide perovskite QDs. For example, $CsPbX_3$ QDs possessed attractive optical properties and exhibited increased cubic-phase stability at RT. Luther et al. reported a synthetic route and purification approach for $CsPbI_3$ QDs, in which the QDs preserved cubic phase for months in ambient air and cryogenic temperature. Solar cells based on $CsPbI_3$ QDs showed enhanced photovoltaic performance with a 10.7% power conversion efficiency, stabilized power output and high open circuit voltage about 1.23 V [26].

One- and two-dimensional perovskites were also used as active materials in solar cells. The reduced dimensionality of the perovskite material is advantageous for the enhanced charge carrier transport. One of the methods used to enhance the charge transport and collection in perovskite solar cells was, by employing the electron injection layer in its one-dimensional structure. In other words, to use one-dimensional perovskite materials as light absorbers in solar cells, the electron injecting layers such as nanostructures of TiO_2 or ZnO were used as scaffolds. Zinc oxide nanorod based perovskite solar cells reported by Park et al. have shown conversion efficiency of 11% [27]. In their work, they have grown the ZnO nanorods on FTO, and the perovskite layer was deposited by two-step spin-coating method, which formed a capping layer on the ZnO nanorod film. The manipulation of the size of the ZnO nanorod is one of the key factors in determining the overall performance of the solar cells. The dimension of the ZnO nanorods, its diameter and length were controlled by adjusting the precursor solution concentration and the immersing time of the film in the solution. This particular device architecture showed a better charge collection and found to be beneficial for better electron transport. In perovskite solar cells based on the perovskite thin films, the charge transport was hindered by the grain boundaries of the NCs, which implies that the carrier transport in the 3D system is not efficient. This, in turn, can affect the charge separation at the n-type oxide/perovskite and the perovskite/hole-transporting layer (HTL) interfaces and lead to an increased recombination rate. Charge transport and the recombination mechanisms are the crucial factors determining the photovoltaic performance of the device, where the charge carrier recombination was faster for mesoscopic structures. Im et al. reported $CH_3NH_3PbI_3$ ($MAPbI_3$) NW solar cell with a power conversion efficiency of 14.71%. $MAPbI_3$ NWs with average diameter 100 nm and length 1 μm were grown using a simple two-step spin-coating method, without using any nanostructure scaffolds for photovoltaic application

[28]. This device structure improved the perovskite/HTL interface and resulted in the better migration of hole from perovskite to HTL and increases the hole collection. NW structures showed better charge separation and improved lateral conductivity than that of the bulk film, found to be promising for photovoltaic applications.

The structural dimensionality of the perovskite materials profoundly influences their physical properties [29]. Due to the structural differences (quantum effects), 2D-perovskite materials were expected to show an environmental stability than that of 3D-perovskite materials. Thus, these materials were also demonstrated for the photovoltaic applications. Photovoltaic devices based on the two-dimensional layered perovskites were first studied by Smith et al. [30]. In their study, they demonstrated the advantages of using the layered perovskite material as a light absorber in the solar cell. Compared to the three-dimensional perovskite materials, the layered structure allows high-quality films that are formed using the spin coating on the substrate without any additional high-temperature processing. Smith et al. synthesized $n=3$ member of the series of $(PEA)_2(MA)_{n-1}[Pb_nI_{3n+1}]$ ($PEA=C_6H_5(CH_2)_2NH_3^+$), $n=$ number of Pb-I sheets in each inorganic layer) by combining (PEA)I, (MA)I ($MA=CH_3NH_3^+$), and PbI_2 in a 2:2:3 stoichiometric ratio in a solvent mixture of nitromethane/acetone. Slow solvent evaporation yielded the red crystals of $(PEA)_2(MA)_2[Pb_3I_{10}]$, and the crystallographic characterization shows the $n=3$ perovskite. They have shown that unlike that of 3D-perovskite materials, film preparation of two-dimensional layered perovskites and the device based on these materials do not require any controlled atmosphere for fabrication. The structure of the photovoltaic device with layered perovskite as the light absorber includes TiO_2, which was used as the electron selective layer and spiro-OMeTAD as the hole selective contact. For the active layer preparation, a solution of *N,N*-DMF containing the stoichiometric quantities of the precursors was deposited using one-step spin-coating method. These films were prepared under ambient air condition, and they formed high-quality films with greater moisture resistance. The photovoltaic device based on layered 2D-perovskite exhibited a PCE = 4.73%, $V_{OC}=1.18$ V, $J_{SC}=6.72$ mA/cm^2, and FF = 0.60. Compared to the 3D metal halide perovskites, layered 2D perovskite shows poor photovoltaic properties, but these devices showed better stability in ambient air conditions. The stability of the 2D films was tested using powder X-ray diffraction (PXRD) by exposing the spin-coated 2D-perovskite film in the controlled-humidity environment. The

PXRD pattern of the 2D film does not show any additional reflections over a period of 46 days. Also there was no significant change in the absorption spectra after the 46 days of humidity exposure.

To combine the structural features of 2D and 3D perovskites, layered perovskites in two-dimensional lead iodide perovskite $(CH_3(CH_2)_3NH_3)_2(CH_3NH_3)_{n-H}Pb_nI_{3n+1}$ series were studied as light absorbers in solar cells [31]. The 2D-perovskite film grown using one-step method showed better film coverage and more moisture resistant compared to the 3D-perovskite materials. The device showed power conversion efficiency 4.02%. Another 2D-perovskite-based solar cell was studied based on the three 2D alkyl diammonium lead iodide materials $[NH_3(CH_2)_4NH_3]PbI_4$ (**BdAPbI$_4$**), $[NH_3(CH_2)_6NH_3]PbI_4$(**HdAPbI$_4$**), and $[NH_3(CH_2)_8NH_3]PbI_4$ (**OdAPbI$_4$**). These devices did not show significant power conversion efficiencies rather; this work shed light to the effect of structural dimensionality on physical properties of the materials [32]. Two-dimensional perovskite crystals of Ruddlesden–Popper phase attained a photovoltaic efficiency of 12.52%. To facilitate the charge transport in the device, they produced perovskite thin films of single crystalline quality in which the alignment of the crystallographic plane of the inorganic perovskite component is in out-of-plane with respect to the contacts [33]. These devices delivered improved stability toward light, humidity, and moisture.

The environmental toxicity and the international waste disposal regulation of Pb-based perovskites is a major concern in the photovoltaic research community. In order to avoid this, Pb-free perovskite materials for photovoltaic applications were developed. Liao et al. reported LD tin perovskites with enhanced air stability compared to its 3D counterpart. They attributed the enhanced stability of the LD tin perovskite to the improved thermodynamic stability resulting from the reduced dimension, encapsulating organic ligands and the compact perovskite film preventing the oxygen penetration. To explore the photovoltaic properties of the LD $FASnI_3$ (formamidinium tin iodide), highly oriented LD $FASnI_3$ perovskite films were grown using the PEA, an organic separating interlayer on nickel oxide (NiO_x) substrates [34]. The orientation of the LD perovskite film on the substrate was controlled by varying the PEA ratio, and introduction of a 20% PEA resulted in the highly oriented perovskite domains perpendicular to the substrate. In the photovoltaic device structure ITO/NiO_x/LD tin perovskite/PCBM/Al, PCBM ([6,6]-phenyl-C_{61}-buteric acid methyl ester) acts as electron-transporting layer. The device gave a

PCE of 5.94%, open circuit voltage 0.59 V, short circuit current density 14.44 mA/cm^2, and fill factor 69%. LD Sn perovskite-based solar cell exhibited enhanced environmental stability sustained over 100 hours. Though the power conversion efficiency of the LD Sn-based solar cell is far below compared to that of the Pb-based 3D perovskites, there are interesting factors about Sn-based solar cells. The highlight of the work by Liao et al. is the synthesis of a lead-free perovskite material that is significantly stable to the humid conditions. Also the highly oriented LD Sn perovskite film was grown perpendicular to the substrate; the perpendicular orientation of the perovskite domains enabled the improved charge carrier transport. As far as the stability of the device is concerned, it can be realized by the effective encapsulation.

7.3.2 Light-Emitting Diodes

The electrical to optical conversion efficiency of LED is called the radiant efficiency (RE) or wall plug efficiency (WPE) and is defined as

$$\mathrm{RE(WPE)} = \mathrm{EQE} \times \mathrm{LE} \tag{7.1}$$

$$\mathrm{RE(WPE)} = \eta_{\mathrm{injection}} \times \mathrm{IQE} \times \eta_{\mathrm{extraction}} \times \frac{L}{IV} \tag{7.2}$$

where EQE is the external quantum efficiency that is defined as the ratio of the number of photons emitted to the number of electrons injected into the device, and light emitting (LE) is the ratio of the number of photons emitted to the energy injected by the source. The most important parameter that determines the LE performance of the materials is its PLQY. PLQY is the ratio of the emitted to the absorbed photons.

Perovskite materials possess high PLQY and are considered to be efficient LE materials. Furthermore, these materials show high color purity (Full width at half maximum (FWHM) ~ 20 nm) [10], broad color tenability over visible and near-infrared (400 nm $\leq \lambda \leq$ 785 nm) wavelength region [35], and low ionization energy forming stable, functional interfaces [36]. The high color purity or the FWHM shown by the perovskite material is insensitive to the crystallite size. In the case of hybrid halide perovskites, it is the Frohlich interaction between the charge carriers and longitudinal optical (LO) phonos, which results in the LO phonon scattering and electron–phonon coupling, determines the homogeneous line width (FWHM ~ 20 nm) in these materials at

RT [37]. It is only the crystal structure that makes a significant effect on the FWHM, not the crystal size, impurities, or the trap states present in the perovskite materials. Because the impurity contributed, line-broadening width is negligible, and trap-assisted recombination was nonradiative decay. Enhanced radiative recombination improves the electron–phonon conversion efficiency, and it depends on the binding energy of the electron–hole pair formed as a result of photoexcitation or with external field injection in LED. These electron–hole pairs can dissociate into free charge carriers or can form excitons with certain binding energy. Perovskite materials possess high dielectric constants, and these materials have Wannier-type excitons. At RT, there is no electron–hole wave function overlap, and no exciton–exciton interaction occurs, resulting in the dissociation of Wannier excitons into free charge carriers. Trapped charge carriers are also turned into free charge carriers or undergo nonradiative decay.

To improve the luminescence efficiency, the exciton formation in materials has to improve over the free charge carrier formation. This can be possible by high excitation density that increases the number of charge carriers and favors the increased rate of wave function overlap. Another method adopted to improve the electron–hole wave function overlap is to reduce the dimensionality of the materials, or in other words confining the perovskite materials to a size <10 nm. The reduced dimensionality results in the confinement of charge carriers and enhances the radiative recombination. In this regard, LD perovskite materials are extensively studied in LED applications.

Perovskite material based LED was first reported by Tan et al. The organic/inorganic hybrid perovskite-based LED used a thin layer of $CH_3NH_3PbCl_{3-x}I_x$ as LE material. The improved light-emission performance of the perovskite-based LEDs showed great promise in the application of these materials in optoelectronic devices. The main drawback of using the bulk material for LED application was the low exciton-binding energies, which hinders the electron–photon conversion efficiency. Zhang et al. developed highly luminescent $CH_3NH_3PbX_3$ QDs using ligand LARP method [10]. They compared the luminance of $CH_3NH_3PbBr_3$ QDs ($\sim$3.3 nm) to that of the corresponding micrometer (2–8 μm) bulk particle. An increased PLQY (up to 70%) was found in the small-sized $CH_3NH_3PbBr_3$ QDs and was attributed to the increases electron–hole binding energy due to the small size and the chemical passivation due to Br rich surface. $CH_3NH_3PbX_3$ QD-based

LED showed narrow FWHM (20–50 nm), high PLQY (50%–70%), and wide wavelength tunability (400–750 nm).

Perovskite LEDs were also studied using the colloidal halide perovskite QDs that show enhanced PL properties. Zhong et al. [38] reported the LE properties of colloidal $CH_3NH_3PbBr_3$ QD-based LEDs. Colloidal metal halide perovskite QDs were synthesized by addition of demulsifier into the $CH_3NH_3PbBr_3$ colloidal solution to control the precipitation in the emulsion. This controlled process of emulsion and demulsion gave monodisperse $CH_3NH_3PbBr_3$ QDs with tunable size 2–8 nm and showed PLQY of 80%–92%. These QDs were extracted from the solution through controlled demulsion process and then purified into QD powder for further studies. The colloidal QD-based EL devices were fabricated with structure ITO/PEDOT:PSS/$CH_3NH_3PbBr_3$ QDs/TPBi/CsF/Al, were colloidal $CH_3NH_3PbBr_3$ QDs dissolved in toluene was spin coated on to the PEDOT:PSS layer. The green LE device showed emission peak at 524 nm, and a maximum current efficiency 4.5 Cd/A, external quantum efficiency 1.1%, and power efficiency 3.5 lm/W. Colloidal perovskite $CH_3NH_3PbX_3$ nanoparticle-based LEDs were further studied in which the emission is tuned over the entire visible region and showed high photoluminescence quantum efficiency. Xing et al. synthesized that colloidal halide perovskite ($CH_3NH_3PbX_3$) nanoparticles with amorphous structure were prepared using the solution method and the LE properties were studied [39]. To enhance the luminescence performance of the perovskite LEDs, the radiation recombination has to be improved. And in organic LEDs, this was achieved by increasing the exciton-binding energy using the amorphous-emitting layers. In the research work by Xiong et al., the synthesis of $CH_3NH_3PbBr_3$ NPs in an amorphous state was synthesized using solution method. The perovskite precursor solutions were prepared by dissolving CH_3NH_3Br, $PbBr_2$, and octyl amine into the DMF and γ-butyrolactone. A certain amount of the precursor solution was added to toluene while stirring. Further, the green colloidal solution was centrifuged at high speed to ovoid the aggregation and to obtain the perovskite colloidal NPs. High performance was shown by colloidal $CH_3NH_3PbBr_3$ NP-based green LEDs. The maximum luminescence efficiency reported was 11.49 cd/A, a power efficiency, and an external quantum efficiency of 7.84 lm/W of 3.8%.

Though organic/inorganic halide perovskite showed high luminescent properties, considering the stability issues, LEDs based on all inorganic perovskite materials were studied widely and showed huge potential.

Song et al. studied the all inorganic cesium halide perovskite QD-based LEDs for the first time [40]. The inorganic perovskite QDs ($CsPbX_3$, X = Cl, Br, I) were prepared using the hot injection method; the average size of the resulting QDs was about 8 nm. They studied the LE properties using a simple LED structure, ITO/PEDOT:PSS/PVK/PeQDs/TBPi/LiF/Al. These LEDs exhibit blue, green, and yellow emission with external quantum efficiencies of 0.07%, 0.12%, 0.09%, respectively. The emission wavelengths of the devices were tuned by changing the size and the halide component (Cl, Br, and I) of the QDs. Considering the fact that the inorganic halide perovskites exhibit more stability under atmospheric environment than that of the organic/inorganic halide perovskites, and by combining the advantage of both QDs and halide perovskites, colloidal all-inorganic halide perovskite QD-based LEDs were fabricated and studied [41]. The colloidal $CsPbX_3$ QDs, synthesized using solution processing, were bandgap tunable and exhibited ultrastable stimulated emission at RT under ambient conditions.

To improve the performance of the perovskite-based LEDs, the most important factor is the optimizing the condition for the formation of a thin, uniform, and pinhole-free film of the active layer. Apart from the synthesis part of the materials, such as lowering the dimension of the materials and utilizing the advantages of the quantum confinement of the materials in optical applications, different studies were carried out in optimizing end in engineering the device architecture and the LE layer of the device. Li et al. [42] fabricated LEDs using the perovskite NCs embedded in a dielectric polymer matrix. For the active layer formation, the perovskite NCs were formed by simple mixing of the perovskite precursor and the polymer followed by annealing. The device structure of the LED was ITO/PEDOT:PSS/Perovskite-PIP/F8/Ca/Ag and the device delivered an EQE of 1.2%. They used $CH_3NH_3PbBr_3$ NCs as green light emitter and the polyimide precursor as a pinhole-free charge blocking layer. These perovskite NCs were uniformly distributed and showed good light emission, whereas the polymer dielectric filled the surrounding voids and avoided the nonradiative recombination. Other than providing a pinhole-free active layer, this modification also provided the benefits of flexibility. In addition to the quality of the active layer, the performance of the optoelectronic devices, including photovoltaic cells and LEDs, depends on the interface of the devices and has to be optimized for proper charge extraction or injection. Inorganic halide perovskite NCs, especially cesium lead halide perovskite NCs showed improved photoluminescence and high quantum yield. The main drawback of the $CsPbX_3$-based LEDs was the presence of the charge carrier injection barrier, which hindered the efficient charge injection to the active layer through the charge-transport layer. Zhang et al. addressed the issue of inefficient hole injection in the $CsPbBr_3$ NCs-based LED by introducing an interfacial layer [43]. In the device structure, they incorporated a thin layer of perfluorinated ionomer (PFI) between the hole-transport layer (HTL) and the perovskite NC active layer. The bright green $CsPbBr_3$ NCs-based LED showed enhanced emissive performance with a peak brightness of 1377 cd/m^2, and FWHM value ~18 nm. The incorporation of a thin layer (~5 nm) of PFI between the HTL and the active layer reduced the luminescence quenching by preventing the NCs from getting charged.

Other than perovskite NCs, one-dimensional perovskite nanorods or NWs and two-dimensional perovskite platelets are also employed in LE devices. First, solution-processed organic/inorganic halide perovskite nanorods were studied for LED application by Yang et al. [15]. In their study, they synthesized $CH_3NH_3PbBr_3$ nanorod arrays by solution processing. Another nanorod array of $CH_3NH_3PbI_3$ was prepared by converting $CH_3NH_3PbBr_3$ using anion exchange reaction. First, hybrid perovskite nanorod based LEDs were fabricated using these nanorod arrays. $CH_3NH_3PbBr_3$ nanorod array based LED exhibited electroluminescence at 533 nm with FWHM of 26 nm. LED fabricated using $CH_3NH_3PbI_3$ as a light emitter exhibited electroluminescence at 782 nm. To produce efficient PeLEDs, two-dimensional perovskite or nanoplatelets that showed excellent photoluminescence properties were also used. $CH_3NH_3PbBr_3$ nanoplatelet based LEDs exhibited emission peak at 530 nm, showed high color purity (FWHM 20 nm) and maximum electroluminescence 10,590 cd/m^2 at a voltage of 12 V. These nanoplatelets were stable and showed good resistances to moisture, which enabled the fabrication of the LEDs to carry out outside the glove box.

Though perovskite materials are found to be promising LE materials with their implementation in LEDs, there are certain issues that need to be addressed for further improvement of the device performance. The charge injection and transport have to be efficient for enhancing the internal quantum efficiency of the devices. The band levels of the charge-transport layers and that of the perovskite active layer have to be matched. The tunability of the emission need to be achieved to yield high color purity and for white light emission.

7.3.3 Photodetector

In a photodetector, the incident light is converted into voltage or current. Examples of photodetectors are photodiodes and phototransistors. Working principle of photodetectors is similar to that of the solar cells, in which the incident solar energy is converted into electrical energy. Metal halide perovskite materials are excellent light harvesters and have shown enhanced photovoltaic properties. The primary requirement of the material in photodetector is the sensitivity of the incident light; therefore, the materials used in the photovoltaic applications can also be used for photodetectors. As far as perovskites are concerned metal halide perovskites are highly sensitive to the visible light, and these are used in photodetectors. The bulk perovskite thin films used in optoelectronic devices contain microsized grain boundaries and defects. The nonuniform films can cause low efficiency and responsivity in photodetectors. To overcome the issues caused by the perovskite thin films, LD perovskites were employed for photodetecting applications. Horváth et al. [14] fabricated $MAPbI_3$ ($CH_3MH_3PbI_3$) NWs by low-temperature solution-processed crystallization using slip-coating method. These NWs showed high sensitivity to visible light and delivered high photovoltage of 1.1 V, which are the necessary criteria to convert solar energy into electrical energy [14]. Horváth et al. explored these optoelectronic properties of the material in the device aspect and used $CH_3NH_3PbI_3$-based NWs in photodetectors for the first time. The devices showed response time less than 500 μs, in other words, $\sim 10^4$ times faster than the state-of-the-art of the photodetectors based on monolayer MoS_2. Perovskite NWs show unique optical and electrical properties that can be attributed to the one-dimensional quantum confinement of the charge carriers, its anisotropic geometry and a large surface-to-volume ratio. These properties make them potential candidate in optoelectronic and photonic devices [44]. In the case of any optoelectronic devices based on NWs, the distribution of the NWs on the substrate is a crucial part. The solution-processed NWs tend to distribute randomly on the surface of the substrate, and this can cause poor performance and also causes a mismatch in the distribution between the random NWs and the electrodes. Therefore to improve the performance of the photodetectors, uniform distribution of NW or ordered NW arrangement by patterning or alignment were employed. Song et al. [12] reported the photodetectors fabricated using organic lead iodide perovskite NWs, in which they have used special strategies for the alignment and distribution of the NWs. For large-area alignment of the NWs, the commonly used method is EISA method. Song et al. fabricated the organic lead iodide perovskite NWs by one-step self-assembly solution method, and they used modified EISA method for the NW alignment and selected area treatment for the NW patterning. Selected area treatment involves the process of treating the substrate surface to make it hydrophilic for better spreading of the solution in the required area. And this is done by UV-ozone treatment combined with a shadow mask. The photodetectors based on the uniformly distributed monolayer $CH_3NH_3PbI_3$ perovskite NWs exhibit enhanced performance compared to the previous ones on the random NWs. The response time reached 0.3 ms, and it shows a responsivity of 1.3 A/W and detectivity of 2.5×10^{12} Jones. In addition to the NWs, the optical and electrical properties of porous NWs are also explored in optoelectronic devices. The unique porous structure of the NWs is advantageous as it reduces the charge-transport distance of electrons, and holes, also it provides a large number of active centers on their surface. Considering the potential of the porous NWs in optoelectronic devices, porous $CH_3NH_3PbBr_3$ NWs were used in the visible photodetectors by Zhang et al. [18]. These porous NWs were prepared using self-template-directed method with a length about 10 μm and diameter of 200 nm. Porous $CH_3NH_3PbBr_3$ NWs were used in visible photodetector that showed a photoresponse time of 0.12 seconds, and decay time 0.086 seconds. The advanced performance of the porous $CH_3NH_3PbBr_3$ NWs in photodetectors was attributed to its enhanced light absorption, porous and one-dimensional geometry, and increased number of active sited on the surface.

Further development in the perovskite photodetector research comes with the study of flexible photodetector arrays. Though the photodetectors (PDs) based on the perovskite thin film and NWs showed good performances, most of them were based on the individual PDs. The practical applications of the photodetectors demand a large number of pixels to create images. To improve the performance of the PDs for the imaging applications, PD arrays were developed. One of the challenges to overcome was, to maintain the uniformity of PD arrays to avoid the addition of the noise signal [45]. Also combining the uniformity and the transparency of the PD arrays expands its application to other fields such as in building integrated photovoltaic devices [46], smart touch screen [47], transparent display [48], and so on. Song et al. reported the flexible and transparent photodetectors based on the organolead triiodide perovskites (OTP) [49]. In their study, the PD arrays with uniformity and transparency were developed via controlling the crystallization. These perovskite PDs were flexible as it was fabricated on polyethylene terephthalate, which showed wide angle (20−80°) and a large number of (~10,000) bending. These flexible transparent PD arrays could detect spatial light intensity distribution. Also, these OTP PDs exhibited enhanced photodetector parameters with reaching the responsivity 0.1 A/W, switching ratio 300, normalized detectivity 1.02×10^{12} Jones, and equivalent dark current derived shot noise 4.73×10^{-15} A/Hz$^{1/2}$.

Inorganic perovskites materials that show excellent photovoltaic properties and stability were also used in photodetector applications. Lee et al. performed the halide exchange reaction in $CsPbBr_3$ colloidal NCs to produce $CsPbBr_3$ and $CsPbI_3$ NCs. Photodetectors were fabricated based on red-emitting $CsPbI_3$ perovskite NCs, which shows longer radiative lifetime. These photodetectors based on $CsPbI_3$ NCs exhibited on/off photocurrent ratio of 10^5 and rise and decay tie of 24 and 29 ms, respectively [50]. To study the application of LD perovskite materials in photodetectors, the unique properties of 2D perovskites were used. These 2D planar materials can also implement in flexible devices. As these materials possess ultrasmooth surface, they are easily compatible with flexible surfaces. Further, compared to the bulk materials, 2D ultrathin layered materials show high optical response, and longer photocarrier lifetime due to the spatial confinement effects. One of the drawbacks of these LD materials is the weak photocurrent generation because of the limited optical absorption cross section. To make use of the unique feature of the LD materials in photodetectors applications, materials with highlight absorptivity and large carrier diffusion length must be produced. Zeng et al. [51] reported the solution-processed flexible photodetectors based on atomically thin 2D $CsPbBr_3$ nanosheets. These nanosheets with thickness 3.3 nm and edge length of about 1 μm were synthesized by solution-processed technique, and they were suitable to apply in the large-area flexible devices. The all inorganic $CsPbBr_3$ nanosheet based visible light photodetectors exhibited enhanced performance with a photosensitivity with a good on/off ratio higher than 10^3 under 442 nm light at a power of 0.34 mW/cm^2, excellent electronic transport properties, and high flexibility (>10,000 cycles). The photodetectors showed high stability under ambient conditions, with a fluctuation of less than 2.6% after irradiating for 12 hours with a 442-nm laser (10 mW/cm^2). Zeng et al. work on high-performance flexible photodetectors based on $CsPbBr_3$ nanosheets show the potential of the all inorganic LD materials in optoelectronic device applications.

7.3.4 Laser

The miniaturized light sources play a crucial role in the advancement of the field of optoelectronics, and therefore extensive studies are carried out to develop highly efficient wavelength tunable and stable light sources on the nanoscale dimension. Semiconductor NW lasers can be considered to represent one of the promising miniaturized light sources. NWs act as both laser cavity and gain medium. Further, NW lasers show efficient wave guiding and highly localized coherent output. The one-dimensional geometry of the NWs and its refractive index facilitates the confinement of the photonic modes guided along the axial directions and giving the low loss wave guiding. There are different NW lasers reported based on inorganic semiconductors including the ZnO, GaAs, CdS, and GaN which spans the UV–vis part of the spectrum. The synthesis of NWs involves procedures to carry out under high-temperature conditions. New efficient materials and designing would be required to implement the RT lasing.

organic–inorganic lead halide perovskite materials were considered to be the potential candidates for the NW lasers. The optoelectronic properties such as long charge carrier lifetime, long diffusion length, high-absorption coefficient, strong photoluminescence, high quantum yield; wavelength tunability, and high gain make them efficient for lasing applications. Sum et al. demonstrated the potential application of the low-temperature solution-processed organic–inorganic lead halide perovskite ($CH_3NH_3PbX_3$ where X = Cl, Br, I) materials in lasing. Their finding shows that the $CH_3NH_3PbX_3$ synthesized through the solution process technique exhibited ultrastable spontaneous emission at a low threshold. It was attributed to their low bulk defect densities, large absorption coefficients, and slow Auger recombination. These materials also showed broad wavelength tunability in the range of 390–790 nm and electrically driven lasing that arises from its ambipolar charge-transport characteristics. One of the barriers in producing highly efficient NW lasers was the high threshold carrier density required for lasing. Lasing threshold describes the operation condition of a laser where the laser emission begins to occur. Zhu et al. demonstrate the fabrication of single-crystal perovskite NW lasers in low-temperature solution-processed method and its room-temperature lasing. The $CH_3NH_3PbI_3$ NWs showed a length about 20 μm with flat rectangular facets. This NW laser showed RT lasing with low threshold ~220 nJ/cm^2 value and high-quality factor ~3600. In this work, they have also shown the tunability of the emission color across the visible spectrum with mixed halides. Enhanced lasing performances of these NW lasers were attributed to the long carrier lifetime and low non-radiative recombination rate. Nanolasers based on organic–inorganic perovskites suffered low stability due to autodegradation and susceptibility to hydrolysis from atmospheric water [52,53], and this limits their utility. Replacing the organic methylammonium cation with cesium was a strategy to improve the stability of the device. Cesium lead halide NWs were favorable materials for lasing applications, and it showed improved stability under ambient conditions. Lasing in all inorganic perovskite NWs (cesium lead bromide and cesium lead chloride) was reported by Yang et al. This single-crystal NWs were grown using the solution processing under low-temperature conditions. Inorganic cesium halide perovskite NW laser exhibited excellent wavelength tunability in the blue, and green region of the visible spectrum through the halide substitution $CsPbBr_3$ NW laser showed enhanced lasing performance with a low threshold value (5 $\mu J/cm^2$) and high Q factor (1009 ± 5).

A different strategy used for increasing the ambient stability of the perovskite-based nanolaser was substituting the methylammonium with less reactive cation such as formamide. Song et al. [54] reported the solution processes single-crystal NW lasers formamidinium lead perovskites with improved lasing performance and better thermal stability. Exchange of organic cation of $MAPbX_3$ from methylammonium to formamidinium resulted in the semiconductor with lower bandgap 1.47 eV and improved temperature and moisture stability [55]. $FAPbX_3$ NW laser exhibited RT lasing in the near IR ~ 820 nm ($FAPbI_3$) region and green ~60 nm ($FAPbBr_3$). FA-based NW laser showed low lasing threshold (~several $\mu J/cm^2$) and high-quality factor (~2000). Compared to the NW lasers based on MA, FA-based ones are widely wavelength tunable from visible to near-IR region and photostable.

7.4 CONCLUSION

LD perovskite materials show unique optical and electronic properties compared to their bulk counterpart. These unique physical and chemical properties of LD perovskite materials show the influence of their structural dimensionality. Organic metal halide perovskite materials with its optical properties, high photovoltaic performance, and emission properties have been attracted the material and device research communities in recent years. To explore the quantum-sized effect of the LD perovskite materials for optical and photovoltaic application, a significant amount of research has been conducted in this field. This chapter on LD perovskite materials offers a summary of the different LD perovskite materials of different sizes and shapes, and their controlled synthesis routes. The synthesis processes of the QDs, NWs, and nanoplatelets were developed in the solution-processing technique under low-temperature conditions to facilitate the application of these materials in optoelectronic devices. The development of the layered perovskite materials including its structure, synthesis, and optical properties were also discussed. In addition to the material synthesis procedures, this chapter gives an overview of the recent studies demonstrating the application of the LD perovskite materials in photovoltaics, LEDs, photodetectors, and lasing applications. The application of the LD-layered perovskite in the photovoltaic devices showed an enhanced environmental stability, which was one of the major issues in the case of devices based on the bulk organic–inorganic halide perovskites. It shows that these materials are promising ones for the future optoelectronic and photonic applications. The extensive study to explore the optical and electrical properties of LD perovskite materials has led to a huge advancement in the performance in various application fields. Though the research on the LD perovskite materials on different optoelectronic devices has achieved a significant improvement in the laboratory research level, practical application is still an issue. Rigorous research progressing to develop high efficient and stable perovskite-based devices, mainly concentrating on the issues of the disadvantage of perovskite materials, is hopeful.

REFERENCES

[1] A. Kojima, K. Teshima, Y. Shirai, T. Miyasaka, Organometal halide perovskites as visible-light sensitizers for photovoltaic cells, J. Am. Chem. Soc. 131 (2009) 6050.
[2] W.S. Yang, J.H. Noh, N.J. Jeon, Y.C. Kim, S. Ryu, J. Seo, et al., High-performance photovoltaic perovskite layers fabricated through intramolecular exchange, Science 348 (2015) 1234.
[3] H. Zhou, Q. Chen, G. Li, S. Luo, T.-B. Song, H.-S. Duan, et al., Interface engineering of highly efficient perovskite solar cells, Science 345 (2014) 542.
[4] J. Burschka, N. Pellet, S.-J. Moon, R. Humphry-Baker, P. Gao, M.K. Nazeeruddin, et al., Sequential deposition as a route to high-performance perovskite-sensitized solar cells, Nature 499 (2013) 316.
[5] Z.-K. Tan, R.S. Moghaddam, M.L. Lai, P. Docampo, R. Higler, F. Deschler, et al., Bright light-emitting diodes based on organometal halide perovskite, Nat. Nanotechnol. 9 (2014) 687.

[6] H. Cho, S.-H. Jeong, M.-H. Park, Y.-H. Kim, C. Wolf, C.-L. Lee, et al., Overcoming the electroluminescence efficiency limitations of perovskite light-emitting diodes, Science 350 (2015) 1222.
[7] Y.I. Dolzhenko, T. Inabe, Y. Maruyama, In situ X-ray observation on the intercalation of weak interaction molecules into perovskite-type layered crystals $(C_9H_{19}NH_3)_2PbI_4$ and $(C_{10}H_{21}NH_3)_2CdCl_4$, Bull. Chem. Soc. Jpn. 59 (1986) 563.
[8] N. Giesbrecht, J. Schlipf, L. Oesinghaus, A. Binek, T. Bein, P. Müller-Buschbaum, et al., Synthesis of perfectly oriented and micrometer-sized $MAPbBr_3$ perovskite crystals for thin-film photovoltaic applications, ACS Energy Lett. 1 (2016) 150.
[9] A. Kojima, M. Ikegami, K. Teshima, T. Miyasaka, Highly luminescent lead bromide perovskite nanoparticles synthesized with porous alumina media, Chem. Lett. 41 (2012) 397.
[10] F. Zhang, H. Zhong, C. Chen, X.-G. Wu, X. Hu, H. Huang, et al., Brightly luminescent and color-tunable colloidal $CH_3NH_3PbX_3$ (X = Br, I, Cl) quantum dots: potential alternatives for display technology, ACS Nano 9 (2015) 4533.
[11] H. Huang, A.S. Susha, S.V. Kershaw, T.F. Hung, A.L. Rogach, Control of emission color of high quantum yield $CH_3NH_3PbBr_3$ perovskite quantum dots by precipitation temperature, Adv. Sci. 2 (2015) 1500194.
[12] H. Deng, D. Dong, K. Qiao, L. Bu, B. Li, D. Yang, et al., Growth, patterning and alignment of organolead iodide perovskite nanowires for optoelectronic devices, Nanoscale 7 (2015) 4163.
[13] P. Zhu, S. Gu, X. Shen, N. Xu, Y. Tan, S. Zhuang, et al., Direct conversion of perovskite thin films into nanowires with kinetic control for flexible optoelectronic devices, Nano Lett. 16 (2016) 871.
[14] E. Horváth, M. Spina, Z. Szekrényes, K. Kamarás, R. Gaal, D. Gachet, et al., Nanowires of methylammonium lead iodide ($CH_3NH_3PbI_3$) prepared by low temperature solution-mediated crystallization, Nano Lett. 14 (2014) 6761.
[15] A.B. Wong, M. Lai, S.W. Eaton, Y. Yu, E. Lin, L. Dou, et al., Growth and anion exchange conversion of $CH_3NH_3PbX_3$ nanorod arrays for light-emitting diodes, Nano Lett. 15 (2015) 5519.
[16] W. Deng, X. Zhang, L. Huang, X. Xu, L. Wang, J. Wang, et al., Aligned single-crystalline perovskite microwire arrays for high-performance flexible image sensors with long-term stability, Adv. Mater. 28 (2016) 2201.
[17] K. Park, J.W. Lee, J.D. Kim, N.S. Han, D.M. Jang, S. Jeong, et al., Light−matter interactions in cesium lead halide perovskite nanowire lasers, J. Phys. Lett. 7 (2016) 3703.
[18] S. Zhuo, J. Zhang, Y. Shi, Y. Huang, B. Zhang, Self-template-directed synthesis of porous perovskite nanowires at room temperature for high-performance visible-light photodetectors, Angew. Chem. Int. Ed. 54 (2015) 5693.
[19] M.J. Ashley, M.N. O'Brien, K.R. Hedderick, J.A. Mason, M.B. Ross, C.A. Mirkin, Templated synthesis of uniform perovskite nanowire arrays, J. Am. Chem. Soc. 138 (2016) 10096.
[20] K. Gauthron, J. Lauret, L. Doyennette, G. Lanty, A. Al Choueiry, S. Zhang, et al., Optical spectroscopy of two-dimensional layered $(C_6H_5C_2H_4-NH_3)_2-PbI_4$ perovskite, Opt. Express 18 (2010) 5912.
[21] W. Niu, A. Eiden, G. Vijaya Prakash, J.J. Baumberg, Exfoliation of self-assembled 2D organic-inorganic perovskite semiconductors, Appl. Phys. Lett. 104 (2014) 171111.
[22] S.T. Ha, X. Liu, Q. Zhang, D. Giovanni, T.C. Sum, Q. Xiong, Synthesis of organic−inorganic lead halide perovskite nanoplatelets: towards high-performance perovskite solar cells and optoelectronic devices, Adv. Opt. Mater. 2 (2014) 838.

[23] Q. Liao, K. Hu, H. Zhang, X. Wang, J. Yao, H. Fu, Perovskite microdisk microlasers self-assembled from solution, Adv. Mater. 27 (2015) 3405.
[24] J. Liu, Y. Xue, Z. Wang, Z.-Q. Xu, C. Zheng, B. Weber, et al., Two-dimensional $CH_3NH_3PbI_3$ perovskite: synthesis and optoelectronic application, ACS Nano 10 (2016) 3536.
[25] J.-H. Im, C.-R. Lee, J.-W. Lee, S.-W. Park, N.-G. Park, 6.5% efficient perovskite quantum-dot-sensitized solar cell, Nanoscale 3 (2011) 4088.
[26] A. Swarnkar, A.R. Marshall, E.M. Sanehira, B.D. Chernomordik, D.T. Moore, J.A. Christians, et al., Quantum dot–induced phase stabilization of α-CsPbI3 perovskite for high-efficiency photovoltaics, Science 354 (2016) 92.
[27] D.-Y. Son, J.-H. Im, H.-S. Kim, N.-G. Park, 11% efficient perovskite solar cell based on ZnO nanorods: an effective charge collection system, J. Phys. Chem. C 118 (2014) 16567.
[28] J.-H. Im, J. Luo, M. Franckevičius, N. Pellet, P. Gao, T. Moehl, et al., Nanowire perovskite solar cell, Nano Lett. 15 (2015) 2120.
[29] C.C. Stoumpos, C.D. Malliakas, M.G. Kanatzidis, Semiconducting tin and lead iodide perovskites with organic cations: phase transitions, high mobilities, and near-infrared photoluminescent properties, Inorg. Chem. 52 (2013) 9019.
[30] I.C. Smith, E.T. Hoke, D. Solis-Ibarra, M.D. McGehee, H.I. Karunadasa, A layered hybrid perovskite solar-cell absorber with enhanced moisture stability, Angew. Chem. 126 (2014) 11414.
[31] D.H. Cao, C.C. Stoumpos, O.K. Farha, J.T. Hupp, M.G. Kanatzidis, 2D homologous perovskites as light-absorbing materials for solar cell applications, J. Am. Chem. Soc. 137 (2015) 7843.
[32] M. Safdari, P.H. Svensson, M.T. Hoang, I. Oh, L. Kloo, J.M. Gardner, Layered 2D alkyldiammonium lead iodide perovskites: synthesis, characterization, and use in solar cells, J. Mater. Chem. A 4 (2016) 15638.
[33] H. Tsai, W. Nie, J.C. Blancon, C.C. Stoumpos, R. Asadpour, B. Harutyunyan, et al., High-efficiency two-dimensional ruddlesden–popper perovskite solar cells, Nature 536 (2016) 312–316.
[34] Y. Liao, H. Liu, W. Zhou, D. Yang, Y. Shang, Z. Shi, et al., Highly oriented low-dimensional tin halide perovskites with enhanced stability and photovoltaic performance, J. Am. Chem. Soc. 139 (2017) 6693.
[35] E.R. Dohner, E.T. Hoke, H.I. Karunadasa, Self-assembly of broadband white-light emitters, J. Am. Chem. Soc. 136 (2014) 1718.
[36] M. Kulbak, D. Cahen, G. Hodes, How important is the organic part of lead halide perovskite photovoltaic cells? Efficient CsPbBr3 cells, J. Phys. Chem. Lett. 6 (2015) 2452.
[37] A.D. Wright, C. Verdi, R.L. Milot, G.E. Eperon, M.A. Pérez-Osorio, H.J. Snaith, et al., Electron–phonon coupling in hybrid lead halide perovskites, Nat. Commun. 7 (2016) 11755.
[38] H. Huang, F. Zhao, L. Liu, F. Zhang, X.-G. Wu, L. Shi, et al., Emulsion synthesis of size-tunable $CH_3NH_3PbBr_3$ quantum dots: an alternative route toward efficient light-emitting diodes, ACS Appl. Mater. Interfaces 7 (2015) 28128.
[39] J. Xing, F. Yan, Y. Zhao, S. Chen, H. Yu, Q. Zhang, et al., High-efficiency light-emitting diodes of organometal halide perovskite amorphous nanoparticles, ACS Nano 10 (2016) 6623.
[40] J. Song, J. Li, X. Li, L. Xu, Y. Dong, H. Zeng, Quantum dot light-emitting diodes based on inorganic perovskite cesium lead halides ($CsPbX_3$), Adv. Mater. 27 (2015) 7162.
[41] Y. Wang, X. Li, J. Song, L. Xiao, H. Zeng, H. Sun, All-inorganic colloidal perovskite quantum dots: a new class of lasing materials with favorable characteristics, Adv. Mater. 27 (2015) 7101.

[42] G. Li, Z.-K. Tan, D. Di, M.L. Lai, L. Jiang, J.H.-W. Lim, et al., Efficient light-emitting diodes based on nanocrystalline perovskite in a dielectric polymer matrix, Nano Lett. 15 (2015) 2640.
[43] X. Zhang, H. Lin, H. Huang, C. Reckmeier, Y. Zhang, W.C. Choy, et al., Enhancing the brightness of cesium lead halide perovskite nanocrystal based green light-emitting devices through the interface engineering with perfluorinated ionomer, Nano Lett. 16 (2016) 1415.
[44] R. Agarwal, C. Lieber, Semiconductor nanowires: optics and optoelectronics, Appl. Phys. A 85 (2006) 209.
[45] A. Rogalski, History of infrared detectors, Opto-Electron. Rev. 20 (2012) 279.
[46] C.J. Emmott, J.A. Röhr, M. Campoy-Quiles, T. Kirchartz, A. Urbina, N.J. Ekins-Daukes, et al., Organic photovoltaic greenhouses: a unique application for semi-transparent PV?, Energy Environ. Sci. 8 (2015) 1317.
[47] J. Song, H. Zeng, Transparent electrodes printed with nanocrystal inks for flexible smart devices, Angew. Chem. Int. Ed. 54 (2015) 9760.
[48] Hsu C.W., Zhen B., Qiu W., Shapira O., DeLacy B.G., Joannopoulos J.D., and et al., presented at the Lasers and Electro-Optics (CLEO), 2014 Conference on, 2014 (unpublished).
[49] H. Deng, X. Yang, D. Dong, B. Li, D. Yang, S. Yuan, et al., Flexible and semitransparent organolead triiodide perovskite network photodetector arrays with high stability, Nano Lett. 15 (2015) 7963.
[50] P. Ramasamy, D.-H. Lim, B. Kim, S.-H. Lee, M.-S. Lee, J.-S. Lee, All-inorganic cesium lead halide perovskite nanocrystals for photodetector applications, Chem. Commun. 52 (2016) 2067.
[51] J. Song, L. Xu, J. Li, J. Xue, Y. Dong, X. Li, et al., Monolayer and few-layer all-inorganic perovskites as a new family of two-dimensional semiconductors for printable optoelectronic devices, Adv. Mater. 28 (2016) 4861.
[52] B. Conings, J. Drijkoningen, N. Gauquelin, A. Babayigit, J. D'Haen, L. D'Olieslaeger, et al., Intrinsic thermal instability of methylammonium lead trihalide perovskite, Adv. Energy Mater. 5 (2015).
[53] G. Niu, X. Guo, L. Wang, Review of recent progress in chemical stability of perovskite solar cells, J. Mater. Chem. A 3 (2015) 8970.
[54] Y. Fu, H. Zhu, A.W. Schrader, D. Liang, Q. Ding, P. Joshi, et al., Nanowire lasers of formamidinium lead halide perovskites and their stabilized alloys with improved stability, Nano Lett. 16 (2016) 1000.
[55] G.E. Eperon, S.D. Stranks, C. Menelaou, M.B. Johnston, L.M. Herz, H.J. Snaith, Formamidinium lead trihalide: a broadly tunable perovskite for efficient planar heterojunction solar cells, Energy Environ. Sci. 7 (2014) 982.

CHAPTER 8

Fabrication and Life Time of Perovskite Solar Cells

P. Mathan Kumar, Abinash Das, Lalu Seban and Ranjith G. Nair
Solar Energy Materials Research and Testing Laboratory (SMaRT Lab), National Institute of Technology Silchar, Cachar, Assam, India

8.1 INTRODUCTION

Photovoltaic (PV) is an emerging technology that converts light into electricity. A device which is working under this principle is known as solar cells. The development of solar cells can be classified into three generations such as wafer-based silicon solar cells, thin-film solar cells, and organic solar cells [1,2]. Organic solar cells consist of dye-sensitized solar cells (DSSCs) [3], organic PVs [4], inorganic cells [5], and quantum dot solar cells [6]. Third generation PV technology provides an efficient photoconversion at low cost of manufacturing, thus providing a promising alternative over the traditional silicon solar cells [7]. First generation PV technologies showed relatively low photoconversion efficiency of around $9.2 \approx 13\%$, later a vast research effort has been made over decades to improve the photoconversion efficiency to Shockley–Queisser limit [8]. Considering the recent energy crisis a new generation solar energy conversion system with high photoconversion efficiency will be a possible alternative over the conventional solar cells [9].

Perovskite solar cell (PSC) discovered by Lev Perovski in the year 2009 as a new class in third generation PV technology, evolved from DSSCs, that holds a major potential for future renewable energy [7]. During its discovery, PSCs showed a photoconversion efficiency $\approx 3.8\%$ [10], later the efficiency of PSCs has been improved to $\approx 22\%$ within this limited period [11]. Therefore this new technology attracts a huge interest in research community. Moreover, its various characteristics such as flexibility, light weight, and semitransparency can unlock many more applications for solar cell. In addition to that, it gives an essential idea to the electronic designers and researchers to achieve its maximum usage by tailoring its properties [12]. Hence, it is strongly believed that the PSCs

Perovskite Photovoltaics
DOI: https://doi.org/10.1016/B978-0-12-812915-9.00008-3

may soon become a promising alternative to conventional silicon solar cells because of its rapid growth in terms of photoconversion efficiency.

For the commercialization of PSCs, there are still many challenges to overcome such as degradation due to moisture, temperature, and ultraviolet (UV) light and resulting degradation of interface layers [10,13]. Moreover, the presence of lead (Pb) in methyl ammonium (MA) lead halides is a major issue. Several attempts have been made to find alternatives to lead and to prevent moisture in the perovskite layer [14]. Due to these reasons, perovskite technology is still routed as technology of the future. Many research efforts on PSCs are still focusing in improving its photoconversion efficiency, stability, and life time by giving attention on tuning the earlier parameters. Researchers are also trying to find the possibilities of combining perovskites with other technologies [15,16]. Meanwhile in advantage to other PV technologies, it can work at different wavelength spectra. Fig. 8.1 represents the growth in photoconversion efficiencies (PCEs) of various emerging solar cells over years [17]. It indicates the drastic rise in PCE of the PSCs compared to other emerging solar cells. The efficiency achieved by the PSCs within the first 3 years is equal to the efficiency achieved by the conventional solar cells in last four decades. This drastic rise in the efficiency of PSCs is significant and impressive compared to all other solar cells. Overall, PSCs are the promising technology to achieve high efficiency at reduced process costs [18].

Perovskite, named after Lev Perovski, is a mineral having the same crystal structure to that of calcium titanium oxide ($CaTiO_3$) and was first discovered in the Ural Mountains [19]. Perovskite structures occur naturally as minerals such as perovskite, loparite, and bridgmanite. Any material in the generic form ABX_3 where X is an anion, and A and B are cations (A being larger than B) are called as perovskite structure. In general perovskite is composed of calcium, titanium, and oxygen in the form $CaTiO_3$. The perovskite crystal structure is shown in Fig. 8.2. It is possible to assemble large number of different elements to form the perovskite structures. The most popular perovskite material namely MA lead iodide has a theoretical efficiency of around 31% [20]. Due to its distinct structure and easy synthesis procedure, it is a suitable, efficient, and low-cost material for PSCs. In addition to that, it allows researchers to selectively design and optimize perovskite structure by its physical, optical, and electrical characteristics.

PSC is a new competitor in the field of solar energy. Multiple researches are streaming on fabricating a high-efficient perovskite material

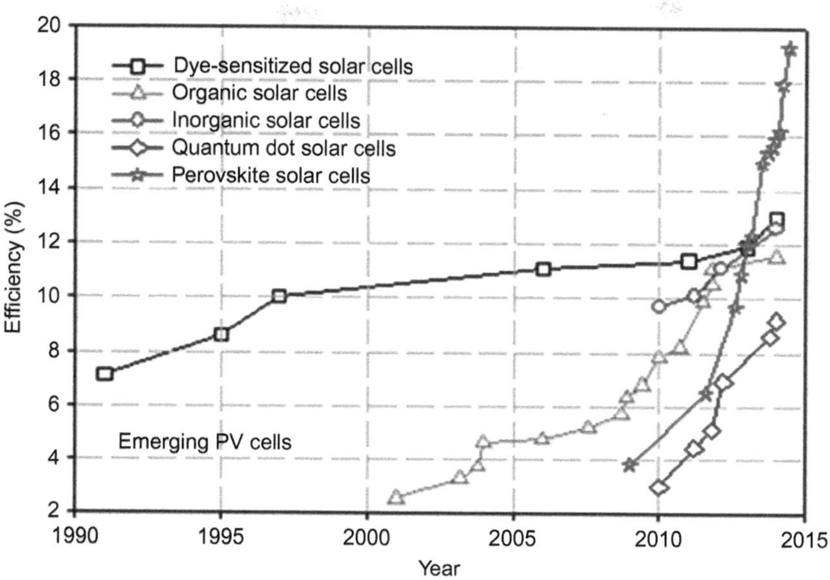

Figure 8.1 Efficiency records vs year of third generation (emerging) PVs. The red line embedded with star indicates that the efficiency of PSCs has improved dramatically in the past 2 years. *M. Batmunkh, C.J. Shearer, M.J. Biggs, J.G. Shapter, Nanocarbons for mesoscopic perovskite solar cells, J. Mater. Chem. A 3 (2015) 9020–9031, doi:10.1039/C5TA00873E reprint permission from RSC Publishers [17].*

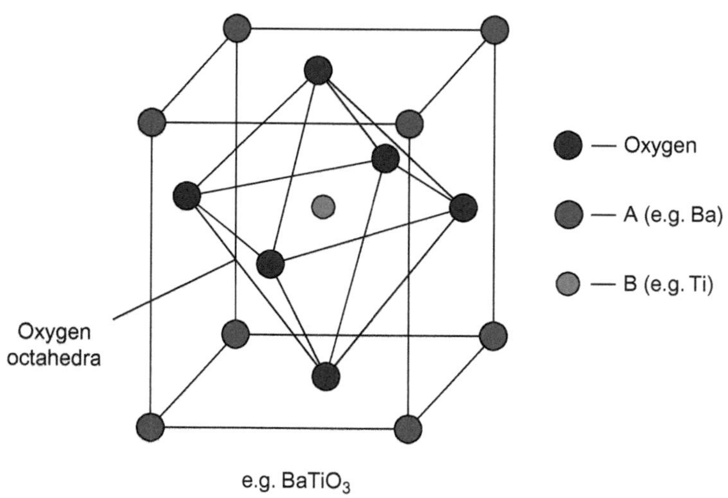

Figure 8.2 Crystal structure of ABX_3 perovskite.

and hence optimizing the process of manufacturing individual components of PSCs. PSC device architectures are similar to DSSC except that the dye molecules used to sensitize the light are replaced by perovskite materials. PSCs can be fabricated using different physical and chemical deposition techniques such as sputtering [21], chemical-assisted spin coating and dip coating [22], chemical vapor deposition [23], coevaporation [24], and low-temperature solution deposition [25]. However, the perovskite layer is usually prepared by following techniques, as shown in Fig. 8.3. These methods are discussed in detail in Section 5.2.1. The difficulties on fabricating an efficient perovskite layer are uniformity in thickness, film quality, and roughness [13]. The photon absorption directly depends on perovskite layer thickness, but for thick perovskite films the efficiency of charge carrier extraction decreases and the recombination become an issue resulting in low photocurrent [26]. Thickness, uniformity, quality, and roughness of the films can be slightly improved by fabricating the perovskite thin film under clean environment condition and optimizing the deposition parameters and the annealing temperature [27]. This chapter gives an outline of the reported fabrication techniques of PSCs in terms of preparation of photoanode and perovskite layer, and selection and optimization of electron transporting material (ETM) and counter electrode (CE). It also gives a brief idea on the factors affecting the stability and life time of the cell.

8.2 FABRICATION OF PEROVSKITE SOLAR CELLS

Recently, PSC has attracted significant attention as one of the leading light harvester in solar cells because of its excellent properties and

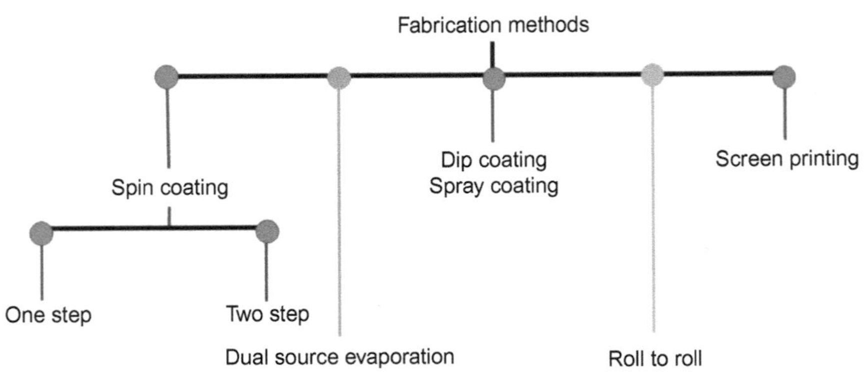

Figure 8.3 Various reported methods to deposit perovskite layers.

advanced device design. Taking this into consideration, many research groups have demonstrated various fabrication methods. PSCs have several advantages in processing over traditional silicon solar cells [23]. Most of the silicon solar cells require expensive multistep processes with high temperatures and high vacuum in special clean room, whereas the perovskite material can be fabricated with simpler wet chemistry techniques in a traditional laboratory environment [18,28,29]. A revolution happens in the PV field in the last 5 years when the new material named organometal halide perovskite (OMHP) entered the PV field. Although OMHP is not a new material, as in 1990s researchers around the world fabricated this material [15]. The PSC technology has also shown great promise in terms of material availability, cost, and ease of handling. Two major issues for the application of PSCs in industry are photo conversion efficiency (PCE) and device stability. To improve the performance in planar perovskite devices the most desired way implemented by different researchers so far is to increase the thickness of the organo-lead halide perovskite layer to improve the photon harvesting [30–32]. Moreover, this is not as direct as it seems because the film quality, the shunting pathways, and the absorption quality of the device are dependent on its crystallization behavior. Thus one of the main challenges in this field is the fabrication of high-quality films with controlled morphology, high surface area, and minimum pinholes for high performance in the solution-processed thin-film perovskite devices. Factors such as charge dissociation efficiency, charge transport, and diffusion length of charge species are dependent on the crystallinity of the film. The performance of these perovskite materials is thus highly dependent on many factors such as deposition methods, composition, type of surfaces, and solvents/additives used [32–34]. Generally a PSC consists of transparent conducting electrode, metal oxide layer, perovskite layer, electron/hole transport layer (ETL/HTL), and conducting CE. In a PSC architecture an appropriate metal oxide–coated transparent conducting electrode is used as a photoanode over which a perovskite layer is deposited as light-harvesting material. Moreover, an ETL/HTL is also deposited along with a suitable conducting electrode as CE. The present research groups in PSCs mainly focus on the high performances by various device structures, fabrication methods, and different analogues.

8.2.1 Photoanode

DSSCs with various compact films prepared by different methods have been reported; however, little research on the compact films for the PSCs has been carried out. Possible preparation methods of tuning the morphology of different photoanode such as TiO_2, ZnO, AL_2O_3, etc. [35–37] range from spin coating [38], spray pyrolysis [39], hydrothermal [40], and microwave-assisted route [40]. Given the widespread use of titania a large variety of different preparation routes for titania nanostructures have been developed in the last decade. Liu et al. reported how the evolution from the DSSCs progressed so far. They have also reported the incorporation of mesoporous Al_2O_3 over mesoporous TiO_2 layer results in significant improvement in PCE [41]. Liu et al. reported that the use of a thin film of ZnO nanoparticles as an ETL in $CH_3NH_3PbI_3$-based solar cells can improve the PCE to more than 10%. The use of ZnO also results in improvements of device performance to a PCE of 15.7% [36]. The most widely used techniques for the preparation of photoanode are discussed in detail later.

Spin coating is a method used to prepare uniform thin films on flat substrates. In this method the substrate is rotated continuously to spread the fluid and a thin film is formed on the substrate. The applied solvent is usually volatile and evaporates simultaneously. The film thickness under this method is directly depend on the angular speed of spinning. Higher speed results in thinner films. Ke et al. reported a thin compact titania film prepared by thermal oxidation (TO) of sputtered titanium (Ti) film and reported a PCE of 15.07%. The titania thin films prepared by TO are dense and reduce interfacial recombination. The thickness of the compact titania thin film in this method is thinner than the films deposited using spin-coating method. One-step sintering process leads to a lower dark current density, a lower series resistance, and a higher recombination resistance than those of two-step sintering. Thus the compact titania film prepared by TO has improved the short-circuit current density (J_{sc}) and the fill factor (FF) of the PSC. And this improvement can be assigned to the roll of compact titania thin film as a blocking layer to avoid the interfacial charge recombination between the perovskite layer and the fluorine-doped tin oxide (FTO) substrate. Furthermore, the one-step sintering method reduces the interface resistance between the compact titania thin film and the porous titania film and requires thinner densified titania layer. The lowest series resistance (R_s) (3.57 Ω cm^2) and the highest shunt resistance (R_{sh}) (1800 Ω cm^2) have been achieved by the device based on the TO-TiO_2 compact film. A maximum PCE of 15.07% has been achieved by the PSC with the TO-TiO_2 compact film, which has a higher short-circuit current density and the FF than that with the TiO_2 compact film. This method has a potential to be applied in high-efficiency, large-area, and low-cost PSCs in the future [38].

Burschka et al. reported that the perovskite pigment can be deposited in a single step onto metal oxide mesoporous films using a PbX_2 and CH_3NH_3X mixture into a common solvent. However, the large morphological variations arise due to uncontrolled precipitation of the perovskite, resulting in a widespread of PV device performance, which impedes the PSC for practical applications. The authors demonstrated a sequential deposition method for the perovskite pigment formation within the porous metal oxide film. Initially, PbI_2 is introduced from solution into a nanoporous titania film and transformed subsequently into the perovskite by exposing it to a solution of CH_3NH_3I. They found that this technique allows a possible control over the morphology of the perovskite material than the previous technique because when the contacts come close to each other, the conversion starts within the nanoporous host. Using this fabrication technique the solid-state mesoscopic solar cells greatly retained the initial performance and enable us to achieve the improved PCE of $\approx$15%. The tests were measured under standard AM1.5G on solar zenith angle, solar light intensity, and cell temperature. This two-step method offers a new opportunity for the solution-processed PSCs due to its unique PCEs and high stability. The fabrication of mesoscopic PSCs through sequential deposition method enables us to retain its maximum initial PCE and enhanced device performance. The highest achieved PCE using solution-processed PVs is of 15% which is in general a new record for organic or hybrid inorganic—organic solar cells. These reports unlock new fabrication routes for perovskite-based PV devices. Since other preformed metal halide mesostructures may be converted into the desired perovskite by the simple insertion reaction detailed here. Thus through these results, it can be expected that a new class of mesoscopic solar cells will provide widespread applications and eventually take over the conventional silicon-based PVs [39]. The thickness of the film also depends on the concentration of the solution. Spin coating is widely used in microfabrication, where it can be used to create thin films with thickness around 100 nm. To get thicker film, multicoating is done [39].

Hydrothermal autoclaving is another widely used method for fabricating electrode in PV devices. However, the conventional hydrothermal method for the fabrication of high-quality nanocrystalline film is time consuming and involves elaborate process. Several compromised solutions, such as use of commercial available powder as starting materials, or non-aqueous synthesis routes have been applied to PSCs' electrodes for delivering high efficiency. Various nanostructures of nanotube, nanofiber, nanorods (NRs), nanocomposite, or hierarchical architectures have been used to fabricate photoanodes for high-efficiency device. Park et al. have reported the fabrication of a high-efficiency ZnO NR-based PSC with PCE of over 11%. Diameter and length of the hexagonal ZnO NRs were manipulated by controlling concentration of the precursor solution and immersion time. Two-step spin-coating of PbI_2 and CH_3NH_3I solutions led to void-filling with $CH_3NH_3PbI_3$ perovskite in ZnO NR film. The ZnO NR-based PSC showed almost ideal external quantum efficiency (EQE) spectral shape with excellent matching of the calculated J_{sc} with the measured J_{sc}. Compared to a limited detection in charge collection due to slow charge collection rate and low PCE for the TiO_2 NR-based PSC, the ZnO NR system showed fast saturation in charge collection, which indicates that ZnO NRs are an effective charge collection system in $CH_3NH_3PbI_3$-based PSCs. Moreover, $CH_3NH_3PbI_3$ perovskite incorporated ZnO NRs achieved a PCE of 11.13% with J_{sc} of 20.08 mA/cm^2, open-circuit voltage V_{oc} of 991 mV, and FF of 0.56. Square spectral feature of EQE was observed, where EQE exceeded 80% covers the wavelength range from 400 to 750 nm, and the integrated current density of 20.03 mA/cm^2 is in well agreement with the observed J_{sc}. Thus a PSC-based on TiO_2 NRs exhibited an integrated current density of 16 mA/cm^2 which is much lower than the measured J_{sc} of 20.9 mA/cm^2 in comparison with the ZnO NRs with a perfect spectral response. Moreover, the time-limited photocurrent response showed that the device signal amplitude was rapidly saturated and charge collection was not fully detected for the ZnO NR-based and TiO_2 NR-based device, respectively. However, the charge collection detection issues for the TiO_2 NR-based devices is due to the slow collection rate. Moreover, the response test was performed under 530 and 700 nm monochromatic beams at 10 Hz and the results concludes that the ZnO NR is an effective charge collection system in $CH_3NH_3PbI_3$-based PSCs [37]. Wang et al. also reported the synthesis of TiO_2 nanoparticles prepared using solution-based deposition as an electron collection layers in mesosuperstructured PSCs and achieved a PCE of 15.6% [33]. However, achieving a high-efficient PSC using solution-processed perovskites and an electron collection layer requires sintering at 500°C which is a limitation for preparing a low cost, plastic substrate applications and multifunction device architectures. Therefore they reported a low-cost electron collection layer prepared through solution-based deposition method using TiO_2 and graphene nanocomposites in mesosuperstructured PSCs. The graphene nanoflakes provide a superior charge-collection in the nanocomposites which enable us to fabricate the devices at higher temperatures more than 150°C. The fabricated devices show an impressing performance and achieved a PCE of 15.6%. Thus this work concludes that graphene/metal oxide nanocomposites can be a possible alternative toward the development of low-cost PV devices. They have also demonstrated the nanocomposites of graphene and organometal trihalide perovskites as a successful combination as an electron collection layer and achieved an enhanced PCE of 15.6%, which is the highest recorded efficiency for a low-temperature processed electron collection layer ever in comparison with the reported efficiencies of a high-temperature processed PSCs. Moreover, this work also achieved a milestone on utilizing graphene incorporation into solar cells. Thus these findings depict the unnecessary usage of high-temperature sintering for preparing electron collection layer on achieving efficient solar cells. By combining the low-temperature sintering with another deposition method, namely the fast printing deposition method, provides the graphene/TiO_2 nanocomposites compatible for reel-to-reel process and clears the pathway to upscale the device fabrication for industrial and commercial applications [33]. Lee et al. reported a low-cost, solution-processed mesosuperstructured PV devices based on high-crystalline perovskite absorber with high visible to near infrared absorptivity and achieved a PCE of 10.9% in a single junction device under simulated sunlight. They stated that the application of mesostructured insulating scaffold thin films assembled with n- and p-type semiconductors is termed as mesosuperstructured solar cells. They suggested that further performance improvement can be achieved by extending the absorption toward 940 nm by using new perovskites or broadening this concept to other solution processable semiconductors. They also suggested that reduced series resistance through the utilization of hole transporters with good mobility or better controlling over the capping layer thickness can also improve the FF and extending to multijunction would further enhance performance [42]. Zhu et al. reported the highest performance of the PSC using a solution-processed SnO_2 ETL and achieved a PCE of 18.8%. They prepared this SnO_2 using hydrothermal method with a thickness of around 120 nm with a high crystallinity and provided a respectable electron transport. Moreover, they also observed an enhanced stability under ambient conditions and retained a 90% of its initial PCE over 30 days with a relative humidity (RH) >70%. They have concluded that SnO_2 improves the performance and stability and have merits in future development of efficient interconnecting layers for PSC-based tandem cells [43]. Controlling the nanoscale structure for mass production using elaborate process is difficult. To achieve an efficient low energy and convenient procedure the conventional hydrothermal procedure has been replaced by microwave heating with a microwave-assisted reactor for the preparation of nanocrystalline electrode. Shen et al. reported that the PCE attained using microwave-assisted synthesized TiO_2 is comparable with that of conventional hydrothermal process. However, the reaction time is reduced from around 12 h to around 25 min using microwave-heating process for TiO_2 nanoparticles formation. Moreover, they have demonstrated a rapid microwave-assisted hydrothermal synthetic technique for the preparation of TiO_2 nanolayer and tested PV performance for dye-sensitized and perovskite-based solar cells. The performance of microwave-based TiO_2 and the one prepared through conventional hydrothermal process are comparable in terms of PCEs. Titania synthesized by microwave-assisted technique-based mesoscopic perovskite–based solar cell showed a PCE over 10%. They have successfully demonstrated the use of microwave-assisted synthesis to fabricate TiO_2 nanoparticles to improve the PV performance of PSC. In the work the material characteristics such as crystal phases, particle sizes, and surface properties are studied in detail. The properties of the prepared material synthesized through microwave-assisted method showed similar quality comparable with commercial product made from conventional hydrothermal method. The photoanodes fabricated using microwave-assisted technique showed an optimal PCE of 8.2% for DSSC. Furthermore, it is the first application of microwave-assisted synthesized TiO_2 nanoparticles in $CH_3NH_3PbI_3$ PSCs with device efficiency as high as 10.7%. These promising results of microwave-heating process open up a facile, cost-effective, and energy saving, potential alternative for the fabrication of PV devices to the conventional hydrothermal method. It offers an energy efficient and low-cost method for the development of photoactive TiO_2 nanocrystallines. Swift heating to the set temperature in microwave-assisted method resulting extremely rapid crystallization is an efficient alternative. Another virtue of the microwave process is the easy adaptation to the conventional heating processes for the industrial usage [40].

Spray pyrolysis is a unique process in which a thin film is deposited by spraying a solution on to a heated surface, where the constituents react to form a chemical compound. Burschka et al. reported the patterned transparent conducting oxide substrates coated with a TiO_2 compact layer by aerosol spray pyrolysis with a PCE of approximately 15% [39]. The chemical reactants are selected such that the products other than the desired compounds are volatile at the temperature of deposition. The process is particularly useful for the deposition of oxides. Spray pyrolysis involves: atomization of the precursor solution, aerosol transport of the droplet, and decomposition of the precursor to initiate film growth. Still lots of researchers are trying to improve the performance of photoanode through some novel approaches.

8.2.2 Perovskite Layer

Several new methods for depositing perovskite film have emerged such as roll to roll, one-step, two-step, vapor-assisted processing, and sequential deposition of perovskite [34,44,45]. These methods enable many routes to process the perovskite film on a variety of substrates with varying thermal treatment requirements. A large number of recent successes on lead halide PSCs are observed in solution-processed thin films due to their facile fabrication, and several strategies have been used to control the crystallization processes. Modification of precursor solution, additives such as hydrogen halide (HX), microwave irradiation, process gas—assisted method, and laser-assisted crystallization has been used for nucleation and to support growth of the perovskite crystals. Till date the strategies reported for the fabrication of hybrid perovskite films are mostly one-step or two-step sequential deposition methods based on solution processing, vacuum deposition, or vapor-assisted solution processing (VASP) [46—48]. The classification of widely used techniques for the deposition of perovskite layer is given in Fig. 8.3.

The optoelectronic properties of the perovskite films decide the processing conditions such as material ratio and growth environment resulting in film quality and device performance. The lattice formations in the perovskite can be determined by its kinetic and thermodynamic factors and it can be tuned by the process parameters such as annealing temperature, time, and ramping rate. Furthermore, it can also be tuned by employing different solvents or additives. One of the important reason for the growth of PCE of PSCs is the understanding of mechanism of film growth and the improved approaches in perovskite material processing. The hybrid perovskite materials crystallize even in low-temperature process, and the final phase of the perovskite formed at relatively high reactions rates between organic and inorganic species. Thus this advantage of hybrid perovskites initiates various fabrication methods like solution processing and gives an idea to achieve new architectures. These advantages substantially increase the varieties of available fabrication methods like thermal evaporation and solution processing. It facilitates the adoption of new and varied device architectures. For example, when using solution deposition techniques, the organic materials soluble in different solvents than those suitable for processing the inorganic framework, which results a poor film quality. For other cases such as both materials are soluble, then the solution techniques must be cautiously controlled due to the fast intercalation reaction and confrontational substrate wetting characteristics [49]. On the other hand the poor heating control of two reactant species and the substrate temperature leads to a dissociation of organic component and decomposition in vacuum deposition techniques. One-step and sequential deposition techniques have been demonstrated in PSCs based on two distinctive architectures: mesoporous scaffold and planar heterojunction configuration [46].

One-step processing is based on the codeposition of both the organic and inorganic components either through solution processing or by thermal evaporation method, which is shown in Fig. 8.4A [45,50]. In solution processing a mixture of MX_2 (M = Pb, Sn; X = Cl, Br, I) and AX (A = MA; form amidinium (FA)) is dissolved in an organic solvent and it will be deposited directly to form a film. Then the film will be subjected to thermal annealing to produce the final perovskite phase. On the other hand the thermal evaporation employs dual sources for MX_2 and AX with different heating temperatures to form a perovskite film. Sequential deposition is mainly depositing a MX_2 (M = Pb, Sn; X = Cl, Br, I) layer

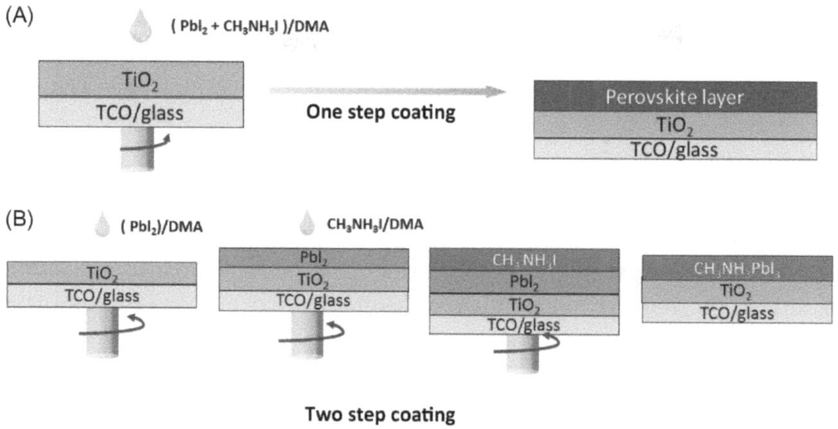

Figure 8.4 (A) One-step solution process based on the mixture PbI_2 and MAI followed by sequential coating of PbI_2 and MAI. (B) Two-step solution process based on the mixture PbI_2 and MAI followed by sequential coating of PbI_2 and MAI [45].

such as PbI_2 and an AX (A = MA, FA) such as methyl ammonium iodide (MAI) sequentially followed by heat treatment to form perovskite film [45]. Perovskite compounds have attracted a great attention in field of PVs. You et al. reported the fabrication of PSCs using mesoporous TiO_2 as an ETL and 2,207,70-tetrakis-(*N,N*-dip-methoxyphenylamine) 9,90-spirobifluorene as the HTL. However, the TiO_2 layer requires high-temperature processing (450°C), restricts the widespread usage of this technology. In this work the authors reported a high efficiency of PSCs prepared through low-temperature processing technique for the devices having both rigid and flexible substrates, with substrate/Indium Tin Oxide (ITO)/poly(3,4-ethylenedioxythiophene)- poly(styrenesulfonate) (PEDOT:PSS)/$CH_3NH_3PbI_{3-x}Cl_x$/phenyl-C61- butyric acid methyl ester (PCBM)/Al device structure, whereas the PEDOT:PSS and PCBM are used as hole and ETLs, respectively. All the layers were solution-processed under 120°C and achieved a PCE of 11.5% for the devices using rigid substrates (glass/ITO), and a 9.2% PCE is achieved for a polyethylene terephthalate/ITO flexible substrate [51]. Mixed halide perovskite such as $CH_3NH_3PbI_{3-x}Cl_x$ was used due to its impressing electrical properties and long carrier lifetime. Thus they have demonstrated a low-temperature solution-processed PSCs and achieved a PCE which is comparable to that of the conventional high-temperature processed devices. They achieved 11.5% PCE for the rigid substrate devices and a 9.2%

efficiency for flexible devices. The reason for the interfacial recombination and V_{OC} loss is due to the imperfection in the perovskite morphology and it can be overcome by tuning the layer morphology. Typically the deposition of MX_2 is achieved using spin coating, while AX can be introduced by, (1) AX solution can be spin-coating on top of the MX_2 layer, (2) dipping the MX_2 layer in the AX solution to induce a solid–liquid reaction, or (3) exposing the MX_2 layer to AX vapor at elevated temperatures [52]. Also, two-step sequential deposition can be carried out in thermal evaporation, by sequentially depositing the inorganic and organic components.

Initially the usage of hybrid perovskite absorbers in solar cells is based on the DSSC structure. On the other hand the perovskite absorber is self-assembled with porous TiO_2 layer formed by sintering the prepared nanoparticles [53]. The typical configuration of this kind of perovskite-based solar cell is FTO/compact TiO_2/mesoporous TiO_2/perovskite/[2,20,7,70-tetrakis-(*N*,*N*-di-4-ethoxyphenylamino)-9,90-spirobifiuorene] (Spiro-OMeTAD)/electrode. It was found that in this structure, perovskite materials are deposited on mesoporous TiO_2 to facilitate transport of charge carriers between the perovskite absorber and the FTO electrode [54]. A succeeding work validated the replacement of relatively conductive porous TiO_2 with an insulating porous Al_2O_3 layer and it indicates that perovskites can be utilized broadly than using it as sensitizers since it is transporting both electrons and holes between cell terminals.

A typical mesoporous scaffold perovskite device fabrication procedure is as follows. Initially a compact TiO_2 blocking layer and a mesoporous oxide layer will be sintered at high temperature (500°C) and deposited on FTO substrates, followed by the perovskite absorber layer deposition over it. For one-step processing, pure iodine-based $MAPbI_3$ is generally deposited from a mixture of PbI_2 and MAI in *N*,*N*-dimethyl formamide (DMF), g-butyro lactone, or a similar solvent system. One-step processing has also received a good achievement in developing mixed halide perovskite systems in which mixed $PbCl_2$ and MAI at 1:3 molar ratio in DMF were deposited onto an insulating Al_2O_3 mesostructured scaffold. Finally, Spiro-OMeTAD and the metal electrode were then deposited by spin-coating and thermal evaporation, respectively. A number of crucial factors are involved in this deposition method that results in poor grain size, poor crystallinity, and low surface coverage [55,56]. For example, in solvent engineering techniques a small amount of additional drop is added during spin-coating, which is recently used to remove the solvent originating from the perovskite solution which greatly enhances the grain size, coverage, and uniformity of the capping layer on the mesoporous structure.

In two-step sequential processing, PbI_2 is deposited initially on a mesoporous substrate, and the perovskite film is deposited either by dipping the substrate in a MAI solution or spin-coating the MAI solution onto the PbI_2 film while using a solvent that does not dissolve PbI_2, which is shown in Fig. 8.4B. The fabricated perovskite film resulted in improved coverage of the capping layer than the film fabricated through one-step process and it also retained the original morphology of the PbI_2 film, and the coating is also uniform throughout the surface. Additionally the incorporation of $PbCl_2$ of a certain amount into PbI_2 offers an opportunity to optimize the morphology [57]. Pellet et al. reported that hybrid organic–inorganic lead halide perovskite $APbX_3$ pigments, such as MA lead iodide, have recently emerged as an excellent light harvester in solid-state mesoscopic solar cells. They have also achieved improved performance by extending its optical-absorption toward the red region in the solar absorption spectrum. They reported a combination of $HN=CHNH_3^+$, FA and $CH_3NH_3^+$, MA cations in the A position of the $APbI_3$ perovskite structure to further improve the optical absorption and this combination improved the short-circuit current resulting in superior device performance compared to only $CH_3NH_3^+$. This opens up a great potential as a versatile tool on tuning the structural, electrical, and optoelectronic properties of the light-harvesting materials. They demonstrated mixed-cation 3D perovskite as a sensitizer for the first time. In this the FA cation has been replaced by MA in lead iodide perovskites, resulting the red-shifted absorption onset of $FAPbI_3$ as compared to that of $MAPbI_3$. Devices based on $FAPbI_3$ were made by sequential deposition and provide a PCE of 11.0% (a lower value than that of $MAPbI_3$ due to the presence of the yellow d-phase). The sequential deposition method and the addition of 20% MA to the FA dipping bath completely avoid the undesirable formation of the d-phase while maintaining the red-shifted band gap of $FAPbI_3$. The mixed-cation perovskite that exhibited superior PV performance to that of the single-cation analogues leads to a greater harvesting and collection of red photons, which resulted in higher short-circuit photocurrents without sacrificing photovoltage. The improved carrier-collection efficiency can be ascribed to longer exciton lifetime of more than 130 ns in the $MA_{0.4}FA_{0.6}PbI_3$ material. Using this technique, they were able to fabricate devices yielding up to 14.9% photon to current conversion efficiency under the AM1.5G simulated solar spectrum. The mixing organic ammonium compounds open up new prospects for further improvement of the PV efficiency of perovskite-sensitized solar cells by tuning the optical, electrical, and morphological properties of the semiconducting sensitizer. It is believed that if the formation of the yellow d-$FAPbI_3$ component could be fully avoided, the performance of pure $FAPbI_3$ would most likely surpass that of $MAPbI_3$-based mesoscopic solar devices. In general, perovskite layer has moderate grain boundary size. In addition, the large grain size is more appropriate for perovskite materials as it can act as a charge carrier scavenger which can reduce the recombination and thus improve the carrier mobility. It has been found that in case of mesoporous structure, grain size of the perovskite layer is limited to pore size of the mesoporous scaffold. Because of the presence of a perovskite thick layer around several hundred nanometer, which is sandwiched between ETL and HTL, planner junction PSC shows better PCE than mesoporous one, which can be ascribed due to the enhanced ambipolar carrier transport and long carrier lifetime. Due to the presence of less number of porous, scaffold structure has a direct influence on the coating as well as of the perovskite layers. Thus an optimization has become essential to convert this deposition method from mesoporous to planar structures. It has been reported that the mesoporous scaffold is a better way where thermal evaporation can be used in a planar configuration which can overcome the difficulty of perovskite precursors penetrating to the nanoporous scaffolds layer.

Planar structures require an uniform layer perovskite film, which is one of the crucial factor as it consists of pinholes that may be a source for severe leakage current. Thus one-step solution processing has been used for the fabrication of planar structure by mixing MAI and $PbCl_2$ in dimethylformamide (DMF) to form $MAPbI_{3-x}Cl_x$ [58]. Moreover, it has been reported that a high PCE of 15.7% is achieved by using sequential deposition procedures instead of one-step deposition and this can be attributed to the effective reduction in the chemical reaction between the perovskite materials and the ETL [36]. Thus the sequential deposition methods have given a variability to ETL options as perovskite films can be prepared effectively at room temperature. Many researchers demonstrated a VASP techniques where reaction takes place between MAI vapor and predeposited PbI_2 to form the perovskite film, as shown in Fig. 8.5B. It has been reported that the resulting $MAPbI_3$ shows excellent film quality. Xiao et al. developed a two-step solution coating method where they used two depositions of PbI_2 and MAI separately, which was followed by

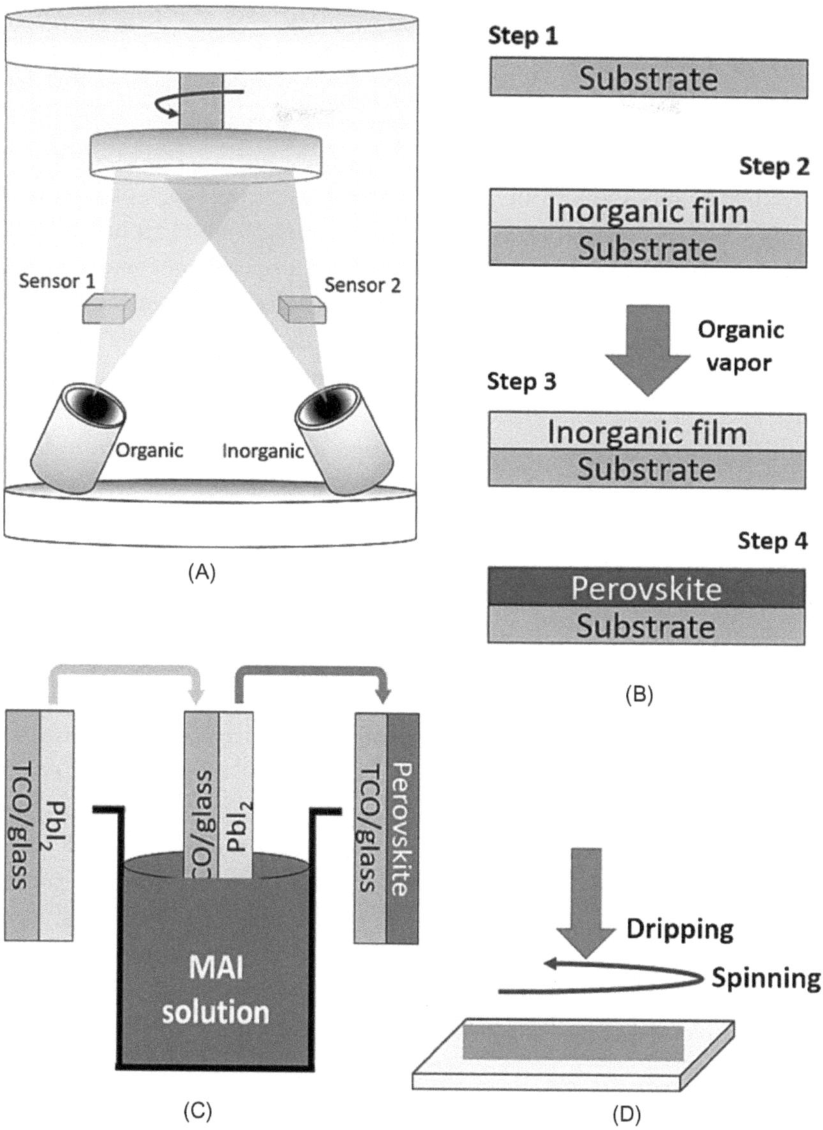

Figure 8.5 (A) Dual source evaporation using $PbCl_2$ and MAI source. (B) Vapor-assisted solution method using the MAI organic vapor to react with the PbI_2 film. (C) Sequential deposition by dipping the PbI_2 film into MAI solution. (D) Dripping and spinning method [45].

an annealing of prepared sample to prepare $MAPbI_3$. The performance of perovskite devices are demonstrated and found very high and which can be ascribed due to the controlled preparation conditions, such as PbI_2 and MAI concentration, speed of spin-coating, and the annealing temperature/duration based on this method [59]. They further reported that the interdiffusion method can be simply applied to other types of perovskite materials for incorporating Cl, Br, F, or other elements from precursor solutions and can also start with thermally evaporated precursor stacking layers to relieve the strict requirements for composition controlling in the coevaporation process. The low temperature used is compatible with plastic flexible substrates. The interdiffusion approach can be potentially scaled up for large-area device fabrication with established solution process techniques, such as die-slot coating, gravure coating, and doctor blade coating. To avoid the problem associated with one-step solution processing, where fiber-shaped crystals form as the $MAPbI_3$ phase grows directly from solution. Sequential procedures have one of the aspects which make it even more useful as their reliance on PbI_2 precursor films as a template to form the eventual perovskite layer.

Spin coating is a well-known, rapid technique used for producing uniform, quality thin films on a variety of substrates, as shown in Fig. 8.5D. Being relatively simple, this coating method is widely used in fabrication of organic electronics and optical device coatings among many others. Spin-coating method comprises of deposition of perovskite solution onto a given substrate and then followed by constant rotation at high speed and consequent drying. Thus it results into formation of a thin perovskite film [60]. Thickness of formed film is controlled through several parameters, such as viscosity of the solution, temperature, and spinning velocity. Postprocessing usually includes annealing to promote evaporation of residual solvent and improve crystallinity. Advantages of spin-coating method to name a few are, low cost, fast processing, possibility of multilayer deposition, and small quantities of solution needed for covering large substrate area. On the other hand disadvantages include significant losses of coating material (over 90%), lack of compatibility with roll-to-roll processing, and sensitivity to the rate of evaporation which can result in numerous defects in the final product [61]. Komikado et al. demonstrated a method to control the optical thickness of perovskite layer according to a quarter of a designed wavelength by adjusting the concentrations of the solutions and rotation speed of the spinner. A multilayered mirror fabricated simply by spin coating of two different kinds of polymers is reported superior compared to the layers fabricated through a simple continuous process of spin-coating solutions. Consequently the increase of the reflectivity at the designed wavelength can be achieved by drying of solvents. Thus the mirrors influenced the reflectivity higher than 99% and here they used the polymeric multilayered mirrors for the first time as the output couplers of a microchip Nd_3^+/YVO_4 laser. Therefore they have demonstrated a polymeric multilayered mirror by spin coating along with the controlled optical thickness and a reflectivity higher than 99% at wavelengths of 633 and 1064 nm. Thus as synthesized, low-cost polymeric multilayered films can be used for many optically and electrically functional applications. They also fabricated a Fabry–Perot wavelength-filter having a sharp bandwidth at a designed wavelength which can be attributed to the intermediate layer with half-wavelength optical thickness. Moreover, they reported the passive Q-switching of a microchip Nd_3^+/YVO_4 laser by means of a saturable dye into one of the stacked layers. Further, they tried to fabricate tunable thin-film filters where the polymeric multilayered films work as the optically and electrically active devices by introducing electro-optic or piezo-electric chromophores into stacked layers and/or in intermediate layers.

Another technique suitable for the fabrication of thin film perovskites is spray pyrolysis, providing easy control over film composition as well as simple experimental setup [62–64]. Formation of a perovskite layer is mostly done by thin-film deposition through spraying solution on the substrate, followed by chemical reaction among the components present in the system. Usually this method is used in synthesis of perovskite oxides. Coating solution is achieved in such a way that all undesirable products are unstable at working temperature. Most used types of spray coating are ultrasonic and electrostatic. Ultrasonic spray deposition used nozzle to supply solution containing droplets of uniform diameter in a form of a spray with the help of various gases acting as carriers and focusing spray onto the substrate. Throughout the electrostatic spray deposition (EDS) technique a positive high voltage is used to generate aerosol from perovskite layer solution. Potential difference between substrate electrode and nozzle is maintained in such a way that it ensures a good coverage with perovskite film. Therefore spray coating method is a very reliable method to deposit the perovskite oxide thin films. Low temperatures may induce cracking of deposited layer, while higher substrate temperatures cause rough and porous films. Furthermore, temperature can mainly influence the texture and crystallinity of the sample, whereas properties and morphology of deposited films can be tuned through various additives. Thus one of the major benefits of using spray coating can be attributed to the low-cost and size selectivity of the droplets of highly diluted solutions [65–67]. Taniguchi et al. reported the deposition of $La_{1-x}Sr_xCo_{1-y}Fe_yO_3$ (LSCF) thin films over the stainless steel by means of EDS technique. The solution has been prepared by dissolving the appropriate amount of strontium chloride, lanthanum nitrate, cobalt nitrate, and iron nitrate into a mixture containing 33 vol.% ethanol and 67 vol.% butyl carbitol, respectively. Prepared LSCF thin films were amorphous or nanocrystalline in nature at the used deposition temperature (573 K). Moreover, the samples were kept for heating at 1173 K for 2 h. Subsequently the crystal structure of the samples gets converted to the anticipated perovskite phase. It has been seen that the surface morphology of LSCF was influenced by these process parameters such as temperatures, deposition time, and liquid flow rates. The reticular layer has been successfully prepared at a deposition temperature of 573 K and liquid flow rate of 1.0 mL/h, while most of the as-prepared films formed three-dimensional irregular networks with micropores. The cross-section picture reveals a very porous superficial layer extending from the layer surface into the layer, and a relative dense layer extending down to the interface of the layer and substrate [65].

Lopez et al. demonstrated the highly porous and macroporous electrodes acquired at temperatures of 250 and 450°C, respectively, having an average grain size of 30–50 nm. Here they have investigated the influence of the postannealing treatment on the microstructure as well as the electrochemical properties of the perovskite material using scanning electron microscopy. Moreover, the different contribution to the polarization was also identified using impedance spectroscopy with air as a function of the oxygen partial pressure. In addition, the samples annealed at 850°C, the area-specific polarization resistance (ASR) values increase up to 0.1–0.6 Ω cm^2, whereas the lowest value corresponding to the film deposited at 250°C can be attributed due to the large porosity and surface area of this film. The degradation performance upon annealing can be ascribed to decreasing reaction. Moreover, the electrochemical measurements were accomplished using impedance spectroscopy on symmetrical cells in air. Three contributions were observed depending on the temperature and the oxygen partial pressure. The high and intermediate frequency contributions were identified as ionic transfer at the electrolyte/electrode interface and ionic diffusion in the LSCF bulk, respectively.

The low-frequency contribution is only visible at high temperatures >700°C and is attributed to gas-phase diffusion of oxygen. All cells annealed at 650°C showed low ASR values of 0.04–0.06 Ω cm^2 measured at 650°C. The ASR is increased due to the low electrode-pore surface area due to higher annealing temperature. However, LSCF electrodes deposited at 250°C with high porosity exhibited lower microstructural changes with the heat treatment, and ASR values of 0.05 and 0.1 Ω cm^2 were obtained after annealing at 650 and 850°C, respectively [67].

Screen printing is another cheap and effective printing technique allowing fabrication of printed layers with 2D patterns. The biggest advantage of this technique in comparison to spin coating is very efficient utilization of material, resulting in virtually minimum loss of paste used for printing. For successful and high-quality screen printing the primary requirements are highly viscous and low volatility solutions along with a large wet film thickness. Mechanism behind this type of printing is relatively simple. In this technique a printing pattern in the screen made of steel mesh or synthetic material is first made impermeable to the coating solution and the screen is then attached to a frame under tension. After bringing the working solution in contact with the substrate, the screen is dragged across it allowing solution leakage through unprotected parts of the mesh reproducing desired pattern [68,69].

Another very simple and effective process (batch or continuous) used for film formation through immersion of substrate in coating solution. After exposure to the solution for a certain period of time, wetting of substrate surface and subsequent film formation takes place. Following this, deposition step is the withdrawal of substrate from the solution bath and evaporation of solvent resulting in formation of film with uniform thickness. Further, posttreatment step may be necessary to obtain final coating. Main forces governing dip-coating process are gravitational force, viscous drag, surface tension, and force of inertia, which plays an important role in processing of perovskites, especially in fabrication of PV devices [70,71]. Garauskis et al. developed perovskite layer through dip-coating technique, employing stable colloidal suspension of perovskite particle dispersed within ethanol media. To study the effect on particle agglomeration and rheological response, two polymer-based surfactants were tested. Subsequently, they reported that through the addition of 2.5 wt.%, addition of selected surfactant can help to obtain dense 15–60 μm thick functional layers. The postannealing has resulted in final particle sizes within submicron range. By employing suspension with pore former, it was possible to significantly increase the surface area of the functional layer. Dense and defect free layers within the thickness range of 15–60 μm were obtained by a single dip-coating deposition. By employing small initial particles the grain size was kept in the submicron range also after the final thermal treatment step. Deposition of a second layer with pore former resulted in a significantly enhanced surface area of the functional layer.

Othman et al. reported the preparation of perovskite membrane precursor using a sol–gel method. It exhibits the Newtonian flow when the perovskite precursor sol was a rheopectic fluid and kept for 2 h of aging. But after 24 h of aging the flow became non-Newtonian. The thickness of the thin layer was found approximately 2 μm. As-prepared precursor was dip-coated to form a perovskite thin layer to demonstrate the permeability experiment conducted at room conditions. The permeability result indicates that it appeared almost impossible to separate oxygen from the oxygen/nitrogen mixture because both have permeability that was nearly identical. However, from the separability experiment, the separation of oxygen factor was found to be higher than that of the theoretical separation factor. This is attributed to the possibility of separation as well as the enrichment of the gas from the binary mixture. Thus the increased separability of oxygen can be ascribed due to the affinitive energy by the perovskite in the range between 11 and 19 kJ/mol. From the permeation test employing pure gas in a single system, the permeability of O_2 and N_2 across the perovskite membrane was observed to be independent of pressure. The permeability values were nearly identical. Separability of oxygen in a binary gas mixture using perovskite membrane seems to have higher theoretical separation factor attributed to the Knudsen diffusion. The higher separability of oxygen can be attributed due to the reasonable adsorption by the perovskite in the range between 11 and 19 kJ/mol.

Fabrication of thin films using thermal evaporation is a method of material deposition directly from the vapor phase and is widely used in magnetic and optical devices. Moreover, during deposition process, temperature maintained to evaporate is high enough to induce vaporization, whereas the substrate temperature is kept at below freezing point of the material. Atoms or molecules being thermally released from evaporant surface, travel to the substrate and they form coating. The substrate follows a straight line without any interaction with residual gas atoms, when the process is conducted under reduced pressure. Moreover, there are two suitable methods to prepare perovskite film using single and dual source evaporation system. After applying large current through the metal sheet, sharp increase of temperature results in ablation of material from metal holder and subsequent deposition onto the substrate. On the other hand, dual source deposition uses organic and inorganic salts that are simultaneously evaporated from two separate sources. In the case of single source ablation, high temperature ensures fast evaporation of inorganic component while avoiding thermal decomposition of the organic part. Furthermore, there is no need for postprocessing because organic–inorganic perovskite self-assembles on the substrate during deposition unlike in dual source method, where annealing is required to obtain fully crystallized perovskite film [41,72].

For PSC with solution-processed silver (Ag), nanoelectrode has the feasibility for the low-cost, large-scale, and high-throughput roll-to-roll production of solid-state devices [73]. Zhang et al. reported the fabrication of a low-temperature–processed planar heterojunction PSC using ZnO nanoparticles as the ETL and silver nanowires (Ag-NWs) as the top electrode. The $CH_3NH_3PbI_3$ perovskite was developed as the light absorber through multistep spin-coating technique. And the fabricated PSC showed the maximum PCE of 9.21% with short-circuit current density of 19.75 mA/cm^2, open-circuit voltage of 1.02 V, and FF value of 0.457. They have reported only a negligible difference on the forward and reverse bias scan in device performance. They also developed a low temperature–processed PSC using Ag-NW as the top electrode and ZnO nanoparticles as the ETL. The solution-processed Ag-NWs electrode was deposited by spray coating without causing any damage to the bottom layers of the device. The prepared PSCs showed a highest PCE of 9.21%. It indicates that the performance PSCs can be improved substantially by an all solution process. Thus roll-to-roll processing of PSCs is feasible in the future [73]. Even then this technique is only possible when the processing temperatures are brought down to 100–150°C. However, most of the high-efficiency PSCs employed a mesoporous TiO_2 scaffold, which requires sintering at high temperature of about 450–550°C. Although further work has been done by replacing mesoporous TiO_2 scaffold with low-temperature process (100–150°C) Al_2O_3 scaffold. A compact TiO_2 layer is still needed and requires high-temperature process. To avoid the high-temperature sintering process and further achieve the fabrication of PSCs by roll-to-roll processing, a low temperature–processed compact ZnO-based ETL on ZnO nanoparticles was introduced.

Various deposition techniques have been used for the fabrication of MA lead halide PSCs. The challenges associated with the one-step precursor solution spin-coating process is the lack of suitable solvents that can dissolve the components in the precursor mixture and the high reaction rate of the perovskite components resulting in incomplete coverage and poor uniformity, leading to low-resistance shunting paths and loss of light absorption in the solar cells. Chloride inclusion, optimization of annealing conditions, and the sequential solution-based deposition method have been demonstrated to increase film coverage and uniformity, as shown in Fig. 8.5C. Dual source vacuum evaporation has also been employed allowing efficient perovskite solar device to be fabricated with excellent film coverage and uniformity but the high-vacuum requirement precludes this method from mass adoption, as shown in Fig. 8.5A [41]. Although the vapor-assisted method has been used to fabricate planar $CH_3NH_3PbI_3$ PSCs, it has not been employed in a cell structure where a mesoporous layer is present. The planar perovskite cell structure without the mesoporous layer has the advantage of a simplified design reducing number of fabrication steps and simplifying the design. This may potentially be beneficial for tandem cell construction and allows the investigation of the underlying device physics without the complication of the mesoporous layer. However, recent research has reported that the presence of mesoporous layer in particular the mesoporous TiO_2 (mp-TiO_2) when its thickness is optimized is beneficial in reducing the hysteresis in PSCs compared with planar structure.

Further, few more attempts have been made to fabricate perovskite layers with improved homogeneity based on planar heterojunctions. Snaith et al. reported the preparation of a uniform mixed perovskite ($CHNH_3PbI_{3-x}Cl_x$) layer through a dual source evaporation process which achieved PCE values up to 15.4%. Although this work represents a significant advancement on the path to high-efficiency planar devices, still it requires a very high vacuum deposition. Moreover, the low-pressure vapor annealing can increase the vapor pressure of CH_3NH_3I, which causes the reduction of annealing temperature. Thus the capability to work with a broader temperature range at a reduced pressure ensures the formation of thin films with different structures, which plays a major role on device performance and minimize the current – voltage hysteresis. It has been reported that the synthesized planar heterojunction solar cells fabricated by low-pressure (LP)-VASP exhibit a record PCE of 16.8% and average efficiency of $15.7 \pm 0.7\%$. Significantly, $I-V$ curves of planar devices show significantly less hysteresis than that of the previously reported planar devices. Moreover, the existence of chloride improves PCE but it has no significant impact on band gap, Fermi level position, and lifetime photo carrier. Therefore these results confirm that the leading role of mixed halide precursors and reduced temperature in LP-VASP have direct impact toward thin films possessing. It has been found that mostly perovskite materials show long-charge carrier lifetimes. However, the solution and coevaporation processing methods showed up to an 11.4% and 15% PCE, respectively, incorporating a planar structure. It has been found that mostly extreme processing conditions could bound the future growth of PSCs as high-quality condensed TiO_2 layer frequently requires high-temperature treatment. Therefore it is difficult to find the possibility of fabricating high-performance perovskite-based solar cells at low temperature. Moreover, low processing temperature offers a broader selection of substrates and electrode materials that could be used in devices and that could be incorporated into PSCs. Thus many groups have demonstrated low temperature–processed PSCs using different transport layer processing. Consequently the resultant device performance is also notably lower compared to high temperature-processed TiO_2. The lower efficiency PSCs can be mainly attributed due to charge recombination at chemical defects in perovskite [19,33,48,58,74,75]. Advantages and disadvantages of popular fabrication methods are given in Table 8.1.

8.2.3 Electron and Hole Transport Layer

To overcome the issues in the traditional DSSCs such as instability due to the lack of suitable electrolyte and sensitivity to moisture, a new class of architecture was introduced that consist of HTL and a hole blocking layer (HBL) which covers the metal electrode and perovskite absorber to transport both the holes and the electrons through the perovskite layer. The absence of liquid electrolyte solution leads to the improvement of the stability of the overall cell [76]. Hence, this architecture is slightly varied from the traditional DSSC architecture known as the mesoporous scaffold structure. Islam et al. reported a blending micrometer long TiO_2 NRs to optimize the mesoporous structure to overcome the criticality on achieving high-efficient PSCs and high loading of perovskite and enhanced light harvesting. They found that a blending of around 10% of NRs promotes the perovskite loading and enhances perovskite connectivity in scaffold and result in good crystalline quality. They also observed that more NR

Table 8.1 Advantages and disadvantages of different fabrication methods

Processing technique	Advantages	Disadvantages
Spin coating	Low cost, fast, easy adjustment of film thickness, multilayer depositing	Nonuniform film thickness, loss of coating material, no patterning
Screen printing	Low cost, adjustable layer thickness	Screen blocking, can be slow
Spray coating	Low cost, use of highly diluted solutions, uniform layer thickness	Difficult to control process parameters
Dip coating	Low cost, simple	Can be slow, nonuniform film surface, no patterning
Vapor deposition	Uniform layer thickness, easy monitoring of deposition rate, patterning	Complex equipment, difficult synthesis of multicomponent materials, can be slow
Dual source evaporation	Significant advancement on the path to high efficiency	It requires high vacuum deposition in a dedicated system

blending will not form a homogenous scaffold thus reduce the performance of PSCs. This result suggests that developing a novel porous architecture with large pores for loading perovskite and results in enhanced perovskite connectivity for efficient carrier transport is a better choice for PSCs [77].

Spiro-OMeTAD is an expensive and most commonly used hole transport material (HTM) for PSCs. Zhou et al. reported a high efficiency of around 19.3% in a planar geometry without using antireflective coating by suppressing carrier recombination and facilitated carrier injection and transport results good carrier extraction at the electrodes. They fabricated the device in air and solution at low temperature and suggested that this will simplify the manufacturing of large-area PSC at low cost and high performance [78]. Later many structures including polymers were developed namely poly[bis(4-phenyl)(2,4,6-trimethylphenyl) amine poly(triarylamine) (PTAA) and achieved a maximum efficiency of 16% [55]. A PSC structure that does not consist of mesoporous structure is called planar heterojunction [79]. It is classified into two: (1) p-i-n and (2) n-i-p. The difference between these two structures is the relative location of charge transport layers. The planar n-i-p heterojunction consists of ETL, usually compact TiO_2 or ZnO thin film, whereas in p-i-n structure poly (3,4-ethylenedioxythiophene): polystyrene sulfonate (PEDOT:PSS) as HTL. Li et al. reported a versatile method to achieve high efficiency using planar $CH_3NH_3I_{3-x}Cl_x$ perovskites. In the work, they found that the domination of chlorine compared to the chemical and electronic doping effects leads to a structural modification, whereas there is no change in the Fermi level. Their entire process is carried out under low pressure in the presence of CH_3NH_3I, an in situ annealing takes place on the lead halide perovskite to make control over intercalation of lowered temperature. They stated that the large hysteresis is not an inherent feature of the lead halide perovskites. They further mentioned that uniform halide perovskite material with optimum composition leads to an efficient charge extraction. Through their physical analysis they found a reduced defect in PSC due to the presence of chlorine under low temperatures. This also enabled a retention of minor secondary phase of lead iodide which is very important for the performance of the PSC. Through their results they found that the dominant impact is not due to chemical or electronic but due to the structure in nature. They concluded that this process may enable to achieve highest efficiency and also lower hysteresis [80]. This p-i-n structure enables the possibility of preparing the structure in low temperature and also compatibility with organic electronics manufacturing which is an advantage over the n−i−p structure. Recently, Zhou et al. [78] reported the perovskite devices using these layers and achieved a maximum reported efficiency of 19.3% using Yttrium-doped TiO_2 as ETL. Then Partk et al. [81] attained a PCE of 17.3% using NiO as a HTL. Partk et al. [81] reported a high-transparent nanostructured NiO films with (111)-oriented and achieved a PCE of 17.3%. Since then the 111-oriented NiO plays a key role in extracting the holes and blocks the electron leakage as HTM. They have the nanostructured transparent NiO films using pulsed laser deposition, by controlling the oxygen partial pressure and deposition time of p-i-n type device having a NiO/$MAPbI_3$/PCBM configuration. They found that the enhanced performance of the prepared film is due to the slow recombination rate during the charge extraction and transport of the dissociated carriers. Through their results they reported that this nanostructured NiO film will be competitive with high-performing $MAPbI_3$-based n−i−p solar cells and an ideal candidate for the HTM in the p-i-n PSC system. Wang et al. demonstrated a small molecule SAF-OMe as an efficient dopant-free HTM for the fabrication of PSC and achieved a PCE of 16.7% and 12.39% with and without dopant. They found that the SAF-OMe displayed a better hold transporting ability than the PTAA and Spiro-OMeTAD. They also found that this SAF-OMe exhibits an improved stability in comparison with later one and stated that it is due to high hole mobility. Thus they concluded that this new SAF-OMe leads the PSC toward the commercialization in future [82]. Xu et al. reported that a device which is covered with a cross-linked HTL retained the active material and the device structure while exposed to both polar and nonpolar solvents. This enhancement not only benefits the single cell but also for multijunction devices. They found that this cross-linked organic HTM coupled with organic metal provides an insoluble and inert physical protection layer combined with high conductance for hole extraction [83]. Domanski et al. reported a triple cation and achieved a PCE of 20%; they fabricated the device using a Spiro-2,2′,7,7′-tetrakis-(N,N-di- 4-methoxyphenylamino)- 9,9′-spirobifluorene (MeOTAD) HTL and an gold (Au) or silver (Ag) back contact and found that the contacts at temperatures exceeding 70°C results in a performance loss. They found that by inserting a thin Cr interlayer between the HTL and Au solves the problem of device degradation at elevated temperatures. Thus they concluded that their results provide a road map for achieving a long-term stability [84]. To facilitate the charge extraction and achieve better performance of these planar heterojunction structures, researchers now focus on finding an alternative choice of new ETL or HTMs. Some of the reported ETL and HTMs were shown in Fig. 8.6.

Another architecture design uses direct deposition of Au on the perovskite layer instead of HTL and in the alternate one does not have electron blocking layer known as HTL-free and ETL-free cells, respectively. In HTL-free cells the built-in field performs the charge separation and the perovskite material act as an hole conductor as well as absorber which forms the heterojunction. Devices using these two architectures achieved a maximum reported efficiency of 16% using ITO/$CH_3NH_3PbI_3$/C60/BCP/Ag (BCP = 2,9-dimethyl-4,7-diphenyl-1,10-phenanthroline) reported by Li [85]. They constructed a hole-conductor-free planar PSC with an inverted-structure by solution process and achieved an efficiency of 16.0% with a better stability which is higher compared to those devices using NiO_x as hole conductor. They also fabricated a reference device with a typical hole conductor layer of NiO_x to compare with their HTM-free device. The excellent performance makes the hole-conductor-free device competitive in PSCs. Ke et al. attained a

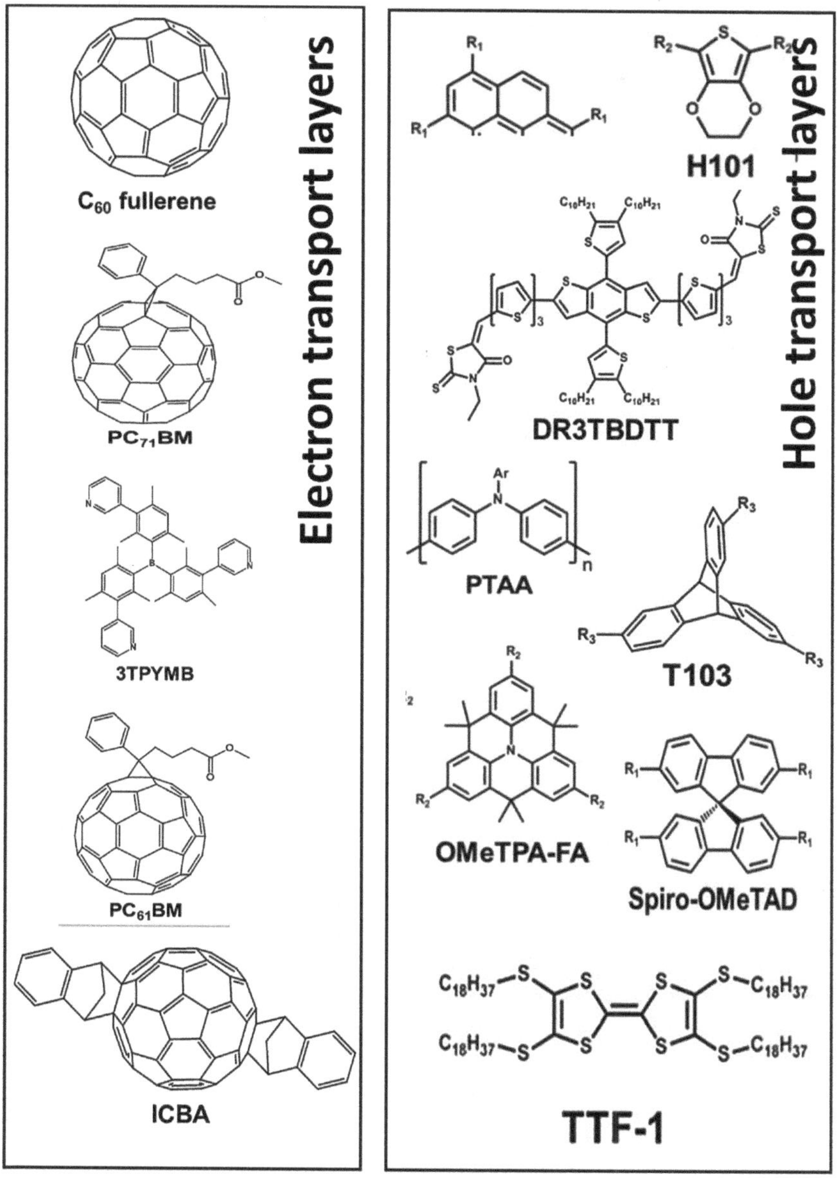

Figure 8.6 Various reported electron/hole transporting materials and its structures [34].

PCE of 14.1% by directly depositing perovskite material on FTO. They also reported a PSC grown directly on FTO without any blocking layers and achieved an efficiency of 14% using solution-processed planar Au/hole-transporting material/$CH_3NH_3PbI_{3-x}Cl_x$/substrate and ultraviolet–ozone treatment for the substrates. In addition, they found that incorporation of chlorine in the synthesis boosts the open-circuit voltage without blocking layers. They suggested that TiO_2 may not be the ultimate HBLs to achieve high PCE for PSCs [86]. Saliba et al. reported a molecularly engineered small molecule fluorene–dithiophene (FDT) HTM and achieved a PCE up to 20.2% compared to Spiro-OMeTAD. They reported the advantages of FDT such as it will dissolve easily in more eco-friendly toluene instead of CB, it can be easily modified and therefore might provide a blueprint for a family of high-performance HTMs. Thus they concluded that FDT has the potential to become a lower cost alternative or possibly replace Spiro-OMeTAD [87]. Also Zhang et al. designed a novel butterfly shaped triphenylamine (TPA)-based small molecule material Z1011 as an HTM for PSC devices and achieved a PCE of 16.3%. They stated that this Z1011 has unique features such as high mobility, suitable energy bands, and render this material as an excellent HTM for $CH_3NH_3PbI_3$-based PSC. It is more stable than that of Sprio-OMeTAD and will be a promising alternative for the future large-scale deployment [88]. Kim et al. reported that a dopant-free polymeric HTM based on benzo[1,2-*b*:4,5:*b*′] dithiophene and 2,1,3-benzothiadiazole, to enhance the stability and performance of PSCs and achieved a PCE of 17.3% and stable over 1400 h with 75% humidity. They reported that this improved performance is due to the deep highest occupied molecular orbital (HOMO) energy level and high hole mobility. The stability is improved by deliquescent or hygroscopic dopants and also avoiding the introduction of hydrophobic polymer layer. The random copolymer devices maintained its initial efficiency over 1400 h with 75% humidity, whereas a device with additives fail after 900 h, respectively [89]. Zhao et al. reported a novel HTM ST1 that prepared using a single step synthesis method and achieved a PCE of 15.4% using undoped ST1 as HTM for PSC. They also tested the stability performance of the ST1 and found more stable in exposure to air than Sprio-OMeTAD. Thus they concluded that this low cost, facile single step synthesis method as well as excellent hole mobility with appropriate energy level makes this new ST1 as a promising alternative for the HTM for the large-scale PSC [90].

8.2.4 Counter Electrode

The CE is one of the crucial component in PSCs. However, it limits the practical application of PSCs because of its high manufacturing cost. The CEs must have a work function higher than the perovskite material and also good electrical conductivity [91]. Platinum (Pt), silver (Ag), and gold (Au) are most widely used CEs typically deposited using vacuum evaporation or E-beam evaporation techniques. Later printable carbon counter electrodes (CCEs) were developed that replaces the familiar Au, Ag, and Pt CEs. The manufacturing of these printable CCEs is bit cheaper and enables us to upgrade the fabrication of large-scale PSCs. Graphite and carbon are the main components of CCEs. Zhang et al. applied a mesoporous graphite/carbon black CEs using different sizes of flaky graphite to hole conductor-free mesoscopic-free PSCs by screen printing technique and achieved a PCE of 11% with optimized graphite thickness of 9 μm. Their results indicated that graphite as a CCE could significantly affect the square resistance of CCEs and which results in differences in the FF and PCE of the devices. They reported that changes in the size of graphite make differences in PbI_2 filling, $CH_3NH_3PbI_3$ precursors, and also the square resistance of CCE resulting in differences in performance of PSCs. They suggested that fully printable hole-conductor-free mesoscopic PSCs based on CCEs have the potential to realize large-scale fabrication in the near future. They concluded that this carbon-based CEs offer a wider prospect to PSC toward further applications due to its availability, low cost, and its properties [92]. However, these materials are quite fragile and require high thickness to achieve low sheet resistance. Thus a novel material is required, which should be cheap and robust. Nickel CE can be a possible alternative over the CCEs or Au/Ag CEs. Xu et al. reported a maximum efficiency of 10.4% PCE using nickel as CE. They replaced the noble metals such as Au, Ag, etc. to nonprecious metals like Cu, Ni, W, and Mo films as the CE for PSC. They prepared the nonprecious metals using sputtering and achieved a PCE of 13.04%, 12.18%, 12.38%, and 11.38%, respectively, which is compared with that of the cells with Ag (15.97%) as a CE. They mentioned that the final PV performance of the prepared PSC is influenced by the morphology and resistance of the as-prepared metal films. Thus they concluded that the low-cost nonprecious metals (Cu, Ni, W, and Mo) can also be the promising CE materials to reduce the fabrication cost of the PSCs [93]. Later Liu et al. employed mesoporous nickel (mp-Ni) film as a printable CE for PSC. They fabricated low-temperature carbon CE-based HTM-free mesoscopic perovskite devices and encapsulated them by low-cost polydimethylsiloxane (PDMS) and achieved a PCE of 10.8%. They found that this technique enhanced the performance to 54% than those without encapsulation. They also found that the involvement of PDMS condensed the mesoscopic layer and improved the $MAPbI_3$/carbon interface condition leads enhanced performance of PSC. Further, they found that the improved performance is due to better charge transfer and slow interfacial recombination and the PDMS acting as shield to protect the PSC from the moisture by isolating water in air. The prepared package devices shown a remarkable stability of testing period of 3000 h. They suggested that this work improves the way for realizing low cost, highly efficient, and stable hybrid PV cells [94]. Still a vast research is required to find novel materials to improve the performance of CE.

The reported growth in the PCE of PSCs is shown in Fig. 8.7. From the figure, it is clear that the PCEs of PSCs have shown a remarkable growth from 10.9% to 22%. The maximum reported efficiency till the date is nearly 22% PCE reported by Ciro et al. [11]. They developed a $CH_3NH_3PbI_3$ p−i−n planar PSC fully fabricated under low temperature and achieved a 10.0% average efficiency with null hysteresis processed in an environment with RH as high as 70%. They found the optimized dipping time for the dip coating with fabrication environment and the selection of the HTL also influences the performance and stability of the PSC. They mentioned that the results were based on a proper treatment of the precursor, implementation of solvent engineering process and adequate solution for HTL. They suggested that the presented technique provides an insight for upscaling the PSC under low-temperature and high-moisture environment. Hence, it is expected that the PSCs technology may further increase in future and will be a promising alternative technology.

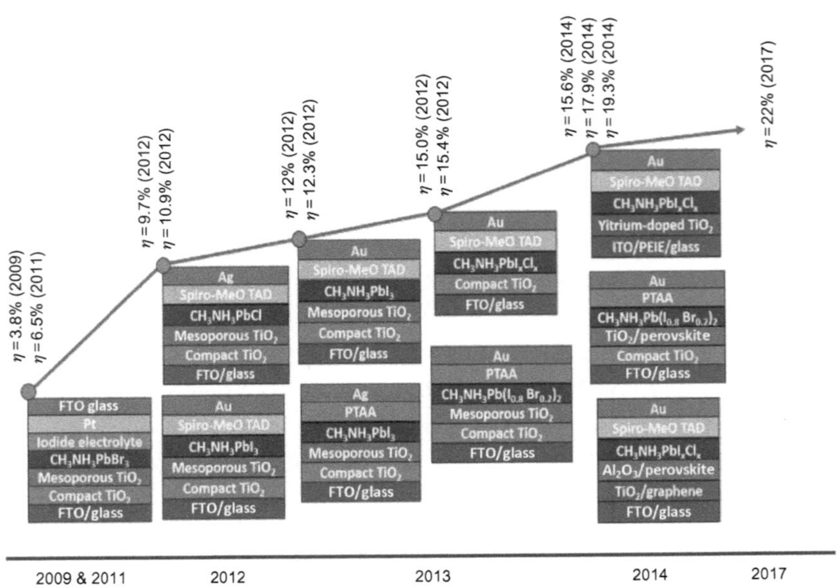

Figure 8.7 Timeline of various reported PCE of perovskite solar cells [17].

8.3 PERFORMANCE OF PEROVSKITE SOLAR CELLS

PSCs are a class of new PV technology which races above all other PV technology in a short duration of time. Besides its unprecedented development, there are still some factors that make undesirable impact on its future growth. Poor stability and short life time are the major problems that keep the PSCs from being commercialized over traditional silicon solar cells and restrict the utilization of this innovative technology. This instability occurs when the fabricated cells react with various environmental factors such as moisture [95], temperature [96], UV light exposure [97] in the perovskite layer, and other factors such as degradation due to electron/hole transporting layer [98], electrodes [99], and buffer layers [100]. These different environmental conditions lead to different problems in PSCs. Hence a vast research is required to improve the stability as well as the PCE of the cell in different possible ways such as device structure, fabrication method, and any new approaches. The major reported factors affecting the stability of PSC are shown in Fig. 8.8.

8.3.1 Moisture Effect

The first and foremost issue is the chemical instability due to interaction of perovskite layer with oxygen and water. Generally, there are two routes where the decomposition of perovskite takes place, as shown in Fig. 8.9 [101]. If PbI_2 and/or CH_3NH_3I combines with other components then the chemical equilibrium reacts in opposite direction leads to decrease in

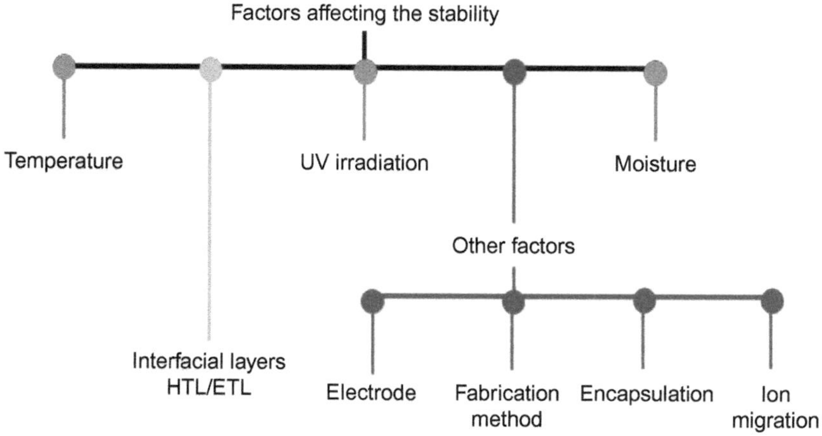

Figure 8.8 Several factors affecting the stability of PSC.

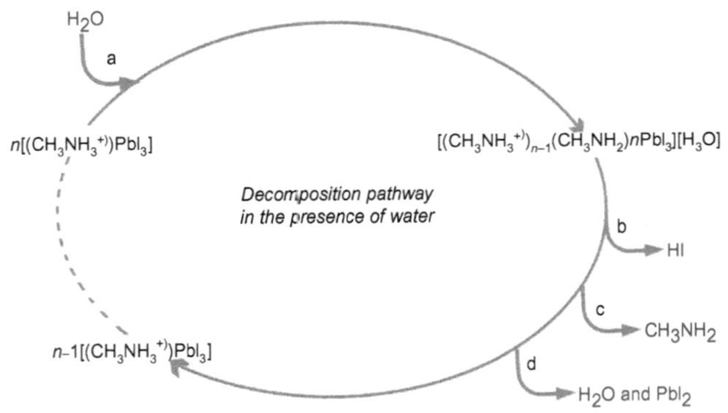

Figure 8.9 Possible decomposition pathway of hybrid halide perovskites in the presence of water. A water molecule (a) is required to initiate the process with the decomposition being driven by the phase changes of both hydrogen iodide ((b) soluble in water) and the methylammonia ((c) volatile and soluble in water). This pathway results in the formation of a yellow solid, which corresponds the experimentally observed PbI_2 (d) [101].

the chemical stability of perovskite layer and secondly the perovskite itself degrade directly into other chemicals. The instability due to moisture occurs when oxygen and moisture directly affect the stability because of the high sensitivity of perovskite to water. Seok and coworkers [102] observed a remarkable change in color from dark (brown) to light (yellow) due to the decomposition of perovskite under 55% humidity as shown in Fig. 8.10. Thus it is clear that degradation cause decreases in the efficiency and lifetime of PSCs and hence limiting this technology for outdoor application.

Later, Yang et al. reported a new phenomenon which boosts the transport property of chemicals resulting an improved PCE up to a maximum of 19.3%. They fabricated the PSCs in controlled humidity conditions, where the hygroscopic organic species could be dissolved by moisture, resulted in accelerated transport property of chemicals [78]. Even though the device fabrication was performed under controlled humidity, they still observed some degradation due to moisture that retains >5% of its original power conversion after 6 days of storage in ambient atmosphere.

Smith et al. studied the device stability where $(PEA)_2(MA)_2(Pb_3I_{10})(PEA = C_6H_5(CH_2)_2NH_3^+)$ as an absorber and they fabricated a high-quality thin film by one-step spin-coating method.

Figure 8.10 Photographs of 3D $TiO_2/CH_3NH_3Pb(I_{1-x}Br_x)_3$ bilayer nanocomposites on FTO glass. This clearly illustrates the change in optical response as the stoichiometry is varied. *Reprinted with permission from J.H. Noh, S.H. Im, J.H. Heo, T.N. Mandal, S. Seok, Chemical management for colorful, efficient, and stable inorganic–organic hybrid nanostructured solar cells. Nano Lett.13 (2013) 1764–1769. doi:10.1021/nl400349b. Copyright 2013 American Chemical Society [102].*

They found that this device shows a better resistance to moisture. With the results, Smith et al. concluded that there is no phase transformation occurs in $MAPbI_3$ due to the decomposition under ambient condition over for 46 days [103]. Bryant et al. reported the stability of MA lead triiodide PSCs when exposed to ambient condition. From this work, they found that light- and oxygen-induced degradation is the main reason for the degradation of the cells. They performed the stability test by exposing the fabricated cells to both light and dry air, and found that the nonencapsulated $MeNH_3PbI_3$ solar cells rapidly degraded in minutes to few hours and observed an enhanced degradation in electrically bias driven current flow in presence of O_2. They also tested the cells in moisture conditions and found a slower degradation at a nitrogen atmosphere with a RH of N_2. They observed that this is due to the light- and oxygen-induced degradation. This can be slowed down by using interlayers which will remove the electrons from the perovskite films before they can react with oxygen to form O^{2-}. Thus they concluded that these ETL properties are critically depend on suitable barrier layers and device configuration [104].

Roose et al. reported a mesoporous Nd-doped titania electrodes and achieved an improved performance. They gradually increased the incorporation of Nd and found an enhanced performance of the fabricated cells especially in the FF. They reported that this improvement is due to the reduction of deep trap states, improved electron transport, and reduced charge recombination. For the 0.3% Nd and for 0.5% Nd doping, they observed a reduced performance of the fabricated device. They also tested the stability under air conditions and found that 18% stabilized power under air for 200 s than that of the undoped devices. They reported that this improved stabilization is due to the Nd passivation of oxygen [105]. Tai et al. reported that the use of lead(II) thiocyanate (Pb $(SCN)_2$) precursor for the fabrication of PSCs under ambient air and achieved a maximum PCE of 15% for a high-quality $CH_3NH_3PbI_{3-x}(SCN)$ than the typical $CH_3NH_3PbI_{3-x}$-based devices. They also reported that the nonencapsulated devices show improved stability in humid air, which is due to the incorporation of thiocyanate ions in the crystal lattice. Thus the authors concluded that these findings provide a pathway to realize efficient and stable PSC in ambient atmosphere [106]. Bi et al. reported a new aliphastic fluorinated amphiphilic additive in $MAPbI_3$ perovskite to enhance the stability without losing the device performance. They achieved an improved PCE of 18% by incorporating 3 mol% 1,1,1-trifluoro-ethyl ammonium iodide additive into the perovskite precursor solution. They also found an enhanced stability due to moisture for some typical optimized devices using these additives and shows a negligible $J-V$ hysteresis on 2D perovskites (sacrificing performance) due to the hydrophobic CF_3 terminal group over the surface of the 3D perovskite, which is a new paradigm of improving environmental stability of PSCs [107]. Arora et al. reported that both intrinsic and extrinsic stability of a fornamidinium lead bromide [CH $(NH_2)_2PbBr_3 = FAPbBr_3$] PSC yields a high photovoltage with a PCE of 8%. They achieved this efficiency by modifying the mesoporous $TiO_2 - FAPbBr_3$ interface using lithium treatment. The emission dynamics studies showed that electron injection from conduction band of $FAPbBr_3$ into that of mesoporous TiO_2 is faster than for the untreated scaffold. They also found that the charge carrier recombination rate within a device based on lithium-treated titania film is an order of magnitude lower than that of the pristine titania. They tested their device both intrinsically, means under nitrogen and ambient conditions and found that the encapsulated devices retained over 95% of its initial over 150 h in nitrogen and dry air and 80% humidity at around 60°C [108]. Deng et al. reported the fabrication of PSC using mixed phase of cation (FA and MA) by doctor blade method and achieved a PCE of 18%. The authors chose mixed phase cation because of its lower band gap and thus tuning the extension absorption of the layer to 850 nm. They also observed an enhanced stability using mixed phase carbon. Then they found that a nonencapsulated device maintained the performance without any PCE loss which is stored in ambient condition of around 25°C with 20%–60% RH for 20–30 days.

Thus they concluded that the Cu can be a more stable cathode than Al or Ag without losing its efficiency and provides a pathway to further improve the performance of the device toward long stability [87]. Back et al. developed a new concept of a chemical inhibition in PSCs using mine-mediated metal oxide systems and succeeded in chemically neutralizing mobile ionic defects through mutual ionic interaction. They found that the degradation process of the planar-type PSC arises from the migration of ionic defects in the perovskite layer yielding insulating Ag–I bonds at the surface of the metal electrode. Then, we can suppress the formation of the insulating Ag–I bonds by adapting a solution-processed metal oxide system as an efficient ion extraction and chemical stabilization layer for the ionic defects between the metallic electrode and PC60BM ETL. Then thereby realizing planar PSCs with long-term stability using an Al (or Ag) electrode and they also observed a long-term stability using Al or Ag electrode. Thus they concluded that their results provided an important step toward air stable and fully printable PSC for next generation renewable energy sources [109].

8.3.2 Ultraviolet Radiation Effect

Another most common factor that affects the performance of PSCs is the degradation due to the illumination of UV light. Photo instability may also occur due to the spatial arrangement of crystals [23]. Multiple research has been undertaken to reduce the instability of PSCs caused under UV light. Hiroshi et al. reported that pristine perovskite layer transformed into PbI_2 after 12 h of light exposure and confirmed the degradation. Wei et al. reported that the degradation of $(R\text{-}NH_3)_2PbX_4$ and this degradation depends on the chemical structure of organic and inorganic part under UV illumination [110]. Leijtens et al. investigated the instability of PSCs and found that the titania layer is susceptible to UV-induced degradation. They identified that the instability does not arises in mesoporous TiO_2-free mesosuperstructured solar cells and its arising only from light-induced desorption of surface-adsorbed oxygen. They also found that TiO_2-free cells can deliver stable photocurrent for over 1000 h continuous exposure and operation under full spectrum–simulated sunlight [111]. Therefore the researchers' attention has turned toward finding a new approach with the conditions such as finding trap sites, avoid UV light from reaching the titania layer and/or replacing the titania with other suitable materials to avoid the instability of PSCs. Wi et al. used cesium bromide (CsBr) to modify the interface between the ETL and the perovskite layer to enhance the performance and stability of the PSCs. They also reported that the CsBr modification improves the stability under the UV light exposure. They concluded that the improved performance is due to the low defects at the heterojunction and improved charge extraction [112].

Ito et al. reported a reduction in the degradation due to UV light by introducing Sb_2S_3 blocking layer and retained 65% of its efficiency after 12 h illumination of light [110]. They fabricated $CH_3NH_3PbI_3$ PSCs using an inorganic hole transporting material (CuSCN) with/without Sb_2S_3 by chemical bath deposition and achieved an efficiency of 5.24%. They inserted Sb_2S_3 layers at the interfaces between TiO_2 and $CH_3NH_3PbI_3$. Then they tested the prepared cell under the light exposure without encapsulation, the $CH_3NH_3PbI_3$ solar cells without Sb_2S_3 deteriorated to zero efficiency in 12 h and were completely changed from black to yellow. They also tested the $CH_3NH_3PbI_3$ solar cells and found that it is stable against light exposure without encapsulation. They concluded that Sb_2S_3 elongates the lifetime of the $CH_3NH_3PbI_3$ perovskite structure during light exposure and that the degradation origin of the $CH_3NH_3PbI_3$ perovskite layer is at the interface between TiO_2 and $CH_3NH_3PbI_3$. They mentioned that further investigation is required to study the degradation mechanism. Hence the incorporated blocking layer acts as a buffer layer and improves the stability. Snaith et al. investigated the stability degradation in PSCs under UV and found that the titania photoanode encapsulated in nitrogen atmosphere shows a rapid decay in PCE. Hence, they reported that nonencapsulated device, in the oxygen atmosphere, could improve the stability because it removes the surface states of titania [111]. Komarala et al. found a new method to prevent the degradation caused by UV light by adding a UV filter before the titania. They coated a down-shifted YVO_4:Eu^{3+} nanophosphorous layer on reverse side of FTO glass by spray coating method and found that the coated layer absorbs the UV light and emit visible light for perovskite layer absorption which significantly improves the stability by retaining 50% of its initial efficiency [113]. They actually reported a simple method to mitigate UV degradation in TiO_2-based PSCs using a transparent luminescent down shifting (DS) $YVO_4:Eu^{3+}$ nanophosphor layer and achieved a PCE of 8.5%. They coated the PSC with DS phosphor which improved the stability under illumination for long time and retaining more than 50% of its initial efficiency, they also tested with the phosphor layer and found that 8.5% enhanced photocurrent. They concluded that the $YVO_4:Eu^{3+}$ layer served a bifunctional role in PSC by reducing photo-degradation also enhancing energy conversion efficiency and suggested that these results may open up further avenues for improving the performance of PSCs.

8.3.3 Temperature Effects

Annealing temperature is an another factor that affects the device stability as during operation the solar modules are subjected to elevated temperature. The temperature from the sun is greater than the transition temperature of perovskite layer; thus phase transition may affect the performance of the device. Various effects cause the thermal instability, such as effects of thermal treatment on the crystal structure and the perovskite decomposition due to annealing temperatures. Annealing is necessary to form the perovskite structure but exposure of PSCs to such elevated temperature makes a critical impact on perovskite crystal structure.

Grastzel et al. studied the thermal behavior of the perovskite and its individual components and reported that the phase transition occurs only in PbI_2 precursor method and not in $PbCI_2$. Therefore these differences in precursors could influence the PV performance and the stability of devices. In their work, they reported that the thermal behavior of the perovskite was explored using thermo gravimetric analysis (TGA) and DSC, and examined the thermal properties of the $CH_3NH_3PbX_3$ ($X = I$ or Cl) perovskite using TGA. They discussed about the sublimation of the excess organic product formed from mixed halide perovskite mixtures and they observed that as a separate process by TGA. The decomposition of organic component of the perovskite by the subsequent loss of HI and CH_3NH_2 where latter species is more tightly incorporated into the perovskite matrix. This behavior likely may affect the long-term stability of these materials when used for solar application. In the case of PbI_2, precursor-based perovskite tetragonal to cubic phase transition was observed. The PV performance of the prepared material is affected by the subtle difference in the formation process of the perovskite. The desired flexibility to fabricate the triiodide perovskite enables to turn the crystallization process. Moreover, it gives a clear view on material's thermal nature and stability, which plays a major role in PV applications [114]. Weber et al. reported a temperature dependent structure, MA tri-halide perovskite $CH_3NH_3PbX_3$ at room temperature that has a tetragonal structure. As the temperature increases, it is transformed to cubic structure which is in line with analogy that, when temperature increases the symmetry increases accordingly. In their work, they reported a temperature-dependent structure of crystalline MA trihalogenoplumbates (1I)-CH_3NH^{3+} PbX^{3-} (X = CI, Br, I) and compared with measurements of the temperature-dependent complex permittivity at frequencies of 50−150 GHz. The dielectric measurements reveal a picosecond relaxation process which corresponds to a dynamic disorder of the MA group in the high-temperature phases of the trihalogenoplumbates [115]. Conings et al. investigated the thermal stability of the PSCs in the presence and absence of HTL to study the interfacial degradations for the samples annealed at 85°C for 24 h under different environment such as N_2/O_2 with pristine as reference. They have reported that both the presence and absence of HTL show a variation in the structure which concludes that the degradation occurs even in low thermal annealing temperature. In their work, they investigated the thermal stability of $MAPbI_3$ perovskite layers in different environmental conditions and found that this type of perovskite is not intrinsically stable while heating at an operating temperature. They suggested that these results are an important message in view of further development of the PV perovskite technology [116]. Bi et al. investigated the effect of thermal annealing on the perovskite films for the perovskite layer annealed at 105°C for 3 h. They have reported an improved mobility; however, annealing for 3 h led to the formation of PbI_2 and concluded that the minimum annealing temperature and time also lead to thermal driven decomposition. In their work, they actually investigated a MA lead iodide perovskite thin films grown by the low-temperature thermal annealing-induced interdiffusion and achieved an efficiency of 13.4%. They found that thermal annealing not only drives the formation of perovskite but also affects the morphology, optoelectronic properties, and correlated device performance. They also found that the reduction of work function in the perovskite films, caused by increased annealing duration, is linearly correlated with the open-circuit voltage loss, which points out a path for further increase of the device efficiency. They mentioned that the thermal annealing at 105°C quickly drives the formation of phase-pure perovskite in a short time of 15 min, and followed thermal annealing up to 2 h continuously increases perovskite crystallinity and grain size without losing film continuity or coverage. They expected that the annealing process can also optimize the performance of PSCs [117]. Phillippe et al. investigated the influence of the temperature on both $CH_3NH_3PbI_3$ and $CH_3NH_3PbI_xCl_x$ films and found that the elevated temperature affects the device stability. They performed the experiment under ultra-high vacuum chamber and studied the effect of temperature in the absence of air and water. For that they studied the stability of methylamine lead halide perovskites and their evolution under diverse conditions such as water exposure, thermal treatment, and time in different environments. They compared $CH_3NH_3PbI_3$ and $CH_3NH_3PbI_{3-x}Cl_x$ with each other and also to $CH_3NH_3PbCl_3$ and found different morphologies and kinetics of formation. They exposed the $CH_3NH_3PbI_3$ to various environments and found that degradation occur at 100°C which is not only related to large humidity even in an inert atmosphere such as argon. They suggested that the obtained results are crucial for a better understanding of this material and will help to improve not only the postconditioning of the cells but also their synthesis process [117]. Saliba et al. reported that a small oxidation of stable Rb^+ can be embedded into a cation cascade and which will create perovskite materials with improved properties. With this implementation, they achieved a stable PCE of 21.6%. They also reported that the polymer-coated cells retained 95% of their initial performance at 85°C of over 500 h under full solar illumination and maximum power point tracking [118]. Yang et al. performed in situ absorption measurement of perovskite film under exposure to 100°C for a long period and 150°C for short periods and found that decomposition of perovskite to PbI_2. They also performed the same treatment of the perovskite films deposited over the ITO substrates and found no change in the films. Thus they reported that the ZnO/perovskite layer interface was responsible for this degradation. In their work, they investigated the use of ZnO as ETL to study the origin of thermal instability. They found that decomposition in the fabricated device is due to the photon transfer at the interfaces which is the unique nature of ZnO surface. They further found the decomposition boosts due to the presence of reduced ligands and surface hydroxyl groups. Calcination of the ZnO layer results in a more thermally stable $ZnO/CH_3NH_3PbI_3$ interface, albeit at the cost of a small decrease in PCE. They expected that these results may help to guide the development of PSCs even in the harsh conditions and large temperatures [119]. Sutton et al. demonstrated a high-efficient cesium-based lead halide PSCs by tuning its band gap and stabilizing the black perovskite phase at lower temperatures. They found that $CsPbI_2Br$ can be used to fabricate solar cells. Thus they employed $CsPbI_2Br$ in a planar architecture device and achieved a PCE of 9.8%. They enhanced the stability using mixed halide composition which offered an enhanced thermal stability over their organic-based counterparts. They reported that further understanding about structural stability is necessary to assess the full application of these inorganic perovskites to solar power generation. Through these results, they concluded that all inorganic perovskite will be a promising new step for PV materials. They mentioned that further understanding about structural stability is necessary to assess the full application of these inorganic perovskites to solar power generation. Through these results they concluded that all inorganic perovskite will be promising next step for PV materials [120,121].

Pisoni et al. reported that the thermal conductive properties of perovskite layer and found a very low thermal conductivity for both the crystals in single and polycrystalline form. The authors depict that the heat caused by light could not spread quickly and causes the mechanical stress that limits the lifetime of the device. They reported the temperature dependence of thermal conductivity of single crystalline and polycrystalline organometallic perovskite $CH_3NH_3PbI_3$. They theoretically demonstrated the importance of the resonant scattering in both specimens. They reported that the interaction between phonon waves and rotational degrees of freedom of $CH_3NH_3^+$ sublattice emerges as the dominant mechanism for attenuation of heat transport and for ultralow thermal conductivity of 0.5 W/Km at room temperature. Recently, Snaith et al. reported a new approach to reduce the thermal degradation and moisture effect by replacing organic HTL material with polymer-functionalized single-walled carbon nanotubes (SWNTs) inside an insulating polymer matrix. In their work, they demonstrated an elegant approach to achieve reduced thermal degradation by using SWNTs embedded with insulating polymer matrix instead of organic HTM and achieved a PCE of 15.3%. They also observed that an enhanced thermal degradation as compared to cells employing organic HTMs and remarkably enhanced the resistance to water ingress. They concluded that these critical developments may lead to achieve an efficient and long-term stable PSCs [122].

8.3.4 Effect of Electron Transport Layer and Hole Transport Layer

In addition to the earlier discussions, there are also many other factors that affect the device stability, such as influence of ETM/HTM on device stability, the impact of fabrication method, and the effect of the structure of metal oxide film on the stability of PSC devices. PSCs made of ETL architecture introduce a nonstoichiometry defects that cause subband gaps which limits the performance. Mesoporous titania is the most commonly used ETL, whereas Spiro-OmeTAD is the most commonly used HTL. It requires additives to improve the conductivity. However, it directly affects the perovskite layer and results in reduced performance of PSC. Pathak et al. reported an improved performance of cell using a low-level doping of Al into titania and found that this low level of doping reduces the number of trap states leads to a reduced nonstoichiometry defects. In detail, they investigated the implications of nonstoichiometric defects in TiO_2 on the dye sensitized and PSCs. They fabricated mesostructured Al-doped anatase TiO_2 and found that the substitution of Al into the TiO_2 lattice reduces the number of subband gap states and dramatically increases electron conductivity. This leads to an overall improvement in the device performance of dye-sensitized and perovskite-based mesosuperstructured solar cells (which employ a compact TiO_2 n-type collection layer). They reported that the oxygen vacancy–induced defects in anatase TiO_2 not only limit the electron transport within the material but also cause of substantial deterioration of device operation upon UV-irradiation in inert condition. They found that the overall performance is attributed to the incorporation of Al in the anatase lattice, "permanently" passivating electronic trap sites in the bulk and at the surface of the TiO_2. They concluded that this study steps forward toward the realization of solid-state DSSCs with practical outdoor operational lifetimes [123]. Song et al. replaced titania with SnO_2 as an ETL. And they fabricated the device using PbI_2 crystallites into the $SnO_2/CH_3NH_3PbI_3$ interface and found an improved stability. However, they also reported that the replacement of titania with SnO_2 results in stable PCE for the nonencapsulated devices stored in ambient conditions which is better than the devices using titania. In their work, they demonstrated a compact SnO_2 thin film prepared by sinter-less spin-coating processes as an electron selective contact for $CH_3NH_3PbI_3$-based planar-heterojunction PSCs. They developed SnO_2 thin films as electron selective contacts for $CH_3NH_3PbI_3$-based PSCs and achieved an efficiency of 13.0% with optimized solvent vapor and annealing time and perovskite layer thickness. They observed that the variation of grain size of PbI_2 layer is dependent on the solvent vapor annealing time resulting in morphological and absorbance changes in the corresponding perovskite films. They found that the fabricated device is highly durable with exposure to the ambient air environment for 30 days. They concluded that this study will help to promote the development of the novel PV technology for perovskite [124].

Another way to improve the stability is by modifying the HTL with additives. Zheng et al. employed a functionalized nanographene as the HTM and improved the efficiency of the pristine film as well as the graphene-doped film. They reported that device shows a high stability under light and 45% humidity without encapsulation [125]. Another way to improve the stability is by introducing some stable materials into the cell structure. To achieve rapid charge extraction, Chen et al. reported a heavily doped planar PSC with inorganic charge extraction layer. They fabricated the inorganic layer of thickness 10–20 nm without pinholes and also eliminating local structural defects over large area. The prepared inorganic layer is of an aperture area $>1\ cm^2$, which has a PCE of 7.5%. Through analysis they found that the fabricated device is stable; since the hysteresis is eliminated in the $I-V$ characteristics, the device retained 90% of initial performance with a testing over 1000 h under light exposure [30]. The report reveals that the coating of Ag over the perovskite layer leads to formation of AgI when it was exposed to moisture results in worse stability than coating with Au [126]. The instability due to moisture is purely depends on the nature of the HTM.

8.3.5 Other Factors

The other factors that affect the stability of the PSCs are fabrication methods, encapsulation, and device testing protocols. Multiple attempts were made on reducing the degradations caused by different factors in the perovskite layer and other interface layer to control the stability. An effective way to control the stability and improve the device stability is making modification on the perovskite films [127]. Most commonly used perovskite layer is $CH_3NH_3PbI_3$, thus modification in this layer may improve the device stability. Noh et al. tuned the stoichiometry of $CH_3NH_3(=MA)Pb(I_{1-x}Br_x)_3$ perovskites by substituting I ions with Br ions and found varied optical property as well as improved stability. They have reported that with increase in the Br content the stability improves further [94]. In their work, they demonstrated the chemical management of $MAPbX_3$ (X = halogen) and chemically tuned the inorganic − organic hybrid materials based on $CH_3NH_3(=MA)Pb(I_{1-x}Br_x)_3$ perovskite and achieved a PCE of 12.3% with improved stability. They controlled the band gap to cover almost entire visible spectrum. They believed that this technique make a step forward toward the realization of low-cost, high-efficiency, and long-term stability with colorful solar cells [102]. Smith et al. used a two-dimensional layered perovskite structure that results in increased band gap as well as improved stability. They developed the 2D layer by slicing the perovskite structure along its crystallographic phase of the most commonly used 3D layers and reported that these high-quality 2D films are more stable to the moisture. They reported the developed 2D-layered perovskite when exposed to a RH of 52%, and observed no decomposition after 46 days even in the presence of moisture [103]. Thus the 2D perovskite structure is more efficient than the commonly used 3D perovskite structure in controlling the stability. Encapsulation is one of the crucial process in improving the performance and commercialization of perovskite devices. Many encapsulation techniques were proposed, which allow one to treat large areas as well as small area cells. Some of the techniques were discussed later. The use of desiccant material filled over the gap of the glass cover to absorb water greatly improved the stability even under high humidity conditions. Han et al. investigated the stability analysis under the elevated temperature (up to 85°C) and humidity (up to 80%) of planar MA lead iodide PSCs. They found better performance at room temperature for the encapsulated devices but degraded at high temperature and humidity. They performed the stability test at one Sun of $100\ mW/cm^2$ illumination at 85°C temperature with 80% humidity. Thus they found that the degradation can be attributed due to the decomposition of $MAPbI_3$ with H_2O and the subsequent reaction of hydroiodic acid with the silver back contact. They concluded that the performance and stability can be improved by using chemically stable conducting materials instead of silver back contact and proper sealing materials [108,128]. The process that takes place at the perovskite decomposition is given as follows: (1) thin silver layer corrosion and void formations in the Spiro-OMeTAD layer due to gaseous products from the decomposition of the perovskite material; (2) void formation in the perovskite layer plus the perovskite layer detaching from the TiO_2 layer and the PbI_2 formation. The HI generation is consumed due to reaction with silver which directly boosts the perovskite materials to decompose further until the cell is fully degraded. Later the advancement in encapsulation technique has been made using roll-to-roll processed polymer solar cells developed by Krebs et al. They reported an optimized inverted polymer solar cells comprising glass/ITO/ZnO/P3HT-PCBM/PEDOT:PSS/silver to exhibit the same performance as model devices with a normal geometry (glass/ITO/PEDOT:PSS/P3HT-PCBM/aluminum) and tested their devices for stability under model atmosphere. The device geometry consists of ITO-ZnO-(substrate) (active layer)-PEDOT:PSS-silver with P3HT-PCBM as the active layer. They compared the prepared devices with a normal geometry and found that both devices have a PCE of 2.7% under 1000/Wm^2, AM1.5G, 72 ± 2°C. They also tested their device with different atmosphere such as dry nitrogen (99.999%), dry oxygen (99.5%), humid nitrogen (90 ± 5% RH), and ambient atmosphere (20% oxygen, 20 ± 5% RH), and both device types were found to be stable in a nitrogen atmosphere during the test period of 200 h. They mentioned that this effect is due to the interaction between the active layer and the aluminum electrode. They also tested that device with a geometry using Al electrode gives stable operation in dry oxygen and not stable in the presence of humidity, and on the other hand the geometry that uses Ag electrode gives an opposite result (stable in humidity and not stable in dry oxygen atmosphere). They fabricated the polymer solar cells with using roll-to-roll process in air and found the PCE of 2.1% for a full module with 120 cm active area (AM1.5G, 393/Wm^2) and up to 2.3% for modules with 4.8 cm active area (AM1.5G, 1000/Wm^2) [129]. Recently, Shao et al. replaced the top electrode with conductive copper tapes that successfully reduced the effects due to air and achieved 12.7% of PCE. The C-tapes can encapsulate the device efficiently; subsequently, it improves the device stability to a great extent. Thus the authors fabricated the top electrodes for planar heterojunction PSCs through a facile lamination process that provides a vacuum-free fabrication method at low temperature. Hence the C-tape enables the low-cost device fabrication with enhanced stability and can be a promising alternative for the commercial application. However, a more detailed study is necessary to optimize its conductivity with softer conductive acrylic adhesive [130]. Nie et al. reported that a slow photocurrent degradation in crystalline large grain size hybrid PSCs is due to the formation of light-activated metastable deep level trap states. Their devices were fabricated in such a way that it can self-heal completely by keeping at dark for less than a minute or degradation can be completely prevented by operating at a temperature of 0°C. They also investigated several mechanisms to explain the microscopic origin for the formation of theses trap states. They suggested that these formations of light-induced defects can be controlled to obtain light stable PSCs by preventing meta-stable electronic state formation and accumulation. Through their investigations, they concluded that these results give a move toward reliable and stable perovskite-based solar cells [128,131].

Another factor that affects the stability is the movement of ions within the perovskite structure. Nagaoka et al. incorporated Zr into titania, found an increase in migration of defect state density in oxide, and reported a significant improvement in PCE. In their work, they investigated the incorporation of zirconium (Zr) into titania (TiO_2) as an ETL used in organometal halide perovskite PVs. They found an improvement in overall PCE and a decreased hysteresis due to the hybrid organometal halide PSCs, which employs $Zr-TiO_2$ electrode and pyridine treatment. Through analysis they achieved a high-carrier lifetime and densities of the fabricated devices in microsecond times in transient photovoltage experiments, and also longer photovoltages over a range from milliseconds to 10 s. They suggested that these results highlight the importance of the perovskite/electrode interface in controlling both hysteresis and overall performance [132]. Thus an innovative research is required to find new pathways to achieve long-term stability in PSCs that could lead these devices toward commercialization.

8.4 CONCLUSION

Research on PSCs has attained a significant attention since its discovery by Lev Perovski in 2009. This chapter highlights the substantial growth of PSCs in terms of PCE (i.e., from 3.9% to 22%), which is comparable with the Shockley–Queisser limit. Various fabrication methods of PSCs in terms of different components and its optimization are also reported. The maximum reported PCE of 22% is obtained by the incorporation of suitable ETL/HTL. Optimization of perovskite layer may be the solution for the improvement of PCE further.

The recent works also reveal that the stability of the PSC is dependent on moisture, temperature, UV light, transport layers, etc. Design and optimization of proper encapsulating material must be performed to reduce the effect of moisture and UV light. Developing proper heterojunctions and understanding the degradation mechanism are the other aspects to improve the device stability. Thus each component of PSC makes a direct impact on the device performance. Therefore optimization of these components is required accordingly to enhance the reported efficiency further, instead of considering it as a whole device. Solving the above issues can trigger the commercialization of PSCs as the promising technology for the growing PV market.

ACKNOWLEDGMENTS

The financial support provided by the Inter University Acceleration Centre (IUAC), New Delhi, vide project no. UFR-58315, and DST-SERB, New Delhi, vide project no. ECR/2016/001186 are gratefully acknowledged.

REFERENCES

[1] L.L. Kazmerski, Photovoltaics: a review of cell and module technologies, Renew. Sustain. Energy Rev. 1 (1997) 71–170. Available from: https://doi.org/10.1016/S1364-0321(97)00002-6.

[2] B. Parida, S. Iniyan, R. Goic, A review of solar photovoltaic technologies, Renew. Sustain. Energy Rev. 15 (2011) 1625–1636. Available from: https://doi.org/10.1016/j.rser.2010.11.032.

[3] M. Grätzel, Dye-sensitized solar cells, J. Photochem. Photobiol. C: Photochem. Rev. 4 (2003) 145–153. Available from: https://doi.org/10.1016/S1389-5567(03)00026-1.

[4] C.J. Brabec, Organic photovoltaics: technology and market, Sol. Energy Mater. Sol. Cells 83 (2004) 273–292. Available from: https://doi.org/10.1016/j.solmat.2004.02.030.

[5] R.W. Miles, G. Zoppi, I. Forbes, Inorganic photovoltaic cells, Mater. Today 10 (2007) 20–27. Available from: https://doi.org/10.1016/S1369-7021(07)70275-4.

[6] A. Salant, M. Shalom, I. Hod, A. Faust, A. Zaban, U. Banin, Quantum dot sensitized solar cells with improved efficiency prepared using electrophoretic deposition, ACS Nano 4 (2010) 5962–5968. Available from: https://doi.org/10.1021/nn1018208.

[7] M.A. Green, A. Ho-Baillie, H.J. Snaith, The emergence of Perovskite solar cells, Nat. Photonics 8 (2014) 506–514. Available from: https://doi.org/10.1038/nphoton.2014.134.

[8] W. Shockley, H.J. Queisser, Detailed balance limit of efficiency of p-n junction solar cells, J. Appl. Phys. 32 (1961) 510–519. Available from: https://doi.org/10.1063/1.1736034.

[9] W.S. Yang, J.H. Noh, N.J. Jeon, Y.C. Kim, S. Ryu, J. Seo, et al., High-performance photovoltaic Perovskite layers fabricated through intramolecular exchange, Science 348 (2015) 1234–1237. Available from: https://doi.org/10.1126/science.aaa9272.

[10] M. Grätzel, The light and shade of Perovskite solar cells, Nat. Mater. 13 (2014) 838–842. Available from: https://doi.org/10.1038/nmat4065.

[11] J. Ciro, R. Betancur, S. Mesa, F. Jaramillo, High performance Perovskite solar cells fabricated under high relative humidity conditions, Sol. Energy Mater. Sol. Cells 163 (2017) 38–42. Available from: https://doi.org/10.1016/j.solmat.2017.01.004.

[12] C.D. Bailie, M.G. Christoforo, J.P. Mailoa, A.R. Bowring, E.L. Unger, W.H. Nguyen, et al., Semi-transparent Perovskite solar cells for tandems with silicon and CIGS, Energy Environ. Sci. 8 (2015) 956–963. Available from: https://doi.org/10.1039/C4EE03322A.

[13] H.S. Jung, N. Park, Perovskite solar cells: from materials to devices, Small (2014) 1–16. Available from: https://doi.org/10.1002/smll.201402767.

[14] J.M. Azpiroz, E. Mosconi, J. Bisquert, F. De Angelis, Defect migration in methylammonium lead iodide and its role in Perovskite solar cell operation, Energy Environ. Sci. 8 (2015) 2118–2127. Available from: https://doi.org/10.1039/C5EE01265A.

[15] H.J. Snaith, Perovskites: the emergence of a new era for low-cost, high-efficiency solar cells, J. Phys. Chem. Lett. 4 (2013). Available from: https://doi.org/10.1021/jz4020162.

[16] G. Niu, X. Guo, L. Wang, Review of recent progress in chemical stability of Perovskite solar cells, J. Mater. Chem. A 3 (2015) 8970–8980. Available from: https://doi.org/10.1039/C4TA04994B.

[17] M. Batmunkh, C.J. Shearer, M.J. Biggs, J.G. Shapter, Nanocarbons for mesoscopic perovskite solar cells, J. Mater. Chem. A 3 (2015) 9020–9031. Available from: https://doi.org/10.1039/C5TA00873E.

[18] K.A. Bush, A.F. Palmstrom, Z.J. Yu, M. Boccard, R. Cheacharoen, J.P. Mailoa, et al., 23.6%-Efficient monolithic Perovskite/silicon tandem solar cells with improved stability, Nat. Energy 2 (2017) 17009. Available from: https://doi.org/10.1038/nenergy.2017.9.

[19] N. Park, Perovskite solar cells: an emerging photovoltaic technology, Biochem. Pharmacol. 18 (2015) 65–72. Available from: https://doi.org/10.1016/j.mattod.2014.07.007.

[20] W.E.I. Sha, X. Ren, L. Chen, W.C.H. Choy, The efficiency limit of CH_3 NH_3 PbI_3 perovskite solar cells, Appl. Phys. Lett. 106 (2015) 1–14. Available from: https://doi.org/10.1063/1.4922150.

[21] K. Wang, P. Shen, M. Li, S. Chen, M. Lin, P. Chen, et al., Low-temperature sputtered nickel oxide compact thin film as effective electron blocking layer for mesoscopic $NiO/CH_3NH_3PbI_3$ Perovskite heterojunction solar cells, ACS Appl. Mater. Interfaces 6 (2014) 11851–11858. Available from: https://doi.org/10.1021/am503610u.

[22] C.-H. Chiang, Z.-L. Tseng, C.-G. Wu, Planar heterojunction Perovskite/PC71 BM solar cells with enhanced open-circuit voltage via a (2/1)-step spin-coating process, J. Mater. Chem. A 2 (2014) 15897–15903. Available from: https://doi.org/10.1039/C4TA03674C.

[23] M. Shahbazi, H. Wang, Progress in research on the stability of organometal Perovskite solar cells, Sol. Energy 123 (2016) 74–87.

[24] Y.S. Lee, T. Gershon, O. Gunawan, T.K. Todorov, T. Gokmen, Y. Virgus, et al., $Cu_2ZnSnSe_4$ thin-film solar cells by thermal co-evaporation with 11.6% efficiency and improved minority carrier diffusion length, Adv. Energy Mater. 5 (2015). Available from: https://doi.org/10.1002/aenm.201401372.

[25] Q. Wang, Y. Shao, Q. Dong, Z. Xiao, Y. Yuan, J. Huang, Large fill-factor bilayer iodine Perovskite solar cells fabricated by a low-temperature solution-process, Energy Environ. Sci. 7 (2014) 2359–2365. Available from: https://doi.org/10.1039/C4EE00233D.

[26] D. Liu, M.K. Gangishetty, T.L. Kelly, Effect of $CH_3NH_3PbI_3$ thickness on device efficiency in planar heterojunction Perovskite solar cells, J. Mater. Chem. A 2 (2014) 19873–19881. Available from: https://doi.org/10.1039/C4TA02637C.

[27] A.T. Barrows, A.J. Pearson, C.K. Kwak, A.D.F. Dunbar, A.R. Buckley, D.G. Lidzey, Efficient planar heterojunction mixed-halide Perovskite solar cells deposited via

spray-deposition, Energy Environ. Sci. 7 (2014) 2944. Available from: https://doi.org/10.1039/C4EE01546K.

[28] Z. Wei, K. Yan, H. Chen, Y. Yi, T. Zhang, X. Long, et al., Cost-efficient clamping solar cells using candle soot for hole extraction from ambipolar Perovskites, Energy Environ. Sci. 7 (2014) 3326–3333. Available from: https://doi.org/10.1039/C4EE01983K.

[29] P.Y. Chen, J. Qi, M.T. Klug, X. Dang, P.T. Hammond, A.M. Belcher, Environmentally responsible fabrication of efficient Perovskite solar cells from recycled car batteries, Energy Environ. Sci. 7 (2014) 3659–3665. Available from: https://doi.org/10.1039/C4EE00965G.

[30] W. Chen, Y. Wu, Y. Yue, J. Liu, W. Zhang, X. Yang, et al., Efficient and stable large-area Perovskite solar cells with inorganic charge extraction layers, Science 350 (2015) 944–948. Available from: https://doi.org/10.1126/science.aad1015.

[31] M. Konstantakou, T. Stergiopoulos, A critical review on tin halide Perovskite solar cells, J. Mater. Chem. A 0 (2017) 1–32. Available from: https://doi.org/10.1039/C7TA00929A.

[32] G. Hodes, Perovskite-based solar cells, Science 342 (2013) 317–318. Available from: https://doi.org/10.1126/science.1245473.

[33] J.T. Wang, J.M. Ball, E.M. Barea, A. Abate, J.A. Alexander-webber, J. Huang, et al., Low-temperature processed electron collection layers of graphene/TiO_2 nanocomposites in thin film perovskite solar cells, Nano Lett. 14 (2014) 724–730. Available from: https://doi.org/10.1021/nl403997a.

[34] T. Salim, S. Sun, Y. Abe, A. Krishna, A.C. Grimsdale, Y.M. Lam, Perovskite-based solar cells: impact of morphology and device architecture on device performance, J. Mater. Chem. A 3 (2015) 8943–8969. Available from: https://doi.org/10.1039/C4TA05226A.

[35] P. Yadav, K. Pandey, P. Bhatt, D. Raval, B. Tripathi, Exploring the performance limiting parameters of Perovskite solar cell through experimental analysis and device simulation, Sol. Energy 122 (2015) 773–782. Available from: https://doi.org/10.1016/j.solener.2015.09.046.

[36] D. Liu, T.L. Kelly, Perovskite solar cells with a planar heterojunction structure prepared using room-temperature solution processing techniques, Nat. Photonics 8 (2013) 133–138. Available from: https://doi.org/10.1038/nphoton.2013.342.

[37] N. Park, 11% efficient perovskite solar cell based on ZnO nanorods: an effective charge collection system, J. Phys. Chem. C 118 (2014) 16567–16573. Available from: https://doi.org/10.1021/jp412407j.

[38] W. Ke, G. Fang, J. Wang, P. Qin, H. Tao, H. Lei, et al., Perovskite solar cell with an efficient TiO_2 compact film, ACS Appl. Mater. Interfaces 6 (2014) 15959–15965. Available from: https://doi.org/10.1021/am503728d.

[39] J. Burschka, N. Pellet, S. Moon, R. Humphry-Baker, P. Gao, M.K. Nazeeruddin, et al., Sequential deposition as a route to high-performance Perovskite-sensitized solar cells, Nature 499 (2013) 316–319. Available from: https://doi.org/10.1038/nature12340.

[40] P. Shen, C. Tseng, T. Kuo, C. Shih, M. Li, Microwave-assisted synthesis of titanium dioxide nanocrystalline for efficient dye-sensitized and Perovskite solar cells, Sol. Energy 120 (2015) 345–356. Available from: https://doi.org/10.1016/j.solener.2015.07.036.

[41] M. Liu, M.B. Johnston, H.J. Snaith, Efficient planar heterojunction Perovskite solar cells by vapour deposition, Nature 501 (2013) 395–398. Available from: https://doi.org/10.1038/nature12509.

[42] M.M. Lee, J. Teuscher, T. Miyasaka, T.N. Murakami, H.J. Snaith, Efficient hybrid solar cells based on meso-superstructured organometal halide perovskites, Science 338 (2011) 648–653. Available from: https://doi.org/10.1126/science.1228604.

[43] Z. Zhu, Y. Bai, X. Liu, C.C. Chueh, S. Yang, A.K.Y. Jen, Enhanced efficiency and stability of inverted perovskite solar cells using highly crystalline SnO_2 nanocrystals as the robust electron-transporting layer, Adv. Mater. 28 (2016) 6478–6484. Available from: https://doi.org/10.1002/adma.201600619.

[44] Rahul, B. Bhattachary, Z.H. Khana, Effect of crystal and powder of CH_3NH_3I on the $CH_3NH_3PbI_3$ based Perovskite sensitized solar cell, Mater. Res. Bull. 89 (2017) 292–296. Available from: https://doi.org/10.1016/j.materresbull.2017.01.035.

[45] T. Song, Q. Chen, H. Zhou, C. Jiang, H. Wang, Y.M. Yang, et al., Perovskite solar cells: film formation and properties, J. Mater. Chem. A 3 (2015) 9032–9050. Available from: https://doi.org/10.1039/C4TA05246C.

[46] V. Gonzalez-Pedro, E.J. Juarez-Perez, W. Arsyad, E.M. Barea, F. Fabregat-Santiago, I. Mora-Sero, et al., General working principles of $CH_3NH_3PbX_3$ Perovskite solar cells, Nano Lett. 14 (2014) 888–893. Available from: https://doi.org/10.1021/nl404252e.

[47] H. Li, S. Li, Y. Wang, H. Sarvari, P. Zhang, M. Wang, et al., A modified sequential deposition method for fabrication of Perovskite solar cells, Sol. Energy 126 (2016) 243–251. Available from: https://doi.org/10.1016/j.solener.2015.12.045.

[48] J. Tian, G. Cao, Design, fabrication and modification of metal oxide semiconductor for improving conversion efficiency of excitonic solar cells, Coord. Chem. Rev. 321 (2016) 193–215. Available from: https://doi.org/10.1016/j.ccr.2016.02.016.

[49] K.D. Karlin, Progress in Inorganic Chemistry, vol. 57, Wiley, Hoboken, NJ, 2012. Available from: https://doi.org/10.1002/9781118148235.

[50] A. Supreeth, Y.S. Shreya, Perovskite solar cells: a review, Int. J. Eng. Dev. Res. 4 (2016) 1974–1977.

[51] J.J. You, Z. Hong, Y.M. Yang, Q. Chen, M. Cai, T. Song, et al., Low-temperature solution-processed perovskite solar cells with high efficiency and flexibility, ACS Nano 8 (2014) 1674–1680. Available from: https://doi.org/10.1021/nn406020d.

[52] N. Ahn, D.Y. Son, I.H. Jang, S.M. Kang, M. Choi, N.G. Park, Supporting information highly reproducible perovskite solar cells with average efficiency of 18.3% and best efficiency of 19.7% fabricated via lewis base adduct of lead(II) iodide, J. Am. Chem. Soc. 137 (2015) 8696–8699. Available from: https://doi.org/10.1021/jacs.5b04930.

[53] S. Ru, D. Cahen, Electron tunneling at the TiO_2/substrate interface can determine dye-sensitized solar cell performance, J. Phys. Chem. B 108 (2004) 17946–17951. Available from: https://doi.org/10.1021/jp047686s.

[54] J.H. Heo, S.H. Im, J.H. Noh, T.N. Mandal, C. Lim, J.A. Chang, et al., Efficient inorganic–organic hybrid heterojunction solar cells containing Perovskite compound and polymeric hole conductors, Nat. Photonics 7 (2013) 486–491. Available from: https://doi.org/10.1038/NPHOTON.2013.80.

[55] N.J. Jeon, J.H. Noh, Y.C. Kim, W.S. Yang, S. Ryu, S. Seok, Solvent engineering for high-performance inorganic–organic hybrid Perovskite solar cells, Nat. Mater. 13 (2014) 897–903. Available from: https://doi.org/10.1038/nmat4014.

[56] J. Ye, H. Zheng, L. Zhu, X. Zhang, L. Jiang, W. Chen, et al., High-temperature shaping Perovskite film crystallization for solar cell fast preparation, Sol. Energy Mater. Sol. Cells 160 (2017) 60–66. Available from: https://doi.org/10.1016/j.solmat.2016.10.022.

[57] N. Pellet, P. Gao, G. Gregori, T.Y. Yang, M.K. Nazeeruddin, J. Maier, et al., Mixed-organic-cation Perovskite photovoltaics for enhanced solar-light harvesting,

Angew. Chemie Int. Ed. 53 (2014) 3151–3157. Available from: https://doi.org/10.1002/anie.201309361.
[58] G.E. Eperon, V.M. Burlakov, P. Docampo, A. Goriely, H.J. Snaith, Morphological control for high performance, solution-processed planar heterojunction Perovskite solar cells, Adv. Funct. Mater. 24 (2014) 151–157. Available from: https://doi.org/10.1002/adfm.201302090.
[59] Z. Xiao, C. Bi, Y. Shao, Q. Dong, Q. Wang, Y. Yuan, et al., Efficient, high yield Perovskite photovoltaic devices grown by interdiffusion of solution-processed precursor stacking layers, Energy Environ. Sci. 7 (2014) 2619. Available from: https://doi.org/10.1039/C4EE01138D.
[60] E.N. Hoggan, D. Flowers, K. Wang, J.M. Desimone, R.G. Carbonell, Spin coating of photoresists using liquid carbon dioxide, Ind. Eng. Chem. Res. 43 (2004) 2113–2122. Available from: https://doi.org/10.1021/ie0308543.
[61] T. Komikado, A. Inoue, K. Masuda, T. Ando, S. Umegaki, Multi-layered mirrors fabricated by spin-coating organic polymers, Thin Solid Films 515 (2007) 3887–3892. Available from: https://doi.org/10.1016/j.tsf.2006.10.119.
[62] D. Perednis, O. Wilhelm, S.E. Pratsinis, L.J. Gauckler, Morphology and deposition of thin yttria-stabilized zirconia films using spray pyrolysis, Thin Solid Film 474 (2005) 84–95. Available from: https://doi.org/10.1016/j.tsf.2004.08.014.
[63] A. Furusaki, H. Konno, R. Furuichi, Perovskite-type lanthanum chromium-based oxide films prepared by ultrasonic spray pyrolysis, J. Mater. Sci. 30 (1995) 2829–2834. Available from: https://doi.org/10.1007/BF00349650.
[64] C.Y. Fu, C.L. Chang, C.S. Hsu, B.H. Hwang, Electrostatic spray deposition of $La_{0.8}Sr_{0.2}Co_{0.2}Fe_{0.8}O_3$ films, Mater. Chem. Phys. 91 (2005) 28–35. Available from: https://doi.org/10.1016/j.matchemphys.2004.10.041.
[65] I. Taniguchi, R.C. Van Landschoot, J. Schoonman, Fabrication of $La_{1-x}Sr_xCo_{1-y}Fe_yO_3$ thin films by electrostatic spray deposition, Solid State Ionics 156 (2003) 1–13. Available from: https://doi.org/10.1016/S0167-2738(02)00615-X.
[66] F. Dong, Y. Huang, S. Zou, J. Liu, S.C. Lee, Ultrasonic spray pyrolysis fabrication of solid and hollow $PbWO_4$ spheres with structure-directed photocatalytic activity, J. Phys. Chem. C 115 (2011) 241–247. Available from: https://doi.org/10.1021/jp108221v.
[67] D. Marrero-López, R. Romero, F. Martín, J.R. Ramos-Barrado, Effect of the deposition temperature on the electrochemical properties of $La_{0.6}Sr_{0.4}Co_{0.8}Fe_{0.2}O_{3-\delta}$ cathode prepared by conventional spray-pyrolysis, J. Power Sources 255 (2014) 308–317. Available from: https://doi.org/10.1016/j.jpowsour.2014.01.021.
[68] F.C. Krebs, H. Spanggard, T. Kjær, M. Biancardo, J. Alstrup, Large area plastic solar cell modules, Mater. Sci. Eng. B 138 (2007) 106–111. Available from: https://doi.org/10.1016/j.mseb.2006.06.008.
[69] J.R.M.J. Cerdà, J. Arbiol, G. Dezanneau, R. Díaz, Perovskite-type $BaSnO_3$ powders for high temperature gas sensor applications, Sens. Actuators B: Chem. 84 (2002) 21–25. Available from: https://doi.org/10.1016/S0925-4005(02)00005-9.
[70] J. Gurauskis, O.F. Lohne, H.L. Lein, K. Wiik, Processing of thin film ceramic membranes for oxygen separation, J. Eur. Ceram. Soc. 32 (2012) 649–655. Available from: https://doi.org/10.1016/j.jeurceramsoc.2011.10.009.
[71] M.R. Othman, J. Kim, Permeability and separability of oxygen across perovskite—alumina membrane prepared from a sol–gel procedure, Ind. Eng. Chem. Res. 47 (2008) 3000–3007. Available from: https://doi.org/10.1021/JZ500059v.
[72] K. Chondroudis, D.B. Mitzi, P. Brock, Effect of thermal annealing on the optical and morphological properties of (AETH)PbX4 (X = Br, I) Perovskite films prepared using single source thermal ablation, Chem. Mater. 12 (2000) 169–175. Available from: https://doi.org/10.1021/cm990516l.

[73] J. Zhang, F. Li, K. Yang, C.P. Veeramalai, T. Guo, Applied surface science low temperature processed planar heterojunction Perovskite solar cells employing silver nanowires as top electrode, Appl. Surf. Sci. 369 (2016) 308–313. Available from: https://doi.org/10.1016/j.apsusc.2016.02.104.

[74] J.M. Ball, M.M. Lee, A. Hey, H.J. Snaith, Low-temperature processed meso-superstructured to thin-film Perovskite solar cells, Energy Environ. Sci. (2013) 1739–1743. Available from: https://doi.org/10.1039/c3ee40810h.

[75] S. Stranks, G. Eperon, G. Grancini, C. Menelaou, Electron-hole diffusion lengths exceeding 1 micrometer in an organometal trihalide Perovskite absorber, Science 342 (2013) 341–344. Available from: https://doi.org/10.1126/science.1243982.

[76] M.M. Lee, J. Teuscher, T. Miyasaka, T.N. Murakami, H.J. Snaith, Efficient hybrid solar cells based on meso-superstructured organometal halide perovskites, Science 338 (2012) 643–646. Available from: https://doi.org/10.1126/science.1228604.

[77] N. Islam, M. Yang, K. Zhu, Z. Fan, Mesoporous scaffolds based on TiO_2 nanorods and nanoparticles for efficient hybrid Perovskite solar cells, J. Mater. Chem. A 3 (2015) 24315–24321. Available from: https://doi.org/10.1039/C5TA06727H.

[78] H. Zhou, Q. Chen, G. Li, S. Luo, T.B. Song, H.S. Duan, et al., Interface engineering of highly efficient Perovskite solar cells, Science 345 (2014) 542–546. Available from: https://doi.org/10.1126/science.1254050.

[79] Q. Chen, H. Zhou, Z. Hong, S. Luo, H.S. Duan, H.H. Wang, et al., Planar heterojunction Perovskite solar cells via vapor-assisted solution process, J. Am. Chem. Soc. 136 (2014) 622–625. Available from: https://doi.org/10.1021/ja411509g.

[80] Y. Li, J.K. Cooper, R. Buonsanti, C. Giannini, Y. Liu, F.M. Toma, et al., Fabrication of planar heterojunction Perovskite solar cells by controlled low-pressure vapor annealing, J. Phys. Chem. Lett. 6 (2015) 493–499. Available from: https://doi.org/10.1021/jz502720a.

[81] J.H. Park, J. Seo, S. Park, S.S. Shin, Y.C. Kim, N.J. Jeon, et al., Efficient $CH_3NH_3PbI_3$ Perovskite solar cells employing nanostructured p-Type NiO electrode formed by a pulsed laser deposition, Adv. Mater. 27 (2015) 4013–4019. Available from: https://doi.org/10.1002/adma.201500523.

[82] Y.K. Wang, Z.C. Yuan, G.Z. Shi, Y.X. Li, Q. Li, F. Hui, et al., Dopant-free spiro-triphenylamine/fluorene as hole-transporting material for perovskite solar cells with enhanced efficiency and stability, Adv. Funct. Mater. 26 (2016) 1375–1381. Available from: https://doi.org/10.1002/adfm.201504245.

[83] J. Xu, O. Voznyy, R. Comin, X. Gong, G. Walters, M. Liu, et al., Crosslinked remote-doped hole-extracting contacts enhance stability under accelerated lifetime testing in perovskite solar cells, Adv. Mater. 28 (2016) 2807–2815. Available from: https://doi.org/10.1002/adma.201505630.

[84] K. Domanski, J.P. Correa-Baena, N. Mine, M.K. Nazeeruddin, A. Abate, M. Saliba, et al., Not all that glitters is gold: metal-migration-induced degradation in perovskite solar cells, ACS Nano 10 (2016) 6306–6314. Available from: https://doi.org/10.1021/acsnano.6b02613.

[85] Y. Li, S. Ye, W. Sun, W. Yan, Z. Bian, Z. Liu, et al., Hole-conductor-free planar Perovskite solar cells with 16.0% efficiency, J. Mater. Chem. A 3 (2015) 18389–18394. Available from: https://doi.org/10.1039/C5TA05989E.

[86] W. Ke, G. Fang, J. Wan, H. Tao, Q. Liu, L. Xiong, et al., Efficient hole-blocking layer-free planar halide Perovskite thin-film solar cells, Nat. Commun. 6 (2015) 6700. Available from: https://doi.org/10.1038/ncomms7700.

[87] M. Saliba, S. Orlandi, T. Matsui, S. Aghazada, M. Cavazzini, J.-P. Correa-Baena, et al., A molecularly engineered hole-transporting material for efficient Perovskite solar cells, Nat. Energy 1 (2016) 15017. Available from: https://doi.org/10.1038/nenergy.2015.17.

[88] F. Zhang, C. Yi, P. Wei, X. Bi, J. Luo, G. Jacopin, et al., A novel dopant-free triphenylamine based molecular "butterfly" hole-transport material for highly efficient and stable perovskite solar cells, Adv. Energy Mater. 6 (2016) 1600401. Available from: https://doi.org/10.1002/aenm.201600401.
[89] G.W. Kim, G. Kang, J. Kim, G.Y. Lee, H.I. Kim, L. Pyeon, et al., Dopant-free polymeric hole transport material for highly efficient and stable perovskite solar cells, Energy Environ. Sci. 9 (2016) 2326–2333. Available from: https://doi.org/10.1039/C6EE00709K.
[90] X. Zhao, F. Zhang, C. Yi, D. Bi, X. Bi, P. Wei, et al., A novel one-step synthesized and dopant-free hole transport material for efficient and stable Perovskite solar cells, J. Mater. Chem. A 4 (2016) 16330–16334. Available from: https://doi.org/10.1039/C6TA05254A.
[91] E.J. Juarez-Perez, M. Wu, F. Fabregat-Santiago, K. Lakus-Wollny, E. Mankel, T. Mayer, et al., Role of the selective contacts in the performance of lead halide perovskite solar cells, J. Phys. Chem. Lett. 5 (2014) 680–685. Available from: https://doi.org/10.121/jz500059v.
[92] L. Zhang, T. Liu, L. Liu, M. Hu, Y. Yang, A. Meia, et al., The effect of carbon counter electrodes on fully printable mesoscopic Perovskite solar cells, J. Mater. Chem. A 3 (2015) 9165–9170. Available from: https://doi.org/10.1039/C4TA04647A.
[93] L. Wang, G. Li, Q. Zhao, X. Gao, Non-precious transition metals as counter electrode of Perovskite solar cells, Energy Storage Mater. 7 (2017) 40–47. Available from: https://doi.org/10.1016/j.ensm.2016.11.007.
[94] Z. Liu, B. Sun, T. Shi, Z. Tanga, G. Liao, Enhanced photovoltaic performance and stability of carbon counter electrode based Perovskite solar cells encapsulated by PDMS, J. Mater. Chem. A 4 (2016) 10700–10709. Available from: https://doi.org/10.1039/C6TA02851A.
[95] M. Salado, L.S. Contrera, L. Calio, A. Todinova, C. Lopez-santos, S. Ahmad, et al., Impact of moisture on efficiency-determining electronic processes in perovskite solar cells, J. Mater. Chem. A 5 (2017) 10917–10927. Available from: https://doi.org/10.1039/C7TA02264F.
[96] A. Dualeh, N. Tétreault, T. Moehl, P. Gao, M.K. Nazeeruddin, M. Grätzel, Effect of annealing temperature on film morphology of organic–-inorganic hybrid Perovskite solid-state solar cells, Adv. Funct. Mater. 24 (2014) 3250–3258. Available from: https://doi.org/10.1002/adfm.201304022.
[97] S.W. Lee, S. Kim, S. Bae, K. Cho, T. Chung, L.E. Mundt, et al., UV degradation and recovery of perovskite solar cells, Sci. Rep. 6 (2016) 38150. Available from: https://doi.org/10.1038/srep38150.
[98] J.C. Blancon, W. Nie, A.J. Neukirch, G. Gupta, S. Tretiak, L. Cognet, et al., The effects of electronic impurities and electron–hole recombination dynamics on large-grain organic–inorganic perovskite photovoltaic efficiencies, Adv. Funct. Mater. 26 (2016) 4283–4292. Available from: https://doi.org/10.1002/adfm.201505324.
[99] E.M. Sanehira, B.J. Tremolet de Villers, P. Schulz, M.O. Reese, S. Ferrere, K. Zhu, et al., Influence of electrode interfaces on the stability of perovskite solar cells: reduced degradation using MoO*x*/Al for hole collection, ACS Energy Lett. 1 (2016) 38–45. Available from: https://doi.org/10.1021/acsenergylett.6b00013.
[100] M.C. Yeh, S.H. Su, W.K. Lin, J.Y. Chen, M. Yokoyama, Efficiency enhancement of Perovskite solar cells with a buffer layer, in: 5th Int. Symp. Next-Generation Electron. ISNE, 2016, https://doi.org/10.1109/ISNE.2016.7543388.

[101] J.M. Frost, K.T. Butler, F. Brivio, C.H. Hendon, M. Van Schilfgaarde, A. Walsh, Atomistic origins of high-performance in hybrid halide Perovskite solar cells, Nano Lett. 14 (2014) 2584–2590. Available from: https://doi.org/10.1021/nl500390f.
[102] J.H. Noh, S.H. Im, J.H. Heo, T.N. Mandal, S. Seok, Chemical management for colorful, efficient, and stable inorganic–organic hybrid nanostructured solar cells, Nano Lett. 13 (2013) 1764–1769. Available from: https://doi.org/10.1021/nl400349b.
[103] I.C. Smith, E.T. Hoke, D. Solis-Ibarra, M.D. McGehee, H.I. Karunadasa, A layered hybrid perovskite solar-cell absorber with enhanced moisture stability, Angew. Chem. Int. Ed. 53 (2014) 11232–11235. Available from: https://doi.org/10.1002/anie.201406466.
[104] D. Bryant, N. Aristidou, S. Pont, I. Sanchez-Molina, T. Chotchunangatchaval, S. Wheeler, et al., Light and oxygen induced degradation limits the operational stability of methylammonium lead triiodide Perovskite solar cells, Energy Environ. Sci. 9 (2016) 1655–1660. Available from: https://doi.org/10.1039/C6EE00409A.
[105] B. Roose, K.C. Godel, S. Pathak, A. Sadhanala, J.P.C. Baena, B.D. Wilts, et al., Enhanced efficiency and stability of Perovskite solar cells through Nd-doping of mesostructured TiO_2, Adv. Energy Mater. 6 (2016) 1–7. Available from: https://doi.org/10.1002/aenm.201501868.
[106] Q. Tai, P. You, H. Sang, Z. Liu, C. Hu, L.W. Helen, Efficient and stable Perovskite solar cells prepared in ambient air irrespective of the humidity, Nat. Commun. 6 (2016) 1–8. Available from: https://doi.org/10.1038/ncomms11105.
[107] D. Bi, P. Gao, R. Scopelliti, E. Oveisi, J. Luo, M. Grätzel, et al., High-performance Perovskite solar cells with enhanced environmental stability based on amphiphile-modified $CH_3NH_3PbI_3$, Adv. Mater. 28 (2016) 2910–2915. Available from: https://doi.org/10.1002/adma.201505255.
[108] N. Arora, M.I. Dar, M. Abdi-Jalebi, F. Giordano, N. Pellet, G. Jacopin, et al., Intrinsic and extrinsic stability of formamidinium lead bromide perovskite solar cells yielding high photovoltage, Nano Lett. 16 (2016) 7155–7162. Available from: https://doi.org/10.1021/acs.nanolett.6b03455.
[109] H. Back, G. Kim, J. Kim, J. Kong, T.K. Kim, H. Kang, et al., Achieving long-term stable Perovskite solar cells via ion neutralization, Energy Environ. Sci. Energy Environ. Sci. 9 (2016) 1258–1263. Available from: https://doi.org/10.1039/c6ee00612d.
[110] S. Ito, S. Tanaka, K. Manabe, H. Nishino, Effects of surface blocking layer of Sb_2 S_3 on nanocrystalline TiO_2 for CH_3 NH_3 PbI_3 perovskite solar cells, J. Phys. Chem. C 118 (2014) 16995–17000. Available from: https://doi.org/10.1021/jp500449z.
[111] T. Leijtens, G.E. Eperon, S. Pathak, A. Abate, M.M. Lee, H.J. Snaith, Overcoming ultraviolet light instability of sensitized TiO2 with meso-superstructured organometal tri-halide Perovskite solar cells, Nat. Commun. 4 (2013) 1–8. Available from: https://doi.org/10.1038/ncomms3885.
[112] W. Li, W. Zhang, S. Van Reenen, R.J. Sutton, J. Fan, A. Haghighirad, et al., Enhanced UV-light stability of planar heterojunction perovskite solar cells with caesium bromide interface modification, Energy Environ. Sci. 9 (2016) 490–498. Available from: https://doi.org/10.1039/C5EE03522H.
[113] N. Chander, A.F. Khan, P.S. Chandrasekhar, E. Thouti, S.K. Swami, V. Dutta, et al., Reduced ultraviolet light induced degradation and enhanced light harvesting using YVO4:Eu3 + down-shifting nano-phosphor layer in organometal halide

Perovskite solar cells, Appl. Phys. Lett. 105 (2014). Available from: https://doi.org/10.1063/1.4891181.
[114] A. Dualeh, P. Gao, S. Seok, M.K. Nazeeruddin, M. Grätzel, Thermal behavior of methylammonium lead-trihalide Perovskite photovoltaic light harvesters, Chem. Mater. 26 (2014) 6160–6164. Available from: https://doi.org/10.1021/cm502468k.
[115] A. Poglitsch, D. Weber, Dynamic disorder in methylammoniumtrihalogenoplumbates(II) observed by millimeter-wave spectroscopy, J. Chem. Phys. 87 (1987) 6373–6378. Available from: https://doi.org/10.1063/1.453467.
[116] B. Conings, J. Drijkoningen, N. Gauquelin, A. Babayigit, J. D'Haen, L. D'Olieslaeger, et al., Intrinsic thermal instability of methylammonium lead trihalide perovskite, Adv. Energy Mater. 5 (2015) 1–8. Available from: https://doi.org/10.1002/aenm.201500477.
[117] B. Philippe, B.W. Park, R. Lindblad, J. Oscarsson, S. Ahmadi, E.M.J. Johansson, et al., Chemical and electronic structure characterization of lead halide perovskites and stability behavior under different exposures – a photoelectron spectroscopy investigation, Chem. Mater. 27 (2015) 1720–1731. Available from: https://doi.org/10.1021/acs.chemmater.5b00348.
[118] M. Saliba, T. Matsui, K. Domanski, J.Y. Seo, A. Ummadisingu, S.M. Zakeeruddin, et al., Incorporation of rubidium cations into Perovskite solar cells improves photovoltaic performance, Science 354 (2016) 206. Available from: https://doi.org/10.1126/science.aah5557.
[119] J. Yang, B.D. Siempelkamp, E. Mosconi, F. De Angelis, T.L. Kelly, Origin of the thermal instability in CH3 NH 3 PbI3 thin films deposited on ZnO, Chem. Mater. 27 (2015) 4229–4236. Available from: https://doi.org/10.1021/acs.chemmater.5b01598.
[120] R.J. Sutton, G.E. Eperon, L. Miranda, E.S. Parrott, B.A. Kamino, J.B. Patel, et al., Bandgap-tunable cesium lead halide perovskites with high thermal stability for efficient solar cells, Adv. Energy Mater. 6 (2016) 1–6. Available from: https://doi.org/10.1002/aenm.201502458.
[121] A. Pisoni, J. Jaćimović, O.S. Barišić, M. Spina, R. Gaál, L. Forró, et al., Ultra-low thermal conductivity in organic–inorganic hybrid Perovskite CH3NH3PbI3, J. Phys. Chem. Lett. 5 (2014) 2488–2492. Available from: https://doi.org/10.1021/jz5012109.
[122] S.N. Habisreutinger, T. Leijtens, G.E. Eperon, S.D. Stranks, R.J. Nicholas, H.J. Snaith, Carbon nanotube/polymer composites as a highly stable hole collection layer in Perovskite solar cells, Nano Lett. 14 (2014) 5561–5568. Available from: https://doi.org/10.1021/nl501982b.
[123] S.K. Pathak, A. Abate, P. Ruckdeschel, B. Roose, K.C. Gödel, Y. Vaynzof, et al., Performance and stability enhancement of dye-sensitized and perovskite solar cells by Al doping of TiO_2, Adv. Funct. Mater. 24 (2014) 6046–6055. Available from: https://doi.org/10.1002/adfm.201401658.
[124] J. Song, E. Zheng, J. Bian, X.F. Wang, W. Tian, Y. Sanehira, et al., Low-temperature SnO_2-based electron selective contact for efficient and stable Perovskite solar cells, J. Mater. Chem. A 3 (2015) 10837–10844. Available from: https://doi.org/10.1039/C5TA01207D.
[125] J. Cao, Y.M. Liu, X. Jing, J. Yin, J. Li, B. Xu, et al., Well-defined thiolated nanographene as hole-transporting material for efficient and stable perovskite solar cells, J. Am. Chem. Soc. 137 (2015) 10914–10917. Available from: https://doi.org/10.1021/jacs.5b06493.

[126] D. Wang, M. Wright, N.K. Elumalai, A. Uddin, Stability of Perovskite solar cells, Sol. Energy Mater. Sol. Cells 147 (2016) 255–275. Available from: https://doi.org/10.1016/j.solmat.2015.12.025.
[127] C. Liu, W. Ding, X. Zhou, J. Gao, C. Cheng, X. Zhao, et al., Efficient and stable perovskite solar cells prepared in ambient air based on surface-modified Perovskite layer, J. Phys. Chem. C 121 (2017) 6546–6553. Available from: https://doi.org/10.1021/acs.jpcc.7b00847.
[128] Y. Han, S. Meyer, Y. Dkhissi, K. Weber, J.M. Pringle, U. Bach, et al., Degradation observations of encapsulated planar CH_3 NH_3 PbI_3 Perovskite solar cells at high temperatures and humidity, J. Mater. Chem. A 3 (2015) 8139–8147. Available from: https://doi.org/10.1039/C5TA00358J.
[129] F.C. Krebs, S.A. Gevorgyan, J. Alstrup, A roll-to-roll process to flexible polymer solar cells: model studies, manufacture and operational stability studies, J. Mater. Chem. 19 (2009) 5442. Available from: https://doi.org/10.1039/b823001c.
[130] Y. Shao, Q. Wang, Q. Dong, Y. Yuan, J. Huang, Vacuum-free laminated top electrode with conductive tapes for scalable manufacturing of efficient Perovskite solar cells, Nano Energy 16 (2015) 47–53. Available from: https://doi.org/10.1016/j.nanoen.2015.06.010.
[131] W. Nie, J.C. Blancon, A.J. Neukirch, K. Appavoo, H. Tsai, M. Chhowalla, et al., Light-activated photocurrent degradation and self-healing in Perovskite solar cells, Nat. Commun. 7 (2016) 11574. Available from: https://doi.org/10.1038/ncomms11574.
[132] H. Nagaoka, F. Ma, D.W. de Quilettes, S.M. Vorpahl, M.S. Glaz, A.E. Colbert, et al., Zr incorporation into TiO_2 electrodes reduces hysteresis and improves performance in hybrid perovskite solar cells while increasing carrier lifetimes, J. Phys. Chem. Lett. 6 (2015) 669–675. Available from: https://doi.org/10.1021/Jz502694g.

CHAPTER 9

Hole Conductor–Free Perovskite Solar Cells

Suresh Maniarasu, Vishesh Manjunath, Easwaramoorthi Ramasamy and Ganapathy Veerappan
International Advanced Research Centre for Powder, Metallurgy and New Materials (ARCI), Hyderabad, Telangana, India

9.1 INTRODUCTION

Crystalline silicon solar cell with an efficiency of 21% is the most developed device in the photovoltaic market. Alternative 3rd generation solar cells, such as organic photovoltaic, dye-sensitized solar cell, and quantum dot solar cells, use only low-temperature process which further reduces the fabrication cost and energy payback time. But still the power conversion efficiency (PCE) of these 3rd generation cells is not able to reach the efficiency of the conventional photovoltaic cells [1]. To overcome these problems, recently evolved organic–inorganic halide perovskite has emerged as an attractive material in the photovoltaic field to make a high-efficiency device with low manufacturing cost. In general, perovskite solar cell (PSC) devices are composed of different layers such as electron and hole transporting layer (HTM), absorber material, cathode and anode for charge transportation. In this chapter, we focus mainly on the influence and effectiveness of HTM-free PSC devices. Till date, most of the highest reported PSC efficiency is achieved using HTM materials, i.e., 2,2',7,7'-Tetrakis[N,N-di(4-methoxyphenyl)amino]-9,9'-spirobifluorene (Spiro-OMeTAD), small organic molecules, polymers, and p-type inorganic semiconducting materials. But most of the reported HTMs are expensive and vulnerable to moisture, oxygen, and other entities, which makes them expensive and unstable [2–4]. Organic–inorganic perovskite has exceptional properties, such as high mobility of charge carriers and long diffusion length for electrons and holes, which are essential for highly efficient solar cells [5–7]. Etgar et al. reported for the first time that a unique new heterojunction device structure (Fig. 9.1) composed of fluorine-doped tin oxide (FTO)/compact TiO_2/mesoporous TiO_2/$CH_3NH_3PbI_3$/Au. In this

Perovskite Photovoltaics
DOI: https://doi.org/10.1016/B978-0-12-812915-9.00009-5

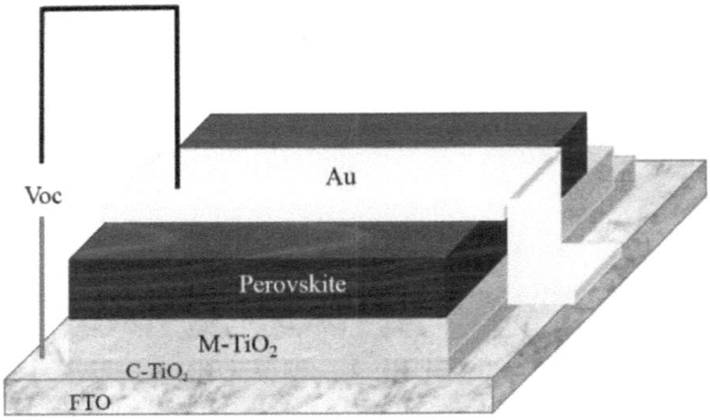

Figure 9.1 Device architecture of hole transporting material (HTM)-free PSC.

device configuration, perovskite acted as p-type as well as an absorber layer, and mesoporous TiO_2 was used as an n-type semiconductor material [8].

Avoiding HTM in the device simplifies the architecture, reduces the cost (94% of the cost in the total cost of fabrication comes from the HTM), enhances the stability of the PSCs, which in turn avoids oxidation and provides consistent results. From the following experiment, they have claimed that perovskite can act as an absorber as well as conduct both the charge carriers, which makes it an efficient material to develop HTM-free heterojunction device [8,9]. This chapter gives a detailed explanation about the PSC without a hole conductor material and its influence on device performance.

9.1.1 Working Mechanism of Hole Transport Material-Free Perovskite Solar Cell

In theory the function of HTM is to transfer the holes and block the electrons in solar cell [10,11]. As we know that, the ambipolar nature of perovskite itself can transfer the charge carriers to the corresponding electrodes [3,12]. Etgar et al. reported for the first time that a device made without any HTM has achieved an efficiency of 5.5%, which paved a breakthrough in the field of HTM-free PSCs [13]. Later, he has improved the device efficiency to around 8.0% by just modifying the mesoporous TiO_2 thickness, as well as by varying the deposition technique to improve the perovskite film quality for efficient extraction of carriers [8]. The schematic diagram depicting the HTM-free PSC device architecture is shown in Fig. 9.1. Device architecture consists of step-by-step coating of transparent conducting substrate coated with semiconducting metal oxide, perovskite material and finally the metal cathode deposition by thermal evaporation (Fig. 9.1).

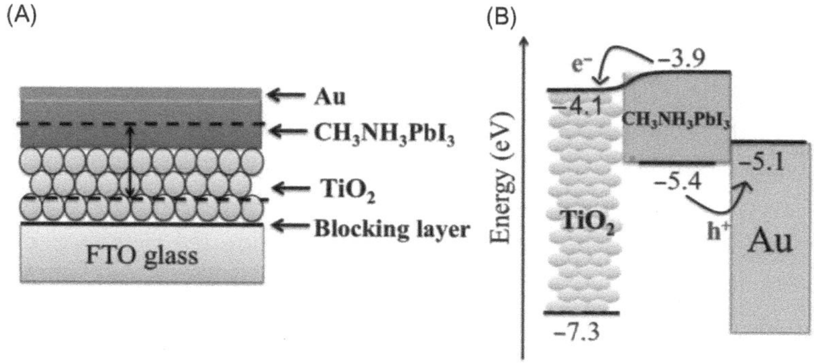

Figure 9.2 (A) The structure of the hole conductor–free $TiO_2/CH_3NH_3PbI_3$ perovskite-based solar cell. The vertical arrow indicates the depletion region observed at the $TiO_2/CH_3NH_3PbI_3$ junction. (B) Energy-level diagram of the discussed solar cell, which shows the charge separation process. *Reproduced with permission from S. Aharon, S. Gamliel, B. El Cohen, L. Etgar, Depletion region effect of highly efficient hole conductor free $CH_3NH_3PbI_3$ perovskite solar cells. Phys. Chem. Chem. Phys. 16 (22) (2014) 10512–10518. Copyright 2014, The Royal Society of Chemistry.*

Eliminating the HTM from the PSC device architecture affects the PV performance, but it also defines that the interfaces of the layer, especially the electron transport layer (ETL)/perovskite interfacial layer, which has more impact on the device performance. In HTM-free device architecture the depletion region exists at the interface of the ETL/perovskite layers (indicated by the arrow line in Fig. 9.2A), which is confirmed by capacitance–voltage measurements and Mott–Schottky analysis. On illumination, perovskite absorbs photons which creates charge carriers that are transferred to the corresponding electrodes. Due to this charge transfer mechanism the depletion region might be formed at the interface of TiO_2/perovskite layers. This depletion region helps to separate the charges from the generated photons and supresses the back reaction at TiO_2/perovskite interface, which consequently improves the PCE of the device [9]. The PV performance of HTM-free device has further improved to 10.85% because of the depletion region width between the TiO_2/perovskite layers [14].

9.1.2 Planar-Type Hole Transport Material-Free Perovskite Solar Cell

Literatures say that few hundred nanometer thickness of perovskite layer itself is good enough to generate more charge carriers. These findings have helped to develop different kind of device architecture in PSC configuration [13,15,16]. The two main types of device architecture that has evolved in PSC are planar and mesoporous [4].

Planar-type HTM-free PSC was fabricated by varying the reaction temperature and thickness of the active layer, the structure of the perovskite was studied [17]. Initially the PbI2 was deposited on top of the compact layer TiO_2, which was converted into uniform and crack-free perovskite layer by passing methylammonium iodide (MAI) vapor at 175°C in N_2 atmosphere. Fig. 9.3 shows the surface morphology and cross-section of PbI_2 and perovskite films. The average thickness of the perovskite film is 330 nm and surface coverage of the film is almost 100%. The

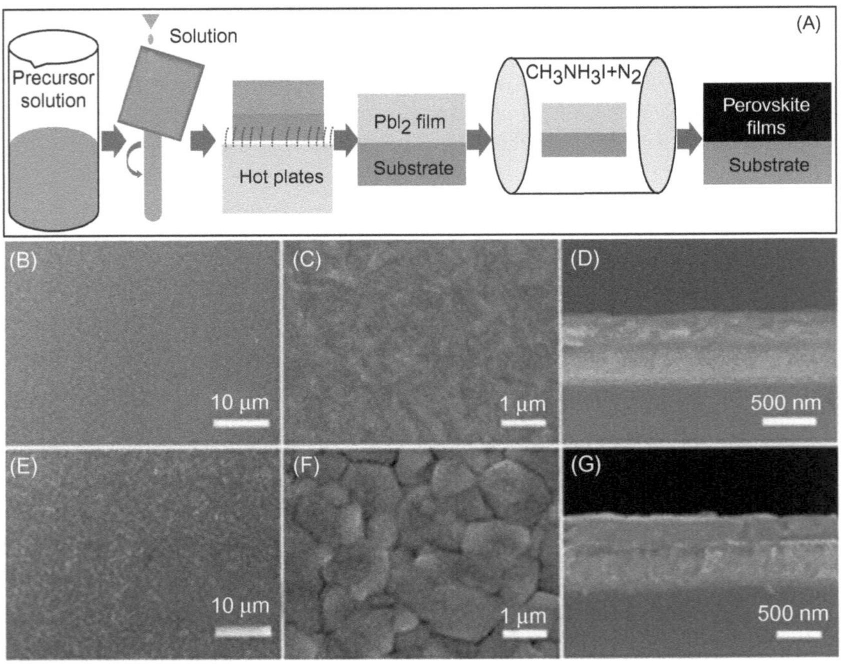

Figure 9.3 (A) Schematic diagram of preparation processes for $CH_3NH_3PbI_3$ perovskite films. (B–D) Scanning electron microscopy (SEM) images of the PbI_2 precursor film deposited on TiO_2-coated FTO by spin-coating at a speed of 2750 rpm and heating at 100°C for 10 min: (B) low-magnification and (C) high-magnification images (top view), (D) cross-sectional image, (E–G) SEM images of $CH_3NH_3PbI_3$ film prepared in the presence of CH_3NH_3I at 175°C in N_2 atmosphere for 3 h: (E) low-magnification and (F) high-magnification images (top view), (G) cross-sectional image. *Reproduced with permission from Y. Liu, S. Ji, S. Li, W. He, K. Wang, H. Hu, et al., Study on hole-transport-material-free planar $TiO_2/CH_3NH_3PbI_3$ heterojunction solar cells: the simplest configuration of a working perovskite solar cell. J. Mater. Chem. A3 (28) (2015) 14902–14909. Copyright 2015, The Royal Society of Chemistry.*

performance of planar thin film solar cell could be improved by making a uniform and complete surface coverage film with reproducibility [18].

The morphology of the $CH_3NH_3PbI_3$ film was investigated by changing the reaction temperature and time (Fig. 9.4). They varied the reaction temperature from 125 to 175°C, and they suggested that a uniform film can be formed with a grain size of 1 μm at 175°C. The films appeared as porous nature at the starting stage of reaction time. As the reaction time increased, the films appeared without any pin holes and became denser. However, cracks formed on the surface of the film due to increased time of reaction.

The photovoltaic performance of the PSC was analyzed by varying the thickness of the active layer. Fig. 9.5 shows the *J*–*V* characteristics of PSC for different thickness range. The thickness of perovskite film even when reduced from 500 to 300 nm showed less absorption. The

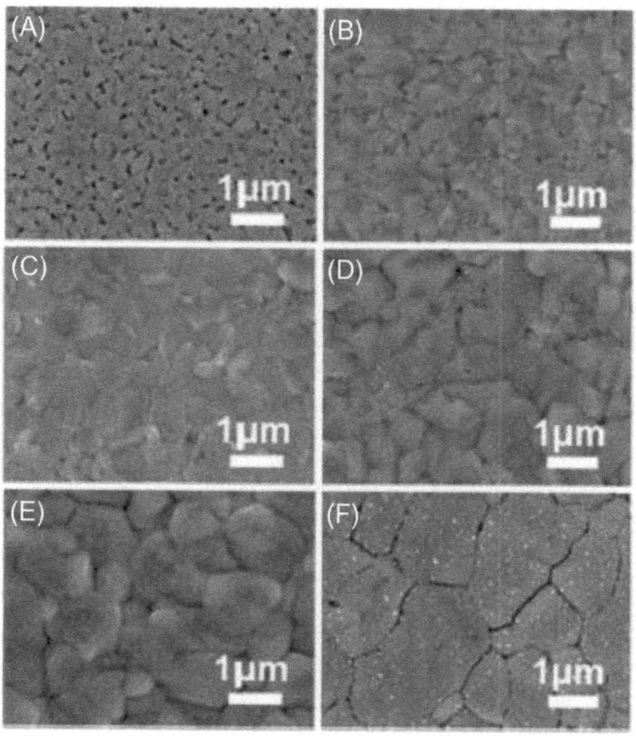

Figure 9.4 SEM images of $CH_3NH_3PbI_3$ films with a reaction time of (A) 0.5 h, (B) 1 h, (C) 1.5 h, (D) 2 h, (F) 3 h, (G) 4 h. *Reproduced with permission from Y. Liu, S. Ji, S. Li, W. He, K. Wang, H. Hu, et al., Study on hole-transport-material-free planar TiO_2/$CH_3NH_3PbI_3$ heterojunction solar cells: the simplest configuration of a working perovskite solar cell. J. Mater. Chem. A3 (28) (2015) 14902–14909. Copyright 2015, The Royal Society of Chemistry.*

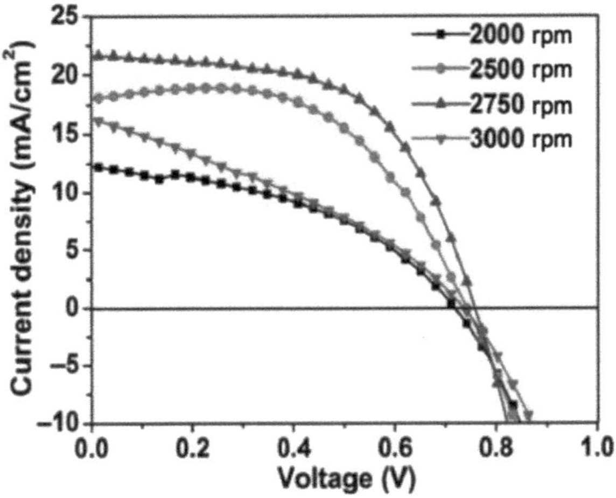

Figure 9.5 *J*–*V* curves of planar HTM-free $CH_3NH_3PbI_3$ solar cells based on $CH_3NH_3PbI_3$ films prepared by reacting PbI_2 precursor films with CH_3NH_3I at 175°C in N_2 atmosphere for 3 h at different spin-coating speed. *Reproduced with permission from Y. Liu, S. Ji, S. Li, W. He, K. Wang, H. Hu, et al., Study on hole-transport-material-free planar $TiO_2/CH_3NH_3PbI_3$ heterojunction solar cells: the simplest configuration of a working perovskite solar cell. J. Mater. Chem. A3 (28) (2015) 14902–14909. Copyright 2015, The Royal Society of Chemistry.*

photovoltaic parameters have shown improvement for the speed range of 2750 rpm. This might be due to low series resistance and reduced thickness of absorber layer which increases the generation of carriers and transportation to the electrodes. Further increment of spin speed reduces the J_{SC} and fill factor (FF) because of reduced shunt path and poor absorption. Fig. 9.5 shows the *J*–*V* curve of planar HTM-free PSC layer, in which the PbI_2 was deposited at 2750 rpm, and the evaporation of MAI was done at 175°C in N_2 atmosphere for 3 h. The value of J_{SC} is higher in this condition, because the suitable thickness of perovskite layer improves the light absorption as well as efficient charge collection.

Dual source coevaporation technique can make a uniform and pin hole–free perovskite film but the evaporation rate for the organic and inorganic precursor should be in proper ratio to get high-quality films. In this work, Cui et al. fabricated a planar HTM-free $MAPbI_3$ by using a sequential-vapor deposition approach to form a high-quality film. The porous PbI_2 film was made by vapor deposition, and then it was reacted with the MAI vapor followed by annealing at 100°C for 30 min to make a textured $MAPbI_3$ film. Fig. 9.6 shows the schematic representation of the sequential-vapor deposition (SVD) process.

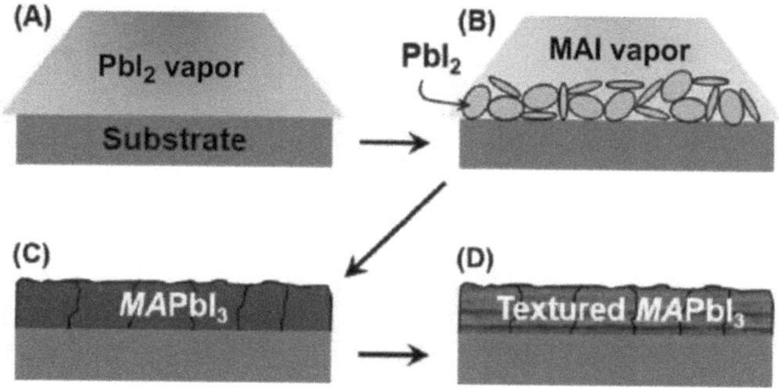

Figure 9.6 Schematic representation of the sequential-vapor deposition (SVD) process for the growth of $MAPbI_3$ perovskite films. (A) PbI_2 vapor onto the substrate, (B) MAI vapor onto the pre-evaporated PbI_2 film, (C) Converted $MAPbI_3$ film, and (D) Heated the $MAPbI_3$ film to get a textured $MAPbI_3$ film *Reproduced with permission from H. Hu, D. Wang, Y. Zhou, J. Zhang, S. Lv, S. Pang, et al., Vapour-based processing of hole-conductor-free $CH_3NH_3PbI_3$ perovskite/C60 fullerene planar solar cells. RSC Adv. 4 (55) (2014) 28964–28967 [19]. Copyright 2014, The Royal Society of Chemistry.*

Usually the PbI_2 film looks like a rough surface when prepared by solution-processed spin-coating technique. But in the case of vapor deposition the PbI_2 film showed uniform coverage and smooth films consisting of some plate kind of morphology. The XRD peak in Fig. 9.7A confirms the orthorhombic phase of PbI_2 with strong texture of 001. Fig. 9.7C and D shows the XRD and surface morphology of the PbI_2 film exposed to MAI vapor for 5 min. It shows the presence of both $MAPbI_3$ and PbI_2 with poor surface and pin holes in the film. When the film was exposed to an additional 10 min with the MAI vapor, the film was completely converted into $MAPbI_3$ with uniform surface and grain size of 500 nm. In addition, XRD also confirms that the pure tetragonal phase of $MAPbI_3$ has formed without any unconverted PbI_2 peak. Fig. 9.7G and H shows XRD and surface morphology of the heat-treated $MAPbI_3$ film with high strong texture 001 and shows individual grains. The vapor-deposited PbI_2 film has porous nature and more surface area, which helps to contact the MAI vapor easily so that pure phase of $MAPbI_3$ uniform film is formed without any pin holes.

In this device architecture, $MAPbI_3$ & C_{60} was deposited by vapor deposition method which acts as an absorber, hole conductor, and electron-acceptor, respectively. C_{60} is a nonoxide material, used here as an ETL since the widely used TiO_2 needs higher annealing temperature so it is eliminated from the device architecture [8,13,20]. Due to the high quality of SVD process of $MAPbI_3$ films, this simple device architecture provides maximum PCE of 5.4% with short circuit current

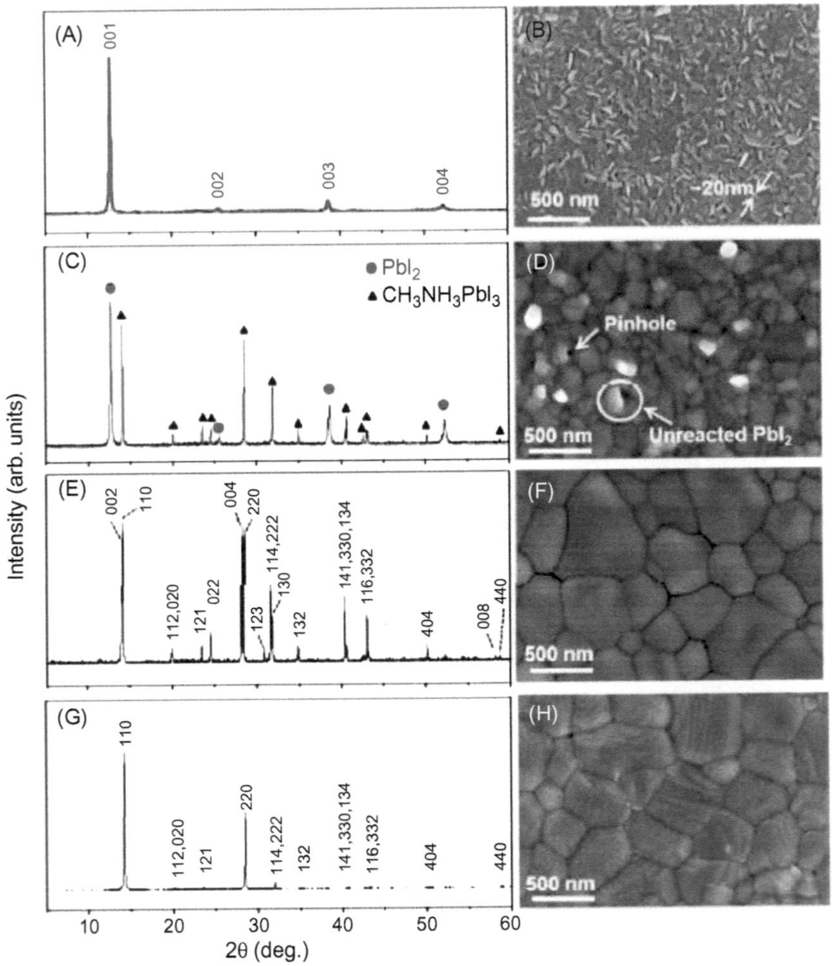

Figure 9.7 Indexed XRD patterns and corresponding SEM images (top surface) of films at the different stages during SVD growth: (A and B) after PbI_2 vapor deposition; (C and D) partially reacted with MAI vapor (5 min); (E and F) fully reacted with MAI vapor (15 min); (G and H) after annealing at 100°C for 30 min. This figure also shows the XRD pattern and surface morphology of each step in fabrication of film. *Reproduced with permission from H. Hu, D. Wang, Y. Zhou, J. Zhang, S. Lv, S. Pang, et al., Vapour-based processing of hole-conductor-free $CH_3NH_3PbI_3$ perovskite/C60 fullerene planar solar cells. RSC Adv. 4 (55) (2014) 28964–28967 [19]. Copyright 2014, The Royal Society of Chemistry.*

density of 13.6 mA/cm^2 and an open circuit voltage of 0.8 V (Fig. 9.8). The fill factor of this device is still low, because of nonideal electrical contacts. But further optimization of Indium tin oxide (ITO) and C_{60}/Ag can improve the device performance.

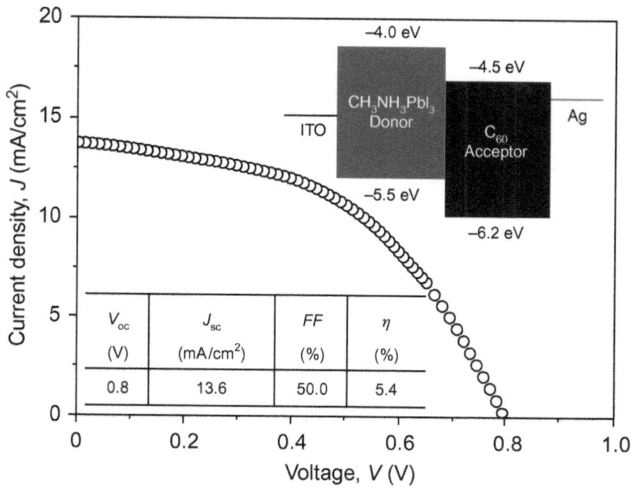

Figure 9.8 Current density (*J*)–voltage (*V*) response of a $MAPbI_3/C_{60}$ solar cell under AM 1.5G one sun illumination (top inset: energy-level diagram; bottom inset: extracted solar cell performance parameters). *Reproduced with permission from H. Hu, D. Wang, Y. Zhou, J. Zhang, S. Lv, S. Pang, et al., Vapour-based processing of hole-conductor-free $CH_3NH_3PbI_3$ perovskite/C60 fullerene planar solar cells. RSC Adv. 4 (55) (2014) 28964–28967 [19]. Copyright 2014, The Royal Society of Chemistry.*

A vapor-assisted solution processing (VASP) technique has evolved as efficient deposition technique to make a uniform and quality film but these techniques are cost efficient which affects the commercialization of this product [15,21].

In this work, Etgar et al. made a unique HTM-free planar perovskite cell with micrometer-size grain perovskite film by using spray deposition method. Fig. 9.9A shows the schematic representation of the spray deposition technique, the one-step solution of perovskite precursor was sprayed on the electrodes and kept on hot plate. The perovskite crystals were immediately formed by the evaporation of solvent in the precursor solution. Fig. 9.9B shows the solar device architecture which consists of FTO/anatase TiO_2/perovskite/Au (as confirmed by XRD; Fig. 9.9C). The thickness of the perovskite layer was controlled by the number of spray passes.

Fig. 9.10 shows the HR-SEM cross-sectional images for various thicknesses of perovskite made by different number of spray passes such as 4, 6, 8, and 10. Depending on the number of spray processes the perovskite gave micrometer-sized grain morphology, which is shown in the

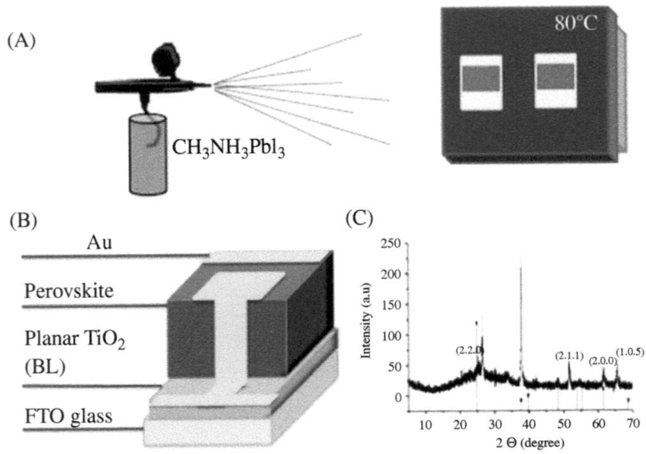

Figure 9.9 (A) Schematic illustration of the spray deposition technique. (B) Planar HTM-free perovskite solar cell structure. (C) XRD of the planar TiO_2 thin film. The anatase TiO_2 crystallographic planes are indexed in the figure. *Reproduced with permission from S. Gamliel, A. Dymshits, S. Aharon, E. Terkieltaub, L. Etgar, Micrometer sized perovskite crystals in planar hole conductor free solar cells, J. Phys. Chem. C 119 (34) (2015) 19722–19728. Copyright 2015, ACS Publications.*

inset image of the cross-sectional image (Fig. 9.10). The number of grain boundaries was less for the micrometer grain size, which is beneficial for photovoltaic performances. The photovoltaic performance of the HTM-free planar device was studied by varying the thickness of the blocking layer and number of spray passes. Table 9.1 shows the PV parameters for different thicknesses of planar TiO_2 with a constant number of passes (10 passes). From the $J-V$ parameters, we can clearly see that the V_{OC} is barely changed with the change in thickness of TiO_2 blocking layer. The device showed better performance for three layers of TiO_2, whereas for 1 and 2 layers of blocking layer, it showed more pin holes in the film, which drastically affects the PV performance. Further increasing the blocking layer upto four and five layers gave a thicker layer, which could not transport the carriers effectively; so the efficiency decreased.

The number of spray passes was varied for further optimization of photovoltaic performance. It can be observed that the high performance was achieved at 10 spray passes because the crystal size is higher than other number of spray passes. The device performance decreased for increased number of spray passes because the thickness of the perovskite became very thick, as a result the layer peeled off from the surface which

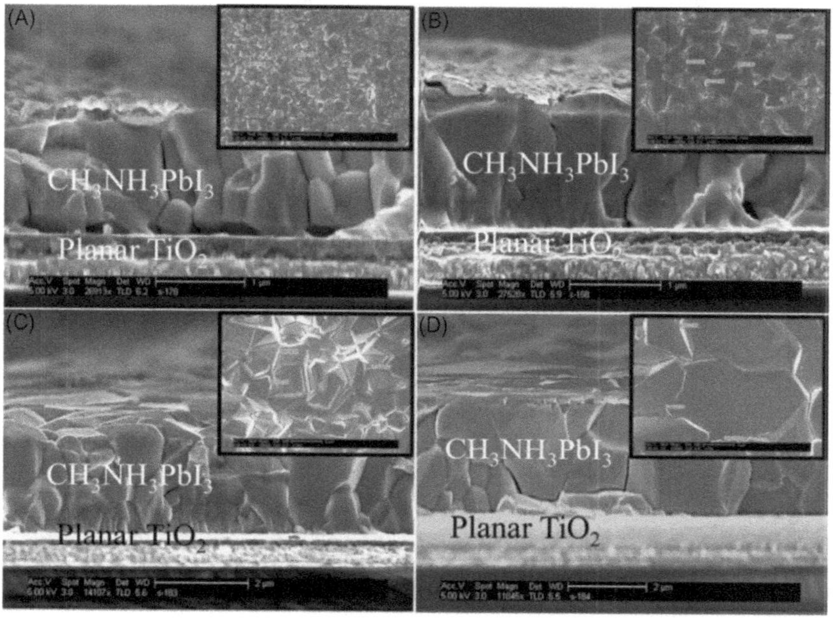

Figure 9.10 (A) Four passes of perovskite spray; (B) 6 passes of perovskite spray; (C) 8 passes of perovskite spray; (D) 10 passes of perovskite spray. (Inset) Top views of different spray passes 4, 6, 8, and 10, respectively. Scale bars in the insets are 2 μm. *Reproduced with permission from S. Gamliel, A. Dymshits, S. Aharon, E. Terkieltaub, L. Etgar, Micrometer sized perovskite crystals in planar hole conductor free solar cells, J. Phys. Chem. C 119 (34) (2015) 19722–19728. Copyright 2015, ACS Publications.*

Table 9.1 PV parameters of cells according to various thicknesses of dense TiO_2 layers (blocking layers); thicknesses as listed in the table[a]

No. of BL	V_{OC} (V)	J_{SC} (mA/cm^2)	FF (%)	Efficiency (%)	BL thickness (nm)
1	0.55	9.35	39	1.99	74.4 ± 11
2	0.66	21.54	32.7	4.63	76.2 ± 15
3	0.69	23.01	43.4	6.93	166.9 ± 19
4	0.65	21.64	39.7	5.58	184.2 ± 13
5	0.68	17.82	42.3	5.11	227.5 ± 42

FF, fill factor; *BL*, blocking layer.
[a]The number of spray passes is constant, equal to 10.
Source: Reproduced with permission from S. Gamliel, A. Dymshits, S. Aharon, E. Terkieltaub, L. Etgar, Micrometer sized perovskite crystals in planar hole conductor free solar cells, J. Phys. Chem. C 119 (34) (2015) 19722–19728. Copyright 2015, ACS Publications.

affected the performance. In the case of 10 spray passes the thickness of the perovskite film is 7.5 ± 1 μm which decreased the recombination of carriers as well as reduced the defects in the crystal [22].

9.1.3 Inverted Planar Type Hole Transport Material-Free Perovskite Solar Cell

In regular PSCs configuration the mesoporous TiO_2 has been used as scaffold for perovskite and ease of electron transfer. This kind of mesoporous TiO_2 used in PSC complicates device fabrication and also affects the device stability under UV illumination [23,24]. The most widely used p-doping HTM such as Spiro-OMeTAD is expensive and difficult to handle, which may limit the future development and application of this type of solar cells. To avoid these problems the inverted planar type has been evolved in the PSCs architecture. The omission of high-temperature TiO_2 from the PSCs helps to make low-temperature solution-processed perovskite device simpler and also offers to fabricate the device on flexible substrate [15–17,25,26].

In this work, Bian and Liu et al. have successfully fabricated a hole conductor–free planar-type PSC with an inverted structure of ITO/$MAPbI_3$/C_{60}/BCP/Ag, where $MAPbI_3$ was deposited on ITO by solution-processed technique. For comparison a reference device with NiO_x as HTM also has been fabricated. In this case the metal halide perovskite is directly deposited onto the ITO or ITO/NiO_x. Then, C_{60} was deposited which acts as ETL followed by BCP that acts as hole blocking layer. Finally, Ag was deposited as cathode.

The devices were characterized by time-resolved photoluminescence and time-integrated photoluminescence spectroscopy to find the effect of HTM on device performance (Fig. 9.11). The perovskite deposited on top of the ITO sample showed high quenching degree. After deposition of C_{60} both the samples showed high quenching degree which means the electron extraction was improved at the interface of perovskite/C_{60}. The PL decay lifetime was decreased from 10.3 and 23.1 ns to 2.1 and 3.5 ns. So they stated that C_{60} has higher electron collection property. The devices with and without HTM were further analyzed by the impedance spectroscopy. The introduction of NiO_x in the device architecture increases the resistance in the interface layers which is confirmed by impedance spectroscopy.

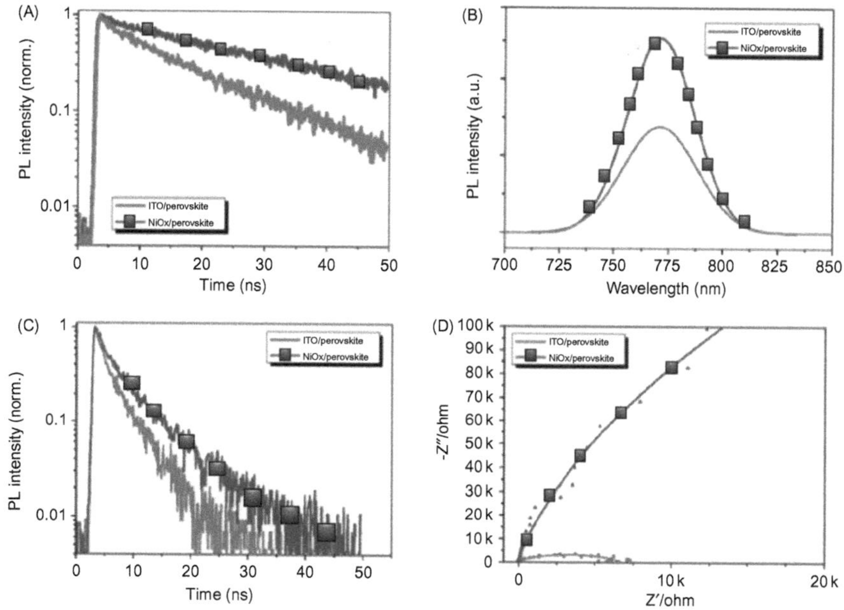

Figure 9.11 Photoluminescence spectral and impedance characterization. (A) Time-resolved PL decay transients measured at 770 ± 10 nm for ITO/$CH_3NH_3PbI_3$ (—) and NiO_x/$CH_3NH_3PbI_3$ (—) under an excitation of 517 nm (1 kHz, 110 nJ/cm^2). (B) Steady-state PL spectra of ITO/$CH_3NH_3PbI_3$ (—) and ITO/NiO_x/$CH_3NH_3PbI_3$ (—) under an excitation of 505 nm. (C) Time-resolved PL decay transients measured at 770 ± 10 nm for ITO/$CH_3NH_3PbI_3$/C_{60} (—) and ITO/NiO_x/$CH_3NH_3PbI_3$/C_{60} (—) under an excitation of 517 nm (1 kHz, 110 nJ/cm^2). (D) Nyquist plots of the perovskite solar cells with NiO_x (—) and without a hole conductor (—), as measured under dark conditions at 0.5 V. *Reproduced with permission from Y. Li, S. Ye, W. Sun, W. Yan, Y. Li, Z. Bian, et al., Hole-conductor-free planar perovskite solar cells with 16.0% efficiency, J. Mater. Chem. A 3 (36) (2015) 18389–18394. Copyright 2015, The Royal Society of Chemistry.*

Fig. 9.12 shows the $J-V$ characteristics of the PSC device with and without NiO_x as the hole conductor material. The detailed photovoltaic parameters are summarized in Table 9.2. As shown in the $J-V$ curve the device with NiO_x shows the high series resistance and shunt resistance which gives results such as low fill factor and J_{SC}. Although the device without any HTM shows low series and high shunt resistance, this device provides better photovoltaic parameters. Moreover, the HTM-free inverted planar PSCs were stable for more than 1000 h because of its less interface and simplified device architecture [27].

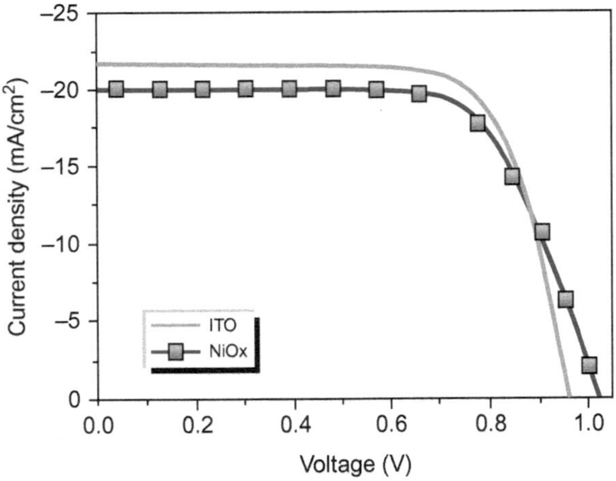

Figure 9.12 The current density–voltage (*J–V*) curves of the champion planar heterojunction solar cells without a hole conductor (—) and with a NiO_x (—) hole conductor under simulated AM 1.5G solar irradiation at 100 mW/cm^2. *Reproduced with permission from Y. Li, S. Ye, W. Sun, W. Yan, Y. Li, Z. Bian, et al., Hole-conductor-free planar perovskite solar cells with 16.0% efficiency, J. Mater. Chem. A 3 (36) (2015) 18389–18394. Copyright 2015, The Royal Society of Chemistry.*

Table 9.2 Performance parameters of solar cells with or without a hole transport layer (HTL)

Device	HTL	V_{OC} (mV)	J_{SC} (mA/cm^2)	FF	PCE (%)
1	NiO_x	994 ± 15	20.4 ± 0.7	0.669 ± 0.022	13.6 ± 0.6
2	–	961 ± 14	21.7 ± 0.5	0.718 ± 0.015	15.0 ± 0.5

Reproduced with permission from Y. Li, S. Ye, W. Sun, W. Yan, Y. Li, Z. Bian, et al., Hole-conductor-free planar perovskite solar cells with 16.0% efficiency, J. Mater. Chem. A 3 (36) (2015) 18389–18394. Copyright 2015, The Royal Society of Chemistry.

9.1.4 Mesoporous-Type Hole Transport Material–Free Perovskite Solar Cell

As discussed earlier, mesoporous-type device is one of the main configuration in PSC where the perovskite deposited on top of the mesoporous TiO_2 can be functionalized as light harvester and HTM.

Initially, PbI_2 was spin coated on top of mesoporous TiO_2 with different waiting time and then annealed at 70°C for 20 min. Then, it was converted into complete $CH_3NH_3PbI_3$ perovskite by dipping the PbI_2 film into CH_3NH_3I solution. In this work the device performance was

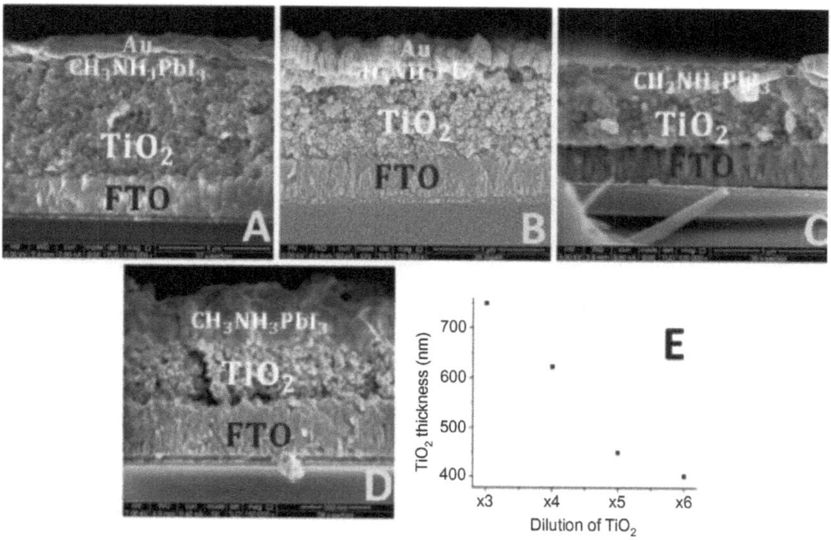

Figure 9.13 HR-SEM of the different TiO_2 thicknesses in the complete TiO_2–$CH_3NH_3PbI_3$ perovskite-based solar cells. (A) The thicker TiO_2 film corresponds to three times dilution (see (E)). (B) Four times dilution. (C) Five times dilution. (D) Six times dilution. Importantly the thickness was measured using a surface profiler, the results of which are shown in (E). *Reproduced with permission from S. Aharon, S. Gamliel, B. El Cohen, L. Etgar, Depletion region effect of highly efficient hole conductor free $CH_3NH_3PbI_3$ perovskite solar cells. Phys. Chem. Chem. Phys. 16 (22) (2014) 10512–10518. Copyright 2014, The Royal Society of Chemistry.*

observed and studied for different waiting time of PbI_2. The device prepared with 3 min of waiting time has shown better performance since the PbI_2 created a uniform film on top of mesoporous TiO_2. On the other hand the precursor solvent evaporated for 5 min of waiting time caused a nonuniform film. Contrarily the PbI_2 is not completely covered for the extremely short wait time.

The mesoporous TiO_2 has an important role as it improves the device performance. The performance of HTM-free device was also studied by varying the thickness of the mesoporous TiO_2 film. The different thicknesses of TiO_2 were achieved by diluting the TiO_2 paste indifferent ratios. Fig. 9.13 shows the cross-sectional image of various thicknesses of TiO_2 in the complete set of TiO_2-$CH_3NH_3PbI_3$ PSC. From the cross-sectional images, we can clearly see that the $CH_3NH_3PbI_3$ layer has formed an overlayer on top of the porous film. The highest performance of the device with high current density and open circuit voltage was achieved for 620 ± 25 nm thickness of mesoporous TiO_2 film. On the

Table 9.3 PV performance of the hole conductor–free TiO_2–$CH_3NH_3PbI_3$ perovskite-based solar cells obtained for various TiO_2 film thicknesses

Thickness of TiO_2 film (nm) ± 25 nm	V_{OC} (V)	J_{SC} (mA/cm^2)	FF	Efficiency (%)
750	0.66	12.47	0.45	3.7
620	0.73	16.03	0.61	7.2
450	0.68	13.27	0.63	5.7
400	0.65	10.47	0.55	3.8

Source: Reproduced with permission from S. Aharon, S. Gamliel, B. El Cohen, L. Etgar, Depletion region effect of highly efficient hole conductor free $CH_3NH_3PbI_3$ perovskite solar cells. Phys. Chem. Chem. Phys. 16 (22) (2014) 10512–10518. Copyright 2014, The Royal Society of Chemistry.

other hand the device fabricated with the thickness of 750 ± 25 nm and 400 ± 25 nm showed lowest efficiency with low FF and current density. Table 9.3 summarizes the photovoltaic parameters of the devices made by different thicknesses of mesoporous TiO_2.

In PSCs the method of deposition is very crucial since it describes the film coverage, thickness of the film and transport properties. There are several deposition methods reported for deposition of perovskite, i.e., vapor deposition, VASP, and solution process (one-step and two-step method) [15,25,28]. Gratzel and his coworkers found out that the two-step deposition technique is a promising method to get high-efficiency PSC [15]. This work described the in-depth study of the parameters (spin velocity, dipping time, annealing temperature, and MAI concentration) influencing the two-step deposition of the PSCs and also showing the effect of these parameters on morphology, optical properties, and photovoltaic performance.

The morphology of the perovskite film made by different spin velocities with same dipping time was studied by HR-SEM. The film had more voids and no proper coverage at 500 rpm. Except at 500 rpm the PbI_2 films showed more uniformity when the spin velocity was increased to 2000, 4500, and 6500 rpm. The reason to get the increment of perovskite crystal size with high spin velocity is directly related to the dipping time of PbI_2 film in the CH_3NH_3I solution. The PbI_2 films showed different thickness with varying spin velocity, and it requires different dipping times to get high quality and uniform perovskite films. When the dipping time is same, the resulting crystal size is dependent on the spin velocity.

The absorption spectra of the PbI_2 and $CH_3NH_3PbI_3$ films prepared at different spin velocities (2000–6500 rpm) which clearly state the difference in the film thickness of PbI_2 and $CH_3NH_3PbI_3$. The absorption of the PbI_2 films at two different spin velocities showed big difference, but the absorption of $CH_3NH_3PbI_3$ is slightly lower than the difference between the PbI_2 samples. This indicates that the final thickness of the complete $CH_3NH_3PbI_3$ is not only dependent on spin velocity but also dependent on the dipping process of PbI_2 electrode. The device prepared with spin velocity of 2000 rpm showed better efficiency of 7.8%. The morphology and PV performance were studied by varying the dipping time (20 s, 1 min, 20 min, and 3.5 h) of PbI_2 electrode in the CH_3NH_3I solution. The best performance of the device was achieved at short dipping time (20 s). The PV performance of the devices was decreased for longer dipping time, because the crystal size is larger which affects the transport properties of the perovskite film and also recombination occurs.

Annealing is the final and important step for perovskite crystallization. The PV performance of the device was studied with different annealing temperatures of 100, 150, 175, and 200°C. At low temperatures the crystals separated from each other. Then the crystals started to mingle together in the range of 150°C. The quality of the crystal was improved with proper sintering at 175°C, which enhances the charge transport properties. Further increment of temperature up to 200°C melts the crystals, which results in lower PCE. The concentration of MAI in the dipping solution also has effects on the device performance. The increased amount of MAI helps to convert the PbI_2 film completely and also increases the crystal size which improves the device performance. On the other hand the excess amount of MAI reduces the conductivity of the perovskite and also decomposes the film. For higher concentration of MAI the current density was low which could be improved by using lower concentration.

The size of the organic cation influences the optical bandgap of perovskite by changing the M–I–M angle (M = Pb, Sn) [29,30]. The most commonly used methylammonium cation can be replaced by larger formamidinium cation which decreases the bandgap, so the optical absorption gets extended into the red region. In this work, both the organic cations (MAI and FAI) are mixed together in the ratio of 1:1 and a hole conductor-free PSC is prepared. It was found that the perovskite with single cation has shown shorter diffusion length than the mixed cation phase perovskite. The PSC was fabricated by the two-step deposition technique, and the effects of composition and annealing temperature on photovoltaic performance were also studied. The PbI_2 electrodes were dipped into the MAI solution and annealed at 100°C/30 min, while in the case of formamidinium iodide (FAI) or FAI:MAI, the samples were annealed at 175°C/15 min.

Fig. 9.14 shows the HR-SEM images of the $MAPbI_3$, $MAPbI_3$:$FAPbI_3$, and $FAPbI_3$ films at different annealing temperatures. Fig. 9.14A, D, and G shows the $MAPbI_3$ film at various annealing temperatures. The film has

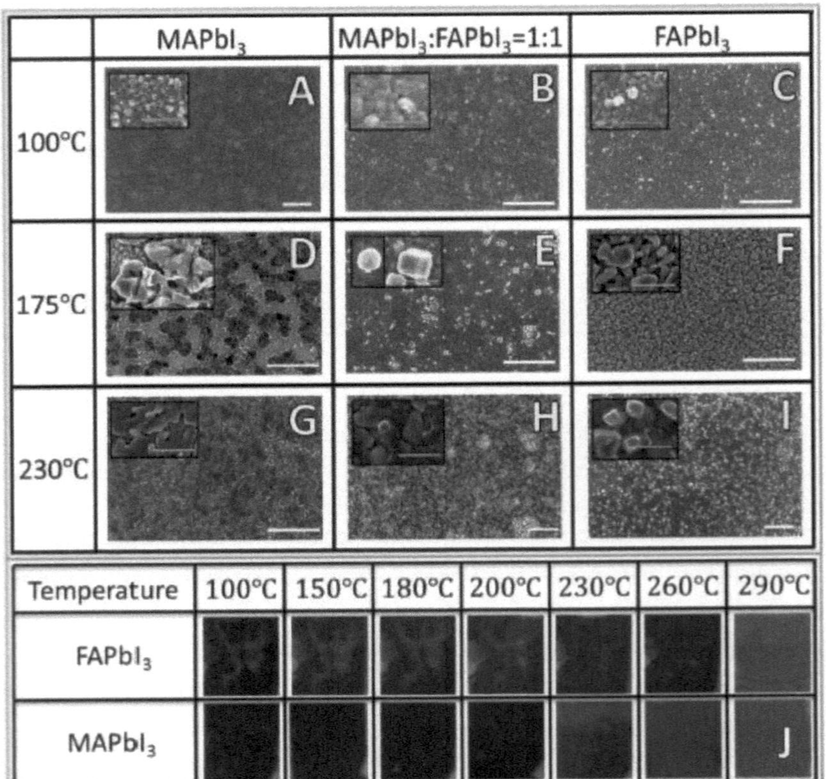

Figure 9.14 HR-SEM images of the perovskite compositions at various annealing temperatures (100, 175, and 230°C). The white scale bar is 5 mm and the orange scale bar (inset) is 1 mm. (A) $MAPbI_3$ annealed at 100°C. (B) $MAPbI_3$:$FAPbI_3$ = 1:1 annealed at 100°C. (C) $FAPbI_3$ annealed at 100°C. (D) $MAPbI_3$ annealed at 175°C. (E) $MAPbI_3$:$FAPbI_3$ = 1:1 annealed at 175°C. (F) $FAPbI_3$ annealed at 175°C. (G) $MAPbI_3$ annealed at 230°C. (H) $MAPbI_3$:$FAPbI_3$ = 1:1 annealed at 230°C. (I) $FAPbI_3$ annealed at 230°C. (J) Pictures of the samples of $MAPbI_3$ or $FAPbI_3$ on glass after annealing at various temperatures. *Reproduced with permission from S. Aharon, A. Dymshits, A. Rotem, L. Etgar, Temperature dependence of hole conductor free formamidinium lead iodide perovskite based solar cells, J. Mater. Chem. A 3 (17) (2015) 9171–9178. Copyright 2015, The Royal Society of Chemistry.*

shown good coverage at the annealing temperature range of 100°C [32]. While increasing the temperature up to 175°C, the coverage of the perovskite film was not as good as 100°C. On the other hand, $MAPbI_3$ was decomposed into lead iodide as can be seen in Fig. 9.14J (~230°C) [33]. The mixed organic cation perovskite showed better coverage at all annealing temperatures (Fig. 9.14B, E, and H), but the crystal can be clearly seen at the temperature range of 175°C (Fig. 9.14E, inset). In case of $FAPbI_3$ the crystals started to grow at 100°C, while at higher temperatures (175 and 230°C), the crystals were clearly visible in the images. This states that the formamidinium-based perovskite films need higher annealing temperatures.

Fig. 9.14J shows the surface images of the $MAPbI_3$ and $FAPbI_3$ films at eight different temperatures. The $MAPbI_3$ films was shown dark in colour up to 200°C (perovskite film), while the colour of the film changed to light colour in the temperature range of 230°C which means the films started to decompose. At 290°C, the entire film was completely changed into PbI_2 film. On the other hand, the $FAPbI_3$ was stable up to 260°C (films with some grain structure in the image). At 290°C, the $FAPbI_3$ film also completely changed to PbI_2 film (compact film) as same as $MAPbI_3$ film. From the earlier results an annealing temperature of 175°C is good enough to make high-quality perovskite crystals, which is in good agreement with the HR-SEM images.

Fig. 9.15 shows the photovoltaic parameters of the hole conductor free device prepared with three different perovskites measured at eight different temperatures (22, 35, 45, 55, 65, 75, 85, and 95°C). From the

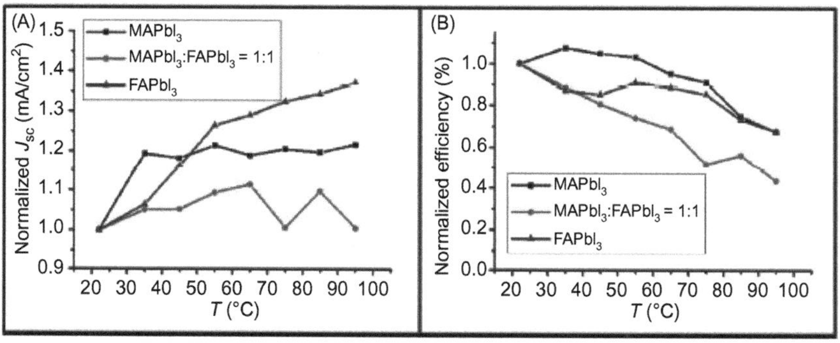

Figure 9.15 (A) The normalized short circuit current density (J_{sc}) of the three different photovoltaic cells at various temperatures. (B) The normalized efficiency of the photovoltaic cells at various temperatures. *Reproduced with permission from S. Aharon, A. Dymshits, A. Rotem, L. Etgar, Temperature dependence of hole conductor free formamidinium lead iodide perovskite based solar cells, J. Mater. Chem. A 3 (17) (2015) 9171–9178. Copyright 2015, The Royal Society of Chemistry.*

results, we can conclude that the value of current density has increased with temperature. With increase in temperature the bandgap of the semiconducting material decreases and this impacts the current density. The $FAPbI_3$ sample has shown more than 30% increment in the J_{SC} because of better stability at high temperature. On the other hand, for mixed cation perovskite the photovoltaic parameters were reduced drastically with increase in temperature. The active layer contains two different cations which get internally affected with changes in temperature due to the formation of strain in the crystalline structure. Fig. 9.15B shows the efficiency results with increased temperature. When temperature increases, the efficiency of the hole conductor free devices reduces with lower FF and V_{OC}. The reason for getting lower V_{OC} with high temperature might be due to the higher intrinsic carrier concentration. The carrier concentration depends on the bandgap of the material. Lower the bandgap gives higher intrinsic carrier concentration which can result in lower V_{OC} [31,34].

9.1.5 Carbon-Based Hole Transport Material–Free Perovskite Solar Cell

As per our previous discussion, the use of HTM in device architecture makes the device costlier and also reduces the stability of the cell, which limits the fabrication of low-cost photovoltaic devices. Secondly, most of the PSCs are fabricated with noble metals such as Ag and Au as counter electrode which are very costly. So, they are not suitable for large scale production as they need high-vacuum deposition methods. Hence, it needs to be replaced with inexpensive and easily available abundant material for making low-cost photovoltaic devices. Carbon-based materials could be an alternative for replacing the metal electrodes because of their low cost, easy preparation, low-temperature process ability, and suitable for screen printing for large scale fabrication. Good connectivity and mechanical performance of the carbon film help to collect the holes effectively and reduce the resistance of the solar devices [35–37].

In this work, formamidinium cation was introduced into the conventional $MAPbI_3$ to extend the light absorption area since formamidinium has a bandgap of 1.47 eV. The perovskite was deposited onto the porous carbon film by two-step sequential deposition method, which is shown in Fig. 9.16. Initially the porous carbon film was infiltrated with PbI_2 by drop-casting and then it was immersed into the mixed solution of MAI and FAI. After 15 min the films were washed with isopropanol to remove the unreacted MAI and FAI, followed by drying the electrodes for 30 min.

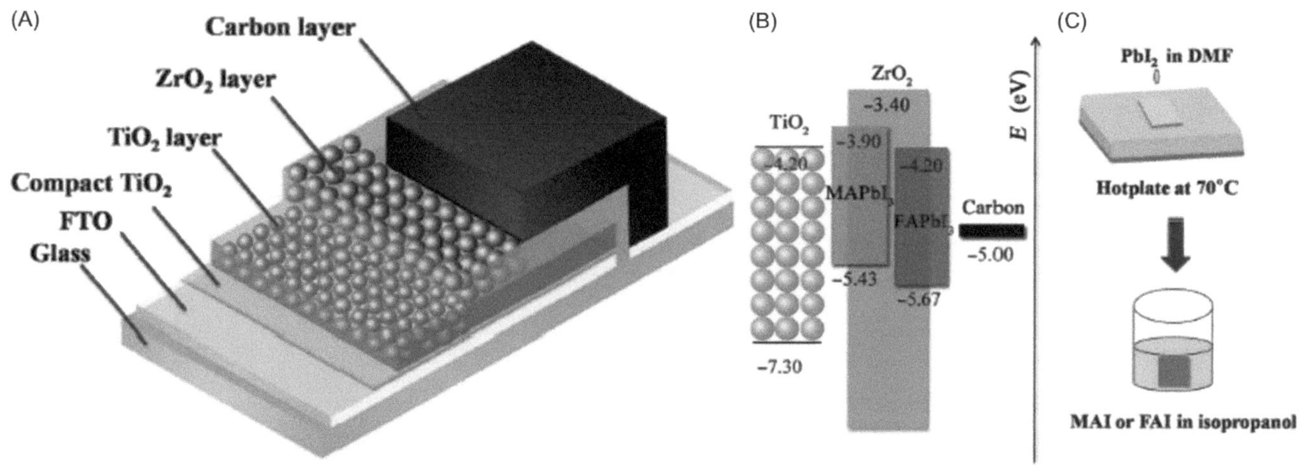

Figure 9.16 (A) The schematic structure of the fully printable mesoscopic perovskite solar cell with a carbon counter electrode; (B) the energy-level diagram of the pure $MAPbI_3$ and $FAPbI_3$ perovskites; (C) the schematic illustration of the sequential deposition method. *Reproduced with permission from M. Hu, L. Liu, A. Mei, Y. Yang, T. Liu, H. Han, Efficient hole-conductor-free, fully printable mesoscopic perovskite solar cells with a broad light harvester $NH_2CH{=}NH_2PbI_3$. J. Mater. Chem. A 2 (40) (2014) 17115–17121. Copyright 2014, The Royal Society of Chemistry.*

The crystallinity of the synthesized $MAPbI_3$ and $FAPbI_3$ was characterized by X-ray diffraction. Generally, $FAPbI_3$ has two phases (i.e., black and yellow phase), with different crystal structure. Among these two phases, black phase belongs to trigonal symmetry, which is similar to that of tetragonal structure of $MAPbI_3$. This black phase absorbs more photons, which consequently increases the PCE of the devices. One-step deposition of $FAPbI_3$ perovskite needs higher annealing temperature and also results in poor quality films. So, sequential deposition method was introduced in this work to make a better film than one-step technique with lower annealing temperature [32,38].

Fig. 9.17 shows the cross-sectional SEM images of TiO_2/ZrO_2/carbon film with infiltration of perovskite and morphology of carbon film.

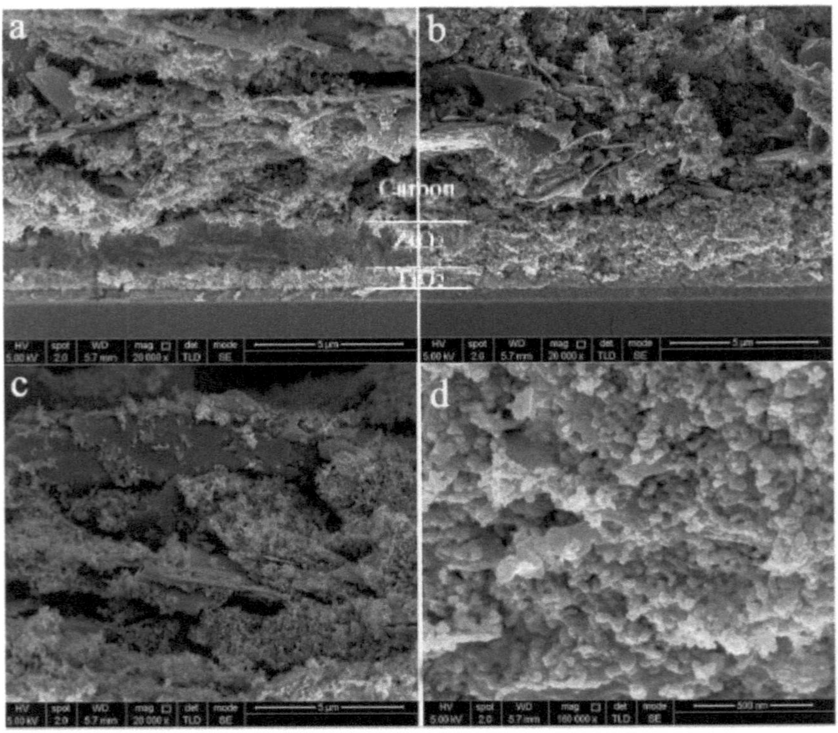

Figure 9.17 (A) SEM image from the cross-section of the blank TiO_2/ZrO_2/carbon film; (B) the cross-section of the mesoscopic solar cell with $FAPbI_3$; (C) the surface morphology of the blank carbon layer; (D) high magnification of the $FAPbI_3$ coated TiO_2/ZrO_2 film. *Reproduced with permission from M. Hu, L. Liu, A. Mei, Y. Yang, T. Liu, H. Han, Efficient hole-conductor-free, fully printable mesoscopic perovskite solar cells with a broad light harvester $NH_2CH=NH_2PbI_3$. J. Mater. Chem. A 2 (40) (2014) 17115–17121. Copyright 2014, The Royal Society of Chemistry.*

Approximately, these TiO_2/ZrO_2/carbon films are having thicknesses of 700 nm, 1.5 μm, and 10 μm, respectively, which have been found out from the cross-sectional images. Fig. 9.17B and D shows the cross-sectional image of the TiO_2/ZrO_2 films with infiltration of $FAPbI_3$ perovskite. It is observed that the TiO_2/ZrO_2 mesoporous layers are fully filled with the $FAPbI_3$ perovskite, which helps in efficient electron transfer from the perovskite to the conduction band of TiO_2. Fig. 9.17C shows the morphology of thick carbon film without perovskite infiltration. It is clearly seen that the small carbon particles are embedded within the large flaky graphite. So the thick film as well as rough mesoporous surface of the carbon films will be suitable for the various sizes of cubic $FAPbI_3$ and $MAPbI_3$ perovskite to infiltrate.

The $J-V$ curves of $MAPbI_3$ and $FAPbI_3$ devices prepared by the sequential method are shown in Fig. 9.18. Both the devices are showing almost the same results, but in the case of current density, $FAPbI_3$ shows higher J_{SC} since it generates photocurrent upto 840 nm when compared with 800 nm achievable by $MAPbI_3$ perovskite. The IPCE spectra of the devices are shown in the figure. The IPCE spectra of $FAPbI_3$ are lower than $MAPbI_3$ at wavelengths longer than 600 nm because of weak light absorption of $FAPbI_3$ at longer wavelengths. To increase the absorption value to longer wavelength region, FAI was mixed with MAI in different ratio. As a result, the absorption value increased when the FAI ratio is

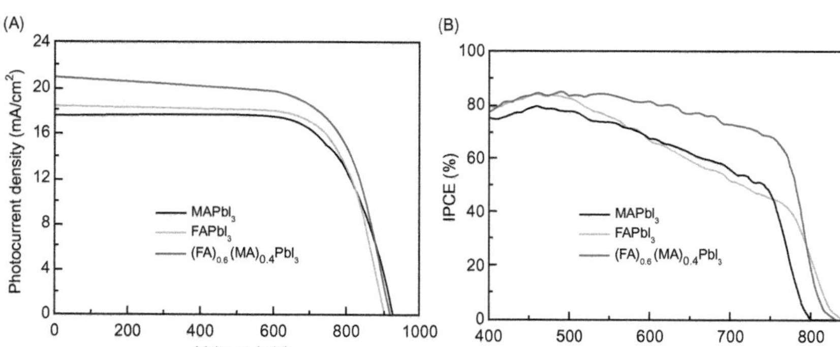

Figure 9.18 (A) $J-V$ curves of mesoscopic perovskite solar cells with $MAPbI_3$, $FAPbI_3$, and $(FA)_{0.6}(MA)_{0.4}PbI_3$ under 100 mW/cm^2 AM 1.5G illumination; (B) the corresponding incident photon to electron conversion efficiency (IPCE) spectra. *Reproduced with permission from M. Hu, L. Liu, A. Mei, Y. Yang, T. Liu, H. Han, Efficient hole-conductor-free, fully printable mesoscopic perovskite solar cells with a broad light harvester $NH_2CH=NH_2PbI_3$. J. Mater. Chem. A 2 (40) (2014) 17115–17121. Copyright 2014, The Royal Society of Chemistry.*

increased. So the device prepared in the ratio of 3:2 (FAI:MAI) showed better performance than pure $FAPbI_3$ and $MAPbI_3$. The increment in PCE for the mixed cation devices might be attributed to the significantly increased current density value than the pristine perovskite devices [35].

The perovskite crystals play a crucial role in the performance of perovskite solar devices [39]. Mei et al. prepared a HTM-free device that uses a double layer of mesoporous TiO_2, ZrO_2, and printed carbon film. The perovskite solution that contained PbI_2, MAI, and 5-aminovaleric acid (5-AVA) was infiltrated into the porous film by drop-casting method. In this work the MA cation was partially replaced with the 5-AVA cation to template the formation of perovskite crystal and to make a crystal growth in normal direction. This mixed cation perovskite has lower defect concentration as well as better surface contact with TiO_2 than the conventional perovskite, resulting in improved efficiency.

The morphology of the two different perovskites was studied by the TEM analysis. Fig. 9.19 shows the morphology of pristine TiO_2, $MAPbI_3$, and $(5\text{-AVA})_x(MA)_{1-x}PbI_3$-covered TiO_2. From images (B) and (C), we can see the morphology difference of the samples prepared from the two different perovskite precursor solutions. As discussed, the

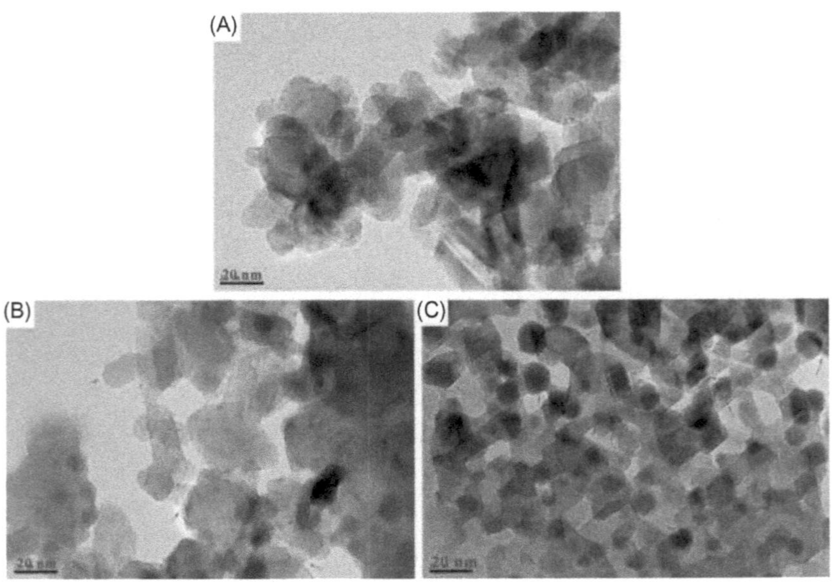

Figure 9.19 Microscopy images. (A–C) TEM images of (A) TiO_2 and mixtures of (B) $MAPbI_3$ and (C) $(5\text{-AVA})_x(MA)_{1-x}PbI_3$, with TiO_2 particles scraped off of a mesoporous film infiltrated with perovskites. *Reproduced with permission from A. Mei, X. Li, L. Liu, Z. Ku, T. Liu, Y. Rong, et al., A hole-conductor-free, fully printable mesoscopic perovskite solar cell with high stability, Science 345 (6194) (2014) 295–298. Copyright 2014, Science.*

$(5\text{-AVA})_x(MA)_{1-x}PbI_3$ perovskite helps to make a template for crystal growth, so that the crystal can grow in length. Also the perovskite had covered most of the TiO_2 surface area. The higher loading of $(5\text{-AVA})_x(MA)_{1-x}PbI_3$ perovskite into the mesoporous TiO_2 probably resulted from the 5-AVA cation which helps in nucleation and growth of perovskite crystal within the scaffold. The $MAPbI_3$ film prepared by one-step process shows partially coated substrate with less surface area where the perovskite absorber is completely absent.

The device prepared with $(5\text{-AVA})_x(MA)_{1-x}PbI_3$ perovskite showed better performance with high value of current density 21.1 mA/cm^2 than $MAPbI_3$ which showed only 13.91 mA/cm^2 (Fig. 9.20). This new cation perovskite achieved a PCE of 12.8% and was stable for more than 1000 h in ambient atmosphere [40].

In carbon-based PSCs the holes are extracted from the perovskite/carbon interface, which strongly depends on the interface engineering, carbon material, particle size/shape, and electronic properties of carbon material [36,41,42]. The FF of nanocarbon-based PSC (<0.75) is lower than the HTM-based devices (>0.78), which limits the performance of the PSC [42,43]. So, Yang et al. used multiwalled carbon nanotubes

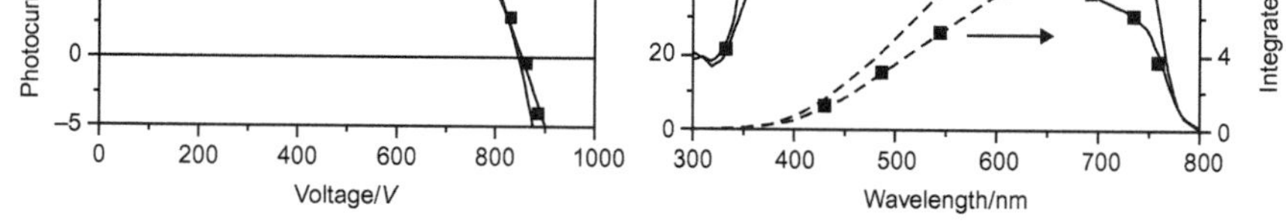

Figure 9.20 (A) *J–V* curves under simulated AM 1.5 solar irradiation at an intensity of 100 mW/cm^2 measured at room temperature. (B) IPCE curves taken with monochromatic light without applied white-light bias for $(5\text{-AVA})_x(MA)_{1-x}PbI_3$ (—) or $MAPbI_3$ (—). The integrated photocurrents calculated from the overlap integral of the IPCE spectra with the AM 1.5 solar emission are also shown. The integrated photocurrent of the $(5\text{-AVA})_x(MA)_{(1-x)}PbI_3$–based photovoltaic is 17.8 mA/cm^2, which agrees closely with the photocurrent density of 18.4 mA/cm^2 measured at the beginning of the testing, which rose to 21.1 mA/cm^2 after 3 min of light soaking. *Reproduced with permission from A. Mei, X. Li, L. Liu, Z. Ku, T. Liu, Y. Rong, et al., A hole-conductor-free, fully printable mesoscopic perovskite solar cell with high stability, Science 345 (6194) (2014) 295–298. Copyright 2014, Science.*

(MWCNTs) as a HTM material to improve the FF and also to reduce the hysteresis effect in device performance. The PSC device was fabricated by two-step sequential deposition method with carbon particle, carbon nanotube, and graphite as HTM for comparison. There are two major steps in the hole extraction; first the hole should be extracted from perovskite to carbon, and second charge should be transported through the carbon film. Fig. 9.21 shows the morphology of the carbon-based

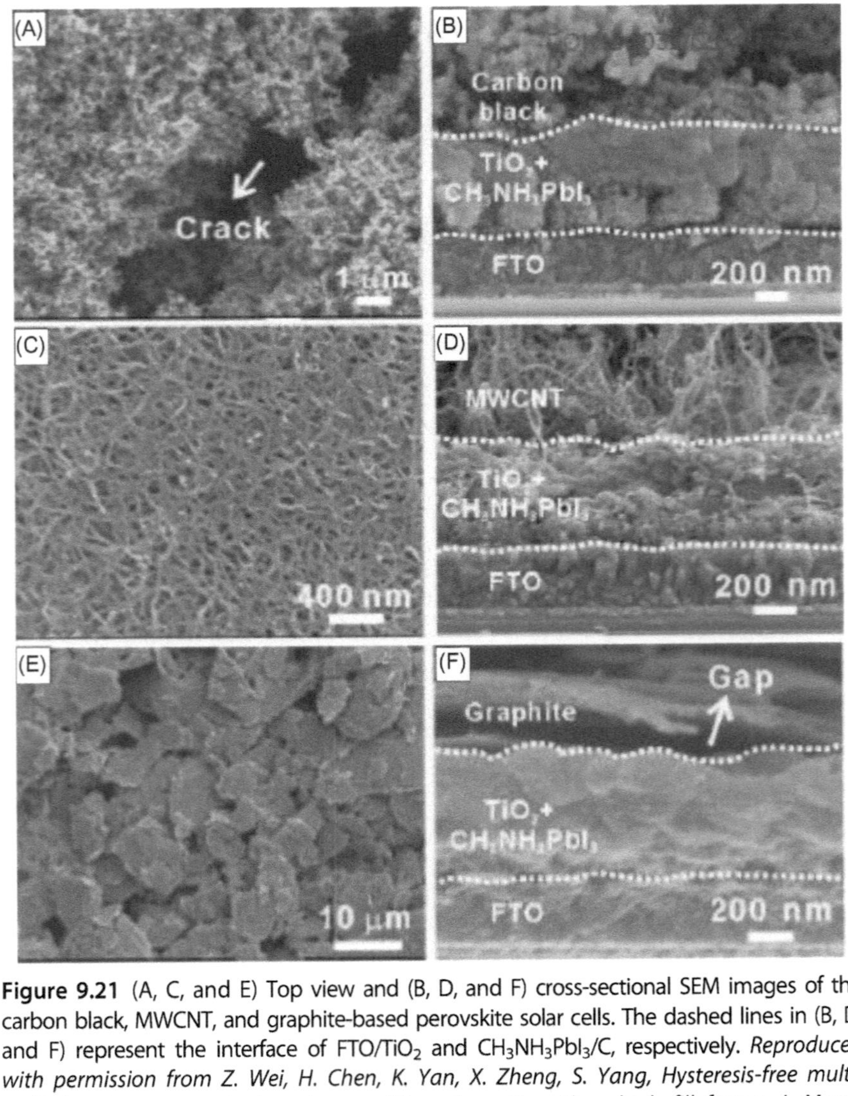

Figure 9.21 (A, C, and E) Top view and (B, D, and F) cross-sectional SEM images of the carbon black, MWCNT, and graphite-based perovskite solar cells. The dashed lines in (B, D, and F) represent the interface of FTO/TiO_2 and $CH_3NH_3PbI_3/C$, respectively. *Reproduced with permission from Z. Wei, H. Chen, K. Yan, X. Zheng, S. Yang, Hysteresis-free multi-walled carbon nanotube-based perovskite solar cells with a high fill factor, J. Mater. Chem. A 3 (48) (2015) 24226–24231. Copyright 2015, The Royal Society of Chemistry.*

PSCs. The device prepared with carbon particle is having high-quality perovskite/carbon interface that helps to extract the holes effectively from perovskite to carbon. However, cracks are formed easily in the carbon film due to the lack of binding energy. So the flow of holes from particle to particle will be slowed down in the cracked film. MWCNTs-based device has excellent interface at the perovskite-MWCNT junction for hole extraction process. The specific property of this MWCNT is that it can act as a highway for charge transport process and improves performances of the photovoltaic devices. On the other hand the device made with graphite flakes has faster charge transportation but is not suitable for making a good contact with the perovskite crystalline film.

Devices were fabricated with these three carbon-based materials and measured under 1 sun condition to evaluate the photovoltaic performance, as shown in Fig. 9.21A and Table 9.4. It is obvious that the graphite-based devices showed very low efficiency of 6.13% with low current density of 10.30 mA/cm^2. The lower current density that is attributed to the gaps between the perovskite and graphite films causes a severe problem with charge transfer resulting in poor performance, whereas the device with carbon black particle shows improvement with an efficiency of 9.35% current density of 15.98 mA/cm^2, V_{OC} of 0.90 V and FF of 0.65. The device with MWCNT is most impressive with FF as high as 0.75 and PCE of 10.30%.

The reason for getting high FF is due to its conductivity (35.60 S/cm), which is 4.3 times higher than that of graphite and 7.43 times higher than that of carbon black. Also the MWCNT device shows low series resistance and high shunt resistance than the other two devices, indicating improved hole extraction and reduced recombination at the interface of perovskite/MWCNT. Fig. 9.22B shows the electrochemical impedance spectra (EIS) of some model devices, i.e., FTO/$CH_3NH_3PbI_3$/MWCNT and FTO/TiO_2/$CH_3NH_3PbI_3$/MWCNT. They were recorded to

Table 9.4 Solar cell parameters of the three as-fabricated carbon-based perovskite solar cells

Solar cells	V_{OC} (V)	J_{SC} (mA/cm^2)	FF	PCE (%)
Carbon black	0.90	15.98	0.65	9.35
MWCNT	0.88	15.60	0.75	10.30
Graphite	0.93	10.30	0.64	6.13

Source: Reproduced with permission from Z. Wei, H. Chen, K. Yan, X. Zheng, S. Yang, Hysteresis-free multi-walled carbon nanotube-based perovskite solar cells with a high fill factor, J. Mater. Chem. A 3 (48) (2015) 24226–24231. Copyright 2015, The Royal Society of Chemistry.

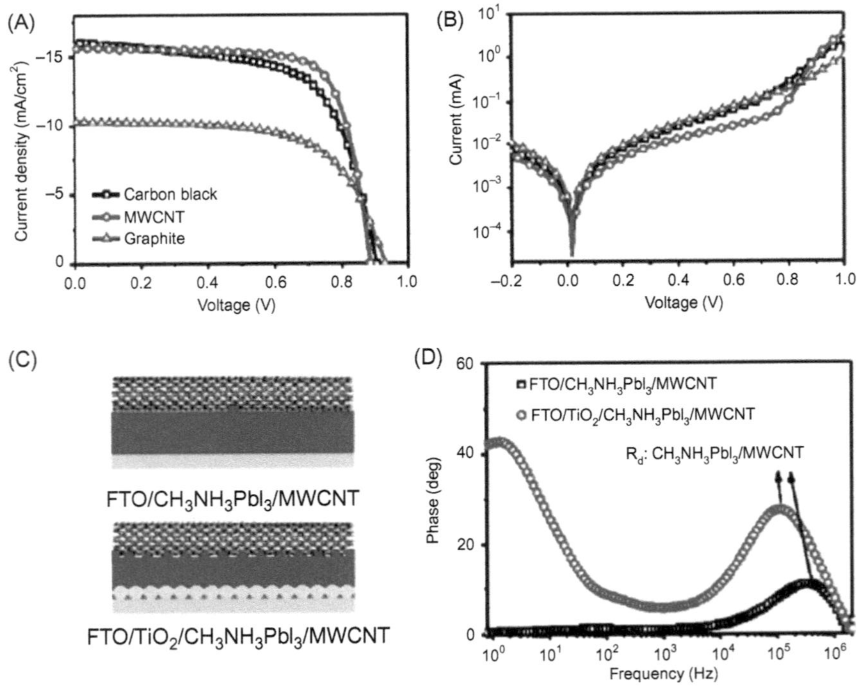

Figure 9.22 (A) *J*–*V* curves under 1 sun illumination and (B) schematic diagrams and (C) EIS spectra (Bode phase plots, bias: 0 V) of two model devices with the configuration of FTO/$CH_3NH_3PbI_3$/MWCNT and FTO/TiO_2/$CH_3NH_3PbI_3$/MWCNT (frequency: 4 MHz to 100 mHz). (D) EIS spectra (Nyquist plot, bias: 0 V). *Reproduced with permission from Z. Wei, H. Chen, K. Yan, X. Zheng, S. Yang, Hysteresis-free multi-walled carbon nanotube-based perovskite solar cells with a high fill factor, J. Mater. Chem. A 3 (48) (2015) 24226–24231. Copyright 2015, The Royal Society of Chemistry.*

analyze the interfacial charge transfer and recombination process. The bode plots show that FTO/$CH_3NH_3PbI_3$/MWCNT device has only one peak in the high-frequency region and FTO/TiO_2/$CH_3NH_3PbI_3$/MWCNT device has one peak in the high-frequency region and an another peak in the low-frequency region. The high-frequency region tells about the charge transfer process at the interface of $CH_3NH_3PbI_3$/MWCNT. In the same way the low frequency attributes to the overall charge recombination in the perovskite thin film.

The difference in PCE mainly depends on the charge transfer process happening at the interface of $CH_3NH_3PbI_3$/carbon. The Nyquist plot of the three carbon-based PSC is shown in Fig. 9.22D. Only one semicircle is observed for each device, which gives information about charge transfer resistance (R_{ct}) of $CH_3NH_3PbI_3$/carbon interface. It can be observed that the device made with MWCNT shows lowest R_{ct} value among the other two devices. From the $J-V$ and EIS results, superior performance was achieved for MWCNT-based device due to its ultra-small R_s and R_{ct} [44].

9.1.6 Stability of Hole Transport Material–Free Perovskite Solar Cell

Though these PSCs are now in a position to compete with the performance of traditional silicon solar cells, long-term stability of the polymeric HTM and moisture sensitivity of perovskite hinder their commercialization. Till date the best performance was achieved by the most commonly used HTM, the spiro-OMeTAD. Generally, these pristine spiro have low conductivity. So, some dopants such as 4-tert-butylpyridine (TBP) and bis(trifluoromethane) sulfonamide lithium salt (Li-TFSI) are used to improve the conductivity and hole mobility of the HTM. Unfortunately, these additives reduce the stability of PSC by corrosion and oxidation of the perovskite material. The degradation mechanism of perovskite with TBP is shown in the following reaction [45,46].

$$MAPbI_3(s) \rightleftharpoons MAI(s) + PbI_2(s) \tag{9.1}$$

$$PbI_2(s) + {}_x TBP(s) \rightarrow PbI_2 - TBP_x(aq) \tag{9.2}$$

$$PbI_2(s) + {}_x TBP(s) \rightarrow [PbI_2 \cdot {}_x TBP](s) \tag{9.3}$$

Moreover, $CH_3NH_3PbI_3$ is very sensitive to moisture due to its hygroscopic amine salts. To overcome this, Noh et al. prepared a HTM-free $CH_3NH_3PbI_3$ device with improved stability, where the large and dense perovskite crystals act as medium to protect the device from oxygen/moisture, thereby improving the stability of the device. Moreover, the stability could be improved by proper sealing of the device in argon atmosphere [17]. The direct exposure of sun light into the PSC device increases the temperature and reduces the stability of the device. There are two kinds of thermal degradation happening in PSC: one due to the intrinsic material property of the perovskite material and the other due to the unstable HTM material. To improve the stability of the PSC device against thermal degradation, Gratzel and his team have prepared an encapsulated triple layer architecture device consisting of $TiO_2/ZrO_2/C$ with high thermal and photo stability [47].

9.2 SUMMARY

HTM-free PSCs have much attention in the photovoltaic field because of the intrinsic property of perovskite material. This chapter gives a detailed information about an operation and parameters that need to be considered for fabrication of HTM-free PSC with different device architecture configuration. Basically, HTM-free PSCs are working with the formation of depletion region between the metal oxide layer and perovskite. High efficiency in HTM-free PSC can be obtained by increasing the depletion region, which suppresses the charge carrier recombination. The metal oxide and absorber layer are having important role in the operation of HTM-free PSC, so this chapter explained clearly about the effect of device performance on various thickness range of these two layers. The metal cathode that used in PSC restricts the commercialization of the product, so carbon-based materials have been used as a cathode to further reduce the cost of the device fabrication. Carbon particle, graphite, and MWCNT were used a cathode. Among these three, MWCNT-based device showed better performance because of its low resistance and high conductivity. Finally, the stability of the HTM free–based devices has been discussed in detail.

ACKNOWLEDGMENTS

Dr. G.V. acknowledges the Department of Science and Technology, New Delhi, India for the financial support through DST-INSPIRE Faculty award (IFA 14-MS-28). G.V. and coauthors extend their acknowledgment to the Director, ARCI.

REFERENCES

[1] T.C. Sum, N. Mathews, Advancements in perovskite solar cells: photophysics behind the photovoltaics, Energy Environ. Sci. 7 (8) (2014) 2518–2534.
[2] S. Ameen, M.A. Rub, S.A. Kosa, K.A. Alamry, M.S. Akhtar, H.S. Shin, et al., Perovskite solar cells: influence of hole transporting materials on power conversion efficiency, ChemSusChem 9 (1) (2016) 10–27.
[3] W. Yan, S. Ye, Y. Li, W. Sun, H. Rao, Z. Liu, et al., Hole-transporting materials in inverted planar perovskite solar cells, Adv. Energy Mater. 6 (17) (2016). 1600474 (1–20).
[4] L. Calió, S. Kazim, M. Grätzel, S. Ahmad, Hole-transport materials for perovskite solar cells, Angew. Chem. Int. Ed. 55 (2016) 14522–14545.
[5] A. Kojima, K. Teshima, Y. Shirai, T. Miyasaka, Organometal halide perovskites as visible-light sensitizers for photovoltaic cells, J. Am. Chem. Soc. 131 (17) (2009) 6050–6051.

[6] S. Lv, S. Pang, Y. Zhou, N.P. Padture, H. Hu, L. Wang, et al., One-step, solution-processed formamidinium lead trihalide ($FAPbI_{(3-x)}Cl_x$) for mesoscopic perovskite–polymer solar cells, Phys. Chem. Chem. Phys. 16 (36) (2014) 19206–19211.
[7] H.S. Jung, N.G. Park, Perovskite solar cells: from materials to devices, Mater.Views 11 (1) (2015) 10–25.
[8] W.A. Laban, L. Etgar, Depleted hole conductor-free lead halide iodide heterojunction solar cells, Energy Environ. Sci. 6 (11) (2013) 3249–3253.
[9] L. Etgar, Hole-transport material-free perovskite-based solar cells, MRS Bull. 40 (08) (2015) 674–680.
[10] W. Yan, Y. Li, W. Sun, H. Peng, S. Ye, Z. Liu, et al., High-performance hybrid perovskite solar cells with polythiophene as hole-transporting layer via electrochemical polymerization, RSC Adv. 4 (62) (2014) 33039–33046.
[11] Y. Zhang, W. Liu, F. Tan, Y. Gu, The essential role of the poly (3-hexylthiophene) hole transport layer in perovskite solar cells, J. Power Sources 274 (2015) 1224–1230.
[12] J.H. Heo, S.H. Im, J.H. Noh, T.N. Mandal, C.S. Lim, J.A. Chang, et al., Efficient inorganic–organic hybrid heterojunction solar cells containing perovskite compound and polymeric hole conductors, Nat. Photonics 7 (6) (2013) 486–491.
[13] L. Etgar, P. Gao, Z. Xue, Q. Peng, A.K. Chandiran, B. Liu, et al., Mesoscopic $CH_3NH_3PbI_3/TiO_2$ heterojunction solar cells, J. Am. Chem. Soc. 134 (42) (2012) 17396–17399.
[14] S. Aharon, S. Gamliel, B. El Cohen, L. Etgar, Depletion region effect of highly efficient hole conductor free $CH_3NH_3PbI_3$ perovskite solar cells, Phys. Chem. Chem. Phys. 16 (22) (2014) 10512–10518.
[15] Q. Chen, H. Zhou, Z. Hong, S. Luo, H.S. Duan, H.H. Wang, et al., Planar heterojunction perovskite solar cells via vapor-assisted solution process, J. Am. Chem. Soc. 136 (2) (2013) 622–625.
[16] D. Liu, T.L. Kelly, Perovskite solar cells with a planar heterojunction structure prepared using room-temperature solution processing techniques, Nat. Photonics 8 (2) (2014) 133–138.
[17] Y. Liu, S. Ji, S. Li, W. He, K. Wang, H. Hu, et al., Study on hole-transport-material-free planar $TiO_2/CH_3NH_3PbI_3$ heterojunction solar cells: the simplest configuration of a working perovskite solar cell, J. Mater. Chem. A 3 (28) (2015) 14902–14909.
[18] G.E. Eperon, V.M. Burlakov, P. Docampo, A. Goriely, H.J. Snaith, Morphological control for high performance, solution-processed planar heterojunction perovskite solar cells, Adv. Funct. Mater. 24 (1) (2014) 151–157.
[19] H. Hu, D. Wang, Y. Zhou, J. Zhang, S. Lv, S. Pang, et al., Vapour-based processing of hole-conductor-free $CH_3NH_3PbI_3$ perovskite/C_{60} fullerene planar solar cells, RSC Adv. 4 (55) (2014) 28964–28967.
[20] J. Shi, J. Dong, S. Lv, Y. Xu, L. Zhu, J. Xiao, et al., Hole-conductor-free perovskite organic lead iodide heterojunction thin-film solar cells: high efficiency and junction property, Appl. Phys. Lett. 104 (6) (2014). 063901 (1–4).
[21] M.M. Tavakoli, L. Gu, Y. Gao, C. Reckmeier, J. He, A.L. Rogach, et al., Fabrication of efficient planar perovskite solar cells using a one-step chemical vapor deposition method, Sci. Rep. 5 (2015). 14083 (1–9).
[22] S. Gamliel, A. Dymshits, S. Aharon, E. Terkieltaub, L. Etgar, Micrometer sized perovskite crystals in planar hole conductor free solar cells, J. Phys. Chem. C 119 (34) (2015) 19722–19728.
[23] W. Li, W. Zhang, S. Van Reenen, R.J. Sutton, J. Fan, A.A. Haghighirad, et al., Enhanced UV-light stability of planar heterojunction perovskite solar cells with caesium bromide interface modification, Energy Environ. Sci. 9 (2) (2016) 490–498.

[24] S. Ito, S. Tanaka, K. Manabe, H. Nishino, Effects of surface blocking layer of Sb2S3 on nanocrystalline TiO_2 for $CH_3NH_3PbI_3$ perovskite solar cells, J. Phys. Chem. C 118 (30) (2014) 16995–17000.

[25] M. Liu, M.B. Johnston, H.J. Snaith, Efficient planar heterojunction perovskite solar cells by vapour deposition, Nature 501 (7467) (2013) 395–398.

[26] B.J. Kim, D.H. Kim, Y.Y. Lee, H.W. Shin, G.S. Han, J.S. Hong, et al., Highly efficient and bending durable perovskite solar cells: toward a wearable power source, Energy Environ. Sci. 8 (3) (2015) 916–921.

[27] Y. Li, S. Ye, W. Sun, W. Yan, Y. Li, Z. Bian, et al., Hole-conductor-free planar perovskite solar cells with 16.0% efficiency, J. Mater. Chem. A 3 (36) (2015) 18389–18394.

[28] J. Burschka, N. Pellet, S.J. Moon, R. Humphry-Baker, P. Gao, M.K. Nazeeruddin, et al., Sequential deposition as a route to high-performance perovskite-sensitized solar cells, Nature 499 (7458) (2013) 316–319.

[29] N. Pellet, P. Gao, G. Gregori, T.Y. Yang, M.K. Nazeeruddin, J. Maier, et al., Mixed-organic-cation Perovskite photovoltaics for enhanced solar-light harvesting, Angew. Chem. Int. Ed. 53 (12) (2014) 3151–3157.

[30] I. Borriello, G. Cantele, D. Ninno, Ab initio investigation of hybrid organic–inorganic perovskites based on tin halides, Phys. Rev. B 77 (2008). 235214 (1–23).

[31] S. Aharon, A. Dymshits, A. Rotem, L. Etgar, Temperature dependence of hole conductor free formamidinium lead iodide perovskite based solar cells, J. Mater. Chem. A 3 (17) (2015) 9171–9178.

[32] S. Pang, H. Hu, J. Zhang, S. Lv, Y. Yu, F. Wei, et al., $NH_2CH = NH_2PbI_3$: an alternative organolead iodide perovskite sensitizer for mesoscopic solar cells, Chem. Mater. 26 (3) (2014) 1485–1491.

[33] B.E. Cohen, S. Gamliel, L. Etgar, Parameters influencing the deposition of methylammonium lead halide iodide in hole conductor free perovskite-based solar cells, APL Mater. 2 (8) (2014). 081502 (1–8).

[34] A. Usami, S. Seki, Y. Mita, H. Kobayashi, H. Miyashiro, N. Terada, Temperature dependence of open-circuit voltage in dye-sensitized solar cells, Solar Energy Mater. Solar Cells 93 (6) (2009) 840–842.

[35] M. Hu, L. Liu, A. Mei, Y. Yang, T. Liu, H. Han, Efficient hole-conductor-free, fully printable mesoscopic perovskite solar cells with a broad light harvester $NH_2CH = NH_2PbI_3$, J. Mater. Chem. A 2 (40) (2014) 17115–17121.

[36] F. Zhang, X. Yang, H. Wang, M. Cheng, J. Zhao, L. Sun, Structure engineering of hole–conductor free perovskite-based solar cells with low-temperature-processed commercial carbon paste as cathode, ACS Appl. Mater. Interfaces 6 (18) (2014) 16140–16146.

[37] H. Zhou, Y. Shi, Q. Dong, H. Zhang, Y. Xing, K. Wang, et al., Hole-conductor-free, metal-electrode-free $TiO_2/CH3_NH3_PbI_3$ heterojunction solar cells based on a low-temperature carbon electrode, J. Phys. Lett. 5 (18) (2014) 3241–3246.

[38] J.W. Lee, D.J. Seol, A.N. Cho, N.G. Park, High-efficiency perovskite solar cells based on the black polymorph of $HC(NH_2)_2PbI_3$, Adv. Mater. 26 (29) (2014) 4991–4998.

[39] J.H. Im, I.H. Jang, N. Pellet, M. Grätzel, N.G. Park, Growth of $CH_3NH_3PbI_3$ cuboids with controlled size for high-efficiency perovskite solar cells, Nat. Nanotechnol. 9 (11) (2014) 927–932.

[40] A. Mei, X. Li, L. Liu, Z. Ku, T. Liu, Y. Rong, et al., A hole-conductor–free, fully printable mesoscopic perovskite solar cell with high stability, Science 345 (6194) (2014) 295–298.

[41] Y. Yang, J. Xiao, H. Wei, L. Zhu, D. Li, Y. Luo, et al., An all-carbon counter electrode for highly efficient hole-conductor-free organo-metal perovskite solar cells, RSC Adv. 4 (95) (2014) 52825–52830.

[42] L. Zhang, T. Liu, L. Liu, M. Hu, Y. Yang, A. Mei, et al., The effect of carbon counter electrodes on fully printable mesoscopic perovskite solar cells, J. Mater. Chem. A 3 (17) (2015) 9165–9170.
[43] W.S. Yang, J.H. Noh, N.J. Jeon, Y.C. Kim, S. Ryu, J. Seo, et al., High-performance photovoltaic perovskite layers fabricated through intramolecular exchange, Science 348 (6240) (2015) 1234–1237.
[44] Z. Wei, H. Chen, K. Yan, X. Zheng, S. Yang, Hysteresis-free multi-walled carbon nanotube-based perovskite solar cells with a high fill factor, J. Mater. Chem. A 3 (48) (2015) 24226–24231.
[45] X. Zhao, N.G. Park, Stability issues on perovskite solar cells, Photonics 2 (2015) 1139–1151.
[46] T. Leijtens, G.E. Eperon, N.K. Noel, S.N. Habisreutinger, A. Petrozza, H.J. Snaith, Stability of metal halide perovskite solar cells, Adv. Energy Mater. 5 (20) (2015). 1500963 (1–23).
[47] X. Li, M. Tschumi, H. Han, S.S. Babkair, R.A. Alzubaydi, A.A. Ansari, et al., Outdoor performance and stability under elevated temperatures and long-term light soaking of triple-layer mesoporous perovskite photovoltaics, Energy Technol. 3 (6) (2015) 551–555.

FURTHER READING

M. Zhang, J.S. Yun, Q. Ma, J. Zheng, C.F.J. Lau, X. Deng, et al., High efficiency rubidium incorporated perovskite solar cells by gas quenching, ACS Energy Lett. 2 (2) (2017) 438–444.
Q. Ma, S. Huang, X. Wen, M.A. Green, A.W. Ho-Baillie, Hole transport layer free inorganic $CsPbIBr_2$ perovskite solar cell by dual source thermal evaporation, Adv. Energy Mater. 6 (2016). 1502202 (1–5).
W. Schottky, Simplified and advanced theory of boundary layer rectifiers, J. Phys. 118 (9–10) (1942) 539–592.

CHAPTER 10

Heterojunction Perovskite Solar Cells

Vidya Raj
IIT Bombay, Powai, Mumbai, India

Organometallic lead halide perovskite solar cells are emerging as a promising material for high performance photovoltaic devices. In 2009, the hybrid organic/inorganic halide perovskites were first utilized as sensitizers in dye-sensitized solar cells (DSSCs), which has given an efficiency of 3.8%. Within a short span of 6 years, the efficiency has reached 22.1% due to the excellent optoelectronic and mechanical properties of the perovskite material. The structural formula of perovskite is AMX_3, where A is an organic cation (ethylammonium, methylammonium, and formamidinium), M is a divalent metal cation (Pb^{2+} and Sn^{2+}) and X is a halide anion (Cl^-, I^-, and Br^-). The MX_6 octahedra forms a three-dimensional (3D) structure and the organic cation A fills the space between the octahedrals. The optical properties of the perovskite depend on the metal halide employed in the structure.

The highest efficiency of 3D perovskites is obtained from methylammonium triiodide ($MAPbI_3$) with a bandgap energy of 1.55 eV, close to the optimum bandgap required for photovoltaic conversion. The factors that can limit the performance of the photovoltaic device are charge recombination and the thickness of the absorber film. The thicker perovskite films can absorb more amount of light, but can reduce the external quantum efficiency and the charge transport. Due to the large absorption coefficient ($1.5 \times 10^4\ cm^{-1}$ at 550 nm) of the perovskite, the thickness of the absorber material can be reduced ($\sim 200-300$ nm). Since the binding energy of the photogenerated electrons and holes is low (~ 50 meV) in 3D perovskites, the charge separation takes place in the absorber layer itself. The diffusion length of the charge carriers in the absorber layer is ~ 100 nm.

The main two barriers for the large scale commercialization of perovskite solar cells are the toxicity due to lead and the environmental instability. The toxicity can be reduced by replacing Pb with Sn. However, Sn

Perovskite Photovoltaics
DOI: https://doi.org/10.1016/B978-0-12-812915-9.00010-1

can easily oxidize to Sn^{4+}, which creates a metal-like behavior in the semiconductor and reduces the photovoltaic performance. The thermal stability of the device can be improved by incorporating formamidinium cation in the structure. Another factor which degrades the photovoltaic performance of perovskite is humidity. The moisture instability is mainly due to the hygroscopic nature of the methylammonium cation. In presence, it easily decomposes to PbI_2 and degrades the optoelectronic properties of the perovskite. The moisture stability of perovskites can be improved by incorporating a hydrophobic and longer chain organic cation at "A" site, which forces it to form a two-dimensional (2D) layered structure.

High performance of the device depends on the quality of the perovskite absorber film. The factors which affect the film quality includes deposition technique, precursor stoichiometry, solvent engineering, and atmosphere. The crystallization behavior of the perovskite film affects the charge separation efficiency, recombination dynamics, and the diffusion lengths of the charge carriers. The large grain size can promote enhanced charge transport. The crystallization dynamics of planar structure is entirely distinct from that of mesoporous scaffold layers. In planar structures, the precursor surface energy is very low and hence the nucleation sites for the growth of the film are very less [1]. The crystallization process of the perovskite film includes nucleation and growth. Volmer-weber growth can occur due to the accelerated growth of the perovskite films, which leads to nonuniform films with lots of pinholes. Therefore, it is essential to control the nucleation and growth of the films in order to attain higher quality films.

Precursor stoichiometry, that is., the ratio of organic to inorganic components is another important factor that can affect the quality of film and charge transport properties. The molar ratios of 1:1 and 3:1 are largely used for making the perovskite precursor solution. Fig. 10.1 depicts the morphology of the perovskite films formed by 3:1 ratio of the MAI and PbI_2 in DMF. The molar ratio of 3:1 for the deposition of mixed halide perovskites $CH_3NH_3PbI_{3-x}Cl_x$ delivers larger grain size and better performance. Therefore for the complete formation of $MAPbI_3$ film, the concentration of methylammonium iodide (MAI) should be three times that of $PbCl_2$. If the precursor stoichiometry is changed from 3:1, either it will excess unreacted MAI or unreacted $PbCl_2$ in the film, which can degrade the optoelectronic performance of the device. The

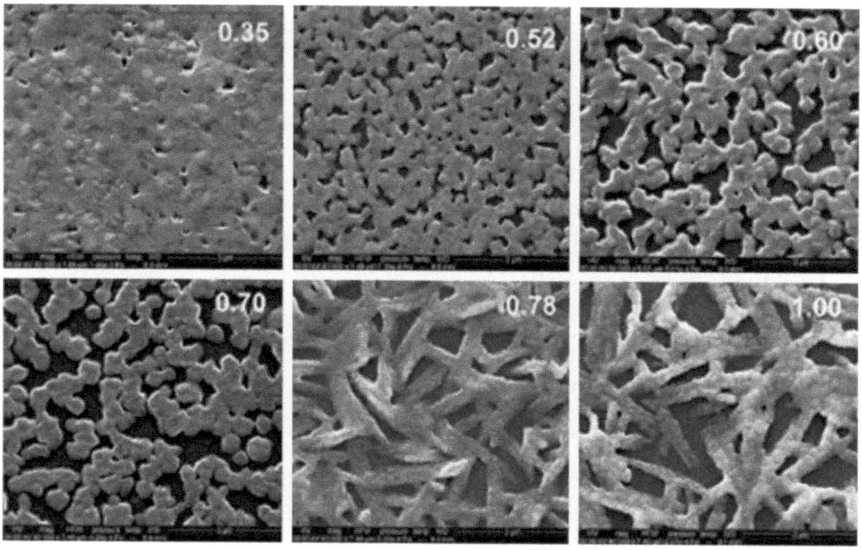

Figure 10.1 SEM images of the perovskite films formed from the nonstoichiometric ratio of PbI_2 to MAI in DMF solvent [3].

XRD patterns obtained for $MAPbI_{3-x}Cl_x$ for molar ratio 3:1 and $MAPbI_3$ for the molar ratio 1:1 are identical. It has been reported that an intermediate phase will form by the introduction of Cl, which enhances the film formation process in one-step process. The carrier lifetime for mixed halide perovskites $MAPbI_{3-x}Cl_x$ (100 ns) is much higher than $MAPbI_3$ (10 ns) [2].

Thermal treatment is required for accelerating the reaction between the organic and inorganic species. If the annealing temperature is very high ($>100°C$), then it will result in the nonuniformity of the perovskite film and can also cause the perovskite film to degrade (Fig. 10.2 (A)). The uniform surface coverage can enhance the light absorption ability of the film, which in turn increases short-circuit current (J_{sc}). Highly uniform and dense $MAPbI_{3-x}Cl_x$ can be obtained by annealing at 80°C for 3 hours or at 100°C for 45 minutes (Fig. 10.2 (B)). Annealing at very low temperatures ($<80°C$) also hinders the formation of perovskite phase. Increasing the annealing time at lower temperature does not improve the film quality. Therefore, it has been reported that the performance of planar perovskite devices can be improved by higher annealing temperature and shorter annealing time [4] (Fig. 10.2).

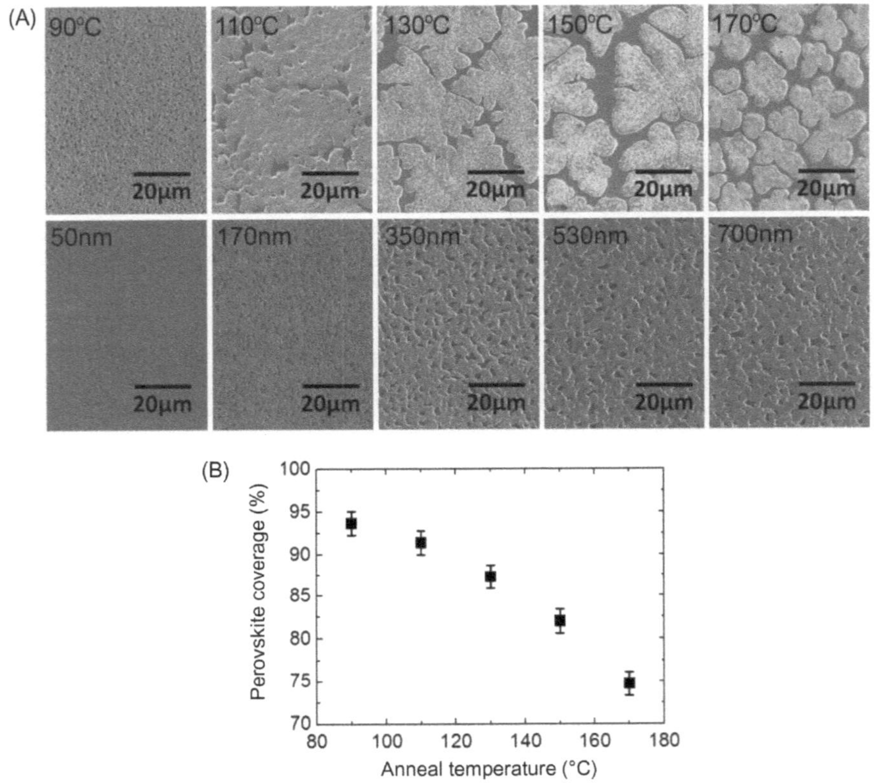

Figure 10.2 (A) SEM images of the one-step solution processed $MAPbI_{3-x}Cl_x$ perovskite films annealed at different temperatures and (B) dependence of surface coverage on annealing temperature [5,6].

10.1 FABRICATION TECHNIQUES OF PEROVSKITE SOLAR CELLS

The techniques for the fabrication of perovskite solar cells includes one-step spin-coating, two-step sequential coating, dip-coating and thermal evaporation (Fig. 10.3).

10.1.1 Single-Step Spin Coating Method

Solution processing is the most widely adopted method for the fabrication of perovskite films due to reduced cost and ease of fabrication. The perovskite precursor solution is deposited on the TiO_2 scaffold by a single-step spin coating. This method minimizes the processing steps; however, result in a nonuniform perovskite films with pinholes due to

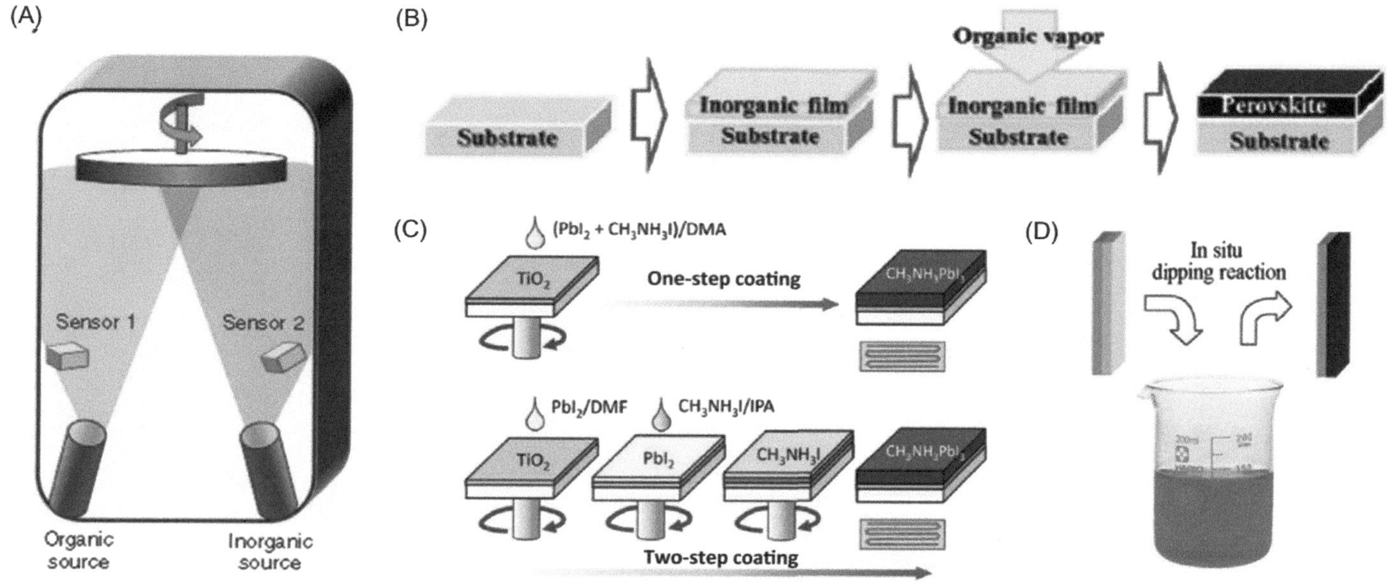

Figure 10.3 Fabrication of perovskite film with various methods (A) co-evaporation of organic and inorganic species, (B) vapor-assisted solution process,(C) one-step spin coating of precursor solution and the sequential deposition of PbI_2 and MAI, and (D) sequential deposition by dipping the PbI_2 film into MAI solution [8–11].

the slow crystallization. The crystallization speed is decreased due to the high boiling point of dimethylformamide (DMF) (153°C). The nonuniformity of the perovskite film is the main drawback of single-step spin coating, which can be overcome by taking the molar ratios of 1: 3 for PbI_2 and MAI [7].

In one-step method, there is a limitation for the efficiency and devices have a broader distribution in performance. This happens due to the uncontrollable growth and variations in the morphology of the grown crystal. These limitations can be overcome by sequential deposition method.

10.1.2 Fast Deposition Crystallization

Antisolvent engineering is used to control the morphology of the film in one-step method. In this method, the perovskite solution is deposited on the TiO_2 film and after a short delay time, an antisolvent is quickly dripped while spinning the precursor solution (Fig. 10.4). The antisolvent enhances the rapid nucleation and growth rate of the perovskite film. This leads to a uniform perovskite film with larger grain size [7]. The carrier lifetime in mixed halide perovskites (100 ns) is much higher than that of $MAPbI_3$ (10 ns).

The different antisolvents which have been tested are chlorobenzene, benzene, xylene, toluene, methanol, ethanol, ethylene glycol, 2-propanol, chloroform, THF, acetonitrile, and benzonitrile [13].

10.1.3 Sequential Deposition Method

This method was first suggested by Mitzi et al. [14]. Here, the PbI_2 film is first deposited over nanoporous TiO_2 by spin coating a solution of PbI_2 dissolved in N,N-DMF at 70°C and subsequently converted to perovskite by the reaction with MAI. The PbI_2 film can be either dipped in the

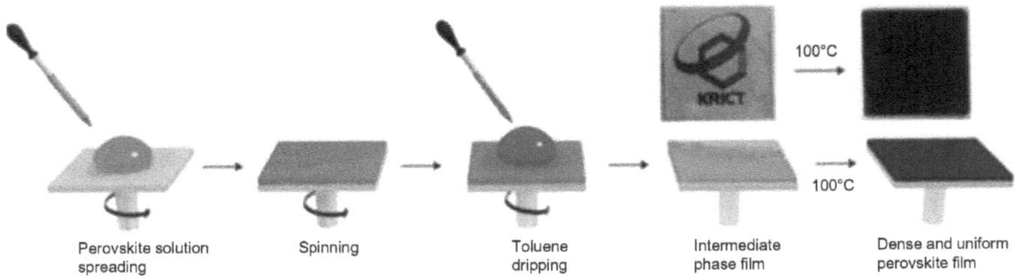

Figure 10.4 Schematic of solvent engineering method [12].

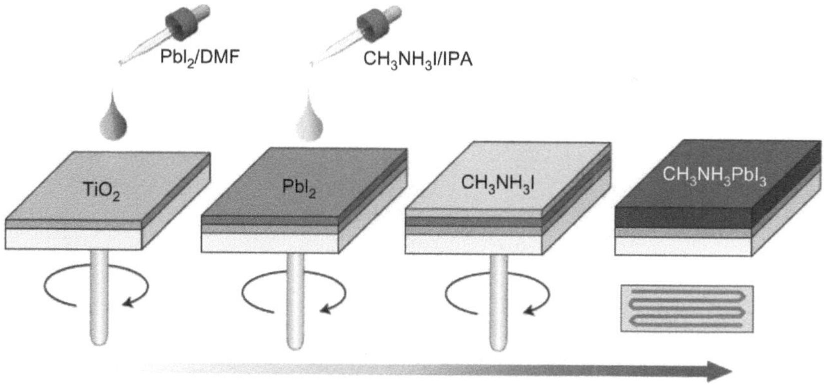

Figure 10.5 Sequential deposition of perovskite film [17].

MAI solution or the MAI solution is spin coated on PbI_2 film (Fig. 10.5). The crucial factors which determine the morphology of the film in the dip-coating process are the dipping time and solution concentration. Another important factor which influences the final morphology of the perovskite film is the substrate preheating process for the deposition of PbI_2. The infiltration of PbI_2 into the porous TiO_2 will be more for heated substrate compared to that of the substrate without heating [15]. The optimum preheating temperature is found to be 50°C. If the PbI_2 films are exposed to vapors of organic solvents like toluene or chlorobenzene, the grain dimension of the PbI_2 layer increases, which in turn increases the reaction sites for MAI [16].

The size of the nanocubic perovskites formed from this method mainly depends on the concentration of the MAI solution and temperature. The growth mechanism can be explained by the Gibbs free energy change [18]. With lesser concentration, larger cuboids and with higher concentration smaller cuboids are yielded. Large cuboids gives enhanced light harvesting and thus increases the J_{sc}. However, medium-sized cuboids (~200 nm) are giving highest open-circuit voltage (V_{oc}) due to higher charge mobility. Also, larger-sized cuboids exhibit lower photoluminescence (PL) intensity and a red shift in the PL peak due to the decrease in the bandgap energy. The solar cells made by this technique give an efficiency of 15% with high reproducibility [19].

10.1.4 Solution: Processed Adduct Method

Lewis acid-base adduct method can be used to yield high quality perovskite films with larger grain size [20]. The carrier mobility in the device

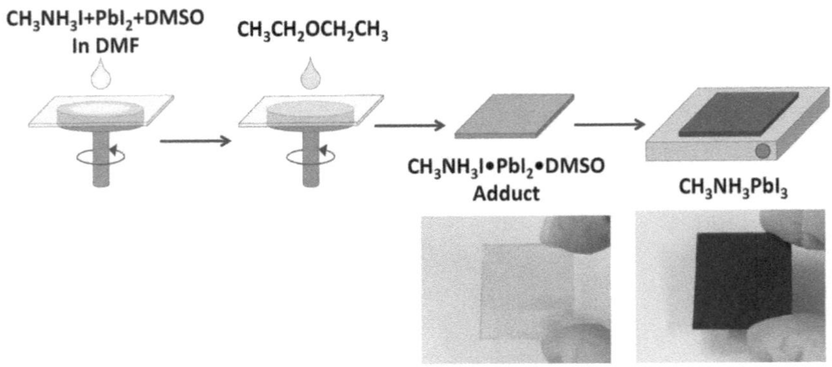

Figure 10.6 Schematic of the solution processed adduct method [20].

using perovskite formed from adduct method is 3.9×10^{-3} cm^2/Vs, which is one orders of magnitude greater than the carrier mobility in the perovskite films made by one-step method (3.2×10^{-4} cm^2/Vs) [21].

The fabrication of perovskite films by Lewis acid-base adduct method is shown in Fig. 10.6. The reaction between Lewis base [dimethyl sulfoxide (DMSO)] and Lewis acid (PbI_2) gives a transparent adduct film. The 1:1:1 molar solution of PbI_2, MAI, and DMSO in DMF solvent is spin coated on the TiO_2 scaffold. While spinning, diethyl ether is splashed for removing DMF and leads to the formation of a transparent 1:1:1 adduct film. By applying mild heat, DMSO can be removed and the adduct film is converted into perovskite [20]. The morphology of the resulted perovskite film is depicted in Fig. 10.7. The highest power conversion efficiency obtained from this method is 19.7%.

10.1.5 Thermal Evaporation

The first vacuum deposited perovskite film was reported by Snaith et al. [22]. Perovskite solar cells fabricated with the co-evaporation of organic and inorganic species yield an efficiency of 15.4% [22]. In this process, PbI_2 and MAI are deposited simultaneously using dual sources. In co-deposition process, it is difficult to monitor the deposition rate of MAI. Therefore, sequential method can be adopted in thermal evaporation where PbI_2 is first coated, followed by the deposition of MAI vapor [23]. The perovsite film deposition by means of thermal evaporation is shown in Fig. 10.8. The best photovoltaic performance was obtained when the substrate temperature was 75°C. It is argued that the film deposited by vacuum deposition is more uniform and flat.

Figure 10.7 SEM image of the perovskite film formed by toluene dripping [12].

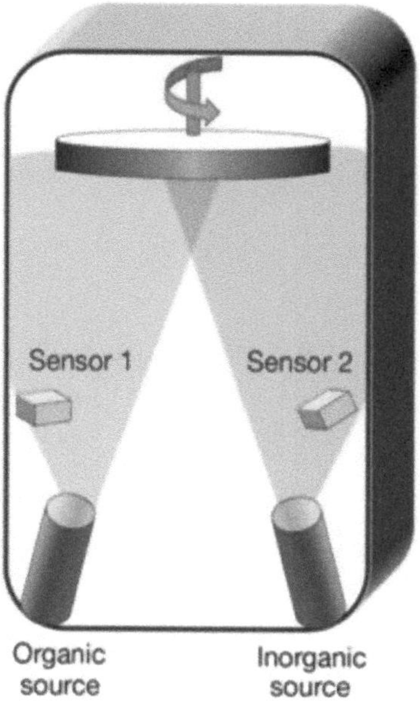

Figure 10.8 Dual-source thermal evaporation [22].

10.1.6 Vapor-Assisted Solution Process Deposition

In vapor-assisted solution process, the PbI_2 film deposited by spin coating is treated with MAI vapor. This method results in the enlargement of the grain size and reduces the surface roughness of the film. The solution processed PbI_2 film is annealed in MAI vapor at 150°C in N_2 atmosphere for 2–4 hours to get a homogeneous perovskite film with larger grain size [24] (Fig. 10.9).

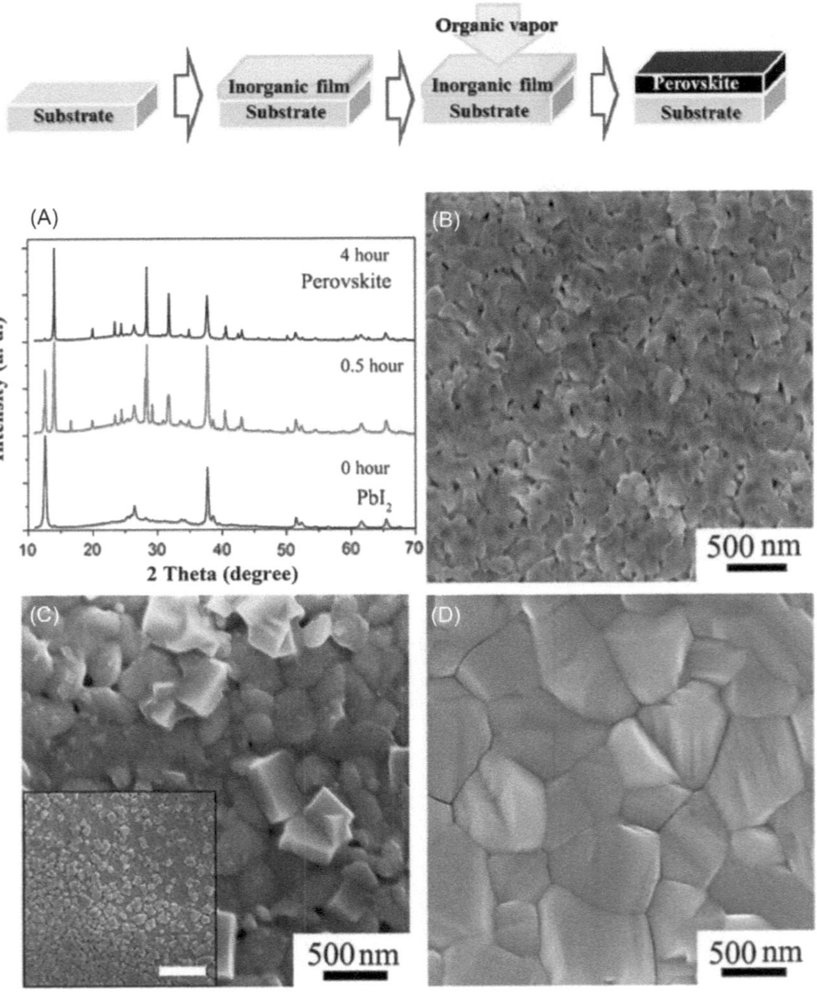

Figure 10.9 (Top) Schematic of vapor-assisted solution process (A) XRD of the perovskite films, (B) SEM image of the film at t = 0 h, (C) SEM image of the film at t = 0.5 h, and (D) SEM image of the film at t = 4 h.

10.2 SOLAR CELL DEVICE ARCHITECTURES

The energy level diagram of different materials that can be used as electron transport membranes (ETMs) and hole transport membranes (HTMs) are depicted in Fig. 10.10.

10.2.1 Electron Transport Layers

The most widely used electron transport layers (ETL) in perovskite solar cells are mesoporous or planar TiO_2 and insulating alumina (Al_2O_3). The

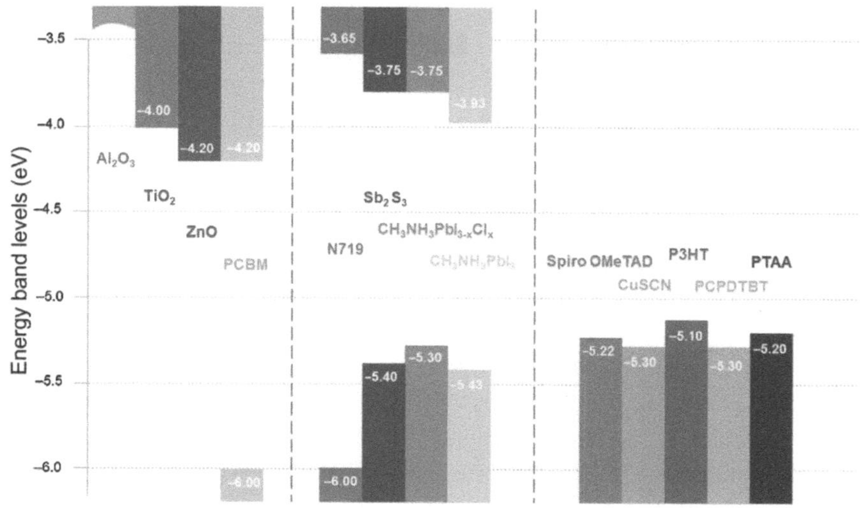

Figure 10.10 Energy levels for various materials that can be used for ETM, absorber, and HTM in solar cells [27].

planar titania layer can allow the electrons to flow and block the transport of holes. The efficiency is improved to 12% by the infiltration of the pores of mesoporous titania [25]. The other ETL that can be used are ZnO, SnO_2, and $SrTiO_3$. ZnO was tried as ETL instead of TiO_2 due to its commensurable energy levels and enhanced transport of electrons. However, the probability of recombination is much higher at the interfaces of perovskite and ZnO layer, which can lower the efficiency of the device. The devices with mesoporous ZnO can achieve efficiency over 12%. The doping of ZnO with Al can increase the charge transport, which in turn can improve the V_{oc} and fill factor (FF).

10.2.2 Hole Transport Layers

The photogenerated holes are injected into the hole transport layer (HTL), which is mostly spiro-OMeTAD [2,2′,7,7′ −tetrakis- (N,N-di-p-methoxyphenylamine) 9,9′ −spirobifluorene]. The doping of spiro-OMeTAD with tris [2-(1H- pyrazol-1-yl) pyridine] cobalt (III) can increase the conductivity, which in turn increases the FF (0.76) [26]. The other hole transporting materials that have been tested are poly-(3-hexylthiophene) (P3HT) and poly {2,6- [4,4-bis- (2-ethylhexyl)-4H- cyclopenta (2,1-b; 3,4-b′)- dithiophene]- alt- 4,7- (2,1,3-benzothiadiazole)} (PCPDTBT). Since the intrusion of the long-chained polymers into the pores of the perovskite is difficult, the performance of the photovoltaic device with polymeric HTMs is inferior to that with spiro-OMeTAD HTMs. The p-type NiO which acts as the hole transport material can improve the hole extraction and leads to an efficiency of 11.4%.

10.3 TYPES OF PEROVSKITE SOLAR CELLS

The cross-sectional SEM micrograph of a perovskite solar cell is given in Fig. 10.11. Regardless of the scaffolding, if the thickness of the perovskite absorber layer is approximately 350 nm, it can absorb sufficient amount of light [28]. However, the thickness of the TiO_2 scaffold can affect the hysteresis behavior [12]. The J_{sc} depends on the morphology of the scaffold layer that affects the loading of the perovskite layer and the scattering of the light.

Based on the deposition technniques, there are three major categories of perovskite solar cells-sensitized, mesoporous and planar heterojunction (Fig. 10.12).

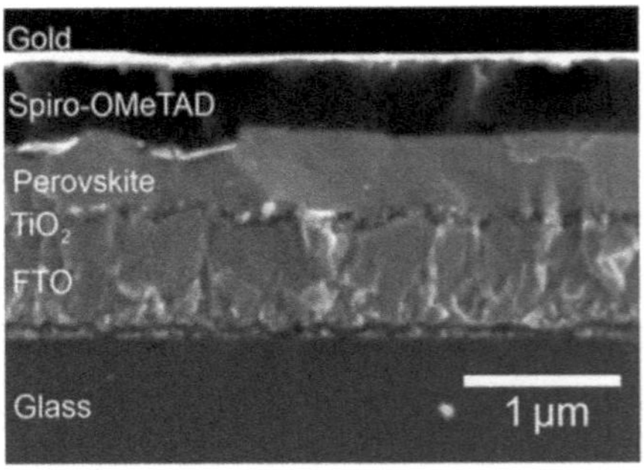

Figure 10.11 Cross-sectional SEM image of a perovskite solar cell [27].

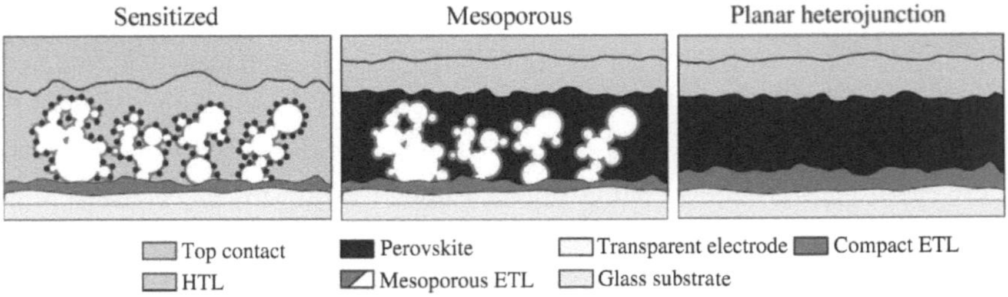

Figure 10.12 The three major categories of perovskite solar cells based on the deposition technique [31].

10.3.1 Mesoscopic Sensitized Solar Cells

In 2009, perovskites were first utilized in DSSCs by Kojima et al. [29], where the perovskite nanoparticles were deposited onto 2 μm thick TiO_2 scaffold. An efficiency of 3.8% is obtained from this configuration. However, there was a problem of poor stability due to the rapid disintegration of perovskite material in the liquid redox electrolyte. Finally, the instability was resolved by replacing the liquid redox electrolyte by a solid-state electrolyte. The first solid-state perovskite solar cell was developed in 2012 by Kim et al. [30], where spiro-OMeTAD was employed as the HTM and meso-TiO_2 ($\sim$600 nm) was used as the ETM. This device architecture promoted the efficiency to 9.7% and this enhanced performance is due to the high absorption coefficient of perovskite material, which is almost 10 times higher than that of the organometallic dye.

10.3.2 Mesoporous TiO_2 Based Perovskite Solar Cells

In this architecture, compact TiO_2 electron transport layer ($\sim$80 nm) is coated on the FTO (flourine doped tin oxide) substrate, followed by the deposition of mesoporous TiO_2 scaffold ($\sim$350 nm). Then, the perovskite absorber layer infiltrates the mesoporous TiO_2 and spiro-OMeTAD ($\sim$200 nm) is deposited as the hole transport layer. At last, gold contact ($\sim$100 nm) is made on top of it [9]. The performance of the device can be upgraded by improving the morphology of the absorber layer and the charge transport through the scaffold layer.

The use of meso-TiO_2 can reduce the hysteresis. The photogenerated electrons are much more easily extracted by using porous TiO_2. The diffusion coefficient of meso- TiO_2 (10^{-4} cm^2/s) is much less than that of $MAPbI_3$ ($>10^{-2}$ cm^2/s) [2]. Thus, it reduces the electron transport speed and it turn balances the charge carrier transport in the device. In the planar configuration, there is an imbalance in the charge transport, that is, one type of charge is accumulated, which can lead to the distortion of the structure of perovskite at the interface and thus increases the hysteresis [32].

10.3.3 Meso-Superstructured Solar Cells

In meso-superstructured architecture, TiO_2 scaffold is replaced by insulating mesoporous alumina and gives a power conversion efficiency of 10.9% [7]. The problem of instability of TiO_2 under UV illumination can be solved by replacing it with alumina. There will not be any electron flow from the perovskite absorber to Al_2O_3 since the conduction band minima of alumina is at a higher position than the conduction band minima of the perovskite. Therefore, the injection of electrons from the absorber layer to alumina is not possible. This indicates that the perovskite not only absorb photons, but also transport charges through it. Thus, it is determined that perovskite is ambipolar in nature. The charge transport is much faster in the case of alumia-based devices compared to that of titania-based devices. The V_{oc} obtained from the Al_2O_3 scaffold is 200 mV higher than obtained from the TiO_2 ETM. In devices encorporating alumina, the surface states at the interface are very less, which lead to higher carrier lifetime of electrons and less loss in the V_{oc}.

10.3.4 HTM-Free Mesoscopic P-N Solar Cells

$MAPbI_3$) is slightly more p-type than n-type. Thus, it can act as both light absorber and hole transporter. It is reported that spiro-OMeTAD which is currently used the HTL is not necessary and can be omitted from the device as this is the most expensive component. This device architecture gives an efficiency of 5.5%. However, the devices without HTM has a poor FF and low V_{oc} [33–40].

10.3.5 Planar Pin Heterojunction Solar Cells

With planar configurations, it is easy to construct multijunction, easy to investigate the underlying physics and there is possibility of mass production. Based on the processing sequence, the planar configurations can be categorized into n-i-p and p-i-n. The n-i-p is the regular structure that follows glass/FTO/ETL/Perovskite/HTL/metal and p-i-n is the inverted with configuration glass/ FTO/HTL/perovskite/ETL/metal [31].

10.3.6 Inverted Planar Structure

In inverted planar configuration, the materials that can be used for hole transport are NiO, WO_3, MoO_3, V_2O_5, and PEDOT:PSS [41]. PEDOT: PSS was used as the hole transport material in the first inverted planar perovskite solar cell. The highest reported efficiency for this material is 15.3% [42]. The inverted structures are hysteresis free and the FF is >0.8. The high FF is contributed by the lower charge transport resistance at the interfaces.

10.4 SUMMARY

The low cost and high efficiency are the two factors that are making the perovskite photovoltaics more attractive compared with today's commercialized photovoltaic systems. Both the regular and inverted configurations have already demonstrated high performance, which is mainly due to the ambipolar and balanced charge transport between the ETMs and HTMs. In order to commercialize perovskite photovoltaics, some of the barriers such as stability issues and toxicity due to the presence of lead must be resolved. This can be achieved by the thorough understanding of the photovoltaic principles and the degradation mechanisms, development of new materials, optimization of device architecture, and associated manufacturing techniques. The semitransparent and colorful windows and flexible modules are the short-term goals of the perovskite research.

REFERENCES

[1] V.M. Burlakov, G.E. Eperon, H.J. Snaith, S.J. Chapman, A. Goriely, Controlling coverage of solution cast materials with unfavourable surface interactions, Appl Phys Lett 104 (2014) 091602.

[2] S.D. Stranks, G.E. Eperon, G. Grancini, C. Menelaou, M.J.P. Alcocer, T. Leijtens, et al., Electron-hole diffusion lengths exceeding 1 micrometer in an organometal trihalide perovskite absorber, Science 342 (2013) 341–344.

[3] Z. Liu, M. Zhang, X. Xu, L. Bu, W. Zhang, W. Li, et al., p-Type mesoscopic NiO as an active interfacial layer for carbon counter electrode based perovskite solar cells, Dalton Trans 44 (2014) 3967–3973.

[4] M. Saliba, K.W. Tan, H. Sai, D.T. Moore, T. Scott, W. Zhang, et al., Influence of thermal processing protocol upon the crystallization and photovoltaic performance of organic–inorganic lead trihalide perovskites, J Phys Chem 118 (2014) 17171–17177.

[5] G.E. Eperon, V.M. Burlakov, P. Docampo, A. Goriely, H.J. Snaith, Morphological control for high performance, solution-processed planar heterojunction perovskite solar cells, Adv Funct Mater 24 (2014) 151–157.

[6] A. Dualeh, N. Tétreault, T. Moehl, P. Gao, M.K. Nazeeruddin, M. Grätzel, Effect of annealing temperature on film morphology of organic–inorganic hybrid pervoskite solid-state solar cells, Adv Funct Mater 24 (2014) 3250–3258.

[7] M.M. Lee, J. Teuscher, T. Miyasaka, T.N. Murakami, H.J. Snaith, Efficient hybrid solar cells based on meso-superstructured organometal halide perovskites, Science 338 (2012) 643.

[8] Q. Chen, H. Zhou, Z. Hong, S. Luo, H.S. Duan, H.H. Wang, et al., Planar heterojunction perovskite solar cells via vapor-assisted solution process, J Am Chem Soc 136 (2013) 622–625.

[9] J. Burschka, N. Pellet, S.-J. Moon, R. Humphry-Baker, P. Gao, M.K. Nazeeruddin, et al., Sequential deposition as a route to high-performance perovskite-sensitized solar cells, Nature 499 (2013) 316–319.

[10] P. Umari, E. Mosconi, F. De Angelis, Relativistic GW calculations on $CH_3NH_3PbI_3$ and $CH_3NH_3SnI_3$ perovskites for solar cell applications, Sci Rep 4 (2014) 4467.

[11] J.-H. Im, H.-S. Kim, N.-G. Park, Morphology-photovoltaic property correlation in perovskite solar cells: one-step versus two-step deposition of $CH_3NH_3PbI_3$, APL Mater 2 (2014) 081510.
[12] N.J. Jeon, J.H. Noh, Y.C. Kim, W.S. Yang, S. Ryu, S.I. Seok, Solvent engineering for high-performance inorganic−organic hybrid perovskite solar cells, Nat. Mater. 13 (2014) 897−903.
[13] N.-G. Park, Methodologies for high efficiency perovskite solar cells, Nano Convergence 3(1) (2016) 15.
[14] K. Liang, D.B. Mitzi, M.T. Prikas, Synthesis and characterization of organic − inorganic perovskite thin films prepared using a versatile two-step dipping technique, Chem. Mater. 10 (1998) 403−411.
[15] H.S. Ko, J.W. Lee, N.G. Park, 15.76% efficiency perovskite solar cells prepared under., J Mater Chem 3 (2015) 8808−8815.
[16] M.I. El-Henawey, R.S. Gebhardt, M.M. El-Tonsy, S. Chaudhary, Organic solvent vapor treatment of lead iodide layers in the two-step sequential deposition of $CH_3NH_3PbI_3$-based perovskite solar cells, J Mater Chem 4 (2016) 1947−1952.
[17] J.H. Im, I.H. Jang, N. Pellet, M. Grätzel, N.G. Park, Growth of CH3NH3PbI3 cuboids with controlled size for high-efficiency perovskite solar cells, Nat. Nanotechno 9 (2014) 927−932.
[18] N. Ahn, S.M. Kang, J.W. Lee, M. Choi, N.G. Park, Thermodynamic regulation of $CH_3NH_3PbI_3$ crystal growth and its effect on photovoltaic performance of perovskite solar cells, J Mater Chem 3 (2015) 19901−19906.
[19] J. Burschka, N. Pellet, S.J. Moon, R. Humphry-Baker, P. Gao, M.K. Nazee ruddin, et al., Sequential deposition as a route to high-performance perovskite-sensitized solar cells, Nature 499 (2013) 316.
[20] N. Ahn, D.Y. Son, I.H. Jang, S.M. Kang, M. Choi, N.G. Park, Highly reproducible perovskite solar cells with average efficiency of 18.3% and best efficiency of 19.7% fabricated via Lewis base adduct of lead(II) iodide, J Am Chem Soc 137 (2015) 8696−8699.
[21] Y. Chen, J. Peng, D. Su, X. Chen, Z. Liang, Efficient and balanced charge transport revealed in planar perovskite solar cells, A.C.S. Appl Mater Inter 7 (2015) 4471−4475.
[22] M. Liu, M.B. Johnston, H.J. Snaith, Efficient planar heterojunction perovskite solar cells by vapour deposition, Nature 501 (2013) 395−398.
[23] W. Chen, H.W. Kang, S.Y. Hsiao, P.F. Yang, K.M. Chiang, H.W. Lin, Efficient and uniform planar-type perovskite solar cells by simple sequential vacuum deposition, Adv Mater 26 (2014) 6647−6652.
[24] J. Emara, T. Schnier, N. Pourdavoud, T. Riedl, K. Meerholz, S. Olthof, Impact of film stoichiometry on the ionization energy and electronic structure of CH_3 NH_3 PbI_3 perovskites, Adv Mater 28 (2016) 553−559.
[25] J.H. Heo, S.H. Im, J.H. Noh, T.N. Mandal, C.S. Lim, J.A. Chang, et al., Efficient inorganic−organic hybrid heterojunction solar cells containing perovskite compound and polymeric hole conductors, Nature Photonics 7 (2013) 486.
[26] J. Burschka, A. Dualeh, F. Kessler, E. Baranoff, N.-L. Cevey-Ha, C. Yi, et al., Tris (2-(1H-pyrazol-1- yl)pyridine)cobalt(III) as p-type dopant for organic semiconductors and its application in highly efficient solid-state dye-sensitized solar cells, J Am Chem Soc 133 (2011) 18042−18045.
[27] P.P. Boix, K. Nonomura, N. Mathews, S.G. Mhaisalkar, Current progress and future perspectives for organic/inorganic perovskite solar cells, Mater Today 17 (1) (Jan/Feb. 2014).
[28] N.-G. Park, Organometal perovskite light absorbers toward a 20% efficiency low-cost solid-state mesoscopic solar cell, J Phys Chem Lett 4 (2013) 2423−2429.

[29] A. Kojima, K. Teshima, Y. Shirai, T. Miyasaka, Organometal halide perovskites as visible-light sensitizers for photovoltaic cells, J Am Chem Soc 131 (2009) 6050−6051.
[30] H.S. Kim, C.R. Lee, J.H. Im, K.B. Lee, T. Moehl, A. Marchioro, et al., Lead iodide perovskite sensitized all-solid-state submicron thin film mesoscopic solar cell with efficiency exceeding 9%, Sci. Rep 2 (2012) 591.
[31] Q. Chen, N. De Marco, Y.M. Yang, T.-B. Song, C.-C. Chen, H. Zhao, et al., Under the spotlight: the organic—inorganic hybrid halide perovskite for optoelectronic applications, Nano Today 10 (2015) 355−396.
[32] B. Wu, K. Fu, N. Yantara, G. Xing, S. Sun, T. Sum, et al., Charge accumulation and hysteresis in perovskite-based solar cells: An electro-optical analysis, Adv Energy Mater 5 (2015) 1500829.
[33] S. Aharon, S. Gamliel, B.E. Cohen, L. Etgar, Depletion region effect of highly efficient hole conductor free CH3NH3PbI3 perovskite solar cells, Phys Chem Chem Phys 16 (2014) 10512−10518.
[34] S. Aharon, B.E. Cohen, L. Etgar, Hybrid lead halide iodide and lead halide bromide in efficient hole conductor free perovskite solar cell, J Phys Chem 118 (2014) 17160−17165.
[35] W.A. Laban, L. Etgar, Depleted hole conductor-free lead halide iodide heterojunction solar cells, Energy Environ Sci 6 (2013) 3249−3253.
[36] B.-E. Cohen, S. Gamliel, L. Etgar, Parameters influencing the deposition of methylammonium lead halide iodide in hole conductor free perovskite-based solar cells, APL Mater. 2 (2014) 081502.
[37] J. Shi, Y. Luo, H. Wei, J. Luo, J. Dong, S. Lv, et al., Modified two-step deposition method for high-efficiency $TiO_2/CH_3NH_3PbI_3$ heterojunction solar cells, ACS Appl Mater Inter 6 (2014) 9711−9718.
[38] Y. Xu, J. Shi, S. Lv, L. Zhu, J. Dong, H. Wu, et al., Simple way to engineer metal−semiconductor interface for enhanced performance of perovskite organic lead iodide solar cells, ACS Appl Mater Inter 6 (2014) 5651−5656.
[39] J. Shi, J. Dong, S. Lv, Y. Xu, L. Zhu, J. Xiao, et al., Hole-conductor-free perovskite organic lead iodide heterojunction thin-film solar cells: high efficiency and junction property, Appl Phys Lett 104 (2014) 063901.
[40] F. Zhang, X. Yang, H. Wang, M. Cheng, J. Zhao, L. Sun, Structure engineering of hole−conductor free perovskite-based solar cells with low-temperature-processed commercial carbon paste as cathode, ACS Appl Mater Inter 6 (2014) 16140−16146.
[41] F. Rui, Z. WenJun, Z. ShaSha, C. Wei, The rising star in photovoltaics-perovskite solar cells: the past, present and future, Sci China 59 (2016) 989−1006.
[42] C.G. Wu, C.H. Chiang, Z.L. Tseng, Md.K. Nazeeruddin, A. Hagfeldt, M. Grätzel, High efficiency stable inverted perovskite solar cells without current hysteresis, Energy Environ Sci 8 (2015) 2725−2733.

CHAPTER 11

Flexible Perovskite Solar Cells

Suresh Maniarasu, Vishesh Manjunath, Ganapathy Veerappan and Easwaramoorthi Ramasamy
International Advanced Research Centre for Powder, Metallurgy and New Materials (ARCI), Hyderabad, Telangana, India

11.1 INTRODUCTION

Terawatt renewable energy challenges could be economically realized when the photovoltaic solar cells are fabricated using high-throughput low-cost manufacturing processes. Light-harvesting layer, considered as the basis of any given photovoltaic technology, plays a critical role in determining the appropriate production method [1]. For example, conventional wafer-based silicon solar cells require energy-intensive vacuum method and therefore usually prepared in a batch process. Second generation thin-film solar cells utilize direct band gap micron thick absorber (CdTe, CIGS) layer for efficient solar to electric energy conversion. The use of thin absorber layer facilitates the adoption of high-throughput device fabrication on a range of substrates. For example, Chirila et al. fabricated CIGS solar cell on flexible polymer film and shown 20.4% power conversion efficiency (PCE) [2]. However, the need of high-vacuum and expensive precursor materials limited the commercial viability of thin-film solar cells. Third-generation solar cells including organic photovoltaics (OPVs) and dye-sensitized solar cells (DSSCs) provided a promising window for low-temperature and non-vacuum processability. Flexible OPV and DSSC modules with moderate PCEs were already fabricated in several research laboratories, and their modern applications were successfully demonstrated [3,4]. Despite continuous efforts on dye and semiconductor polymer development, PCE of third-generation solar cell technologies had stagnated around 12% and need to be improved for competing with existing technologies.

Organometal halide perovskites recently received notable attention from photovoltaic community due to their high light absorption coefficient, fast carrier transport, and low-temperature solution processability [5]. A state-of-the-art perovskite solar cell (PSC) fabricated with a

Perovskite Photovoltaics
DOI: https://doi.org/10.1016/B978-0-12-812915-9.00011-3

configuration of fluorine doped tin oxide (FTO) glass/comp-TiO_2/meso-TiO_2/$MAPbI_3$/spiro-OMeTAD/Au exhibits about 22.7% PCE under one sun conditions. The high PCEs of organometal trihalide perovskite—based solar cells are attributed to ambipolar charge carrier transport and long range electron—hole diffusion lengths. Organometal trihalide perovskite absorbers are initially synthesized by wet chemical methods. Freshly synthesized methylammonium iodide (MAI) and PbI_2 salts are mixed in a common solvent such as γ-butyrolactone or dimethylformamide and then deposited over the substrate followed by the heat treatment to evaporate the solvent. As the nucleation and crystal growth of $MAPbI_3$ takes place in the bulk of solution, it is difficult to obtain a uniform and pinhole-free perovskite absorber film in this approach. To solve this issue, a two-step sequential deposition method was being developed, in which first the PbI_2 film was deposited on a desired substrate and then exposed to MAI solution. In this method, perovskite crystals are formed by interdiffusion of precursor elements in PbI_2 matrix whose kinetics control the perovskite crystallization and film morphology. Being one-step or two-step solution methods, polycrystalline perovskite film with excellent requisite properties over large area substrate could be fabricated below 100°C [6], which makes this wonder material ideal choice for next-generation low-cost high-performance flexible solar cells.

Flexible PSCs were fabricated on either polymer or thin metallic substrates, and very recently ultrathin glasses also entered into the fray. Fig. 11.1 shows the schematic cross-sectional diagram of typical flexible PSCs. Organometal halide perovskite absorber film with suitable charge extraction layers in sandwich configuration is fabricated on a flexible substrate. The notable difference between the flexible and glass

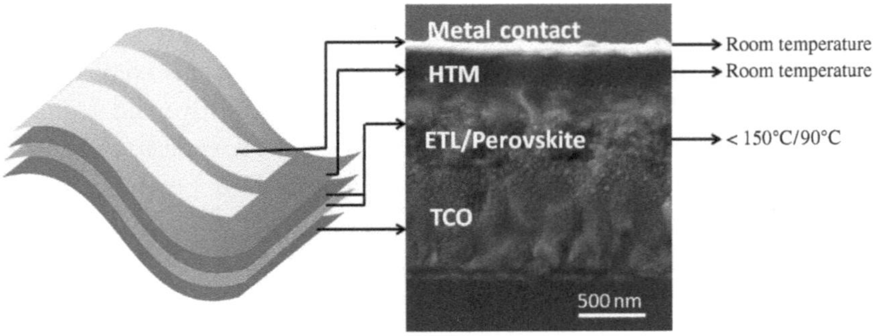

Figure 11.1 Schematic cross-sectional diagram of flexible PSC. Various active layers with required typical processing temperatures also provided for clarity.

substrate–based PCS is the selection and deposition methods of electron acceptor/hole barrier layers. Compact and mesoporous TiO_2 layers, employed as hole barrier and electron extracting layer in glass substrate–based high-efficiency PSCs, require high-temperature (>400°C) annealing treatment. However, flexible substrates are incompatible with such high-temperature treatment and hence a range of new materials (ZnO, Zn_2SnO_4, PCBM) and alternative methods (photonic curing, chemical sintering, UV irradiation) were developed. As hole extraction layers of conventional PSCs are fabricated using small-molecule organic polymers or inorganic *p*-type semiconductors at room temperature, the same materials and methods can be adopted for flexible PSCs. Finally, a metal cathode is deposited by thermal evaporation to complete the device fabrication process.

In this chapter, we review in detail the progress made from the material development, thin-film deposition, device architecture, and long-term stability pertinent to flexible perovskite solar cells. Section 11.2 provides a comprehensive overview of various device architectures adopted for the fabrication of high-efficiency and stable flexible PSCs. High-throughput production methods of flexible PSCs are elaborated in Section 11.3. Section 11.4 emphasizes on stability issues and reliability aspects of flexible PSCs, while Section 11.5 will present a roadmap for potential modern applications of flexible PSCs.

11.2 DEVICE ARCHITECTURE

PSCs benefit from the decade long research expertise built on DSSCs. Organometal halide perovskites indeed made their debut in traditional liquid electrolyte DSSCs by replacing a ruthenium sensitizer [7]. Despite having 3.8% initial PCE, devices suffered from poor stability due to the dissolution of methylammonium lead iodide ($MAPbI_3$) in redox electrolyte. Park et al. adopted a solid-state DSSC approach to address the dissolution of $MAPbI_3$ in which corrosive iodine/tri-iodide redox couple electrolyte is replaced with a small-molecule organic polymer 2,2′,7,7′-tetrakis-(N,N-di-4-methoxyphenylamino)-9,9′-spirobifluorene (abbreviated as spiro-OMeTAD) [8]. The evolution of PSC device architecture was originated from the typical solid-state DSSC, where the compact TiO_2 and mesoporous TiO_2 layers were employed as hole blocking and electron accepting layers, respectively. However, mesoporous TiO_2 needs higher annealing temperature, which makes the device fabrication expensive as well as increasing the processing time. It has been reported that organometal trihalide perovskite has long charge carrier diffusion length and ambipolar property, indicating that perovskite itself can transport the electrons and holes to the respective electrodes. All of these results indicated that the efficient PSCs could be prepared without mesoporous TiO_2 layer and planar PSC was developed. The simple planar PSC can be further divided into regular *n-i-p* and inverted *p-i-n* structure based on the order of carrier selective contacts. The inverted *p-i-n* structure provides a number of processing advantages over regular *n-i-p* configuration and stands as a front runner for flexible PSC technology. This section provides a comprehensive overview on three major device architectures (Fig. 11.2) adopted for the realization of high-efficiency PSCs with more emphasize on their relevance to flexible PV technology.

11.2.1 Mesoporous Flexible PSCs

In general, PSCs with mesoporous TiO_2 or Al_2O_3 scaffold on compact TiO_2 layer–coated FTO glass substrate have shown higher PCE and operational stability. Such scaffolds are typically prepared by solution approach followed by high-temperature treatment to remove the binders and improve the interconnectivity between the particles. In PSC, the development of thin hole blocking compact layer is indispensable, which reduces the carrier recombination at the interface between transparent conducting oxide (TCO) and perovskite layer. So far, the compact layer was deposited by spray pyrolysis, sputtering, electrodeposition, and spin coating of nanoparticle (NP) dispersion. However, the same approach is not applicable to flexible polymer substrate due to the temperature sensitivity. To overcome with the high-temperature annealing, Brown et al. proposed a UV irradiation technique for developing mesoporous TiO_2

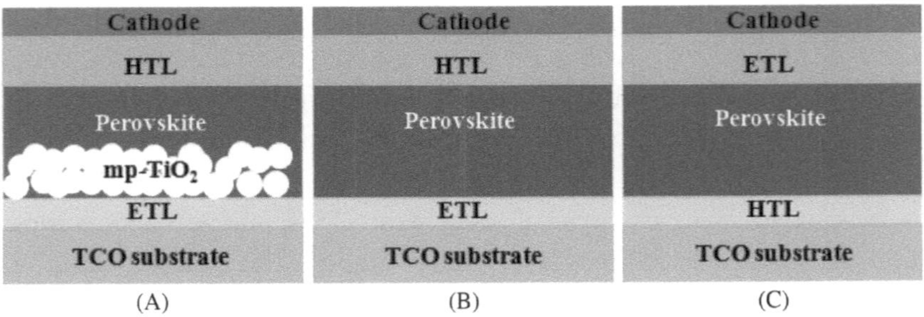

Figure 11.2 Evolution of PSC architecture: (A) mesoporous, (B) planar (*n-i-p*), and (C) inverted planar (*p-i-n*) configuration.

scaffold on plastic substrates [9]. Plasma-assisted atomic layer deposition (ALD) technique was introduced to make an effective compact layer with lower pinhole density on temperature sensitive indium tin oxide (ITO)–coated plastic substrates. Very low reverse current density of 0.004 $mA \cdot cm^{-2}$ at −1 V applied bias in the dark indicates efficient hole blocking at ALD compact TiO_2/perovskite interface. The mesoporous TiO_2 scaffold was deposited on aforementioned substrate by spin coating or screen printing method and subjected to an UV irradiation for 1 hour. Absorber layer was fabricated by spin coating $CH_3NH_3PbI_{3-x}Cl_x$ precursor in a nitrogen atmosphere and annealing at 95°C. Fig. 11.2 shows the typical schematic of mesoporous flexible PSC with required processing temperature for various active layers. Laboratory-scale flexible PSC with an active area of 0.12 cm^2 and spin-coated mesoporous TiO_2 scaffold showed 8.4% PCE, whereas similar devices with screen printed scaffold showed 4.4%. The UV irradiation effectively removed the organic binders and also promoted interparticle bonding without much affecting the underneath plastic substrate. These successful laboratory-scale devices were scaled up over the large area module that consists of four series connected cells on a 5.6 × 5.6 cm PET/ITO substrate. A maximum PCE of 3.1% with excellent mechanical stability was demonstrated in 7.92 cm^2 active area flexible PSC module (Fig. 11.3).

There are some limitations associated with the use of plastic substrates for flexible PSC applications. For example, crack formation in TCO coating during bending cycles drastically reduces the device performance.

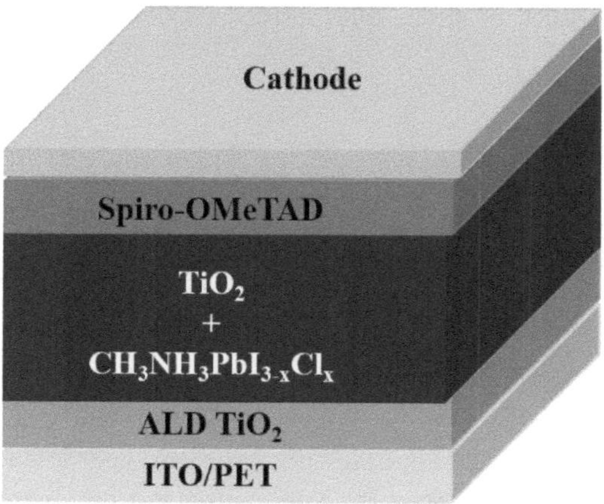

Figure 11.3 Schematic of mesoporous flexible PSC with ALD compact-TiO_2 and UV-irradiated mesoporous TiO_2 scaffold layer.

Thin metal substrates could provide an alternative solution to aforementioned issues due to their high inherent conductivity and mechanical stability. As most of the metal foils are opaque to light, they cannot enter into the active layer through front side illumination device, and back contact metal cathode needs to be surface engineered for carrier collection as well as light transmittance. Moreover, metal foil–based flexible PSCs could tolerate high-temperature sintering required for mesoporous scaffold layer, and one can expect similar photovoltaic performance and operational stability to those of conventional glass–based PSCs.

Lee et al. fabricated a mesoporous flexible PSC on Ti foil substrate, in which all of the active layers were processed at similar experimental conditions to those of glass-based PSCs [10]. Here notable difference is sunlight entered into the device thorough back contact Ag metal cathode, as shown in Fig. 11.4A. The Ag cathode with various thicknesses (8, 12, 18, and 20 nm) was deposited on top of the hole transporting layer (HTL) by thermal evaporation technique. The transmittance of Ag ultrathin metal film (UTMF) deposited on spiro-OMeTAD was measured to estimate the amount of light penetrated the cathode and reach the perovskite absorber film. The bare glass substrate coated with spiro-OMeTAD did not show any notable decrease in visible light transmittance. On the other hand, the transmission has come down to 20%–25% at the wavelength range of 450–750 nm for thin layer of Ag-coated sample (Fig. 11.4B). The device with different thickness range of UTMF was prepared to find the statistics of the device performance with respect to the thickness. As per the transmittance data, the device prepared with thick Ag film (20 nm) limits the penetration of photon into the active area results in less value of

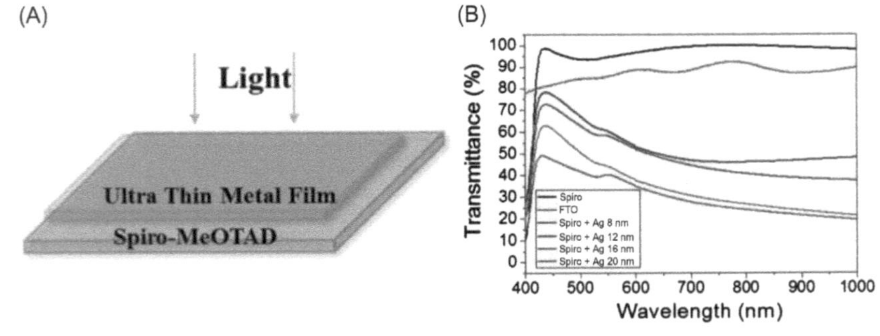

Figure 11.4 (A) Schematic diagram for light harvesting to the perovskite layer from top illumination. (B) Transmittance of the UTMF/spiro-MeOTAD with UTMF top illumination. *Reproduced with a permission from Ref. [10].*

J_{SC}(6.1 $mA \cdot cm^{-2}$). On the other hand, the device fabricated with thin Ag film (8 nm) also showed low J_{SC} and PCE, which might be due to nonuniform and discontinuous film and high-series resistance. Among these studies, the highest PCE of 3.85% was achieved for 12-nm Ag cathode; still there is a room for high PCE through judicial surface engineering of metal cathode.

Wong et al. introduced another interesting variant of metal foil–flexible PSC. By using facile electrochemical anodization, vertically aligned TiO_2 nanotube (TNT) arrays with desired features were directly grown on Ti foil [11]. Grown amorphous TNT arrays are converted into anatase phase by thermal annealing at 450°C and employed as scaffold for perovskite absorber deposition. Novel transfer printing approach was adopted to fabricate highly transparent CNT film on Ti foil/TNT/$CH_3NH_3PbI_3$/spiro-OMeTAD stack, as depicted in Fig. 11.5. The CNT film used here has dual advantage: High transmittance to incoming solar irradiation and its porous nature facilitates the infiltration of spiro-OMeTAD HTM into CNT networks. It has been found that $TiCl_4$ treatment improves the voids between the compact layers thereby greatly reduces charge carrier recombination at substrate/perovskite interface. Surface roughness of Ti foil also found to have notable influence on the photovoltaic performance of the device through flat perovskite layer and reducing the unfavorable shunting paths. Current–voltage characteristics and incident photon to current conversion efficiency (IPCE) of flexible PSCs fabricated with different thickness Ti foil substrate are shown in Fig. 11.6. Flexible PSC fabricated using thin and ultra-smooth Ti foil showed 8.3% PCE under standard test

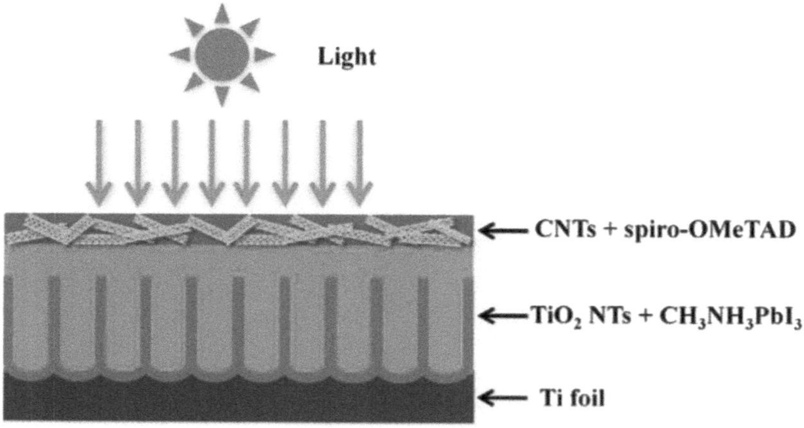

Figure 11.5 Schematic of solid-state PSCs based on Ti foil/TiO_2 nanotubes and carbon nanotubes. *Reproduced with a permission from Ref. [11].*

condition. These devices retain more than 90% of the initial performance after completing 100 bending cycles, exhibiting their suitability for modern portable power applications.

Transparent conductive adhesive (TCA)—coated PET substrate embedded with Ni mesh has a potential for semitransparent cathode application in metal foil—based flexible PSCs [12]. This counter electrode avoids the use of expensive noble metals and hence metal ion induced performance degradation. A schematic diagram of flexible PSC with laminated cathode is shown in Fig. 11.7. In this device architecture, they newly introduced PEDOT:PSS layer on top of spiro-OMeTAD plays a crucial role by providing a conductive pathway to collect the charges from HTL and transport them to the conductive sites within the TCA network. Ni mesh—embedded PET substrate also acts as a barrier for

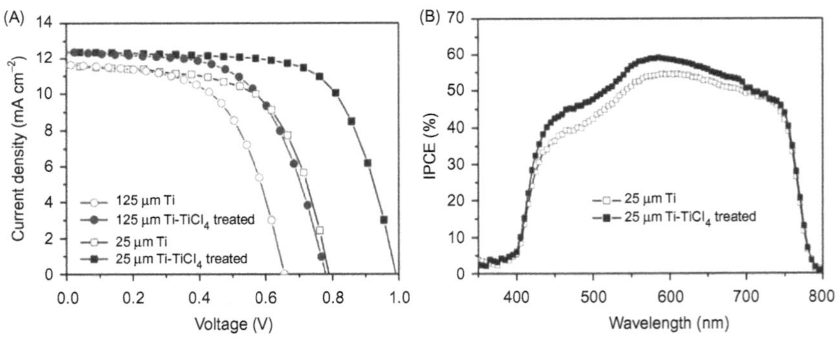

Figure 11.6 (A) Photocurrent—voltage characteristics of TNT and CNT-based flexible PSCs prepared on different Ti foil thickness and with/without $TiCl_4$ treatment; (B) IPCE of 25-μm-thick Ti-based PSCs with/without $TiCl_4$ treatment. *Reproduced with a permission from Ref. [11].*

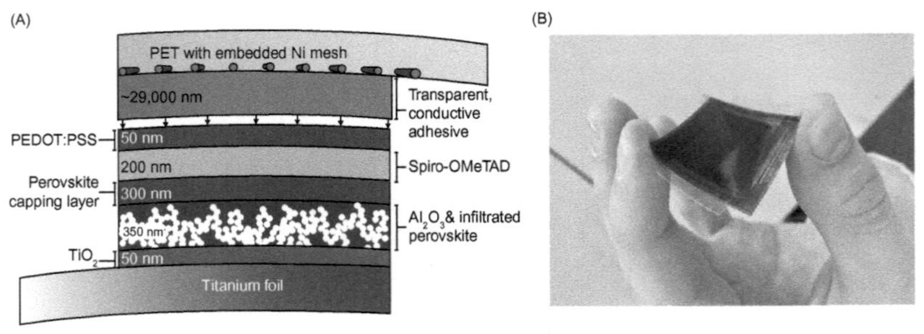

Figure 11.7 (A) A schematic representation of a metal-mounted PSC and (B) Photograph of flexible PSC fabricated on Ti foil and laminated with TCA cathode. *Reproduced with a permission from Ref. [12].*

atmospheric moisture and enhancing the device life time. High transmittance of ex situ fabricated TCA ensured minimum loss in J_{SC} compared with front side illuminated devices, and a remarkable PCE of 10.3% was achieved. Performance of these devices dropped 7% of its initial value after 200 bend cycles, demonstrating their better stability and flexibility.

11.2.2 Planar Flexible PSCs

The previous section emphasized the challenges associated with low-temperature processing of mesoporous scaffold layer on temperature sensitive flexible substrates. Planar device architecture has been evolved as a low-temperature processable alternative to conventional PSC design. In this configuration, perovskite absorber material is sandwiched between the compact electron and hole extraction layer. Simple configuration, ease of fabrication, and enhanced versatility made this architecture more attractive for fabricating high-performance flexible PSCs. Poor crystallinity associated with low-temperature processing of inorganic (TiO_2) compact electron transporting layer seems a major bottleneck for realizing high efficiency in planar flexible PSCs. In this context, binary metal oxides with high intrinsic carrier mobility (ZnO, SnO_2) or ternary metal oxides (Zn_2SnO_4, $BaSnO_3$) with tunable carrier mobility emerged as a natural

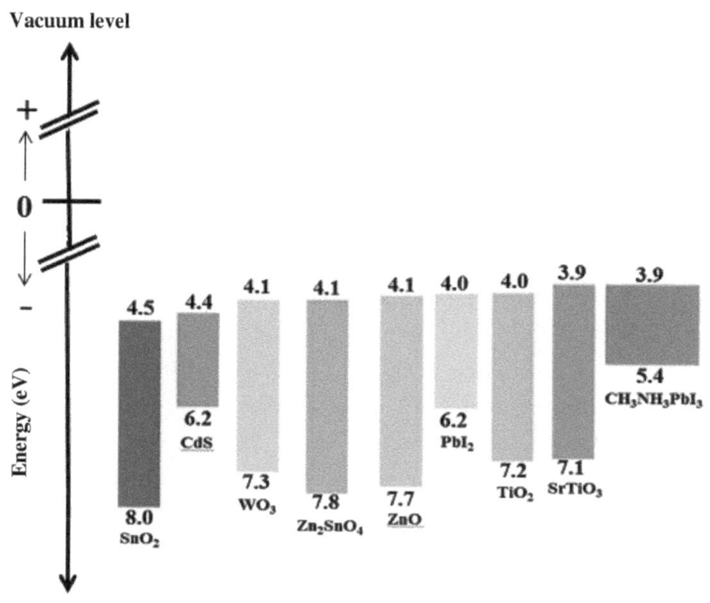

Figure 11.8 Band energy diagram of various inorganic electron transporting materials.

choice for electron transport layer (ETL) application. Fig. 11.8 shows the band energy diagram for various inorganic ETL materials. Apart from this, organic materials such as 6,6-phenyl C61-butyric acid methyl ester (PCBM), C60 and solid-state ionic liquids also explored as low-temperature processable ETL for hysteresis-free and stable planar flexible PSCs.

Synthesis and solution deposition of highly crystalline and uniform electron transporting material over large area plastic substrate is a major issue. Mali and Hong introduced RF magnetron sputtered TiO_2 thin film as an annealing-free ETL for planar flexible PSCs [13]. Defect-free and amorphous TiO_2 ETL with various thicknesses (30–100 nm) was sputter deposited at room temperature on PET/ITO substrate. $MAPb(I_{1-x}Br_x)_3$ and PTAA were employed as an absorber and hole transporting material, respectively. Flexible PSC prepared with 30-nm thickness TiO_2 ETL showed less fill factor (FF) and attributed to partial isolation of PTAA from underneath ITO substrate. The FF was notably increased once the ETL thickness approaches 50 nm because of perfect isolation between PTAA and ITO substrate. Further increasing ETL thickness drastically reduces the J_{SC} value from to 20 to 17 $mA \cdot cm^{-2}$, mostly caused by hindered electron transport in thick amorphous TiO_2 layer. By carefully optimizing between partial and large isolation conditions (Fig. 11.9), a champion PCE of 15.88% achieved, which is almost similar to that of control device prepared on a glass substrate.

Zinc oxide (ZnO) received considerable interest for low-temperature processable ETL because of its high electron mobility ($100-200\ cm^2 \cdot V^{-1} \cdot s^{-1}$ vs $0.1-4\ cm^2 \cdot V^{-1} \cdot s^{-1}$ of TiO_2) and long diffusion coefficient. Liu and Kelly fabricated thin and compact ETL on ITO/PET substrate using spin coating a colloidal solution of ZnO NPs [14]. Surface smoothness of ETL layer and thickness was systematically controlled by multistep deposition approach. The effect of ZnO ETL thickness (0–70 nm) on the photovoltaic performance of resulting device was evaluated in detail. The device prepared without any ZnO layer showed less V_{OC} and FF because of charge carrier recombination between the perovskite and ITO layer. At the optimized ETL thickness of 25 nm, flexible device showed PCE of 10.2% with FF of 0.74, an open-circuit voltage of 1.03 V and short-circuit current density of 13.4 $mA \cdot cm^{-2}$. This work made a substantial step towards room-temperature volume reduction of flexible PSCs. Motivated by these impressive results, high electron mobility SnO_2 is also explored as ETL for high-efficiency PSCs [15].

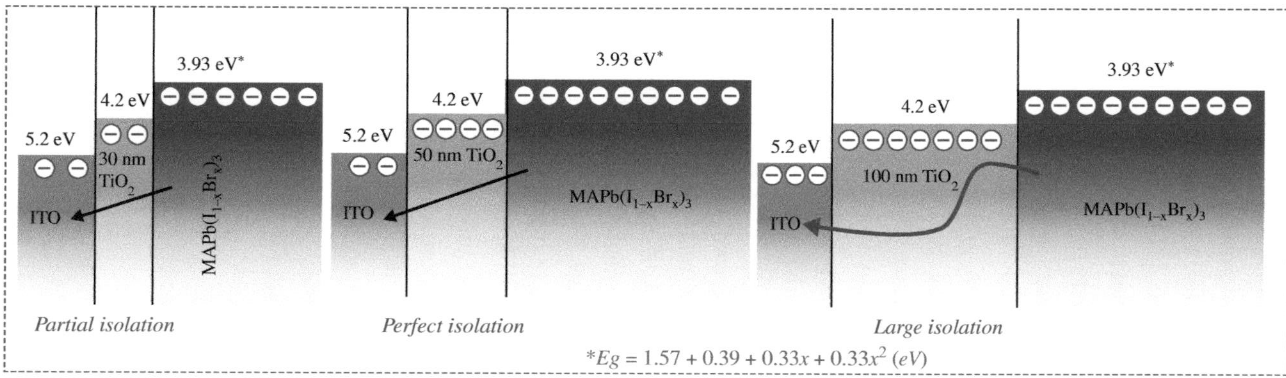

Figure 11.9 Energy band diagrams of ITO, BI-TiO_2 and MAPb $(I_{1-x}Br_x)_3$ perovskite absorbing layer. The band gap of the mixed halide perovskite MAPb $(I_{1-x}Br_x)_3$ has been calculated using equation $E_g(x) = 1.57 + 0.39x + 0.33x^2$, where E_g is the band gap of the perovskite in eV and x is the amount of Br relative to the amount of I. *Reproduced with a permission from Ref. [13].*

Higher PCEs obtained for ZnO and SnO_2 ETL planar flexible PSCs naturally extended the interest towards ternary metal oxides. Ternary metal oxides, apart from high intrinsic carrier mobility, provide flexible approach to custom-tailor the electrical and microstructural properties through composition engineering. Yang et al. for the first time introduced Zn_2SnO_4(ZSO) as an ETL in flexible PSC [16]. ZSO had previously been used in DSSC and OPVs as photo electrode because it has a conduction band edge at similar level to that of TiO_2 and ZnO. Moreover, it has high chemical stability with respect to acid/base solution and polar organic solvents. Generally, the synthesis of highly crystalline metal oxide NPs needs high temperature and high pressure. But formation of ZSO does not require high temperature and pressure, which made them highly suitable to use on plastic substrates. The prepared ZSO NPs was deposited on top of ITO/PEN substrate by spin coating method and then followed with an active layer by two-step spin coating deposition process. The typical device cross-sectional SEM image was shown in Fig. 11.10A, and energy level diagram of Zn_2SnO_4 ETL PSC is provided in Fig. 11.10B. The flexible PSC was characterized under AM 1.5G $100\ mW \cdot cm^{-2}$, and it showed an impressive PCE of 15.3%.

The presence of deep trap states in binary and ternary metal oxide—based ETL materials and severe hysteresis problem always affect the operational stability of flexible PSCs. Recently it has been discovered that organic ETLs including fullerene derivatives provide hysteresis-free planar PSCs. Choi and coworkers introduced C_{60} as an ETL, these C_{60} helps to reduce the density of trap states and passivate the grain boundaries of the absorbing layer [17]. In this study, the C_{60} was deposited on flexible substrates by vacuum-based thermal evaporation method. The effect of device performance on various thickness of C_{60} was studied. The

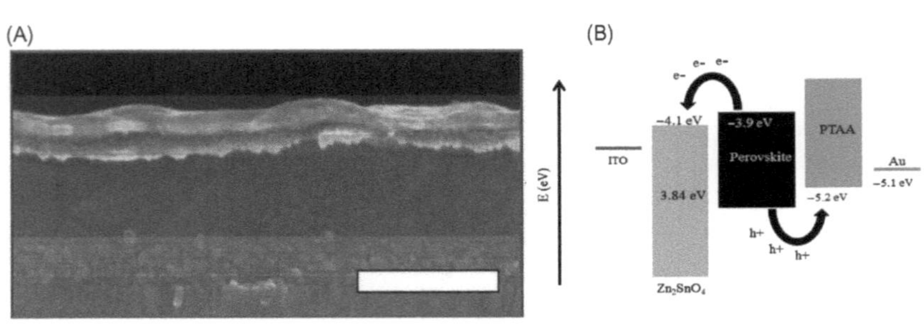

Figure 11.10 (A) Cross-section SEM image and (B) energy-level diagram of the Zn_2SnO_4 ETL flexible PSCs. *Reproduced with a permission from Ref. [16].*

deposited C_{60} layer with thickness of less than 35 nm showed a number of pin holes which attributes to the charge carrier recombination. At the optimized thickness of 35 nm, they achieved a device efficiency of 19.1% on rigid substrate and 16.0% of efficiency without any hysteresis problem. To demonstrate the device stability against mechanical force bending, they performed the bending test up to 1000 bending cycles with different radius of curvature. At radii of 10 nm the devices did not show any decrement from its initial efficiency up to 1000 cycles. On the other hand, device bended at radii of 5 nm provided 95% of its initial efficiency after 100 cycles, and PCE got decreased 20% from its initial PCE after 1000 bending cycles, which exhibits that the device with C_{60} has good stability against with the mechanical bending (Fig. 11.11).

11.2.3 Inverted Planar Flexible PSCs

Inverted planar PSC architecture is adopted from OPV configuration; and therefore, most of the carrier selective contacts were already explored for their carrier selectivity in organic solar cells. Initial studies on the photoluminescence quenching of perovskite emission highlighted the suitability of various organic carrier selective contacts for PSC fabrication. In inverted planar architecture, organic *p*-type semiconductors (i.e., PEDOT:PSS, PTAA) HTM is directly coated on top of the substrate followed by perovskite deposition. Electron selective contact (PCBM, C60) is prepared on top of the perovskite absorber layer and device fabrication is completed by cathode deposition. Inverted planar architecture is relatively easy to incorporate on flexible substrate because most of the organic layers are prepared in low temperature. However, organic materials are susceptible to ambient (oxygen, moisture, UV irradiation, and temperature) and therefore operational stability of inverted planar flexible PSC also a point of concern [18].

The first inverted planar architecture with *p-i-n* configuration for flexible PSC was developed by Jeng et al., in which PEDOT:PSS and fullerene derivate were employed as hole and electron-selective contacts respectively [19]. By introducing PCBM/TiO_x bilayer ETL and systematically optimizing the various interfaces, PCE of inverted planar flexible PSC was improved to 6%. Surface roughness of underneath TCO substrate influences the perovskite formation and surface coverage, thereby affecting light absorption and device performance. Fig. 11.12 shows the representative SEM image of the inverted device and corresponding

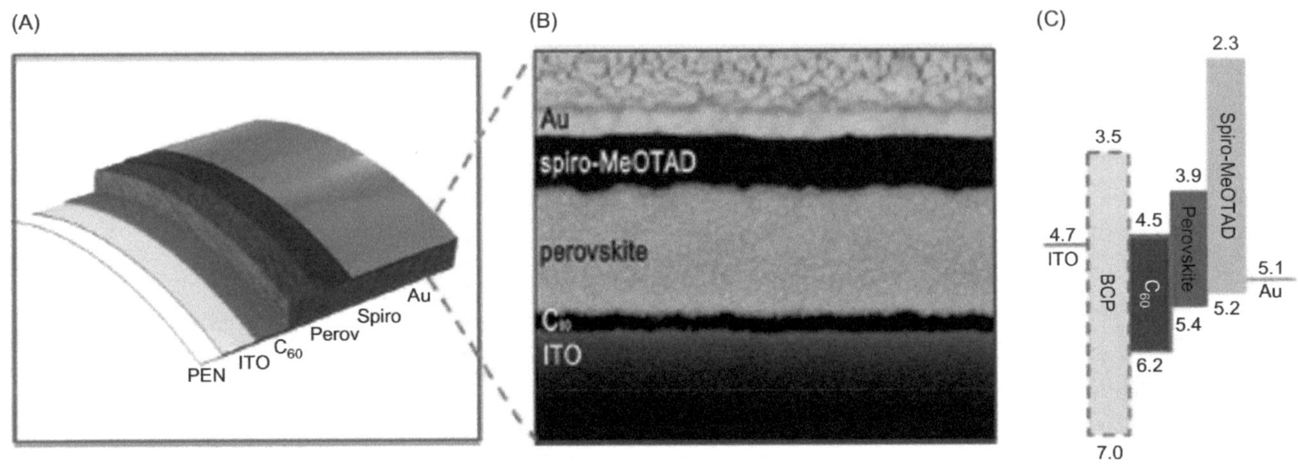

Figure 11.11 (A) The structure of a flexible $MAPbI_3$ PSC on a PEN/ITO substrate. (B) Cross-sectional FESEM image of a flexible $MAPbI_3$ PSC. (C) The relative energy level diagram of the $MAPbI_3$ PSC adopting C_{60} as an ETL with or without the BCP layer. *Reproduced with a permission from Ref. [17].*

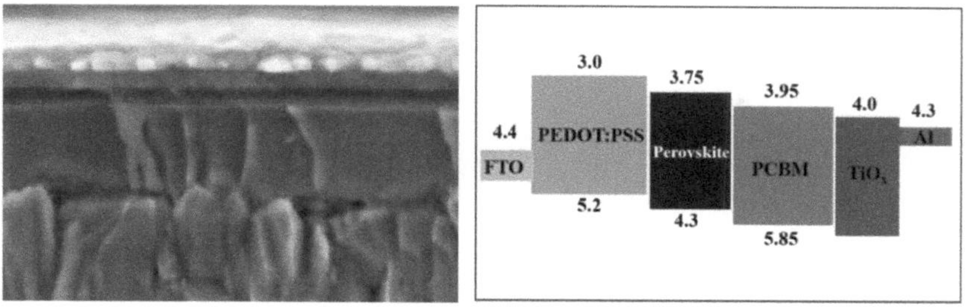

Figure 11.12 (A) Cross-sectional SEM image of inverted planar PSC and (B) energy-level diagram of the corresponding device. *Reproduced with a permission from Ref. [19].*

energy level diagram. It is worthwhile to mention here that a similar device prepared on ITO glass substrate also exhibited essentially similar device parameters, highlighting the suitability of inverted architecture for flexible devices [20].

The PCE of inverted planar PSC could be further improved by reducing the charge trap states in perovskite and eliminating the photocurrent hysteresis. In this perspective, Carmona et al. used flexible PET coated with a thin layer of silver sandwiched between aluminum doped ZnO (AZO) as substrate for realizing robust and moderately efficient inverted planar flexible PSCs [21]. Poly-TPD and PCBM were introduced as an effective electron and hole blocking layer, respectively (Fig. 11.13). A maximum PCE of 7% with V_{OC}: 1.04 V, J_{SC}: 14.3 $mA \cdot cm^{-2}$ and FF of 0.47 were achieved under one sun condition. Similar device made on glass substrate showed a PCE of 12%, primarily because of high FF and slight improvement in the photocurrent. The decrement of FF in flexible device might be due to the higher thickness of hole and electron transporting layer, so it needs to be optimized to get an improved PCE. These devices were very stable even after 50 bending cycles.

The most of the inverted type flexible device uses PEDOT:PSS as a HTL because it can be processed at a very low temperature. However, unfortunately, the device based on this PEDOT:PSS is not stable in long term due to its high acidity and hygroscopic nature. To overcome with this problem, some of the inorganic *p*-type semiconductor materials such as CuI and NiO_x are explored as a HTL in inverted planar flexible PSCs. Usually NiO_x films were deposited by pulsed laser deposition followed by high-temperature annealing. However, the same approach cannot be extended to flexible substrate. Yin et al. adopted a novel approach where presynthesized crystalline NiO_x NPs were used to make thin HTM layer

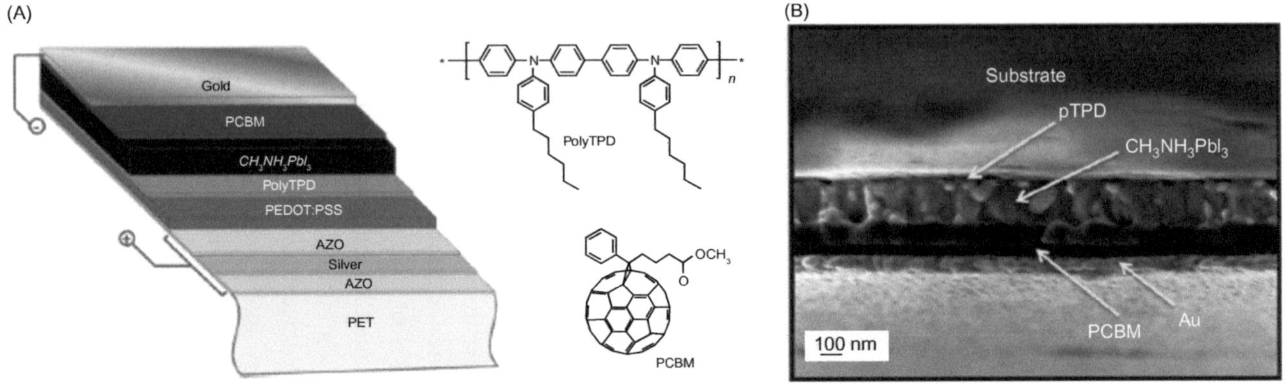

Figure 11.13 (A) Schematic layout of inverted planar flexible PSC and chemical structure of materials used as the electron and hole blocking layer. (B) SEM cross-section of the perovskite device on the PET foil. *Reproduced with a permission from Ref. [21].*

on flexible substrate at room temperature [22]. The NiO_x was deposited on to the ITO-coated PEN substrate by spin coating method, and interestingly there is no much difference in the visible light transmittance of bare and NiO_x layer—coated ITO-PEN substrate. An impressive PCE of 13.43% was observed with V_{OC}: 1.04 V, J_{SC}: 18.74 $mA \cdot cm^{-2}$ and FF of 0.69, whereas the similar device prepared on rigid substrate showed 16.47% efficiency.

Cuprous iodide is one of the potential material for HTL application in inverted planar PSCs because of its high hole mobility and simple fabrication. Nejand et al. for the first time prepared an inverted planar flexible PSC using 10-μm thick copper foil because of its favorable work function [23]. Cuprous iodide thin film with thickness of 200 nm was grown on Cu foil in zinc blend structure. The complete schematic of the device and work function of various layer are shown in Fig. 11.14. According to the band diagram, the generated electron can be easily moved to the ZnO layer because of its high electron affinity at the interface of ZnO layer and perovskite. The Ag nanowire network was deposited by spray coating method shows high uniformity, good connectivity with the ZnO layer, and good transparency for passing the light from top side. The complete device measured under one sun illumination showed an efficiency of 12.8 %. Based on the knowledge gained in a laboratory-scale device, large area device with an active area of 0.8 cm^2 was fabricated, and a power output of 7.07 mW was demonstrated. These inorganic ETL- and HTL-based flexible PSC showed good durability over a period of 60 days.

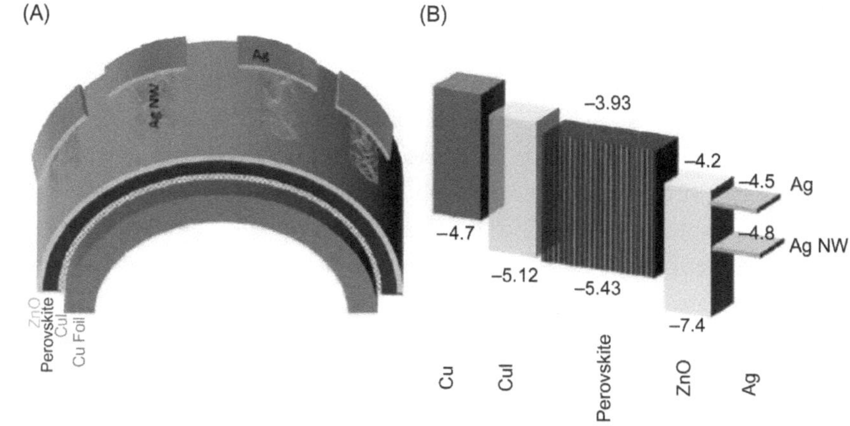

Figure 11.14 (A) Schematic of the device architecture (B) and energy level diagram of $Cu/CuI/CH_3NH_3PbI_3/ZnO/Ag$ stack. *Reproduced with a permission from Ref. [23].*

11.3 MANUFACTURING OF FLEXIBLE PSCS

Crystalline silicon solar cells that currently dominate the photovoltaic market with >90% employ energy-intensive manufacturing process. Although cost of the raw materials decreased drastically in recent years, price of the renewable electricity produced by first generation solar cell is more expensive than traditional grid power. PSCs, as extensively discussed in previous sections and chapters, provide multiple advantages including low-temperature fabrication, less material consumption, flexible, and light weight. Certified power conversion efficiencies of above 20% made PSCs commercially competitive with c-Si solar cells. The recent cost–performance analysis carried out on rigid perovskite modules with moderate efficiency–cheap material or high efficiency–expensive material combination concludes that even a moderate module efficiency of 12% and life time of 15 years could lead to the levelized cost of electricity (LCOE) generated from PSC technology in the range of 3.5–4.9 US cents/kWh [24]. It is worth to mention here that LCOE for grid power is 7.04–11.90 US cents/kWh and for c-Si solar cell is 9.78–19.33 US cents/kWh. It has been pointed out that raw materials, module design, and manufacturing process play critical role in arriving LCOE estimation. Flexible PSCs provide multiple manufacturing advantages over their glass-based counterpart. This section describes the potential scale-up approaches suitable for high-throughput manufacturing flexible PSC modules.

11.3.1 Batch Processing

The batch process involves deposition of active layers at one go on a large number of substrates and perform postdeposition treatments (if any) such as temperature annealing, UV treatment, and laser scribing. Fig. 11.15 schematically depicts various stages involved in the fabrication of monolithically integrated perovskite solar modules [25]. Note that the module configuration and processing steps are more or less similar between the glass and flexible substrate and does not significantly affect the process flow. Transparent conductive substrate and metal cathodes can be prepared on a temperature-sensitive substrate by sputtering process. Electron-selective ZnO contact and hole-selective NiO contacts may be screen printed on a semi-processed device in a desired sequence and will be subjected to low-temperature annealing for evaporating the solvent and also to induce the interparticle connectivity. Single-step deposition of perovskite ink made up of PbI_2 and MAI dissolved in dimethyl formamide

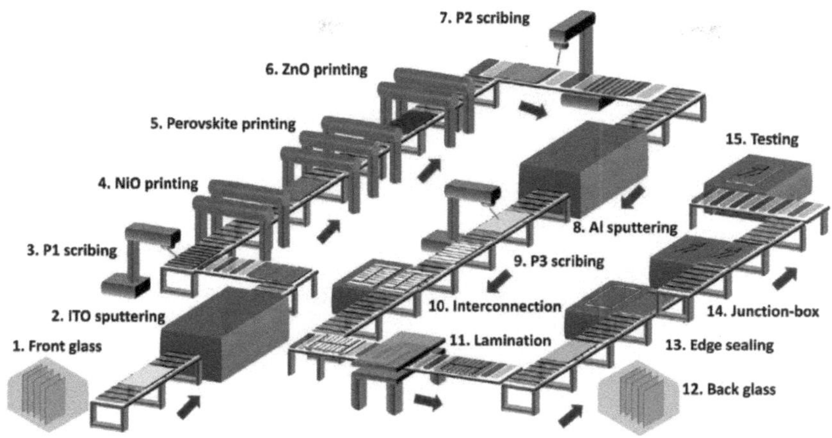

Figure 11.15 Schematic diagram showing the various steps involved in the batch production of PSCs. *Reproduced with a permission from Ref. [25].*

solvent is preferred for screen printing a $MAPbI_3$ absorber layer over a large area substrate. Laser or mechanical scribing will be utilized at an appropriate place to create stripe pattern cells and interconnect the adjacent stripe cells in a series-monolithic configuration.

11.3.2 Roll-to-Roll Processing

Roll-to-roll (R2R) is a continuous manufacturing process meant for the deposition of various thin and thick films on rollable substrates over a long-length scale. The R2R process provides immense cost-reduction vis-à-vis traditional batch processing of various electronic devices. Most of the high-throughput coating (slot-die, meniscus) and printing (gravure, rotary screen printing) technologies can be adopted for the roll-to-roll fabrication of solar cells. First attempt of roll-to-roll fabrication of large area solar cells was performed on OPVs and impressive PCE in the range of 5%–7% was achieved [26]. Solution processing and low-temperature compatibility naturally extended the potential of roll-to-roll to PSC manufacturing. Fig. 11.16 shows the typical roll-to-roll process for the fabrication of PSC on rollable substrate. Hwang et al. introduced the gas quenching on slot die-coated perovskite absorber layer and improved the PCE of roll-to-roll processed PSC to 12% [27]. Very recently, Solliance developed an inline roll-to-roll coating, drying, and annealing process below 120°C for both the ETL and perovskite absorber at a linear speed of 5 m/minute on a 30-cm wide ITO/PET foil. The entire process was performed under ambient conditions and maximum PCE of 12.6% demonstrated.

Figure 11.16 Roll-to-roll fabrication of PSCs on plastic substrate [28].

11.4 STABILITY OF FLEXIBLE PSCS

Long-term stability of flexible PSC needs to be addressed for the successful commercialization of this technology. In addition to the routine factors such as thermal stability, photo-induced perovskite degradation, and metal ion migration, flexible PSCs may suffer from additional factors caused by the flexible nature of the substrate. The flexible PSC devices are made up of conductive layer—coated PET or PEN substrates which are very sensitive to ambient environmental factors. First, high oxygen and moisture permeability of polymer substrates in comparison with conventional glass substrate can cause unprecedented performance degradation. Second, these plastic substrates are more sensitive to UV light and temperature; both of them largely affect the photovoltaic performance of flexible PSCs even in mild indoor or outdoor environment. Third, as all of the active layers are sequentially deposited on flexible polymer substrate, any stress caused by flexibility of underneath substrate can damage the TCO coatings and also weaken the various interfaces, thereby induce the photovoltaic performance degradation and eventually cell failure. Fourth, lack of mechanical rigidity possesses additional challenges for hermitic sealing and installation of the device.

Still, the PCE and stability of flexible PSCs lag behind with the conventional device on rigid substrate. Because, these flexible PET substrates have low conductivity and poor mechanical robustness. Moreover, these TCO electrodes are having high sheet resistance of 30 ~ 50 ohms/square, which increases the series resistance of the complete device. To overcome this issue, Li et al. [29] made a Ag mesh by nano-imprinting technique over the flexible PET substrate to improve the flexibility and robustness, transmission, and also these substrate showed less sheet resistance ~3 ohms/square, which suggest its future applications in flexible solar cell (Fig. 11.17). The resultant substrate with total thickness of 57 μm shows a high transmission of 82%–86% in the visible region without any light diffraction and scattering. A 150-nm thick PH1000 was coated on top of the PET/Ag-mesh substrate by spin coating method to collect the charge carriers along both the lateral and vertical direction. An inverted, planar flexible PET/Ag-mesh/PH1000/PEDOT:PSS/$MAPbI_3$/PCBM/Al device showed a highest efficiency of 14.2% with 0.80 FF. The short-circuit current density of these devices was shown very less compared with rigid substrates because of its less transmission in the visible region. In addition, ultrathin Ag-mesh substrate has given a high specific power of 1.96 kW/kg, which helps to use this solar device in unmanned aerial vehicles.

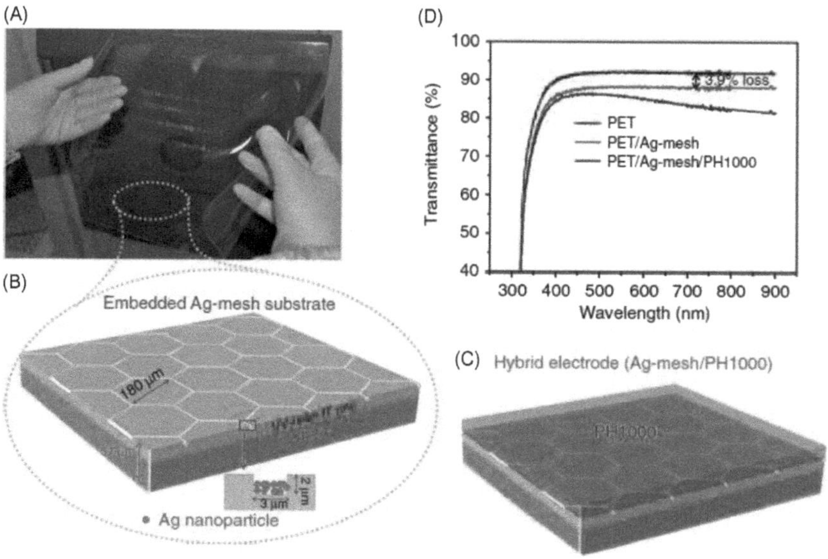

Figure 11.17 Schematic illustration of the FEAM substrate and hybrid electrode. (A) An image of the large-area FEAM substrate. (B) The structure of the FEAM substrate with detail parameters. (C) The diagram for the hybrid electrode (PET/Ag-mesh/PH1000). (D) Transmission spectra of bare PET, PET/Ag-mesh, PET/Ag-mesh/PH1000-based substrates. *Reproduced with a permission from Ref. [29].*

The device prepared with PET/Ag-mesh/PH1000 substrates shown 95.4% of its initial efficiency even after 5000 cycles of bending test (Fig. 11.18). On the other hand, the efficiency of the conventional device made on ITO substrate decreased 30% of its initial efficiency after 1000 bending cycles which demonstrates that PET/Ag-mesh/PH1000 could be a promising candidate for making a large-scale flexible PSC with high thermal stability.

Because the most commonly used PEDOT:PSS corrodes the TCO substrates and perovskite layer because of its acidic nature (pH ~1). So, an alternative material with high crystalline and air stability needs to be found as an HTL to fabricate a large area flexible PSC. Park et al. [30] used Li doped P3HT nanofibrils (LN-P3HT) as HTL with improved air stability than the PEDOT:PSS and Spiro-OMeTAD. Fig. 11.19 shows the schematic and energy level diagram of the device architecture used in this work. Here, the ITO substrate was modified with poly(ethylenimine)-ethoxylated to increase the work function of ITO from −4.6 to

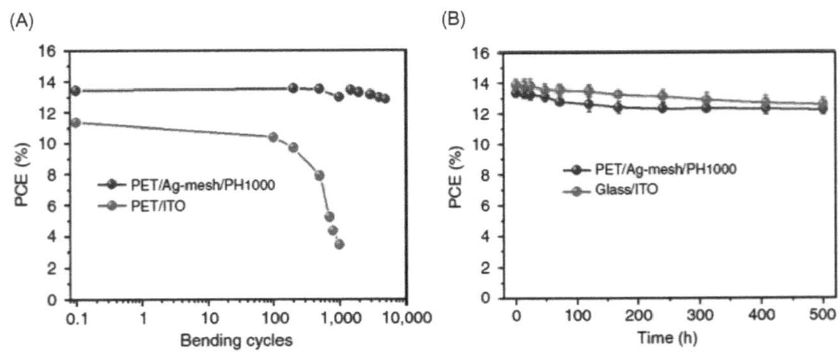

Figure 11.18 Bending and stability test of flexible PSCs. (A) PCEs of flexible PSCs based on both PET/Ag-mesh/PH1000 and PET/ITO electrodes as a function of bending cycles at a radius of 5 mm. *Reproduced with a permission from Ref. [29].*

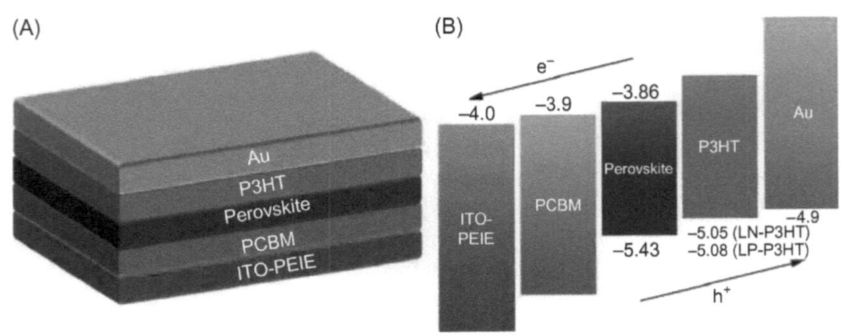

Figure 11.19 (A) Schematic of the PSC structure. (B) Energy diagram of the device showing the flow of electrons and holes. *Reproduced with a permission from Ref. [30].*

−4.0 eV, which increase the transfer of electrons from the lowest unoccupied molecular orbital to the ITO substrate. In addition, there was an upward shift in the highest occupied molecular orbital, i.e., −5.08 eV and −5.05 eV for Li doped pristine P3HT and LN-P3HT, respectively. The prepared devices were measured, and it showed an efficiency of 13.41% and 15.18% for LP-P3HT- and LN-P3HT-based HTL devices, respectively. The improvement of efficiency for LN-P3HT-based HTL device might be due to its high conductivity and lesser grain boundary which improves the charge carrier transportation, and also these devices showed less hysteresis than the LP-P3HT-based HTL device

The LN-P3HT-based HTL device prepared with an active area of 1 cm^2 showed an initial efficiency of 13.12% with high mechanical stability (Fig. 11.20). These devices were shown only the little decrement of efficiency even after 500 bending cycles for a bending radius of up to 10 mm. Due to the plastic deformation of the substrate, the device performance was drastically decreased after 100 bending cycles for bending

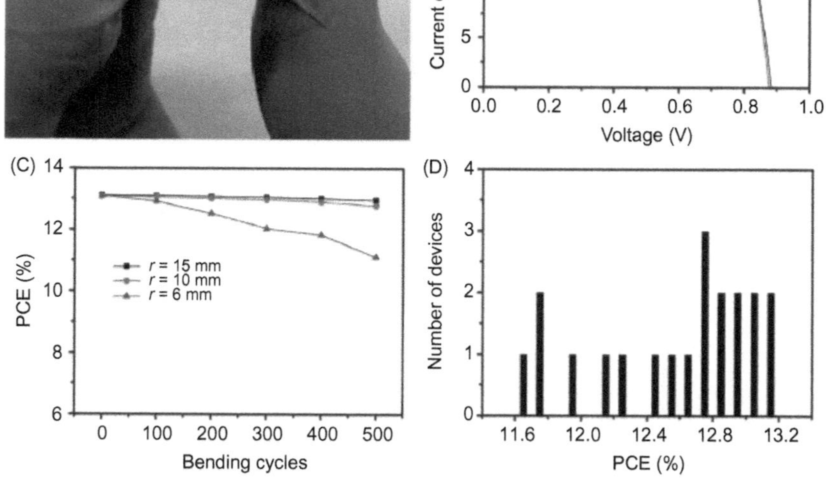

Figure 11.20 (A) Photograph of the flexible PSC, with an active area of 1 cm^2, based on LN-P3HT. (B) J–V curves of the flexible PSCs measured during a backward–forward scan, for 200 Ms of scan-delay time. (C) Bending test: flexible PSCs were subjected to 500 bending cycles at $r = 15$ mm, 10 mm, and 6 mm. (D) Histogram of PCEs for 20 devices. *Reproduced with a permission from Ref. [30].*

radius at 6 mm (Fig. 11.21). Generally, the amorphous region and grain boundaries of P3HT allow the oxygen to the inner layer, which degrades the thiophenes. In case of LN-P3HT-based device, the tight packing and long length of nanofibrils acts as barrier for oxygen molecules, which limits the degradation of perovskite. These devices were stored under ambient conditions showed high air stability and photo stability for 30 days with 87% of its initial efficiency.

Flexible PSC made on conventional PET or PEN substrates are not stable with higher temperature processing unlike polyimide (PI) substrates. Because, the glass transition temperature is very high for PI substrate while compared to PET (T_g ~78°C) or PEN (T_g ~123°C), but these PI substrate is light brownish color so that cannot be used as transparent electrode for flexible PSCs. Kim et al. used 60-μm thick colorless PI (CPI) substrate as an transparent electrode for flexible PSC application [31]. These ITO/CPI substrates can be annealed up to 300°C because of its high glass transition temperature. Moreover, these substrates can be exposed to light or plasma during sputtering due to its better thermal stability, and also ITO/CPI substrate showed better electrical properties than the ITO film on PET or PEN substrate. The optical measurement of CPI-based substrate showed an average transmittance of 81% between 400 and 800 nm wavelength range (Fig. 11.22), which is higher while compared with ITO film on PET substrate (Table 11.1).

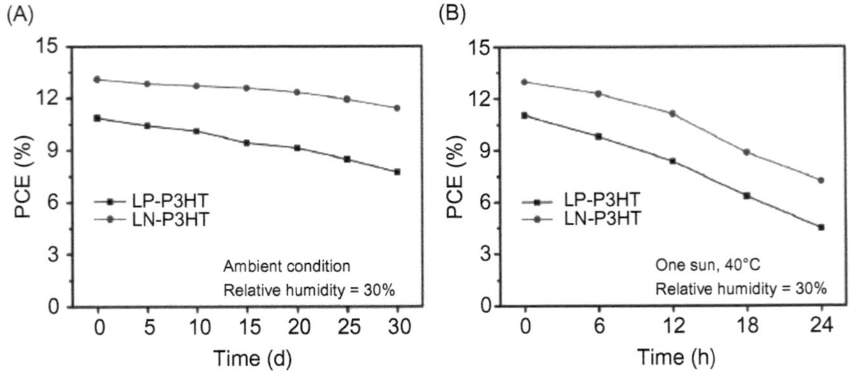

Figure 11.21 Air- and photo-stability test of the flexible devices (A) PCEs of the flexible LP-P3HT and LN-P3HT devices (active area = 1 cm^2) obtained under ambient conditions and a relative humidity of 30% for 30 days. (B) PCEs of the flexible LP-P3HT and LN-P3HT devices obtained at one sun, 40°C and a relative humidity of 30% for 12 hours. *Reproduced with a permission from Ref. [30].*

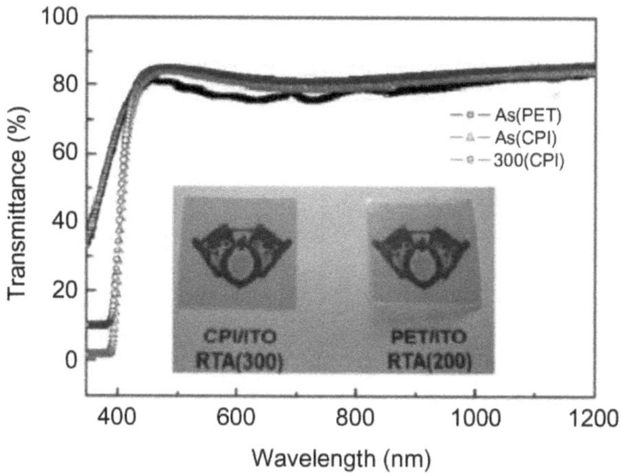

Figure 11.22 Optical transmittance of as-deposited ITO/PET, as-deposited ITO/CPI, and annealed ITO/CPI. The inset showed the transparency and color of rapidly annealed ITO/CPI (300°C) and ITO/PET (200°C). *Reproduced with a permission from Ref. [31].*

Fig. 11.23 shows the transmission spectra of $MAPbI_3$ coated on ITO/PET and ITO/CPI substrate. Among these two, the $MAPbI_3$ coated on ITO/CPI substrate showed more transmission of light over 800-nm wavelength range due to the absence of absorption by conventional perovskite. The external quantum efficiency results showed that ITO/CPI-based device cuts off the light below 400 nm due to UV—cut-off functionality of the film. Both of the devices showed same transmittance in the whole region of up to 400- to 800-nm wavelength, but the device with ITO/CPI showed 5% improvement in the light absorption than the PET substrate due to its reduced reflection. The ITO/CPI-based device achieved the highest efficiency of 15.5% with better FF than the PET-based solar cell (14.8%) due to its lower sheet resistance. Finally, the device performance was analyzed after bending cycle test to demonstrate the mechanical stability of both the devices. Among from the two devices, the CPI substrate−based device was showed better stability than the ITO/PET substrate (Fig. 11.24).

11.5 MODERN APPLICATIONS OF FLEXIBLE PSCS

Rapid progress made in PCE and stability of PSCs increased its commercial viability for various modern applications. The notable intrinsic features such as band gap tunability and ultrathin absorber layers pave the

Table 11.1 Representative device performance of mesoporous, planar, and inverted planar flexible PSCs

Substrate	Device architecture	V_{OC} (V)	J_{sc} (mA/cm^2)	FF	PCE (%)	Reference
Mesoporus						
PET	ITO/compact-TiO_2/scaffold-TiO_2/Perovskite/Spiro-OMeTAD/Au	0.83	12.6	0.71	7.4	[9]
PET	ITO/compact-ZnO/scaffold-ZnO/Perovskite/Spiro-OMeTAD/Au	0.8	7.5	0.43	2.62	[32]
Ti	Compact-TiO_2/scaffold-TiO_2/Perovskite/Spiro-OMeTAD/Ag (thin film)	0.89	9.5	0.73	6.15	[10]
Ti	Compact-TiO_2/scaffold-TiO_2/Perovskite/Spiro-OMeTAD/Ag & ITO	1.0	18.5	0.61	11.0	[11]
Ti	Compact-TiO_2/scaffold-Al_2O_3 /Perovskite/Spiro-OMeTAD/PEDOT:PSS /PET-Ni	0.98	17	0.61	10.3	[12]
Planar						
PET	IZO/TiO_X/Perovskite/Spiro-OMeTAD/Ag	0.99	17.6	0.70	12.3	[33]
PEN	ITO/C_{60}/Perovskite/Spiro-OMeTAD/Au	1.01	23.6	0.64	15.5	[17]
PEN	ITO/ALD-TiO_2/Perovskite/Spiro-MeOTAD/Ag	0.95	21.4	0.60	12.2	[34]
PET	ITO/ionic liquid/Perovskite/Spiro-OMeTAD/Au	1.07	22.72	0.66	16.0	[35]
PDMS /Willow glass	ITO /ZnO/Perovskite/Spiro-MeOTAD/Au	0.98	19.3	0.69	13.1	[36]
PET	PEDOT:PSS/Perovskite /PCBM/Cr_2O_3 & Cr/Au/polyurethane	0.93	17.5	0.76	12.1	[37]
Inverted Planar						
PET	FTO/PEDOT:PSS/Perovskite/$PC_{60}BM$/TiO_X/Al	0.88	14.4	0.51	6.4	[20]
PET	AZO/Ag + AZO/PEDOT:PSS/Poly TPD/Perovskite/PCBM/Au	1.04	14.38	0.47	7.0	[21]
PET	ITO/PEDOT:PSS/Perovskite/PCBM/Bis-C_{60}/Ag	0.86	14.6	0.75	9.43	[38]
PET	ITO/PEDOT:PSS/Perovskite/PCBM/Al	0.86	16.5	0.64	9.20	[39]
Cu	CuI/Perovskite/ZnO/Ag	0.95	22.5	0.59	12.8	[23]

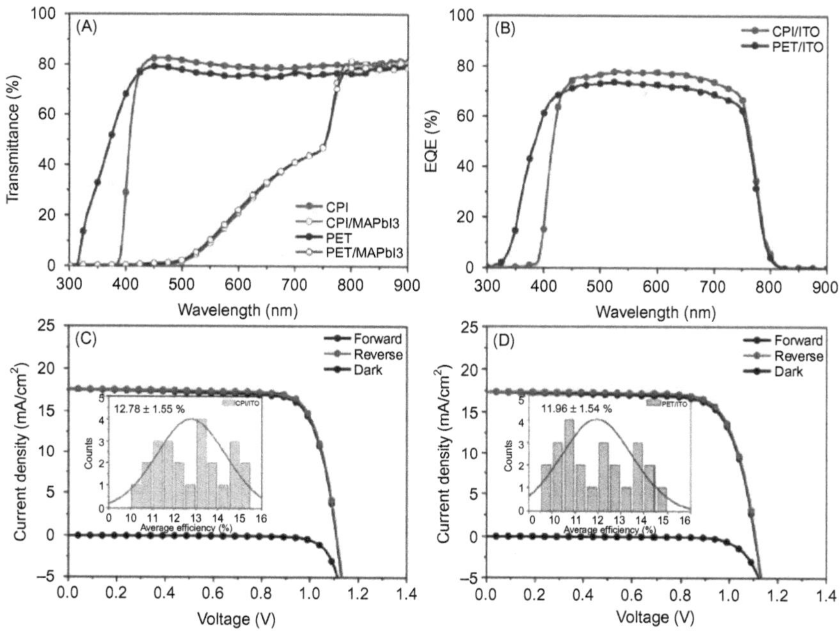

Figure 11.23 (A) Transmittance spectra of ITO/CPI, ITO/PET (polyethylene terephthalate), $MAPbI_3$/ZnO/ITO/CPI, and $MAPbI_3$/ZnO/ITO/PET samples. (B–D) Photovoltaic properties of flexible $MAPbI_3$ PSCs: (B) EQE (external quantum efficiency) spectra and (C and D) J-V (current density–voltage) curves of flexible $MAPbI_3$ PSCs comprised (C) Au/PTAA/$MAPbI_3$/ZnO/ITO/CPI substrate and (D) Au/PTAA/$MAPbI_3$/ZnO/ITO/PET. Insets: Efficiency deviations of 24 samples. *Reproduced with a permission from Ref. [31].*

way for the fabrication of multicolor and see-through solar cells for building integrated photovoltaic and vehicle integrated photovoltaic applications. Low-temperature processability and ultrathin absorber layer could help to realize lightweight flexible solar cells for portable electronic applications. Indeed some of startup companies dry to build smart products for unconventional applications where established Si wafer-based PV technology cannot be adopted. Fig. 11.25 shows some of the modern applications of flexible PSCs expected in the near future.

11.6 SUMMARY

In this chapter, we have discussed in detail the progress made towards flexible PSCs. Selection, processing, and optimization of materials play an important role in fabrication of high-performance flexible PSCs. Engineered low-temperature deposition techniques such as sputtering and

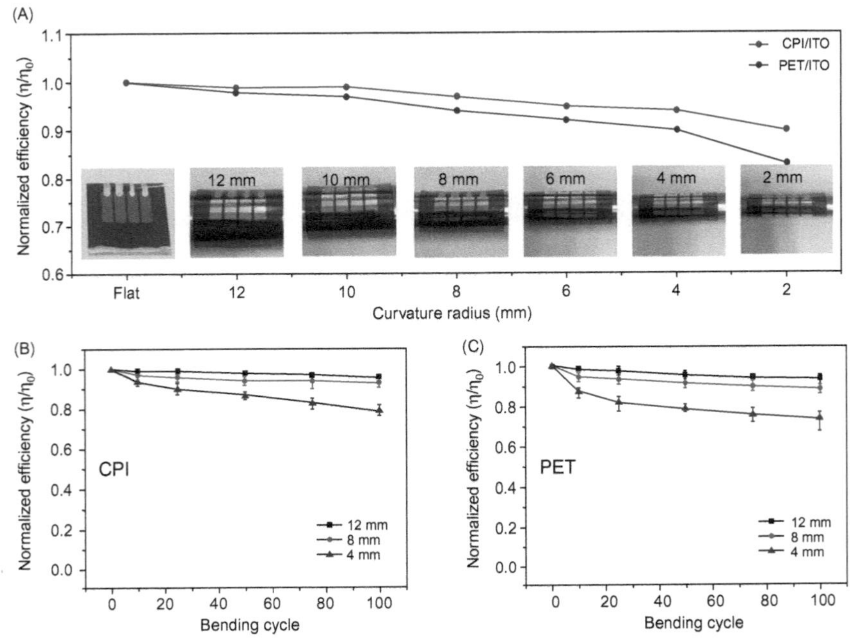

Figure 11.24 (A) Normalized PCE of flexible Au/PTAA/$MAPbI_3$/ZnO/ITO/CPI and Au/PTAA/$MAPbI_3$/ZnO/ITO/PET devices versus bending radius (R); (Band C) normalized PCE versus bending cycle number of flexible MAPbI3 devices on (B) CPI substrate and (C) PET substrate. *Reproduced with a permission from Ref. [31].*

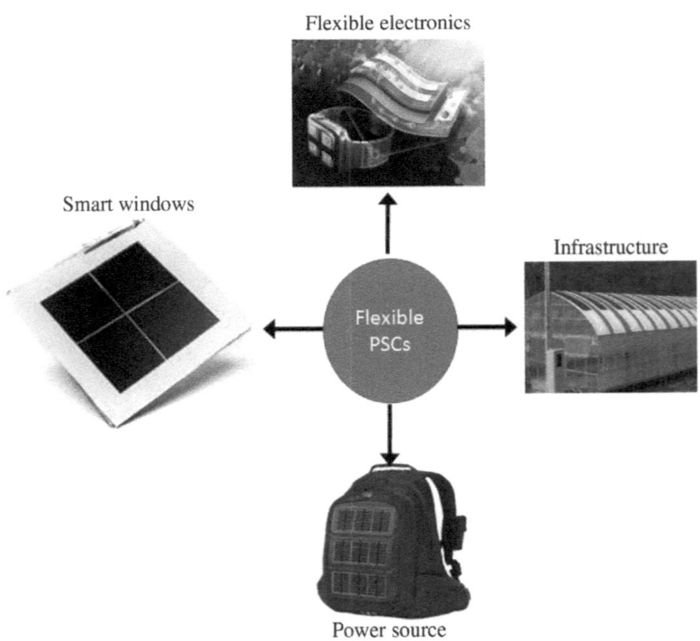

Figure 11.25 Potential modern applications of flexible PSCs.

ALD required to deposit transport layers have been discussed. Various types of flexible substrates used to fabricate PSCs and their influence on overall performance of the devices are compared. High-throughput industrialization techniques namely batch processing and roll-to-roll processing to fabricate commercial modules are highlighted. Finally, potential applications of these flexible PSCs and their deployment in niche markets are discussed. To realize the application of these flexible PSCs, the problem related to their long-term stability and low throughput has to be solved.

REFERENCES

[1] N.M. Haegel, R. Margolis, T. Buonassisi, D. Feldman, A. Froitzheim, R. Garabedian, et al., Terawatt-scale photovoltaics: trajectories and challenges, Science 356 (6334) (2017) 141–143.

[2] A. Chirila, P. Reinhard, F. Pianezzi, P. Bloesch, A.R. Uhl, C. Fella, et al., Potassium-induced surface modification of Cu (In, Ga) Se2 thin films for high-efficiency solar cells, Nat. Mater. 12 (12) (2013) 1107.

[3] N. Gasparini, L. Lucera, M. Salvador, M. Prosa, G.D. Spyropoulos, P. Kubis, et al., High-performance ternary organic solar cells with thick active layer exceeding 11% efficiency, Energy Environ. Sci. 10 (4) (2017) 885–892.

[4] C. Wu, B. Chen, X. Zheng, S. Priya, Scaling of the flexible dye sensitized solar cell module, Sol. Energy Mater Sol. Cells 157 (2016) 438–446.

[5] J.P. Correa-Baena, A. Abate, M. Saliba, W. Tress, T.J. Jacobsson, M. Grätzel, et al., The rapid evolution of highly efficient perovskite solar cells, Energy Environ. Sci. 10 (3) (2017) 710–727.

[6] J. Yoon, H. Sung, G. Lee, W. Cho, N. Ahn, H.S. Jung, et al., Super flexible, high-efficiency perovskite solar cells utilizing graphene electrodes: towards future foldable power sources, Energy Environ. Sci. 10 (1) (2017) 337–345.

[7] A. Kojima, K. Teshima, Y. Shirai, T. Miyasaka, Organometal halide perovskites as visible-light sensitizers for photovoltaic cells, J. Am. Chem. Soc. 131 (17) (2009) 6050–6051.

[8] H.S. Kim, C.R. Lee, J.H. Im, K.B. Lee, T. Moehl, A. Marchioro, et al., Lead iodide perovskite sensitized all-solid-state submicron thin film mesoscopic solar cell with efficiency exceeding 9%, Sci. Rep. 2 (2012) 591.

[9] F. Di Giacomo, V. Zardetto, A. D'Epifanio, S. Pescetelli, F. Matteocci, S. Razza, et al., Flexible perovskite photovoltaic modules and solar cells based on atomic layer deposited compact layers and UV-irradiated TiO_2 scaffolds on plastic substrates, Adv. Energy Mater. 5 (8) (2015). 1401808.

[10] M. Lee, Y. Jo, D.S. Kim, Y. Jun, Flexible organo-metal halide perovskite solar cells on a Ti metal substrate, J. Mater. Chem. A 3 (8) (2015) 4129–4133.

[11] X. Wang, Z. Li, W. Xu, S.A. Kulkarni, S.K. Batabyal, S. Zhang, et al., TiO_2 nanotube arrays based flexible perovskite solar cells with transparent carbon nanotube electrode, Nano Energy 11 (2015) 728–735.

[12] J. Troughton, D. Bryant, K. Wojciechowski, M.J. Carnie, H. Snaith, D.A. Worsley, et al., Highly efficient, flexible, indium-free perovskite solar cells employing metallic substrates, J. Mater. Chem. A 3 (17) (2015) 9141–9145.

[13] S.S. Mali, C.K. Hong, A.I. Inamdar, H. Im, S.E. Shim, Efficient planar n-i-p type heterojunction flexible perovskite solar cells with sputtered TiO_2 electron transporting layers, Nanoscale 9 (9) (2017) 3095−3104.
[14] D. Liu, T.L. Kelly, Perovskite solar cells with a planar heterojunction structure prepared using room-temperature solution processing techniques, Nat. Photon. 8 (2) (2014) 133.
[15] J.Y. Chen, C.C. Chueh, Z. Zhu, W.C. Chen, A.K.Y. Jen, Low-temperature electrodeposited crystalline SnO_2 as an efficient electron-transporting layer for conventional perovskite solar cells, Sol. Energy Mater Sol. Cells 164 (2017) 47−55.
[16] S.S. Shin, W.S. Yang, J.H. Noh, J.H. Suk, N.J. Jeon, J.H. Park, et al., High-performance flexible perovskite solar cells exploiting Zn_2SnO_4 prepared in solution below 100°C, Nat. Commun. 6 (2015) 7410. Available from: https://doi.org/10.1038/ncomms8410.
[17] H. Yoon, S.M. Kang, J.K. Lee, M. Choi, Hysteresis-free low-temperature-processed planar perovskite solar cells with 19.1% efficiency, Energy Environ. Sci. 9 (7) (2016) 2262−2266.
[18] L. Meng, J. You, T.F. Guo, Y. Yang, Recent advances in the inverted planar structure of perovskite solar cells, Acc. Chem. Res. 49 (1) (2015) 155−165.
[19] J.Y. Jeng, Y.F. Chiang, M.H. Lee, S.R. Peng, T.F. Guo, P. Chen, et al., $CH_3NH_3PbI_3$ perovskite/fullerene planar-heterojunction hybrid solar cells, Adv. Mater. 25 (27) (2013) 3727−3732.
[20] P. Docampo, J.M. Ball, M. Darwich, G.E. Eperon, H.J. Snaith, Efficient organometal trihalide perovskite planar-heterojunction solar cells on flexible polymer substrates, Nat. Commun. 4 (2013) 2761. Available from: https://doi.org/10.1038/ncomms3761.
[21] C. Roldán-Carmona, O. Malinkiewicz, A. Soriano, G.M. Espallargas, A. Garcia, P. Reinecke, et al., Flexible high efficiency perovskite solar cells, Energy Environ. Sci. 7 (3) (2014) 994−997.
[22] X. Yin, P. Chen, M. Que, Y. Xing, W. Que, C. Niu, et al., Highly efficient flexible perovskite solar cells using solution-derived NiO_X hole contacts, ACS Nano 10 (3) (2016) 3630−3636.
[23] B.A. Nejand, P. Nazari, S. Gharibzadeh, V. Ahmadi, A. Moshaii, All-inorganic large-area low-cost and durable flexible perovskite solar cells using copper foil as a substrate, Chem. Commun. 53 (4) (2017) 747−750.
[24] Molang Cai, Yongzhen Wu, Han Chen, Xudong Yang, Yinghuai Qiang, Liyuan Han, Cost-performance analysis of perovskite solar modules, Adv. Sci. 4 (2017) 1600269.
[25] Z. Song, C. Mcelvany, A.B. Phillips, I. Celik, P. Krantz, S.C. Watthage, et al., A technoeconomic analysis of perovskite solar module manufacturing with low-cost materials and techniques, Energy Environ. Sci. 10 (6) (2017). Available from: https://doi.org/10.1039/c7ee00757d.
[26] R. Sondergaard, et al., Roll-to-roll fabrication of polymer solar cells, Mater. Today 15 (2012) 36−49.
[27] K. Hwang, Y. Jung, Y. Heo, F. Scholes, S. Watkings, J. Subbiah, et al., Toward large scale roll-to-roll production of fully printed perovskite solar cells, Adv. Mater. 27 (2015) 1241−1247.
[28] Available from: https://www.csiro.au/en/Research/MF/Areas/Innovation/Flex-Electronics/Printable-perovskite-solar-cells.
[29] Y. Li, L. Meng, Y. Yang, G. Xu, Z. Hong, Q. Chen, et al., High-efficiency robust perovskite solar cells on ultrathin flexible substrates, Nat. Commun. 7 (2016) 10214. Available from: https://doi.org/10.1038/ncomms10214.

[30] M. Park, J. Park, I. Han, J. Oh, High-performance flexible and air-stable perovskite solar cells with a large active area based on poly(3-hexylthiophene) nanofibrils, J. Mater. Chem. A 4 (29) (2016) 11307–11316. Available from: https://doi.org/10.1039/c6ta03164a.
[31] J. Park, J. Heo, S. Park, K. Hong, H. Jeong, S. Im, et al., Highly flexible InSnO electrodes on thin colourless polyimide substrate for high-performance flexible $CH_3NH_3PbI_3$ perovskite solar cells, J. Power Sources 341 (2017) 340–347.
[32] M. Kumar, N. Yantara, S. Dharani, M. Gratzel, S. Mhaisalkar, P. Boix, et al., Flexible, low-temperature, solution processed ZnO-based perovskite solid state solar cells, J. Chem. Soc., Chem. Commun. 49 (2013) 11089–11091.
[33] Y. Dkhissi, F. Huang, S. Rubanov, M. Xiao, U. Bach, L. Spiccia, et al., Low temperature processing of flexible planar perovskite solar cells with efficiency over 10%, J. Power Sources 278 (2015) 325–331.
[34] B.J. Kim, D.H. Kim, Y.Y. Lee, H.W. Shin, G.S. Han, J.S. Hong, et al., Highly efficient and bending durable perovskite solar cells: toward a wearable power source, Energy Environ. Sci. 8 (3) (2015) 916–921.
[35] D. Yang, R. Yang, X. Ren, X. Zhu, Z. Yang, C. Li, et al., Hysteresis-suppressed high-efficiency flexible perovskite solar cells using solid-state ionic-liquids for effective electron transport, Adv. Mater. 28 (26) (2016) 5206–5213.
[36] M.M. Tavakoli, K.H. Tsui, Q. Zhang, J. He, Y. Yao, D. Li, et al., Highly efficient flexible perovskite solar cells with antireflection and self-cleaning nanostructures, ACS Nano 9 (10) (2015) 10287–10295.
[37] M. Kaltenbrunner, G. Adam, E.D. Głowacki, M. Drack, R. Schwödiauer, L. Leonat, et al., Flexible high power-per-weight perovskite solar cells with chromium oxide-metal contacts for improved stability in air, Nat. Mater. 14 (10) (2015) 1032–1039.
[38] J.W. Jung, S.T. Williams, A.K.Y. Jen, Low-temperature processed high-performance flexible perovskite solar cells via rationally optimized solvent washing treatments, RSC Adv. 4 (108) (2014) 62971–62977.
[39] Q. Chen, H. Zhou, Z. Hong, S. Luo, H.S. Duan, H.H. Wang, et al., Planar heterojunction perovskite solar cells via vapor-assisted solution process, J. Am. Chem. Soc. 136 (2) (2013) 622–625.

CHAPTER 12

Fabrication of Semitransparent Perovskite Solar Cells

Jayeeta Chattopadhyay[1] and Tara Sankar Pathak[2]
[1]Amity University Jharkhand, Ranchi, Jharkhand, India
[2]Surendra Institute of Engineering and Management, Siliguri, West Bengal, India

12.1 INTRODUCTION

The transformation to a fossil-fuel civilization occurred more rapidly than any other change in energy regimes in world history. We heat our homes and businesses with fossil fuels, run our factories with fossil fuels, light our cities, and communicate our distances with electricity derived from fossil fuels, power our transportation with fossil fuels, grow our food with the help of fossil fuels, construct our buildings with materials produced from fossil fuels, treat our sickness with pharmaceuticals made from fossil fuels derivatives, store our surpluses with plastic containers and packaging made from fossil fuels, and produce our home appliances with petrochemicals [1]. In this way, the global fossil-fuel energy demand is expecting to be doubled within next 50 years. This demand cannot be fulfilled by the combined usage of crude oil, natural gas, coal, and nuclear energy. At the same time, the reliance of civilized society on oil and gas makes the world economy dependent on a limited number of exporting countries [2]. But global warming represents the darkest side of the commercial ledger of the fossil-fuel civilization. From the past several hundred years, and especially in the 20th century, human beings burned massive amounts of "stored sun" in the forms of coal, oil, and natural gas to produce the energy made an industrial way of life possible. That spent energy has accumulated in the Earth's atmosphere and has started affecting adversely the climate of the planet and the workings of its ecosystems [1]. The global average air temperature near the Earth's surface raised $0.74°C \pm 0.18°C$ ($1.33°F \pm 0.32°F$) during the last 100 years. The Intergovernmental Panel on Climate Change (IPCC) concludes, "Most of the observed increase in globally averaged temperatures since the mid-20th century is very likely due to the observed increase in anthropogenic

Perovskite Photovoltaics
DOI: https://doi.org/10.1016/B978-0-12-812915-9.00012-5

greenhouse gas concentrations" [3] via the greenhouse effect, as a result, *climate change is the most severe problem that we are facing today, more serious even than the threat of terrorism*. In response to the above stated two problems, governments across the world are making continuous effort in exploration of clean, renewable alternatives, and stimulating the utilization of renewable energies and resources such as, solar, wind, water, and biomass [2,4]. Nowadays, photovoltaics (PVs) can be considered as an established technology that can contribute to lower energy consumption and massive reduction in greenhouse gas emissions.

PV modules actually based on crystalline silicon cells (c-Si) are predominant on the market; however, its conversion efficiencies are 15% for polycrystalline and 20% for monocrystalline silicon cells [5]. Moreover, these cells are mostly rigid, opaque, and flat, which are not suitable in integrating the system due to the lack of transparency. In this process, many attempts have been made to encapsulate c-Si cells in laminated glass, with the application of transparent spacing between them. Despite this lack of efficiency, it is still one of the preferred solutions, due to its lower costs and consistent, long-lived mass production.

In recent days, many researchers are working on amorphous silicon solar cells (a-Si) [6,7]. Among them, Green et al. have currently reached the best laboratory efficiency of 10.2% [7]. Organic PVs represent an interesting technology for building integration of photovoltaic, which use thin, flexible layers of organic light-harvesting molecules to generate power. As per the report, this technology can attain 11.5% of efficiency for semitransparent devices [8]. However, this device cannot be commercialized due to the lack of durability [8,9].

Now-a-days, the emergence of perovskite-based solar cells has revolutionized the field of new generation of PVs. These devices are easily processable, solid-state high conversion efficiency solar cells [10], most commonly based on hybrid organic–inorganic metal halides (ABX_3), with $A = (CH_3NH_3, NH_2CHNH_2, Cs)$, $B = (Pb, Sn)$ and $X = Br, Cl, I$ enabling accurate tuning of band gaps between ~ 1.2 and 3 eV [11–13]. A conversion efficiency of $20.1\% \pm 0.4\%$ has been achieved by this recent developed technology [7]. Several strategies have been proposed to realize highly transparent perovskite cells. In this type of device, a layer of perovskite placed in between two different electron and hole transporting materials, which are used as anode and cathode, respectively. In this case, thick layer of perovskite layer acts to absorb all incidents light, which results in an opaque device. The enhancement of transparency can be possible, by making the perovskite layer thinner or by fabricating the perovskite morphology [14]. Usually, perovskite contains discontinuous micro-islands, which can become invisible to the human eye by suitably designed and results into the formation of neutral tinted films, with minimal impacts in the spectral properties of light entering way.

The considerable efforts are being devoted by many research groups to form highly efficient PVs via improved crystallization of perovskite and searching for low-cost interface materials [15–17]. Esthetic semitransparent perovskite solar cells have been simultaneously receiving growing attention because of their specific application in transparent architectures [18–20] such as windows, rooftops, greenhouses, and other fashion elements. To achieve efficient semitransparent perovskite devices, both the anode and the cathode of the devices should be highly transparent and conductive to minimize the optical and resistance losses. In addition to low-cost materials, the cost reduction of PVs substantially depends on the ability to use high-throughput coating methods combining with roll-to-roll processing techniques [8]. Till now, most of the devices reported are employed with thin metal films (Al, Ag, and Au) as top electrodes which were fabricated based on energy-intensive evaporation processes [19–21].

To attain the low-cost and high-stability criterion, carbon-based materials have received much attention by utilizing conducting electrodes for perovskite solar cells [22,23]. Li et al. have worked on semitransparent perovskite solar cells using carbon nanotubes (CNTs) as top electrodes. However, the sheet resistance of CNTs films (2–25 $k\Omega \cdot sq^{-1}$) limits the device performance, results in low fill factors (FFs) due to the high series resistance of the device [24]. Moreover, the involved transfer and lamination procedures increased the complexity in device fabrication which is incompatible with large-scale, continuous roll-to-roll manufacturing. In this regard, Mei et al. have reported the application of carbon black/graphite composite material as top electrode [22]. Despite all these negative factors, the device with carbon electrodes exhibits high efficiency and long-term stability; to achieve the desired sheet resistance, the carbon black/graphite layer had to be made thicker which makes the device nontransparent. Recently, many researchers are working on the alternative materials of Indium tin oxide (ITO) for optoelectronic devices, such as touch screen panels, solar cells, and light-emitting diodes [25]. Among these proposed materials, silver nanowires (AgNWs) represent one of the most promising materials with benchmark optoelectronic properties of

~90% transmittance at 550 nm and <20 $\Omega \cdot \text{square}^{-1}$ sheet resistance that are comparable with the commercial ITO electrode [26,27].

This chapter will mainly discuss on the difficulties of neutral and multicolored semipermanent perovskite solar cells, due to the presence of toxic metals, and problems faced during the fabrication of large area and/or flexible devices. Here we will emphasis on the ITO-free alternative materials, viz. silver nanowires and thin Au bottom electrodes, carbon-based materials The future aspects in wide-scale applications of these devices will also be taken into account with accountability of their stability.

12.2 CNT AND GRAPHENE-BASED PEROVSKITE SOLAR CELLS

The new age solar cell structures utilize perovskite material as both absorber and charge-transporting material; the hole transport properties of $CH_3NH_3PbI_3$ have enabled perovskite/Au solar cells with power conversion efficiency (PCE) of 8% [28]. To decrease the cost of solar cell fabrication, it is necessary to replace the evaporated metal (Au/Ag) electrode with a low-cost material. CNTs with their excellent electrical conductivity, chemical stability, and unique nanostructure have been applied as electrodes in CdTe [29,30] dye-sensitized solar cells [31,32], and organic solar cells [33,34]. Not only did the CNT electrode show comparable performance to conventional architectures, it also provided the solar cell with more features such as flexibility and semitransparency for tandem solar cells [35].

The laminated CNT network has been reported as the top electrode of the semipermanent perovskite solar cells by Li et al. [36]. Fig. 12.1A represents the architecture of the perovskite solar cell by this group, in which a thin and dense TiO_2 layer has deposited on top of the fluorinated tin oxide (FTO) substrate using spray pyrolysis method, followed by the mesoporous nanocrystalline TiO_2 layer. Further light-harvesting perovskite semiconductor layer was deposited by employing a sequential deposition method [37]. Finally, thin CNT network film was placed on the perovskite active layer to complete the device structure [38]. This CNT film was synthesized using chemical vapor deposition (CVD) technique. In this case, the adhesion between the perovskite layer and the CNT film was improved by adding a few droplets of toluene. This happens as the surface tension occurring during vaporization of the toluene

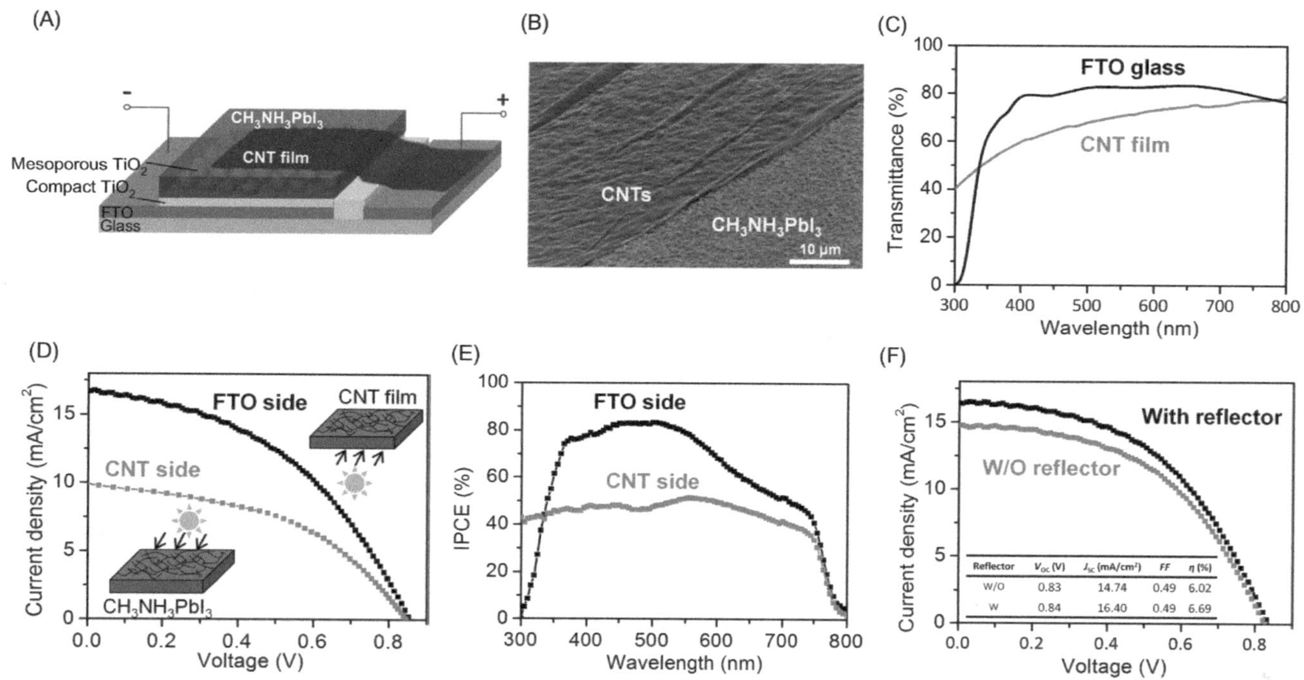

Figure 12.1 (A) Schematic diagram of $CH_3NH_3PbI_3$ perovskite solar cell with CNT film electrode. (B) Tilted SEM image of $CH_3NH_3PbI_3$ perovskite substrate (blue) partly covered by CNT film (purple). (C) UV-vis transmittance of a CNT film and FTO glass. (D) Light J-V curves of $CH_3NH_3PbI_3$ perovskite/CNTs solar cell with illumination from the FTO and from the CNT side, under conditions of AM1.5 100 $mW \cdot cm^{-2}$. (E) IPCE of a $CH_3NH_3PbI_3$ perovskite/CNTs solar cell with illumination from the FTO and from the CNT side. (F) Light J-V curves of a $CH_3NH_3PbI_3$ perovskite/CNTs solar cell with and without an Al reflector. *Reproduced with permission from Z. Li, S.A. Kulkarni, P. P. Boix, E. Shi, A. Cao, K. Fu, et al., ACS Nano, 8 (2014) 6797–6804. Copyright 2015, American Chemical Society.*

pulled the CNT network film toward the active layer due to the presence of van der Waals force without dissolving the active layer. This phenomena result into the enhanced electrical contact, as CNT network successfully transferred onto the perovskite layer in a conformal manner. Fig 12.1B represents the SEM image of the CNT film. This image has shown that the film was very thin (20–50 nm); the optically transparent property was achieved due to its lower thickness. The PCE of the device with the FTO side illumination is 6.29% ($V_{oc} = 0.86$ V, $J_{sc} = 16.7\ mA \cdot cm^{-2}$, and FF = 44%), while the corresponding PCE is 3.88% ($V_{oc} = 0.85$ V, $J_{sc} = 9.9\ mA \cdot cm^{-2}$, and FF = 46%) when illuminated from the CNT side. This lower PCE with the CNT side illumination primarily arises from the small J_{sc} value, which is attributed to the slightly lower transmission of the CNT film, leading to the limited adsorptions in the active layer. Fig. 12.1F shows J_{sc} value as $16.4\ mA \cdot cm^{-2}$ due to the introduction of the opaque layer at the backside of the device.

Graphene materials become popular because of their attractive features as compared to ITO and FTO, such as high optical transparency, high electrical conductivity, mechanical robustness, chemical stability, and low cost, thus providing a new pathway to remarkably enhanced performance characteristics of various types of solar cells. With the help of its fascinating properties described above, the semitransparent perovskite solar cells have been demonstrated by laminating CVD-grown multilayer graphene as the top electrode as shown in Fig. 12.2A [39].

The device architecture comprises a thin and compact TiO_2 layer (electron transport layer), perovskite light-absorbing semiconductor (photoactive layer), Spiro-OMeTAD/PEDOT:PSS (hole transport layers) and single-layer CVD graphene on top of the FTO glass substrate. A thin PEDOT:PSS layer is predeposited to improve the electrical conductivity of the graphene film, which is ascribed to the fact that the Fermi level of the PEDOT:PSS (5.0 eV) is higher than the Dirac point of graphene (4.6 eV), thus leading to an electrostatically doped graphene layer for more hole injections. Fig. 12.2B is showing the images of the fabricated graphene-based semitransparent perovskite solar cell devices with different thickness of 150–350 nm. It is to be notified here that the two-layer graphene film are synthesized by layer-by-layer stacking process to further improve the electrical conductivity and are used in all semitransparent devices. It is clear from the results that optical transparency becomes lower as the thickness of the perovskite layer increases, which is represented in Fig. 12.2C. With the illumination of FTO and graphene side,

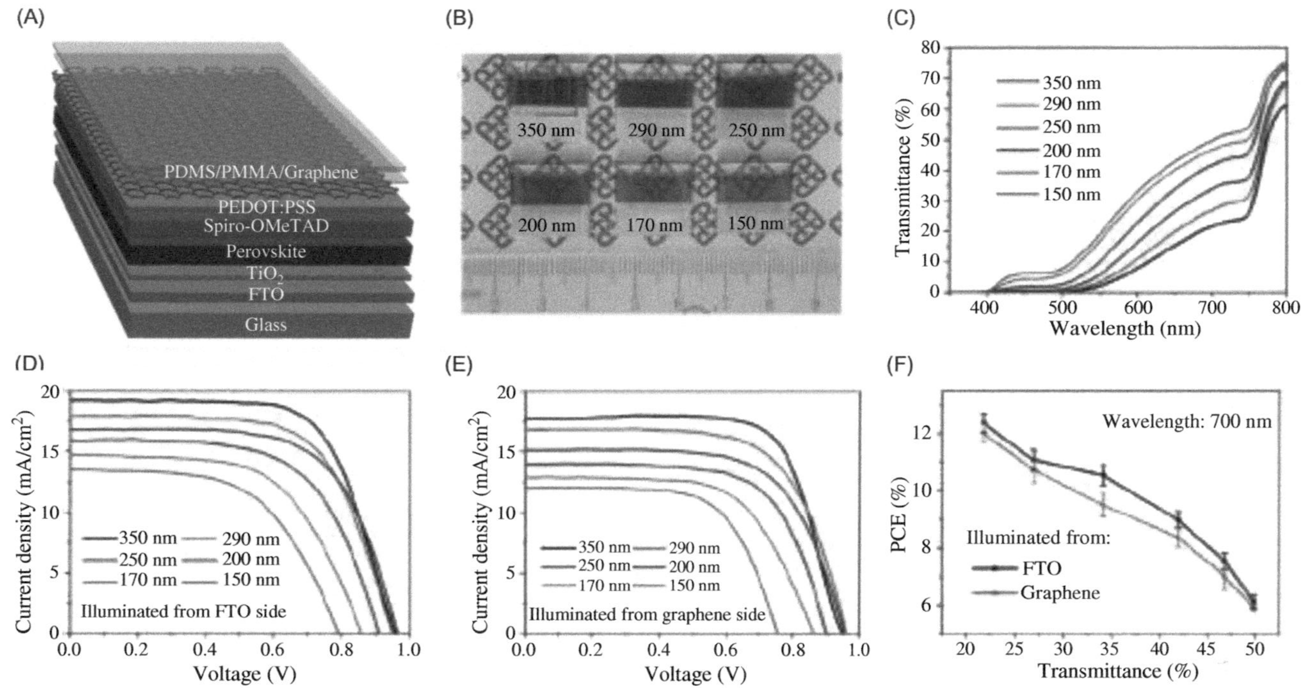

Figure 12.2 (A) Schematic view of a semitransparent perovskite solar cell employing a graphene-based transparent top electrode. (B) Photographs of semitransparent perovskite solar cells with transparent graphene electrodes. The thicknesses of the perovskite layers in the six devices are approximately 350, 290, 250, 200, 170, and 150 nm, respectively. (C) Transmittance spectra of the above devices. J-V characteristics of the semitransparent solar cells illuminated from (D) the FTO side and (E) the graphene side. (F) The average PCEs of the semitransparent perovskite solar cells as a function of the transmittance at the wavelength of 700 nm. PDMS, polydimethylsiloxane; PMMA, poly(methyl methacrylate); Spiro-OMeTAD: 2,2′,7,7′-tetrakis(N,N-di-p-methoxyphenylamine)-9,9-spirobifluorene. *Reproduced with permission from S.L. Hellstrom, M. Vosgueritchian, R.M. Stoltenberg, I. Irfan, M. Hammock, Y.B. Wang, et al., Nano Lett. 12 (2012) 3574–3580. Copyright 2015, Wiley.*

J-V curves are showing different results with variation of perovskite layer thickness (Fig. 12.2D and E). It can be concluded from the results that, with a device with 350 nm thick perovskite layer, the optical and electrical performance can be reached at its highest level with a PCE of 12.37% combined with a transmittance of ~23% at 700 nm. It is clear from the results that with rise in the thickness of the perovskite absorber layer, higher amount of photocurrent have been generated, which is enabled by the strong optical absorptions in the active layer. When the perovskite thickness became less than 200 nm, V_{oc} value reduced to 0.76 V from 0.95 V, which had been explained as the deficient coverage of the perovskite layer. Fig. 12.1F had employed the relationship between the average PCE of the semitransparent perovskite solar cells, where the graphene film has been used as the top electrode during the illumination from both the FTO and graphene sides at the optical transmittance at 700 nm with the transmission efficiency of ~50%. The device with 150 nm thick perovskite layer had shown relatively higher PCE value than the device with similar average transparency.

12.3 SILVER NANOWIRES

The carbon electrodes usually exhibits high efficiency and long-term stability in achieving the desired sheet resistance, but the CNTs and graphene layers have to be much thicker, which makes the device nontransparent in normal cases. There comes the significance of solution processable transparent electrodes in development of semitransparent perovskite devices. Recently, several such conducting materials have been processed extensively as being used as alternatives of ITO for optoelectronic devices, such as touch screen panels, solar cells, and light-emitting diodes [40]. However, Au is not an ideal material for the transparent electrode due to its high cost and low transparency in the near infrared spectral region, the latter of which notably hinders tandem solar cell construction. Among these proposed materials, silver nanowires (AgNWs) represent one of the most promising materials with benchmark optoelectronic properties of ~90% transmittance at 550 nm and <20 $\Omega \cdot sq.^{-1}$ sheet resistance that are comparable with the commercial ITO electrode [41,42]. However, the poor compatibility of Ag with the perovskite semiconductor resulted that Ag-based electrodes remained unexploited in the perovskite solar cells. To conquer this challenge, Guo et al. had investigated on ZnO nanoparticles–based thin interfacial layer

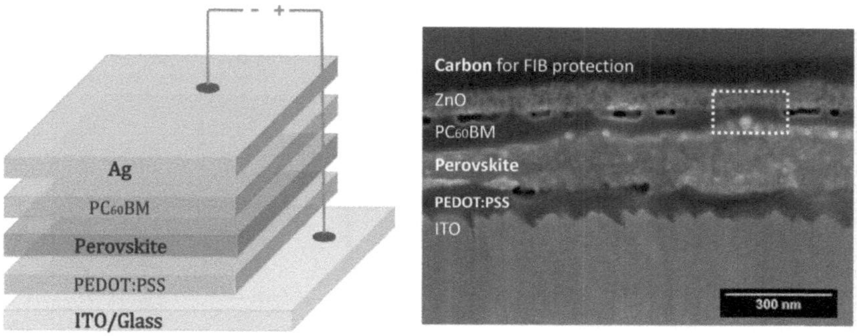

(A) (B)

Figure 12.3 (A) Schematic diagram of the devices with solution-processed top AgNWs. (B) Cross sectional SEM image of the whole device stack. The white rectangle represents the area where perovskite is not covered by [6,6]- phenyl C_{61}-butyric acid methyl ester ($PCBM_{60}$). *Reproduced with permission from F. Guo, H. Azimi, Y. Hou, T. Przybilla, M. Hu, C. Bronnbauer, et al., Nanoscale 7 (2015) 1642–1649 [45]. Copyright 2015, Royal Society of Chemistry.*

between AgNWs and the electron transporting layer, which can efficiently prevent the diffusion of the Ag atoms into the perovskite material and also ensure ohmic contact at the top electrode [43,44]. The device structure, which consists of PEDOT:PSS/$CH_3NH_3PbI_{3-x}Cl_x$/$PCBM_{60}$/ZnO/AgNWs on the ITO/glass substrate, is schematically depicted in Fig. 12.3A.

PEDOT:PSS:poly(3,4-ethylenedioxythiophenone):poly(styrenesulphonic acid); ITO. Without inserting an additional ZnO electron transport layer between AgNWs and PCBM60, it is difficult for the charges to be efficiently extracted to the electrode, which is ascribed to a large energy barrier between $PCBM_{60}$ and AgNWs. Moreover, ZnO nanoparticles are compatible with evaporated Ag as well as solution-processed AgNWs; therefore, it is served not only as a capping layer but also as an ohmic contact formation at the ZnO-Ag interface for efficient electron extraction, both of which are responsible for achieving high values of V_{oc} and FF. The optimal thickness of the ZnO layer had been reported between 60 and 90 nm. When the thickness of the ZnO layer was too thin, there was a possibility that Ag atom diffusion occurs through the thin ZnO layer, reducing the built-in potential and hence the V_{oc} of the solar cell devices. A spray coating method had been used as it could create a uniform thin film without damaging the underlying layer due to its very fast drying

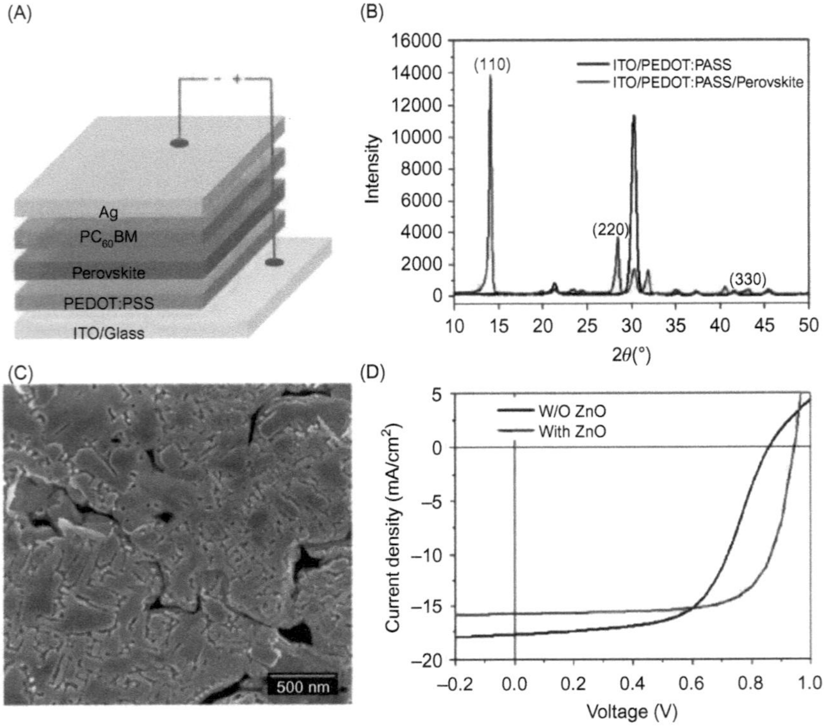

Figure 12.4 (A) Schematic architecture of the perovskite reference device. (B) XRD pattern of the perovskite crystals grown on PEDOT:PSS/ITO (gray line), with blank sample of the PEDOT:PSS-coated ITO substrate for comparison (black line). (C) Top-view SEM image of the as-grown perovskite film. (D) J–V curves of the two devices with and without a layer of ZnO between the $PC_{60}BM$ and Ag top electrode. *Reproduced with permission from F. Guo, H. Azimi, Y. Hou, T. Przybilla, M. Hu, C. Bronnbauer, et al., High-performance semitransparent perovskite solar cells with solution-processed silver nanowires as top electrodes, Nanoscale, 7 (5):1642–1649. doi:10.1039/c4nr06033d. Copyright 2014, Royal Society of Chemistry.*

characteristics for the deposition of AgNW electrode. Fig. 12.3B is showing a cross-sectional view of the fabricated semitransparent perovskite solar cell device. This result clearly depicted that the interface between AgNWs and ZnO is fairly distinct. This implies that spray-coated AgNWs can be well supported by the underlying ZnO layer. It is important to note that the semitransparent perovskite solar cells with the AgNW top electrode described in this study show greatly improved stability as compared to what has been observed so far. This is because having a compact ZnO interface layer can physically separate the perovskite light-absorbing medium from the AgNW top electrode, thereby hindering the Ag atoms from forming Ag halide by chemical reactions, which causes performance degradation. This finding provides the possibility of realizing the tandem solar cells by exploiting the semitransparent perovskite solar cells with a highly efficient and highly transparent AgNW top electrode as a top cell, which can be mechanically stacked or placed on top of the bottom cell with a tunnel junction. Fig. 12.4A represents the similar architecture of the same research group. In this case, the perovskite absorber is sandwiched between the PEDOT:PSS and $PC_{60}BM$, which serves as an electron and hole-blocking layer, respectively. They grew the perovskite crystals from a 3:1 molar ratio of NH_3CH_2I and $PbCl_2$ precursor solution on PEDOT:PSS-coated ITO substrates. Fig. 12.4B has shown the sharp diffraction peaks of the perovskite in XRD pattern.

12.4 CONCLUSIONS

In recent years, the perovskite solar cell becomes huge popular as a strong candidate in solar cell technology for the upcoming days due to its superior characteristics, whose performance can be comparable with conventional Si-based solar cell technology, with economical options. This alternate device has already been reached over 20% of its conversion efficiency. However, there are many factors that control this technology from commercialization, e.g., the toxicity of materials commonly utilized to produce high performance, the long-term stability, and applicability to large area fabrication process without any loss of performance. The researchers should approach these issues to improve the performance further. It is clearly seen that the fabrication in the device architecture actually enhances the photocharge generation with the minimum usage of light in the device. It is expected to be one of the most promising candidates in future solar cell technology.

REFERENCES

[1] Jeremy Rifkin, The Hydrogen Economy, Tarcher Penguin, New York, 2002, p. 64. **Chapter 4**.
[2] Jean-Paul Lange, Lignocellulose conversion: an introduction to chemistry, process and economics, Biofuels, Bioproducts and Biorefining 1 (2007) 30–48.
[3] Summary for policymakers. Climate change 2007: the physical science basis. In *Contribution of Working Group I to the Fourth Assessment Report of the Intergovernmental Panel on Climate Change*. Intergovernmental Panel on Climate Change (2007).
[4] M. Ni, Dennis Y.C. Leung, Michael K.H. Leung, K. Sumathy, An overview of hydrogen production from biomass, Fuel Process. Tech. 87 (2006) 461–472.

[5] D.O. Akinyele, R.K. Rayudu, N.K.C. Nair, Distributed photovoltaic *power* generation for *energy*-poor households: The Nigerian perspective, Renew. Sustain. Energy Rev. 48 (2015) 112–139.

[6] D.E. Carlson, C.R. Wronski, Amorphous silicon solar cell, Appl. Phys. Lett. 671 (1976) 1–4.

[7] A. Green Martin, E. Keith, H. Yoshihiro, Solar cell efficiency tables (version 46), Prog. Photovolt. Res. Appl. 23 (2015) 1–9.

[8] C.J.M. Emmott, A. Urbina, J. Nelson, Environmental and economic assessment of ITO-free electrodes for organic solar cells, Sol. Energy Mater. Sol. Cells 97 (2012) 14–21.

[9] H. Tan, A. Furlan, W. Li, K. Arapov, R. Santbergen, M.M. Wienk, Highly efficient hybrid polymer and amorphous silicon multijunction solar cells with effective optical management, Adv. Mater. 28 (2016) 2170–2177.

[10] M.A. Green, A. Ho-Baillie, H.J. Snaith, The emergence of perovskite solar cells, Nat. Photon 8 (2014) 506–514.

[11] K. Yang, F. Li, J. Zhang, C.P. Veeramalai, T. Guo, All-solution processed semi-transparent perovskite solar cells with silver nanowires electrode, Nanotechnology 27 (2016) 95202.

[12] W. Zhang, M. Anaya, G. Lozano, M.E. Calvo, M.B. Johnston, H. Miguez, Highly efficient perovskite solar cells with tunable structural color, Nano Lett. 15 (2015) 1698–1702.

[13] G.E. Eperon, S.D. Stranks, C. Menelaou, M.B. Johnston, L.M. Herz, H.J. Snaith, Formamidinium lead trihalide: a broadly tunable perovskite for efficient planar heterojunction solar cells, Energy Environ. Sci. 7 (2014) 982.

[14] G.E. Eperon, V.M. Burlakov, A. Goriely, H.J. Snaith, Neutral Color Semitransparent Microstructured Perovskite Solar Cells, ACS Nano 8 (2014) 591–598.

[15] P.W. Liang, C.Y. Liao, C.C. Chueh, F. Zuo, S.T. Williams, X.K. Xin, et al., Additive enhanced crystallization of solution-processed perovskite for highly efficient planar-heterojunction solar cells, Adv. Mater. 26 (2014) 3748–3754.

[16] Z. Xiao, C. Bi, Y. Shao, Q. Dong, Q. Wang, Y. Yuan, et al., Efficient, high yield perovskite photovoltaic devices grown by interdiffusion of solution-processed precursor stacking layers, Energy Environ. Sci. 7 (2014) 2619–2623.

[17] J.Y. Jeng, K.C. Chen, T.Y. Chiang, P.Y. Lin, T.D. Tsai, Y.C. Chang, et al., Nickel oxide electrode interlayer in CH_3 NH_3 PbI_3 perovskite/PCBM planar-heterojunction hybrid solar cells, Adv. Mater. 26 (2014) 4107–4113.

[18] Z. Li, S.A. Kulkarni, P.P. Boix, E. Shi, A. Cao, K. Fu, et al., Laminated carbon nanotube networks for metal electrode-free efficient perovskite solar cells, ACS Nano 8 (2014) 6797–6804.

[19] G.E. Eperon, V.M. Burlakov, A. Goriely, H.J. Snaith, Neutral Color Semitransparent Microstructured Perovskite Solar Cells, ACS Nano 8 (2013) 591–598.

[20] C. Roldán-Carmona, O. Malinkiewicz, R. Betancur, G. Longo, C. Momblona, O. Malinkiewicz, et al., High efficiency single-junction semitransparent perovskite solar cells, Energy Environ. Sci. 7 (2014) 2968–2973.

[21] L.K. Ono, S. Wang, Y. Kato, S.R. Raga, Y. Qi, Pinhole-free hole transport layers significantly improve the stability of $MAPbI_3$-based perovskite solar cells under operating conditions, Energy Environ. Sci. 7 (2014) 3989–3993.

[22] A. Mei, X. Li, L. Liu, Z. Ku, T. Liu, Y. Rong, et al., A hole-conductor-free, fully printable mesoscopic perovskite solar cell with high stability, Science 345 (2014) 295–298.

[23] L. Zhang, T. Liu, L. Liu, M. Hu, Y. Yang, A. Mei, et al., The effect of carbon counter electrodes on fully printable mesoscopic perovskite solar cells, J. Mater. Chem. A 3 (2014) 9165–9170. Available from: https://doi.org/10.1039/C4TA04647A.

[24] Z. Li, S.A. Kulkarni, P.P. Boix, E. Shi, A. Cao, K. Fu, et al., Laminated carbon nanotube networks for metal electrode-free efficient perovskite solar cells, ACS Nano 8 (2014) 6797–6804.
[25] D.S. Hecht, L. Hu, G. Irvin, Emerging Transparent Electrodes Based on Thin Films of Carbon Nanotubes, Graphene, and Metallic Nanostructures, Adv. Mater. 23 (2011) 1482–1513.
[26] D.S. Leem, A. Edwards, M. Faist, J. Nelson, D.D. Bradley, J.C. de Mello, Efficient organic solar cells with solution-processed silver nanowire electrodes, Adv. Mater. 23 (2011) 4371–4375.
[27] A. Kim, Y. Won, K. Woo, C.H. Kim, J. Moon, Highly Transparent Low Resistance ZnO/Ag Nanowire/ZnO Composite Electrode for Thin Film Solar Cells, ACS Nano 7 (2013) 1081–1091.
[28] W. Abu Laban, L. Etgar, Depleted hole conductor-free lead halide iodide heterojunction solar cells, Energy Environ. Sci. 6 (2013) 3249–3253.
[29] T.M. Barnes, X. Wu, J. Zhou, A. Duda, J. van de Lagemaat, T.J. Coutts, et al., Single-wall carbon nanotube networks as a transparent back contact in CdTe solar cells, Appl. Phys. Lett. 90 (2007) 243503.
[30] A.B. Phillips, R.R. Khanal, Z.N. Song, R.M. Zartman, J.L. DeWitt, J.M. Stone, et al., Wiring-up Carbon Single Wall Nanotubes to Polycrystalline Inorganic Semiconductor Thin Films: Low-Barrier, Copper-Free Back Contact to CdTe Solar Cells, Nano Lett. 13 (2013) 5224–5232.
[31] T. Chen, L.B. Qiu, Z.B. Cai, F. Gong, Z.B. Yang, Z.S. Wang, et al., Intertwined Aligned Carbon Nanotube Fiber Based Dye-Sensitized Solar Cells, Nano Lett. 12 (2012) 2568–2572.
[32] S. Zhang, C.Y. Ji, Z.Q. Bian, P.R. Yu, L.H. Zhang, D.Y. Liu, et al., Single-Wire Dye-Sensitized Solar Cells Wrapped by Carbon Nanotube Film Electrodes, ACS Nano 6 (2012) 7191–7198.
[33] S. Chaudhary, H.W. Lu, A.M. Muller, C.J. Bardeen, M. Ozkan, Hierarchical Placement and Associated Optoelectronic Impact of Carbon Nanotubes in Polymer-Fullerene Solar Cells, Nano Lett. 7 (2007) 1973–1979.
[34] R. Ulbricht, S.B. Lee, X.M. Jiang, K. Inoue, M. Zhang, S.L. Fang, et al., Transparent carbon nanotube sheets as 3-D charge collectors in organic solar cells, Sol. Energy Mater. Sol. C 91 (2007) 416–419.
[35] X.Y. Xia, S.S. Wang, Y. Jia, Z.Q. Bian, D.H. Wu, L.H. Zhang, et al., Surface modification of colloidal CdX-based Quantum dots for biomedical applications, J. Mater. Chem. 20 (2010) 8478–8482.
[36] Z. Li, S.A. Kulkarni, P.P. Boix, E. Shi, A. Cao, K. Fu, et al., Laminated carbon nanotube networks for metal electrode-free efficient perovskite solar cells, ACS Nano 8 (2014) 6797–6804.
[37] J. Burschka, N. Pellet, S.-J. Moon, R. Humphry-Baker, P. Gao, M.K. Nazeeruddin, et al., Sequential deposition as a route to high-performance perovskite-sensitized solar cells, Nature 499 (2013) 316–319.
[38] Z. Li, Y. Jia, J. Wei, K. Wang, Q. Shu, X. Gui, et al., Large area, highly transparent carbon nanotubespiderwebs for energy harvesting, J. Mater. Chem. 20 (2010) 7236–7240.
[39] S.L. Hellstrom, M. Vosgueritchian, R.M. Stoltenberg, I. Irfan, M. Hammock, Y.B. Wang, et al., Strong and stable doping of carbon nanotubes and graphene by MoOx for transparent electrodes, Nano Lett. 12 (2012) 3574–3580.
[40] D.S. Hecht, L. Hu, G. Irvin, Emerging Transparent Electrodes Based on Thin Films of Carbon Nanotubes, Graphene, and Metallic Nanostructures, Adv. Mater. 23 (2011) 1482–1513.

[41] D.S. Leem, A. Edwards, M. Faist, J. Nelson, D.D. Bradley, J.C. de Mello, Efficient organic solar cells with solution-processed silver nanowire electrodes, Adv. Mater. 23 (2011) 4371–4375.

[42] A. Kim, Y. Won, K. Woo, C.H. Kim, J. Moon, Highly Transparent Low Resistance ZnO/Ag Nanowire/ZnO Composite Electrode for Thin Film Solar Cells, ACS Nano 7 (2013) 1081–1091.

[43] F. Guo, P. Kubis, T. Stubhan, N. Li, D. Baran, T. Przybilla, et al., Fully solution-processing route toward highly transparent polymer solar cells, ACS Appl. Mater. Interfaces 6 (2014) 18251–18257.

[44] F. Guo, P. Kubis, N. Li, T. Przybilla, G. Matt, T. Stubhan, et al., Solution-Processed Parallel Tandem Polymer Solar Cells Using Silver Nanowires as Intermediate Electrode, ACS Nano 8 (2014) 12632–12640.

[45] F. Guo, H. Azimi, Y. Hou, T. Przybilla, M. Hu, C. Bronnbauer, et al., High-performance semitransparent perovskite solar cells with solution-processed silver nanowires as top electrodes, Nanoscale 7 (5) (2015) 1642–1649. Available from: https://doi.org/10.1039/c4nr06033d.

CHAPTER 13

Commercialization of Large-Scale Perovskite Solar Energy Technology and Scaling-Up Issues

Madhulika Bhati and Radhika Rai
National Institute of Science, Technology and Development Studies [CSIR-NISTADS], New Delhi, India

13.1 INTRODUCTION

The industrial revolution that began in the second half of the 18th century was primarily based on the more conventional energy sources like coal and later oil and natural gases. These are the finite resources on the earth and continuously depleting. Several studies analyzed the approximate lifetime of world's petroleum, natural gas, and coal reserves with current consumption rate, which is 42 years, 56 years, and 118 years, respectively (Fig. 13.1). Apart from pressure on total resources and availability, the burning of vast quantities of coal and oil and other fossil fuels are responsible for environment deterioration and also significantly contributed to global warming. Thus the challenge before us to harness the energy forms in such a way that they decrease the CO_2 emission (footprint) and also preserve the environment and the ecological balance [1].

Green technologies such as wind technology, solar technology, small hydropower, biomass and waste to energy technologies are developed and promoted as alternative sources that make little or no contribution to climate change and provide sustainable energy services to meet the global energy demands. These green technologies have strong potential to meet the future energy demand without compromising the environment quality as they have very less carbon. At first in 1980, United Nations general assembly put it the need of energy transition as "The international community will have to make substantial and rapid progress in the transition

Perovskite Photovoltaics
DOI: https://doi.org/10.1016/B978-0-12-812915-9.00013-7

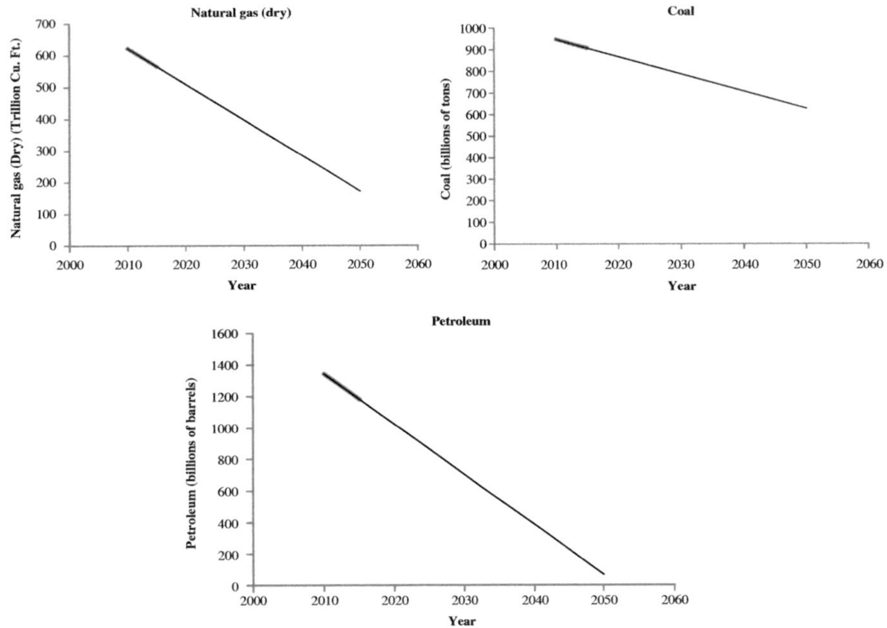

Figure 13.1 Graphs showing depleting natural resources with time in years. *From Green Technology Report. Available from: http://www.nistads.res.in/images/ISTIP/report/green%20technology.pdf.*

from the present international economy based primarily on hydrocarbons. It will have to rely increasingly on new and renewable sources of energy, seeking to reserve hydrocarbons for non-energy and non-substitutable uses" [1,2].

The conception of renewable energy was further emphasized by the 150 nations of the world at the New Energy and Renewable Energy conference by United Nations at Nairobi in August 1981. Renewable energy was defined clearly that the renewable energy was developed and utilized by using new technology and new materials. Renewable energy is different from other fossil energy for its sustainable development, restoring ability, and environment friendliness. Some main types of green technologies are solar, wind, small hydro, biomass, and waste to energy. This renewable or green technology helps in meeting the present energy requirement without compromising future energy needs and environmental factors [3].

To reduce the burden from the limited resources like fossil fuel and further emphasis on the reduction of the carbon footprint, solar technologies is an effective solution. Total estimated potential of renewable power generation includes various sources like solar, wind, small-hydro power, biomass power, bagasse-based cogeneration in sugar mills, and waste to energy (Table 13.1).

According to International Renewable Energy Agency report [5], the energy sector accounts for more than two-thirds of global greenhouse gas emissions. As such, energy must be our priority in bringing down CO_2 emissions. Countries accounting for 90% of global CO_2 emissions have submitted their climate plans (INDCs) in advance of the Paris negotiations later this year.

Table 13.1 Various renewable energy generation options and their contribution

Renewable energy generation option	Contribution	Percentage share
Solar	748990	62.48
Wind (80 m height + 100 m height)	405023	33.78
Small hydro power	19749	1.65
Biomass	17538	1.46
Bagasse-based cogeneration in sugar mills	5000	0.42
Waste to energy	2556	0.21

Source: Energy Statistics, Central Statistics Office, Government of India. 2017. Available from: http://www.cbgaindia.org/wp-content/uploads/2017/04/Energy_Statistics_2017.pdf [4].

There is a gap (15–17 gigatons) between what countries have promised to do, and the emission reductions needed to peak emissions by 2020 and stay on the 2°C trajectory. Even if all INDCs are fully implemented, the temperature will rise roughly 2.7°C by the end of the century.

Renewable energy and energy efficiency are the most effective way to close that gap for two reasons:

- Renewable energy can deliver half of all emission reductions needed to keep temperature rise below 2°C. Energy efficiency measures can deliver the rest.
- Renewable and efficiency are the only technologies that can be deployed fast enough and at a sufficient scale to close this gap in time.

Renewable power capacity has accounted for more than half of capacity additions in the global power sector since 2011 [6]. Renewable energy already accounts for more than 22% of total global electricity generation—reducing total power sector emissions by 20%. Many countries are already well on their way to meeting "100% renewable" objectives, as are a number of leading businesses. They are doing this because renewable energy is not only good for the climate, it's good for communities, it's good for the economy, it's good for job creation, and it's good for business. In just the last 3 years, the world added more than 100 gigawatts of new renewable energy capacity every year—a number equivalent to the total installed generation capacity of Brazil. Of note, 164 countries now have radiant efficiency targets, up from 43 countries in 2005 [7] As of October 2015, 133 countries have expressed support for the long-term goal of decarbonization, net zero emissions, or carbon neutrality.

Energy technology innovation will be a key in achieving the 2°C scenario. A comprehensive portfolio of low-carbon technologies, including solutions for decarbonization, will be needed to achieve policy climate goals. Renewable are now not in some Indian future, but today the most affordable source of power in many parts of the world, a trend that will continue to grow [8].

Solar photovoltaic (PV) module costs have fallen as much as 80% since 2009. Renewable energy technologies are creating more jobs on average than fossil fuel technologies and are therefore beneficial for society and the economy. Of note, 7.7 million people work in the renewable energy sector (excluding large hydro), an 18% increase from the previous year. If a share of 36% renewable energy is achieved, employment levels could exceed 24 million jobs by 2030. Solar PV creates more than twice the number of jobs per unit of electricity than coal or natural gas. China employs more than 3.4 million people in its renewable energy sector, which represents around 20% of the total workforce in the country's energy sector, and is significantly more than the number of jobs in the oil and gas sectors combined (2.6 million) [9].

Solar energy is an attractive and feasible option to traditional resources such as coal and fossil fuel for the present growing energy demand. In this perspective, developing solar cells is one of the best approaches to convert solar energy into electrical energy based on the PV effect. Over the years, silicon-based cells have been used for industrial purposes due to their efficient solar-to-power generation (~30%), particularly crystalline silicon. However, the cost of Si-based PV cells is relatively high and difficult to utilize in large-scale industries. An alternative to silicon solar cells is third-generation excitonic PV devices, which have been developed based on various dye sensitizers, organic and hybrid (organic–inorganic) materials. There is also a chance to enhance perovskite cell efficiency to 31%. And even higher efficiencies might be achieved by combining silicon and perovskite devices [10]. Among these materials, perovskites (organic–inorganic) have reached top position (~20.1%) within ~5 years, due to substantial improvement of power conversion efficiency (PCE) and low processing costs. As the efficiency of the device increases, the number of publications is also rapidly increasing since 2009 (Fig. 13.2). Significant aspects of perovskites are synthetic feasibility, strong optical absorption, charge recombination rate, and ease of fabrication. Moreover, hybrid perovskites (HPs) can be prepared by simple synthetic methods and are easy to capitalize when compared to the existing excitonic PV technologies such as dye-sensitized solar cells (DSSCs) and

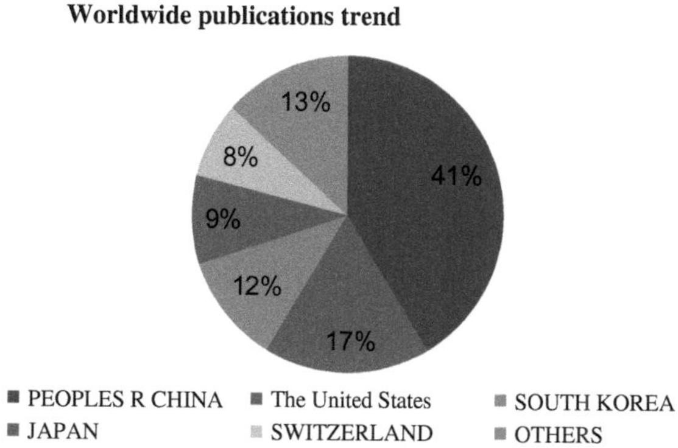

Figure 13.2 Major countries' players. *From Web of Science database.*

organic solar cells (OSCs). Another important aspect is high charge-carrier mobility, which is more useful for developing high-performance solar cell devices [11]. However, toxicity of lead is a major concern, which easily degrades on exposure to humidity and ultraviolet (UV) irradiation. Present-day research mainly focuses on the commercialization of perovskite solar cells (PSCs) by controlling degradation and toxicity. In this chapter, we highlight the fundamental aspects of perovskites and recent progress in perovskite-based device fabrication. Still, there exist some issues which need to be resolved in the commercialization of perovskites.

Solar cell technologies are traditionally divided into three generations. First-generation solar cells are mainly based on silicon wafers and typically demonstrate a performance about 15%–20%. Crystalline silicon (c-Si) cells (both single crystalline silicon and multi-crystalline silicon) and gallium arsenide (GaAs) cells belong to the wafer-based PV. C-Si-based solar cell dominates the market. This first generation of solar cell dominate the market and comprises the 90% of the total share [12,13]. The good performance and high stability are the important factors for their successful commercialization. The second-generation solar cells are based on amorphous silicon, CIGS and CdTe, where the typical performance is 10%–15%. These types of cell, although used less material, require energy intensive treatments like vacuum processes and high temperature treatments. Furthermore, the second-generation solar cells are based on scarce elements and this is a limiting factor in the price (Fig. 13.3).

Third-generation solar cells are made of organic materials such as small molecules or polymers. Thus, polymer solar cells are a subcategory of OSCs. The third generation also covers expensive high-performance experimental multi-junction solar cells which hold the world record in solar cell performance [15,16]. This type has only to some extent a commercial application because of the very high production price. However, third-generation approaches to PVs aim to decrease costs to well below the \$1/W level of second-generation PVs to \$0.50/W, potentially to \$0.20/W or better, by significantly increasing efficiencies but maintaining the economic and environmental cost advantages of thin-film deposition techniques [17].

13.2 PSC ARCHITECTURE

In recent years, a new class of thin film solar cell, perovskite-based solar cells have raced to the front of emerging PVs and show a huge potential of beyond 20% PCE (Fig. 13.4). These emerging technologies have

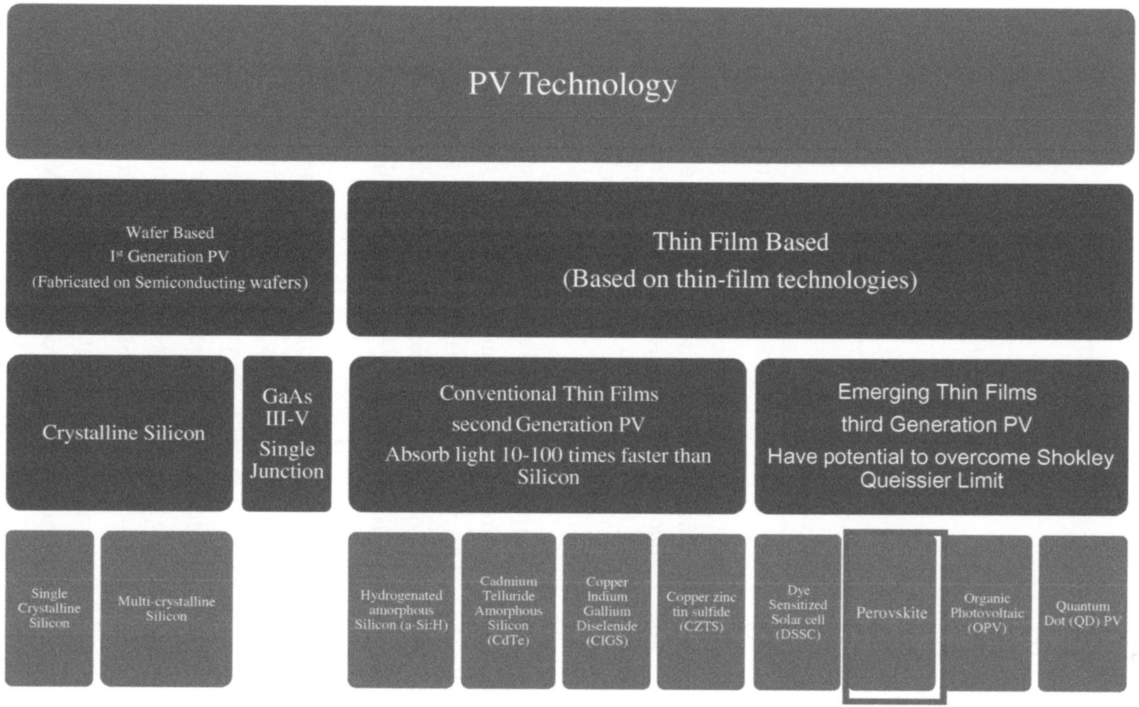

Figure 13.3 Solar PV generations. *From Perovskite Photovoltaics 2016–2026: Technologies, Markets, Players, IDTechEx, 2016 [14].*

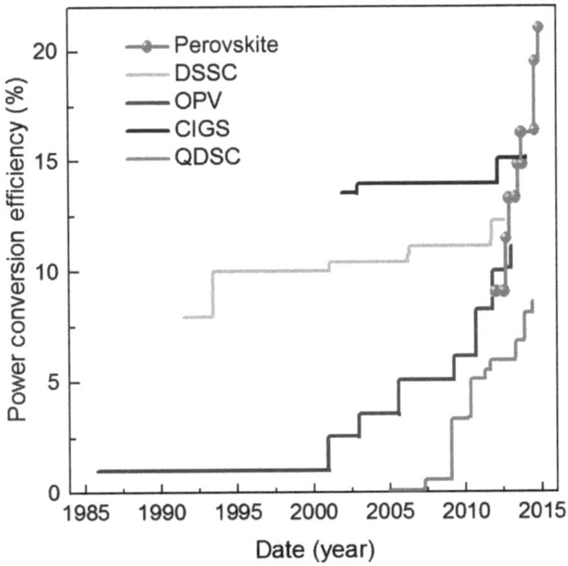

Figure 13.4 Solar conversion efficiency progress for solution-processed solar cells. *From P.P. Boix, T.M. Koh, S. Agarwala, G.S. Mahisalkar. Perovskite solar cell: beyond methylammonium lead iodide. J. Phys. Chem. Lett. 6 (2015) 898–907, doi: 10.1021/jz502547 [18].*

recently come up and are competing with well-established solar technologies such as the inorganic thin-film and multi-crystalline silicon used in solar panels worldwide. This technology has the potential to solve flying demands of energy production and has created thriving ideas of devices and material development, consequently increasing patent filing activities. Filing early patents is advantageous because it reduces the competition and also creates understanding of what to file. Although the performance and stability of third-generation solar cells is still limited compared to the first- and second-generation solar cells, they have great potential and are already commercialized. In recent years several review paper has been published, which demonstrate the recent progress in this emerging technologies (Table 13.2) but none of the paper map the progress of this technology through publication and patent analysis so far. This paper is aimed at providing an insight of development in the perovskite solar PV field through publication and patent activities worldwide. It will provide the knowledge about the leading public and private players in prolific countries.

Based on the Web of Science database search for review papers, interest in the area started around 2008; however, patent search was done by

Table 13.2 Review papers related to PSC Research Development [19–33]

S. No.	Title	Authors	Journal	Year	Country
1.	A perspective of mesoscopic solar cells based on metal chalcogenide quantum dots and organometal-halide perovskites	JH Rhee, et al. [19]	NPG Asia Materials, 5	2013	Taiwan
2.	Current progress and future perspectives for organic/inorganic perovskite solar cells	PP Boix, et al. [20]	Materials Today, 17 (1)	2014	Singapore
3.	Organometal perovskite based solar cells: sensitized versus planar architecture	S Gamliel & L Etgar [21]	RSC Advances, 4 (55)	2014	Israel
4.	Third generation solar cells: a review and comparison of polymer: fullerene, hybrid polymer and perovskite solar cells	JF Yan & BR Saunders [22]	RSC Advances, 4 (82)	2014	England
5.	Perovskite photovoltaics: a high efficiency newcomer to the solar cell family	BH Wang; XD Xiao, & T Chen [23]	Nanoscale, 6 (21)	2014	China
6.	Perovskite as Light Harvester: A Game Changer in Photovoltaics	S Kazim, et al. [24]	Angewandte Chemieinternational Edition 53 (11)	2014	Spain & Switzerland
7.	Organolead halide perovskites: a family of promising semiconductor materials for solar cells	WH Zhang, & B Cai [25]	Chinese Science Bulletin, 59 (18)	2014	China
8.	The emergence of perovskite solar cells	MA Green, et al. [26]	Nature Photonics, 8 (7)	2014	Australia & England

(Continued)

Table 13.2 (Continued)

S. No.	Title	Authors	Journal	Year	Country
9.	Organohalide lead perovskites for photovoltaic applications	P Gao, et al. [27]	Energy & Environmental Science 7 (8)	2014	Switzerland & Saudi Arabia
10.	Advancements in perovskite solar cells: photophysics behind the photovoltaics	TC Sum, & N Mathews [28]	Energy & Environmental Science 7 (8)	2014	Singapore
11.	Recent progress and perspective in solution processed Interfacial materials for efficient and stable polymer and organometal perovskite solar cells	CC Chueh, et al. [29]	Energy & Environmental Science 8 (4)	2015	The United States
12.	Perovskites: transforming photovoltaics, a mini-review	AK Chilvery, et al. [30]	J. Photon. Energy. 5, (1), 057402	2015	The United States
13.	Perovskite solar cells join the major league	M Sessolo, & HJ Bolink [31]	Science. 350, (6263), 917−917 (2015)	2015	Spain
14.	Planar heterojunction organometal halide perovskite solar cells: roles of interfacial layers	H Kim, et al. [32]	Energy & Environmental Science 9 (1)	2016	South Korea
15.	A review on triphenylamine (TPA) based organic hole transport materials (HTMs) for dye sensitized solar cells (DSSCs) and perovskite solar cells (PSCs): evolution and molecular engineering	P Agarwala, & D Kabra [33]	Journal of Materials Chemistry A, 5 (4)	2017	India

the authors and found that a patent filed in India by the Netherland's patentees in 2003 [34] have mentioned in the description that perovskite material could be used in solar cell application by quoting "it is now possible for the electrode to be printed as a layer onto another component and, in this way, to add a function to this component. This component may, for example, form part of a *photovoltaic solar cell or of an electrochromic window*." As the electrode made was compared to those which are known previously can be less expensive, can have a higher round trip efficiency, in particular at relatively high current intensities, can be produced more easily, can be used in the form of a thin film or a thick layer, and may optionally be enclosed in a matrix which may also comprise a lightweight, inexpensive plastic material of moderate conductivity. This patent indicates that it was already conceptualized although went unnoticed. We discuss our findings and provide an overview of patenting activity from a historical, organizational, geographical, and technological point of view. This exercise is instrumental for private as well as public stakeholder aiming at both internal and external technology creation.

Technology development is still following an exponential growth path, with the majority of patents coming from the Asian continents like China, Japan, and Korea. One interesting fact has been observed among prolific Asian countries that key players are found in university in the case of China and rest of the Asian countries private companies are the applicant. Technology focus is traced back using the Espacenet database; whereas for devices, the main technological focus can be traced back to the "H01L-031" international patent classification (IPC) main group. Finally, encapsulation is found to be a less mature research field given the smaller number of patent families. It shows that the technology has not matured to the level where processing is carried out on a commercial scale.

13.3 LANDSCAPE OF PSC: BIBLIOMETRIC ANALYSIS

Scholarly publications and patent applications can serve as starting objective material to figure out success or failure of any new promising technology research and development (R&D). The scholarly publications as objects of a bibliometric analysis have been increasingly used in evaluating the technology development in any field. During the evaluations, we have reviewed the current situation worldwide with regard to the evolving research under competitive large-scale universities and companies, introducing multidimensional parameters key indicators based on research productivity and laboratory to market approach, the patents as the technological birth and the management activities of the large-scale and competitive technology applications. Because patents often considered as output indicators for science and technology (S&T), do actually function as input into the economy primarily to provide cutting edge solutions. We conducted joint bibliometric analysis of patents and scholarly publications for perovskite application in solar cells. Below, we briefly introduce analysis approach and parameters to highlight the technology trend worldwide.

13.3.1 Publication Analysis

Web of Science database being an online subscription-based scientific citation indexing service incorporates citations from science and social science journals and proceedings. Those refer cross-disciplinary research delivering detailed reports of their citation performance, their citation network, and the count and type of materials published. It may help to track the rising stars and influencers—user can directly compare journals using quartile, percentile, or trend analyses to understand the publishing activities, rising stars, and impactful groups influencing the direction of research. Sorting journal and category rankings by clearly defined metrics helps to easily export the results into a custom report. Impact assessment at the journal and category levels could be identified. Also visualize the relationship between citing and cited journals. Trace a journal's impact over time. Here we have used some of its key services like tracking publishers, their countries, year-wise publications, country-wise publication analysis using key string devised on the basis of prior analysis research that also presents an analysis of the processes, procedures, and outcomes of the novel research publication.

Publication trends were analyzed for the period from 2010 using Web of Science database [35] using string ((perovskite* solar* OR perovskite* photovoltaic*). A steep rise in the publications has been observed, from 36 publications in 2013 to 1182 in 2016, published from different geographical locations worldwide (Tables 13.2 and 13.3). Among the top five players, three are the Asian countries like China, Japan, and Korea. Asian countries are the major contributor and cover approximately 62% of the total publications. The perovskite research is a recent phenomenon in India. Indian Institute of Technology and Council of Scientific and Industrial Research have emerged as major players in this field.

An examination global publication trend of perovskite solar research publications highlights the growth of this area for the period of 2010–16. It is clear that PSC research has become a significant research discipline in recent years and becoming a rapidly expanding and advancing research field. Fig. 13.5 indicates increase and a continuous rise in publications in energy sector applications from 2011 (1) to 2016 (1182). It can also be observed in Fig 13.6 that the positions of top 10 players are changing since 2011 due to change in share of publications.

Table 13.3 PSC year-wise publications

Year	Number
2011	1
2012	7
2013	36
2014	309
2015	747
2016	1182

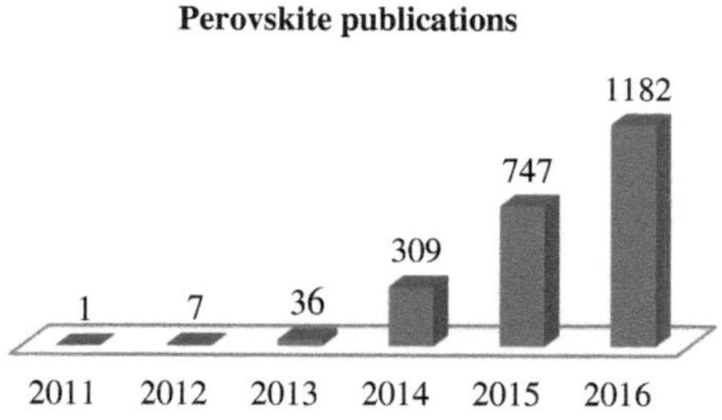

Figure 13.5 PSC publications. *From Web of Science database.*

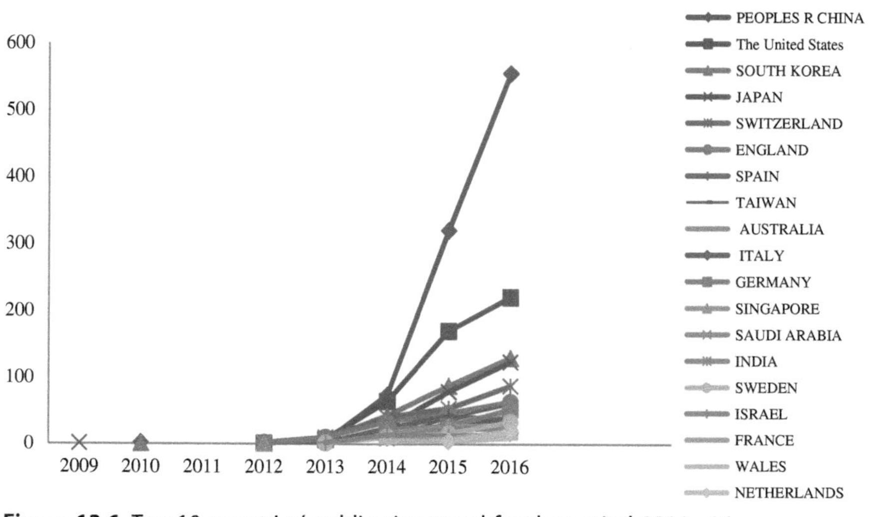

Figure 13.6 Top 10 countries' publication trend for the period 2011–16.

China is continuously increasing publishing in this field and surpassed the other comparator countries and became the one of the most prolific country by acquiring a global rank of first in this field in year 2016. China is contributing 41% of publication of total share followed by the United States, South Korea, and Japan.

China and the United States are the most prolific countries in research publications in this area and other Asian countries like Japan and Korea is also emerging as a key player in the global publications in this area. It shows Asian countries are continuously working on pressing concerns for economic development and to develop sustainable path to achieve self-sufficiency in energy.

Publication trends can visualize the hotspots of research and can help to find out the players at the market level (Table 13.4).

Emerging technologies and the interest of scientific community to provide cutting edge solutions at market level is the key for the societal development, for which many research organizations, governmental, and private are working to make a mark. Chinese universities are rigorously working in the field of material science, chemistry, and nanoscience/nanotechnology.

Table 13.4 Top 20 countries in PSC research

Rank	Top countries/territories	Publications	Total publications, %
1	People's Republic of China	1099	41.32
2	The United States	459	17.26
3	South Korea	301	11.32
4	Japan	241	9.06
5	Switzerland	207	7.78
6	England	167	6.28
7	Spain	118	4.44
8	Taiwan	118	4.44
9	Australia	106	3.99
10	Germany	106	3.99
11	Italy	77	2.90
12	Singapore	76	2.86
13	Saudi Arabia	65	2.44
14	Sweden	65	2.44
15	India	64	2.41
16	Iran	44	1.65
17	Israel	41	1.54
18	France	24	0.90
19	Netherlands	23	0.87
20	Wales	21	0.79

Chinese researchers being a part of the most prolific developing country never underestimate the seeds of novel research and bring out a plan to work out all possible scenarios, finally making the product out of it, whereas universities in the United States are working in similar fields as material science, chemistry, and nanoscience/nanotechnology disciplines, whose strategy is to start early for standing at market level. They have sector expertise, manufacturing prowess, and project pipeline to bring new solar technology to market. Japanese researchers take a problem-solving approach; at first they define the problem clearly finding the area left for exploring which can also be seen by their elaborate patent classification system they have subcategorized even the subcategories of cooperative classification system. Korean universities are working in the area of chemistry and material science for new PSC development. The strategies of these countries can help to work out the best possible ways for R&D in India where researchers have just started working in the area of PSCs (Table 13.5).

13.3.2 Patents Analysis

Patent is one of the significant bibliometric indicators. When a research output is having elements like novelty, non-obviousness, and a useful purpose than according to intellectual property laws, it is patentable. Patenting provide exclusive rights to patentee, can build strong market position, provides higher returns on investments, provides licensing and cross-licensing rights for the patentee, and helps in building positive image for your enterprise, WIPO [36].

Patents are the valuable asset in analyzing technology know-how's of a country. There are governmental and intergovernmental organizations managing the databases containing legal as well as technical information. Most of the countries have patent offices of their own. Database of these offices provides a user-friendly method to search for the required patents. Keyword-based and class-based searches are the most favored ones. An IPC system is a hierarchical system of language-independent symbols for the classification of patents and utility models according to the different areas of technology to which they pertain.

For retrieval of patent documents in "prior art" searches, classification search plays an important role. A database of PSC publications was compiled from the Espacenet database by using a series of class-based, keyword-based, and a combination of two using class strings (derived by the authors) as well as cooperative classes like these mentioned in the Table 13.6.

Table 13.5 Top universities in the most prolific countries

People's Republic of China	The United States	Japan	Korea	India
Chinese Academy of Sciences	United States Department of Energy (DOE)	Toin University of Yokohama	Sungkyunkwan University	Indian Institute of Technology (IIT)
Huazhong University of Science Technology	University of Washington, Seattle	Kyushu Institute of Technology	Seoul National University	Council of Scientific Industrial Research (CSIR)
Peking University	University of California System	Japan Science Technology Agency (JST)	Korea University	Pandit Deendayal Petroleum University
Suzhou University	National Renewable Energy Laboratory, The United States	Kyoto University	Korea Institute of Science Technology	
Tsinghua University	University of California Los Angeles	National Institute of Materials Science (NIMS)	Kyung Hee University	
North China Electric Power University	University of Toledo	University of Tokyo	Korea Research Institute of Chemical Technology	
Dalian University of Technology	University of Nebraska Lincoln	University of Hyogo	Ecole Polytechnique Federale De Lausanne	
Hong Kong University of Science Technology	University of Nebraska System	Ecole Polytechnique Federale De Lausanme	Pohang University of Science Technology	
Shanghai Jiao Tong University	Stanford University	Okinawa Institute of Science Technology Graduate University	PostechYonsei University	
Wuhan University	University of Oxford	Shanghai Jiao Tong University	Ulsan National Institute of Science Technology	

Table 13.6 Description of classification codes using Espacenet

Y	General tagging of new technological developments
Y02	Technologies or applications for mitigation or adaptation against climate change
Y02E	Reduction of greenhouse gases emission, related to energy generation, transmission, or distribution
Y02E10/00	Energy generation through renewable energy sources
Y02E10/50	PV energy
Y02E10/52	PV systems with concentrators
Y02E10/54	Material technologies (not used; see subgroups)
Y02E10/541	$CuInSe_2$ material PV cells
Y02E10/542	DSSCs
Y02E10/543	Solar cells from Group II–VI materials
Y02E10/544	Solar cells from Group III–V materials
Y02E10/545	Microcrystalline silicon PV cells
Y02E10/546	Polycrystalline silicon PV cells
Y02E10/547	Monocrystalline silicon PV cells
Y02E10/548	Amorphous silicon PV cells
Y02E10/549	Organic PV cells
H	Electricity
H01	Basic electric elements
H01L	Semiconductor devices; electric solid-state devices, not otherwise provided for
H01L21/00	Processes or apparatus adapted for the manufacture or treatment of semiconductor or solid-state devices or of parts thereof
H01L22/00	Testing or measuring during manufacture or treatment
H01L23/00	Details of semiconductor or other solid-state devices
H01L24/00	Arrangements for connecting or disconnecting semiconductor or solid-state bodies; methods or apparatus related thereto
H01L25/00	Assemblies consisting of a plurality of individual semiconductor or other solid-state devices; multistep manufacturing processes thereof
H01L27/00	Devices consisting of a plurality of semiconductor or other solid-state components formed in or on a common substrate

Source: European Classification Search. Available from: https://worldwide.espacenet.com/ [37].

Classes mentioned above were all incorporated in the search string to maximize the coverage of patents applications, and even the subclasses under these classes were covered in the search by optimizing the search string using Espacenet database search platform tools (Fig. 13.7).

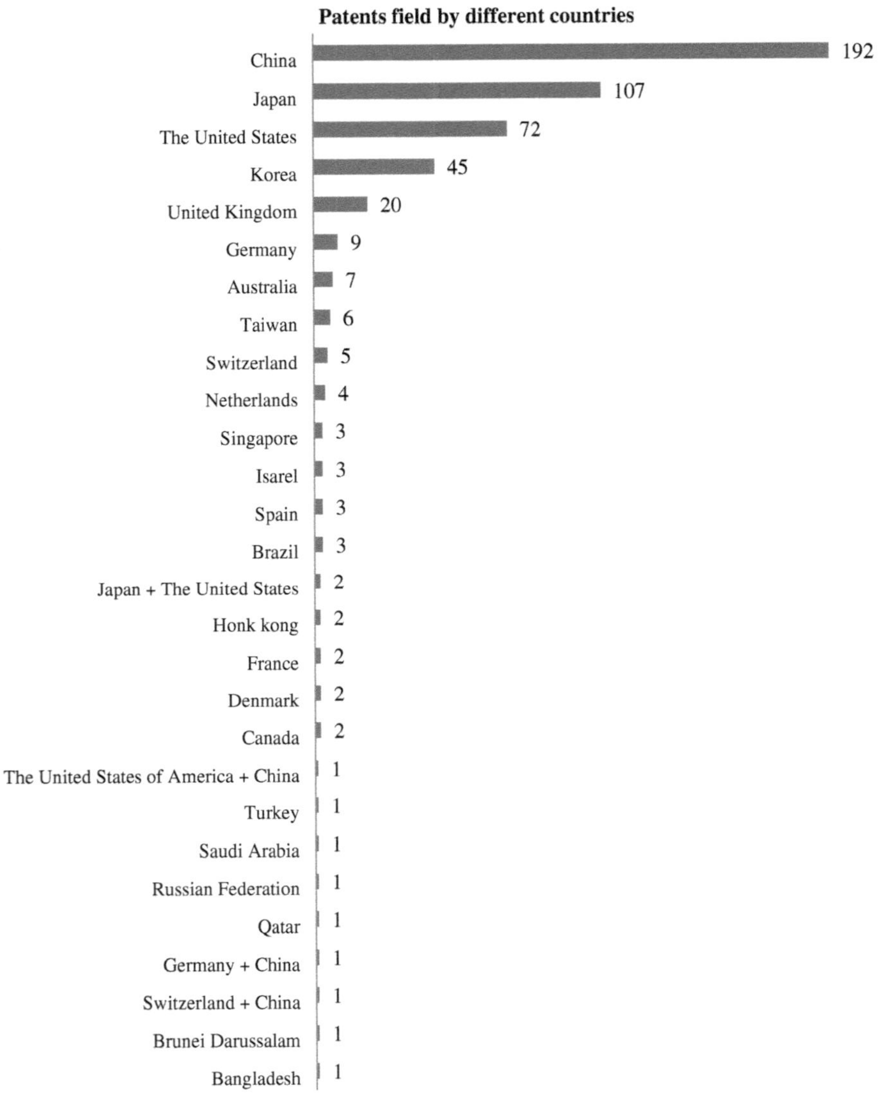

Figure 13.7 Country-wise number of published patents and collaborative patent.

Energy production and supply being necessity of time and the area of PSC being a new hope in the third-generation PVs; researchers across the world are making efforts in the area (Fig. 13.8).

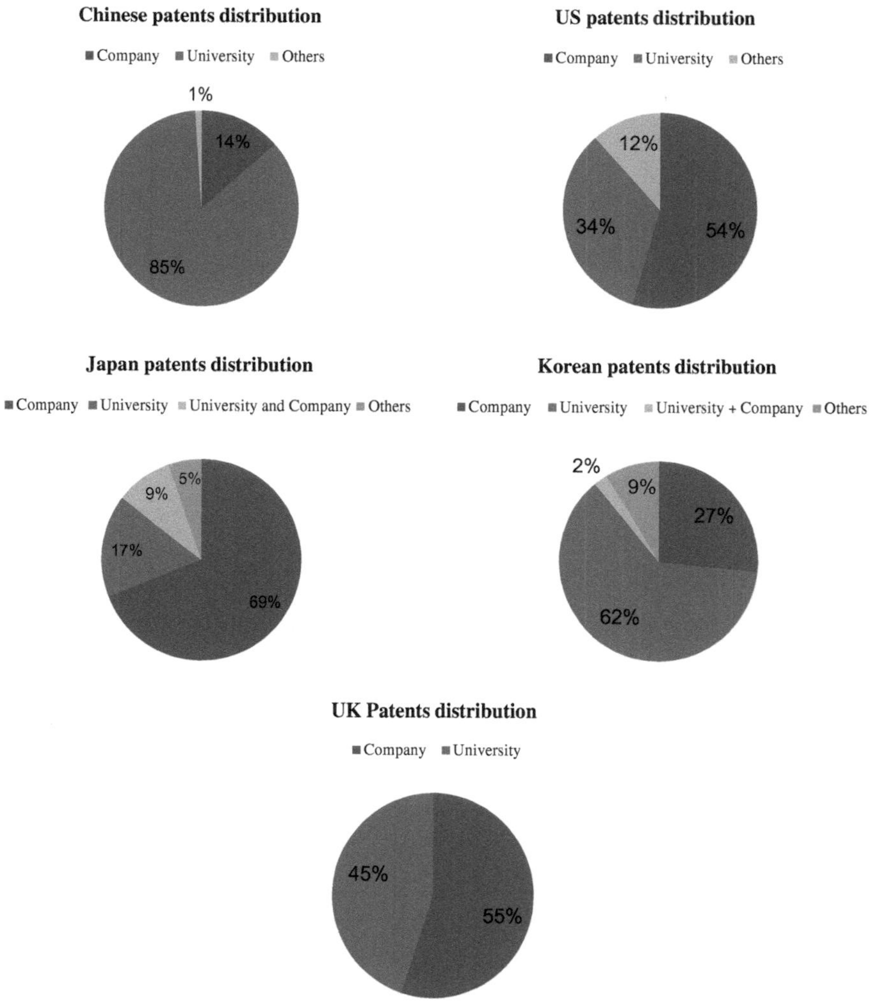

Figure 13.8 Patent distribution in top five countries. *From Espacenet Database.*

13.4 PV TECHNOLOGY STATUS

13.4.1 Perovskite Market

In the recent years, PSC are started to emerge in the market. Worldwide different companies are constructing several devices competing with the conventional Solar PV-based technologies. The effort is continuously enhancing the efficiency of the solar cell. Patents are already filed from different part of the world, which indicate the technology that has the potential of scaling up which will further make it cost-effective. NREL recent chart about research cell efficiency indicates a continuous enhancement in solar cell efficiency (Fig. 13.9) through various materials and methodologies which we have traced back from various patents [38].

13.4.1.1 Commercialization Overview Through Patenting Activity in Different Countries

China established itself among the top in patenting the perovskite-based solar cell. China filed approximately 183 patents out of which 156 filed by universities. Universities are emerging as a major player and comprising the approximate 85% of the total patent and remaining 14% has done by industry. Technology focus has been trace backed by the abstract of patent; it has been found that most of the patents related to the method of preparation and assembly (Tables 13.7 and 13.8).

13.5 TECHNOLOGY FOCUS OF TOP COUNTRIES

Technology development in PSCs basically has these areas of research like DFT (density functional theory) & band structure, device configuration & hysteresis, new material for solar cell; synthesis and characterization of perovskite nanostructure, perovskite device physics & engineering, energetics at perovskite interface, molecular dynamics photophysics of perovskite-based nanomaterial, light emission water splitting novel applications. Whereas, module of PSCs, and device and tandem cells with stability and reliability are the current areas in development (Tables 13.9 and 13.10).

Through the analysis of patent claimed by universities and companies in the different countries (Table 13.11), their technology's focus has been analyzed. The list of technology focus of top countries is mentioned in Table 13.12.

According to Rui et al. (2016), PSCs would one day dominate the world PV market which is currently dominated by crystalline Si-solar cells The first-generation crystalline silicon (C-Si) solar cells with most worked area and advanced market application currently dominates roughly 90% of the PV market. Its price has fallen dramatically from $76.67 per watt in 1977 to $0.30 per watt in 2015. The second generation of thin-film PVs, including cadmium telluride (CdTe) and copper indium gallium selenide (CIGS) solar cells, shared roughly 10% of the PV market. The emerging PVs reported to date include DSSC, quantum dot solar cell, CZTS solar cell, OSC and PSC. Most of them have reached efficiencies of more than 10%. Although it is still lower than the threshold level to

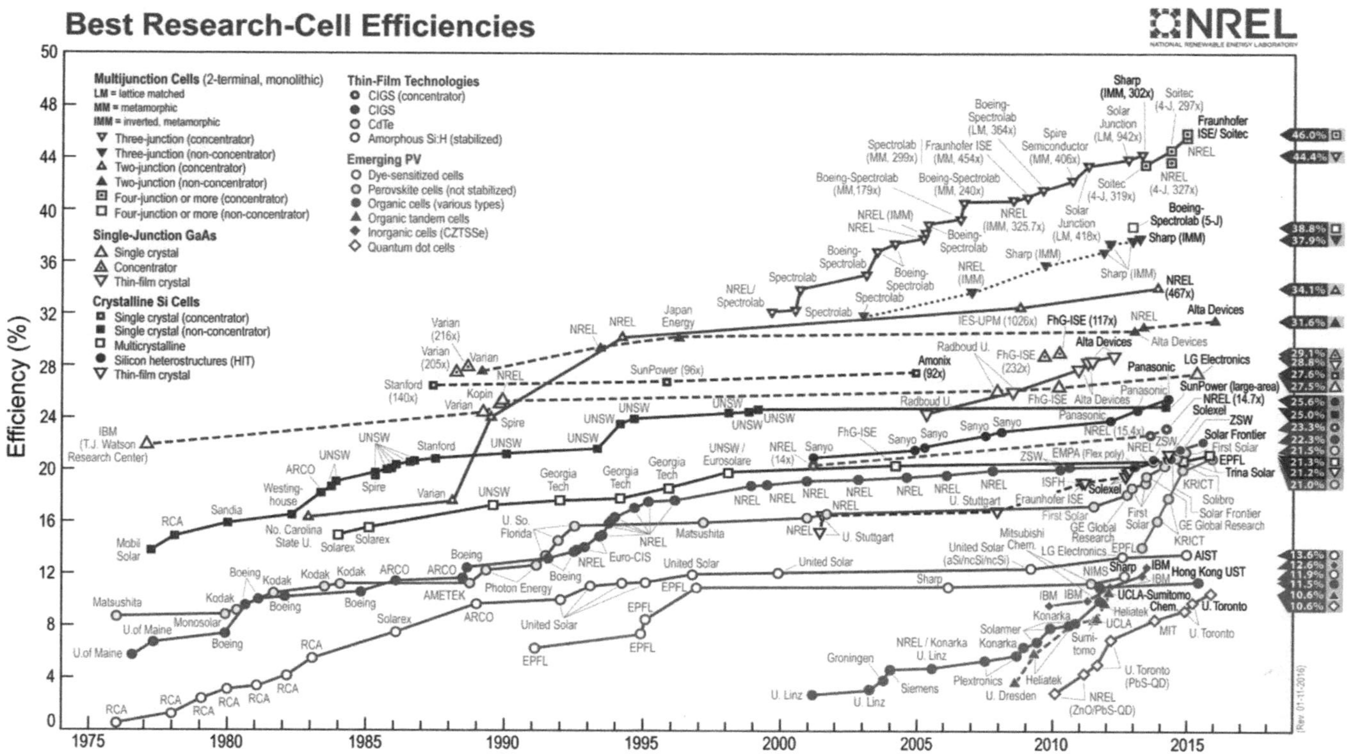

Figure 13.9 Solar cell efficiency chart. *From NREL. Available from: https://www.nrel.gov/pv.*

Table 13.7 China main universities and companies

Universities		Company
• Univ Changzhou • Huazhong Science Tech • Univ Soochow • Univ Wuhan Tech • Univ Beijing • Univ Bohai • Univ Liaoning Technology • Univ North China Elec Power • Univ Tsinghua • Chinese Acad Physics Inst • Nw Inst Non Ferrous Metal Res • Univ Xiamen • China Petroleum & ChemicalSinopec Fushun Res Inst Pet • Dalian Chemical Physics Inst • Nat Ct Nanoscience NCNST China • Shanghai Tech Physics Inst • Univ Xidian • Univ Yunnan Normal • Wuhan Inst Technology • Beijing Inst Technology	• Univ Hebei Technology • Univ Nanchang Hangkong • Univ Zhejiang • Changzhou Trina Solar Energy • Univ Hohai Changzhou • Chongqing Inst Green & Intelligent Technology CAS • Inst Semiconductors CAS • Qingdao Inst Bioenergy & Bioprocess Technology CAS • Shanghai Res Inst • Microelectronics Shrime Peking Univ • Univ Beijing Chemical • Univ Beijing Technology • Univ China Mining • Univ East China Normal • Univ Hubei • Univ Jilin • Univ Nanjing Inf Sci & Tech • Univ Peking Shenzhen Grad Scho • Univ Shanghai • Univ Tianjin Technology	• Wuhan Xinshen Optoelectronics Technology Co Ltd • China Lucky Group Corp • Xiamen Weihua Solar Co Ltd • 48th Res Inst China Electronics Technology Group Corp • Dongguan Rizhen Film PV Technology Co Ltd • Valiant Co Ltd • Basf Se [DE] • Basf China Co Ltd • Beijing Kingdom Network Technologies Co Ltd • Beijing Shouqiyi Science & Technology Co Ltd • Kailuan Energy Chemical Co Ltd • Nanjing Yousibeite Photoelectric Material Co Ltd • Sh Nat Eng Res Ct Nanotech Co • Suzhou Creative Carbon Nanotechnology Co Ltd • Wuhan Xinshen Photoelectric Science & Technology Co Ltd • Zhenjiang Shuangli Optoelectronic Technology Co Ltd

Table 13.8 Korean patents filed by companies and industries

University		Industry
• Korea Inst Sci & Tech • Korea Res Inst Chem Tech • Univ Sungkyunkwan Res & Bus • Univ Sungkyunkwan Res & Bus with Global Frontier Center For Multiscale Energy Systems • Daegu Gyeongbuk Inst of Science and Tech • Univ Seoul Ind Coop Found • Univ Sungkyunkwan Res & Bus Global Frontier Ct For Multiscale Energy Systems • Global Frontier Center For Multiscale Energy Systems Univ Seoul Nat R & Db Found	• Global Frontier Center For Multiscale Energy Systems with Univ Seoul Nat R & DB Found with Univ Sungkyunkwan Res & Bus • Korea Energy Research Inst • Korea Univ Res And Business Found • SNU R&DB Foundation • Univ Hongik Ind Acad Coop Fdn • Univ Korea Nat Transp Ind Acad Coop Found • Univ Korea Res & Bus Found • IUCF-HYU (Industry-University Coop Found Hanyang University)	• LG Chemical Ltd • Dongjin Semichem Co Ltd • Hyundai Motor Co Ltd • Kceracell • Samsung Electronics Co Ltd • Samsung SDI Co Ltd

Table 13.9 United Kingdom patents filed by these companies and industries

University	Company
• Swansea University and Univ Swansea • Univ St Andrews • Univ Swansea • Bangor Univ • The University Court of The University of St. Andrews • Univ Court, Irvine John Thomas Sirr, and Tao Shanwen	• Oxford Photovoltaics Ltd • Isis Innovation • Ceres Power Ltd

Table 13.10 Japan patents filed by these companies and industries

Company		University	Collaboration
• Fujifilm Corp • Panasonic Corp • Matsushita Electric Ind Co Ltd • Nippon Catalytic Chem Ind • Peccell Tech Inc • Seiko Epson Corp • Sekisui Chemical Co Ltd • Toyota Motor Co Ltd • Kaneka Corp • Tosoh Corp • Hitachi Shipbuilding Eng Co • Revo Internat Inc • Toshiba Corp • Fujitsu Ltd • Chemical Materials Evaluation and Res Base • Fujikura Ltd • Goto Taichi, Honda Motor Co., Ltd • Hunan Province Tianci Sunshine Solar Energy Co Ltd • Kabushiki Kaisha Toshiba • Kyocera Corp • Mitsubishi Materials Corp Ishihara Tatsumi Takita Yusaku	• Seimi Chemical Co., Ltd • Revo International Inc • Ngk Insulators Ltd • Nippon Telegraph & Telephone • Nissan Motor • Nissan Motor Seimi Chem Kk • Noritake Co Ltd • Ricoh Co Ltd • Mitsubishi Heavy Ind Ltd • Toho Gas Kk Nippon Catalytic Chem Ind • Tosoh Corp • Toto Ltd • Toyota Central Res & Dev • Toyota Motor Co Ltd Japan Fine Ceramics Center Nag • Toyota Motor Corp • Yokohama Rubber Co Ltd • The Japan Science and Technology Agency • Sanyo Electric Co • Seimi Chem Kk • Sumitomo Chemical Co	• Japan Science & Tech Agency • Japan Science & Tech Agency California Inst of Techn • Japan Science & Tech Corp • Nat Inst of Advanced Ind Scien • Okinawa Inst of Science and Tech School Corp • Nat Inst for Materials Science • Nat Inst Natural Sciences Tokyo Yogyo Kk • Nat Institute For Materials Science • Tokyo Inst TechUniv Kyoto • Univ Kanazawa • Univ Kyushu	• Konica Minolta Holdings IncUniv Shinshu • Equos Res Co LtdUniv Kyushu • Fuji Electric Co Ltd Santoku Corp Univ Hokkaido Nat Univ Corp • Nat Inst of Advanced Ind Scien Ngk Insulators Ltd • Tanaka Precious Metal IndUniv of Science Tokyo • Toyota Motor CorpUniv of Science TokyoToyota Ind Corp • Univ KyotoNippon Catalytic Chem Ind • Univ Kyoto Showa Co Ltd • Univ Nagoya Nat Univ Corp • Univ KyotoShowa:Kk • Univ Shizuoka

Table 13.11 United States patents filed by these companies and industries

Company		University
• Hunt Energy Enterprises LLC • IBM • Nissan Motor • Nutech Ventures • Sharp Lab of America Inc • Alliance Sustainable Energy • Amoco Corp • Bloom Energy Corp • BOC Group Inc • The Boc Group, Inc • Zaf Energy Systems Inc	• Cabot Microelectronics Corp • Ceramatec, Inc • Solar-Tectic LLC • Solar-Tectic LLC • Corning Inc • Kateeva Inc • Los Alamos Nat Security LLC • Recapping Inc • Saint Gobain Ceramics • Zinc Air Fuel Cells Inc	• Univ Northwestern • Massachusetts Inst Technology • Midwest Research Inst • California Inst of Techn • Univ Akron • California Institute of Technology • Univ California • Univ California with Cosmo Chemical Co Ltd [KR] and Cosmo AM&T Co Ltd [KR] • Univ Chicago • Univ Leland Stanford Junior • Univ Leland Stanford Junior and Massachusetts Inst Technology • Univ Nebraska • Univ of Washington Though Its Center for Commercialization

Table 13.12 Renewable energy consumption versus generation statistics of top countries and percentage share worldwide

S. No.	Country	Renewable energy–solar consumption share, %	Cumulative installed PV power share, %
1.	China	19.9	25.9
2.	The United States	17.1	13.4
3.	Japan	14.9	14.2
4.	India	3.6	3.0
5.	South Korea	1.6	1.4
6.	The United Kingdom	3.1	3.9

Source: BP Statistical Review of World Energy. 2017. Available from: https://www.bp.com/content/dam/bp/en/corporate/pdf/energy-economics/statistical-review-2017/bp-statistical-review-of-world-energy-2017-full-report.pdf [39].

enter market competition, no one could deny that those emerging technologies have big potentials in the future, because their developing histories are much shorter than the commercialized PVs. Among them, PSC outshines others, at least in the efficiency evolution. Though PSC only appeared 6 years ago, its efficiency record has been dramatically improved from 3.8% to 21.0% or more. Prospects for PSCs towards industrialization include improving the efficiency of PSCs to >25% that will strengthen its economic competitiveness. Research in areas of PCE of PSC module, stability, costs, and toxicity is still progressing.

13.6 SHORT-TERM INDUSTRIALIZATION PROSPECTS

Instead of long-term goals, we can also apply this perovskite technology in the development of solar cells which are flexible enough to be fabricated as semitransparent and colorful power windows, the feasibility of which has already been demonstrated in laboratory. This can be applied in building integrated PV applications—their application can help not only for sunshade but can generate electricity in c-Si solar modules that are nontransparent and can only be equipped on the roofs of the buildings, which are scarce resources in countries like China. Second, PSCs can be made on flexible substrates. Flexibility and portability are notable features for certain applications, such as wearable and portable powers required for many civilian and military roles [40].

US Department of Energy's Office of Energy Efficiency and Renewable Energy started SunShot's next-generation PV projects [41] that investigate transformational PV technologies with the potential to meet SunShot cost targets. The goals increase efficiency, reduce costs, improve reliability, and create more secure and sustainable supply chains. On October 22, 2014, SunShot awarded more than $14 million to 10 research institutions to meet or exceed SunShot targets by improving performance, efficiency, and durability of solar PV devices. The R&D projects will explore a spectrum of leading-edge solutions, from new high-performance materials like perovskites to novel techniques for creating solar cells with high efficiency and lower manufacturing cost. Grants were awarded to Duke University (Pb-halide-based perovskites), Sandia National Laboratories, Stanford University (develop and deposit thin films of perovskite semiconductors with band gaps ranging from 1.6 to 2.3 eV on top of silicon solar cells to advance cell efficiency and reduce cost), The Regents Of The University Of Michigan, National Renewable Energy Laboratory (managed and operated by The Alliance For Sustainable Energy, LLC) (develop high-efficiency single-junction perovskite (top and bottom) solar cells), University of Nebraska-Lincoln (developing tandem junction solar cells with organo-lead trihalide perovskites based on high-efficiency perovskite cells as the top cell), University of Houston, University of Washington (developing high band gap HP materials and a novel two-terminal monolithic tandem device architecture that are capable of >25% PCE under one-sun conditions).

Duke University team was working for Pb-halide-based PSCs, emphasizing that PV properties of absorbers depend on the structural and electronic dimensionalities which are same for some absorbers but different for others. They reported that some double perovskite such as Ag- and Bi-containing halide; double perovskites are structurally 3D, they are electronically 0D. High electronic dimensionality is preferred for high-performance PVs materials. The electronic dimensionality can account for many of the trends observed for PV properties and device performances of perovskite and non-perovskites reported in literature. Their results can provide critical guidance for searching for Pb-free halide perovskite PV absorbers and for fabricating efficient devices [42].

Stanford University goal was for optimizing film deposition conditions to reduce defects and obtain the best crystal size and orientation. A valuable insight for nature of perovskite and PSCs fracture was presented, and the influence of material selection, deposition techniques' environmental variables, and molecular additives on these fracture processes are reported. Because mechanical stability of PSCs is crucial for field applications, to determine the mechanical stability, the fracture resistance of a multitude of solution-processed organometal trihalide perovskite films and cells utilizing these films were studied. In all cases, the perovskites offered negligible resistance to fracture, failing cohesively below 1.5 J/m2. The solar cells studied featured these perovskites and a variety of organic and inorganic charge transporting layers and carrier-selective contacts. These ancillary layers were found to significantly influence the overall mechanical stability of the PSCs and were repeatedly the primary source of mechanical failure, failing at values below those measured for the isolated fragile perovskite films [43].

One of the project by NREL was to develop high-efficiency single-junction PSCs; later their team found that more refined preparation of heterojunctions can possibly be done by good lattice matching, and continuous interface at the molecular level between perovskite and carrier collectors is essential to ensure high stability of the device [44].

13.7 JAPAN

According to Asia Biomass energy cooperation promotion office, 2016 report, Japanese researchers have developed a new solar cell. In 2009, a group led by Professor Tsutomu Miyasaka from Toin University of Yokohama identified the ability of compound thin films with a crystalline structure known as perovskite to function as solar cells, and verified that its generating efficiency was 3.9%. Later, in 2012, as a result of advances made through joint research between Toin University of Yokohama and the University of Oxford, a maximum generating efficiency of 10.9% was achieved, which has been earning attention from around the world. The National Renewable Energy Laboratory (NREL) in the United States investigates and publicly announces the efficiency of solar cells from around the world. It announced that as of June 2015, a PSC developed by the Korea Research Institute of Chemical Technology had demonstrated a maximum efficiency of 20.1%. A mere 6 years since they were first announced, PSCs have achieved a generating efficiency that surpasses not only that of crystalline silicon solar cells but also amorphous silicon solar cells, organic thin-film solar cells, and DSSCs. Perovskite is a type of crystalline structure and shares the same name with another type of crystalline structure. Those that are currently being studied as solar cells can be expressed by the chemical formulas $CH_3NH_3PbBr_3$ and $CH_3NH_3PbI_3$ The solar cells can be formed simply by applying a solution that serves as the raw materials over a porous membrane of titanium oxide. As a result, not only do the solar cells have a high generating efficiency but it is also expected that they will be of low cost as well [45].

13.8 SOUTH KOREA

A new record for PSC efficiency was made. South Korean researchers have developed a highly stable PSC that has a photo-conversion efficiency of more than 20% [46].

13.9 CHINA

Chinese researchers taking the lead are heading towards optimizing factory layouts, strengthening supply chains for the perovskite market (Table 13.13).

Table 13.13 Chinese companies' areas of expertise in PSC development

University and private industry	Area of research
Wuhan Xinshen Optoelectronics Technology Co Ltd (4)	• Preparing a PSC through a low-temperature full-solution method • Copper indium sulfur/perovskite heterojunction solar cell • Uses a method of rotating coating of a silver electrode for perovskite cell • Solar battery with silver sulfide/perovskite heterojunction and a preparation method thereof
China Lucky Group Corp (3)	• Preparation of planar heterogeneous PSC • "Planar heterojunction PSC (utility model granted)" • Ultra-thin orientation layer
Xiamen Weihua Solar Co Ltd (3)	• PSC with the conducting polymer as the back electrode • Method for a perovskite thin film solar cell assembly • PSC preparation method based on the full-coating process one-step annealing method
48th Res Inst China Electronics Technology Group Corp (2)	• Preparation of the PSC by a solution method • Preparation method-process
Valiant Co Ltd (2)	• Derivative is applied to PSCs.-stability /life expectancy • Carbazole diarylamine dendritic compound-hole-transporting material of a PSC
Basf Se [DE]Basf China Co Ltd [CN] (1)	• Preparation with conductive inter layer
Beijing Kingdom Network Technologies Co Ltd (1)	• Supercapacitor
Kunshan Innovation Res Inst Of Nanjing UniversityKunshan Sunlaite New Energy Co Ltd (1)	• PSC with titanium dioxide nanotube layer
Sh Nat Eng Res Ct Nanotech Co (1)	• Perovskite cell with spin-coated titanium oxide and nanocrystalline perovskite
Suzhou Creative Carbon Nanotechnology Co Ltd (1)	• Method of manufacturing fiber-shaped PSC
Wuhan Xinshen Photoelectric Science & Technology Co Ltd (1)	• PSC with a cadmium sulfide-doped perovskite layer
Zhenjiang Shuangli Optoelectronic Technology Co Ltd (1)	• Tin perovskite structure absorbing layer with low toxicity

Material synthesis and manufacturing are the key strengths of Chinese industries. PSC is developed using nanotube and nanocrystalline material. Methods like spin coating, rotation coating, and low-temperature full-solution were devised in manufacturing process. Structural changes for efficiency and stability were addressed by researchers as fiber-shaped PSC and back electrode conductive polymer. Advancement is done by making devices with heterojunctions. Lead was also replaced by tin for toxicity issue. In China companies like Wuhan Xinshen Optoelectronics Technology, China Lucky Group Corporation, and Xiamen Weihua Solar Corporation Ltd are working actively in the area of PSCs development as well as manufacturing for the application in larger areas. Wuhan Xinshen Photoelectric Science & Technology Corporation Ltd developed cadmium sulfide–doped perovskite layer, and solar cells with tin perovskite structured absorbing layer were devised by Zhenjiang Shuangli Optoelectronic Technology Corporation Ltd (Table 13.14).

Table 13.14 Chinese universities: areas of expertise in PSC development

University and private industry	Area of technology
Univ Huazhong Science Tech (7)	• Mesoscopic solar cell based on a perovskite light absorption material • Nanoparticles of the mesoporous electron collection layer • Flexible PSC production technology-dropwise addition of perovskite to the carbon counter electrode so that it penetrates from the carbon counter electrode onto the ZnO nanowire • Method for producing a lead halide perovskite nanowire • Binary metal composite perovskite material • A trans-planar PSC based on a doped NiO hole transport layer • Solar cell-$M_zA_yBX_{z+y+2}$ perovskite-based photoelectric functional material modified with an organic amphipathic molecule
Univ Soochow (7)	• Method—black cubic crystal system perovskite film-solvent annealing process • Method—unannealed perovskite type solar cell to get electron transport layer • Method—inverted thin film solar cell • Hole transport layer-Spiro-OMeTAD triarylamine derivative

(Continued)

Table 13.14 (Continued)

University and private industry	Area of technology
	• Method—multioriented ordered crystallization–methylammonium iodide, lead chloride, indium chloride, or iodination • Method—water doping in perovskite precursor solution-for high crystallinity and good stability • Method—orderly arranged small titanium dioxide balls
Univ Changzhou (6)	• Overlapped structure sunlight effective absorption • Method—preparation of battery with different deposition layers-perovskite-structured $CH_3NH_3PbI_3$ layer is prepared by a solution method or a co-evaporation method; The metal electrode is prepared by thermal evaporation or sputtering • Preparation—Solar cell of a perovskite structure, a sandwich protection structure to improve the stability and photoelectricity • Spin-coated solution of perovskite-structured $CH_3NH_3PbI_3$ layer doped with graphene quantum dots with toluene for more pores • Method for depositing the perovskite layer filled in the ITO barrier layer with dimethylacetamide and ultrapure water in a solvent • Solar batteries—stretchable organic–inorganic HP
Univ Wuhan Tech (5)	• Method—light absorbing layer is a two- $(C_6H_5CH_2NH_2)_2(CH_3NH_3)_{n-1}Pb_nI_{3n+1}$ (n = 1, 2) material of the layered structure–sustained high humid conditions • Perovskite thin-film PV cell • Chlorine–bromine–iodine mixing perovskite thin film and a PV battery • SnO_2 porous structure perovskite PV cell • Method—bilayer nanometer mesoporous electron-transporting layer
Univ Liaoning Technology (5)	• Preparation method of methylamine lead bromide for light absorbing material • Organic/inorganic hybridized tin-lead mixed perovskite material subjected to low-speed spin-coating • Tin perovskite inorganic hole solar cell- inorganic hole transport layer is cuprous iodide • Nanoarray titanium dioxide perovskite

(*Continued*)

Table 13.14 (Continued)

University and private industry	Area of technology
	• Improvement type nano-zinc oxide piece array perovskite type solar cell
Univ North China Elec Power (5)	• Method of preparation of perovskite film • PSC—plane heterojunction • Perovskite-lead sulfide quantum dot stacked solar cell-a dot layer of thickness 15 to 400 nm • PSC—lead acetate and CH_3 NH_3 I is spin-coated as a precursor solution • PSC—skeleton layer is a barium stannate film
Univ Tsinghua (3)	• Method—wholly iodine perovskite material $CH_3NH_3PbI_3$, a chlorine/bromine-doped perovskite material • Method of preparing a matrix deposited with a precursor BX_2 film by adopting a vapor deposition method • Nanometer perovskite oxygen electrode powder • Electrolytic perovskite-low temperature self-propagating method
Univ Beijing (3)	• PSC and a preparation method steps improving surface morphology • PSC with a hole transport system free of ionic additive • Perovskite solar energy battery having organic skeleton structure
Univ Xiamen (3)	• Flexible sensitized solar cell—a perovskite absorption layer • $SrTiO_3$ of perovskite oxide in an OSC • Method—fullerene in perovskite
Chinese Acad Physics Inst (3)	• Perovskite-type organic lead halide—thin-film solar cells interface structure of controllable thickness • Thermoplastic carbon counter electrode for a PSC • A perovskite thin film solar cell comprising a hole transport layer with a polymer comprised graphite alkynes
Nat Ct Nanoscience Ncnst China (3)	• Battery comprising hole transport layer includes PEDOT: PSS, P3HT, PTAA, PThTPTI, metal oxide, and oxide graphene, said electron transport layer comprising 58π-electron fullerene PCBM, 56π-electron fullerene OQMF, 54π-electron fullerene OQBMF, PFN, PEIE, ZnO, TiO_2 • Improving the performance of the absorber layer of a PSC additive to the perovskite reactant • Perovskite solar battery and the bulk heterojunction-organic blend layer is a mixed film of the bulk heterojunction solar cell donor material and the acceptor material

(Continued)

Table 13.14 (Continued)

University and private industry	Area of technology
Shanghai Tech Physics Inst (3)	• Zinc telluride serving as hole transporting layer • Metal (nanoparticle) pellet-based perovskite thin film solar cell structure • Hole transport layer of inorganic compound as copper sulfide, etc.
Univ Hebei Technology (2)	• Solid perovskite microcrystalline silicon composite film • Method—thin film crystalline silicon perovskite heterojunction solar cell
Univ Nanchang Hangkong (2)	• Method—solar cell mesoporous layer is mesoporous zirconia • Preparation method for Back electrode is magnetron sputtering and spin coating.
Univ Zhejiang (3)	• CdSe nanocrystals • Self-assembled monolayer
Changzhou Trina Solar Energy (2) Univ Hohai Changzhou	• Z-type serial PSC assembly by hole transporting layer • Large-area PSC module
Chongqing Inst Green & Intelligent Technology Cas (2)	• Semitransparent PSC-conductive layer is a super-thin metal Ag or a conductive polymeric material • Flexible PSC where flexible substrate is PET/PEN
Inst Semiconductors Cas (2)	• Perovskite solar battery and the manufacturing method-p-type silicon-based alloy thin film comprising a microcrystalline silicon carbon alloy and n-type silicon-based thin film having a film thickness of 5–50 nm and a deposition method for chemical vapor deposition • HIT battery comprises a back metal electrode
Qingdao Inst Bioenergy & Bioprocess Technology Cas (2)	• Planar structures preparation—preparing an organic–inorganic perovskite film by a one-step solution method • Controllable technical process for film formation
Shanghai Res Inst Microelectronics Shrime Peking Univ (2)	• PSC based on a full-sol-gel process and a preparation method • Large-area perovskite battery with Palladium material contact layer of back electrode
Univ Beijing Chemical (1)	• Fibrous solar battery with titanium wire and the carbon nanotube as battery electrodes
Univ Beijing Technology (1)	• Flexible all-solid-state array type perovskite solar battery–titanium dioxide nanotube array with the controllable structure
Univ East China Normal (2)	• Organic perovskite film preparation method • Method—aluminium alloy back electrode
Univ Hubei (2)	• Sandwich structure

(Continued)

Table 13.14 (Continued)

University and private industry	Area of technology
	• ZnO nanorods are also used as the skeleton layer and Perovskite layer synthesized by the two organic hole transport layers is composed of organic Spiro-MeOTAD film
Univ Jilin (2)	• PSC perovskite layer by the solid fog interface reaction • Method of preparing the electrode is a vapor deposition coating
Univ Peking Shenzhen Grad Scho (2)	• PSC-ultrasonic spraying method • Cuprous oxide heterojunction solar cell
Univ Shanghai (2)	• Method—single-source flash evaporation method • Method—glass substrate coated with FTO (SnO_2:F) and TiO_2-ZnO nanorod arrays
Univ Xidian (2)	• Method—plane PSC • Planar PSC, photoelectric conversion
Univ Yunnan Normal (2)	• Two-sided thin film solar cell • Silica-based perovskite laminated solar cell
Nw Inst Non Ferrous Metal Res [CN] (1)	• Interclude layer $NdGaO_3$
Univ Tianjin Technology (1)	• Method—hybrid solar cell—the zinc oxysulfide electron transfer layer, an inorganic HP material • Method—organic optoelectronics—three-dimensional ordered mesoporous support layer
Beijing Inst Technology (1)	• Light emitting diode (LED)—structural formula of the perovskite is $R1NH_3AB_3$
California Inst Of Techn [US] (1) Haile Sossina M [US]Shao Zongping [CN]	• Cathode active materials comprise perovskites of the general form ABO_3, where A is a cation (+2 charge), and B is a cation (+4 charge)
Ecole Polytech [Ch] (1)Univ Huazhong Science Tech [Cn]	• Method of fabrication-mesoporous P-I-N framework
Guangzhou Inst Energy Conv Cas (1)	• Perovskite structure heterocomplex low-temperature film formation
Hefei Inst Physical Sci Cas (1)	• Framework layers of the quasi-one-dimensional TiO_2 nanostructure arrays of solar perovskite battery
Inst Electrical Eng Cas (1)	• Method for preparing an organic/inorganic HP membrane

(Continued)

Table 13.14 (Continued)

University and private industry	Area of technology
Inst Plasma Physics Cas (1)	• DSSCs–perovskite oxide counter electrode
Ningbo Inst Materials Technology & Eng Cas (1)	• PSC preparation by spin coating method, a doctor coating method
Univ Beifang Nationalities (1)	• Ceramic dielectric laminated supercapacitor
Univ Beijing Posts & Telecomm (1)	• Perovskite/nanowire hybrid solar cell with perovskite being filled between i regions adjacent to said nanowires
Univ Beijing Science & Tech (1)	• Perovskite/P-type quantum dot composite solar cell
Univ Central South (1)	• Method—quantum dot cosensitization type PSC–the hole-transporting material poly-3-hexylthiophene is replaced by Spiro-OMeTAD
Univ Changzhou (1) Univ Jiangsu	• Solar batteries—stretchable organic–inorganic HP
Univ Dalian Tech (1)	• Method—superimposed PSC preparation method
Univ Electronic Science & Tech (1)	• Flexible substrate solvothermal annealing process of the DMF or DMSO
Univ Fudan (1)	• Flexible braided coaxial linear PSC
Univ Fuzhou (1)	• The quality and performance control using spraying technology
Univ Hefei (1)	• Method—preparing a vapor deposition of a PSC
Univ Henan (1)	• Method—vacuum deposition of a layer of silver electrode
Univ Huaqiao (1)	• Preparation of hybridization solar battery for perovskite-like sensitized photoanode
Univ Nankai (1)	• High-efficiency perovskite thin-film solar cell–PVP-prepared perovskite film
Univ Ningbo (1)	• The methylamine lead iodine polycrystalline film is modified by dodecyltrimethoxysilane
Univ Qingdao (1)	• Perovskite nanometer fiber film prepared–electrostatic spinning method
Univ Shaanxi Normal (1)	• Method—perovskite solar battery photoanode
Univ Shaanxi Normal (1)Dalian Chemical Physics Inst	• Method—large-area perovskite thin sheet which is convenient to machine
Univ Shanghai Jiaotong (1)	• Method—SOFC-resists carbon deposition and has sulfur resisting property
Univ Shanghai Jiaotong (1) Jiangsu Sunshore Solar Energy Industry Co Ltd	• Solar heat storing and collecting device-laminated perovskite or polyalcohol

(*Continued*)

Table 13.14 (Continued)

University and private industry	Area of technology
Univ Shanxi (1)	• Method—PSC at low temperature and without calcination
Univ Shijiazhuang Tiedao (1)	• Method—organic halide thin-film solar cell photoanode preparing
Univ South China Normal (1) Shenzhen Guohua Photoelectric Science & Technology Co Ltd	• Method—semitransparent perovskite and crystalline silicon tandem laminated solar cell
Univ South China Tech (1)	• Method—organic–inorganic planar heterojunction solar cell–inactive photoactive layer
Univ Sun Yat Sen (1)	• Method—silica-based perovskite heterojunction solar cell
Univ Xi An Jiaotong (1)	• Perovskite—Homogenous layer
Univ Zhengzhou (1)	• Plane-mesopore mixed PSC-stacking
USTC Univ Science Tech CN [CN] (1)	• Two-phase composite compact-ceramic membrane

Chinese universities' researchers had rarely left an area for exploration in PSC development. Perovskite thin-film solar cell materials were devised with a hole transport layer comprising inorganic compounds like graphite alkynes, doped NiO, zinc telluride, copper sulfide, cuprous oxide, metal oxide, and oxide graphene and organic compounds like Spiro-OMeTAD single-layered or double-layered and spiro-OMeTAD triarylamine derivative worked by Soochow University. Development of quantum dots composited solar cells by the Beijing Science & Technology University.

Deposition methods like chemical vapor deposition sol-gel process, film-forming process like spin coating process, a spraying process, a printing process, and scratch coating process. Steps were formulated for improving surface morphology and efficiency of excitons transport by making micro-nano-interpenetrating structure of electron transport layer.

Huazhong Science Technology University, Soochow University, Changzhou University, Wuhan Technology University, North China Electric Power University, and Liaoning Technology University are working rigorously in the area of photosensitive material development, formalizing novel methods for manufacturing at large scale and improving the stability and photo-electricity.

Devices like ceramic dielectric laminated supercapacitor and flexible all-solid state array type solar cells were developed by the researchers. PSC comprised mesoporous P-I-N framework. Large area PSC module comprising a plurality of monomeric PSCs were developed by the team of Changzhou Trina Solar Energy Company and Hohai Changzhou University. Because the application of the technology for the field trail is the primary goal, South China University in collaboration with Shenzhen Guohua Photoelectric Science & Technology Cooperation Limited has developed semitransparent perovskite and crystalline silicon tandem laminated solar cell (Table 13.15).

Japanese researchers are working for the development of high temperature superconductor, crystal particle like tantalate for DSSC, optically pumped semiconductor, solid junction solar cell with high durability, and also issues like mass production as well as high efficiency were taken into account.

Chemical structure and composition with organic as well as inorganic compounds were used in manufacturing process. Konica Minolta Holdings, Inc. in collaboration with Shinshu University is working in area of tantalate particles for DSSCs, Tosoh Corporation, Panasonic Corporation, Peccell Technologies, Inc., and Seiko Epson Corporation are working in area of PSC development (Table 13.16).

National Institute for Materials Science has developed multicomponent oxide photocatalyst and paint luminescent nanosheet having perovskite octahedral crystals combined together in a planar configuration, where photo filter of solar cell comprises this luminescent nanosheet. National Institute of Advanced Industrial Science working for crystal structure and in collaboration with NgK Insulators Ltd is working for the phosphor microparticles and Shizuoka University developed a semiconductor element with epitaxially grown gallium nitride single crystal layer on the Aluminium Nitride (AlN) foundation layer at a specific temperature (Table 13.17).

Premier companies of the United States are Hunt Energy Enterprises LLC, IBM, Solar-Tectic LLC, and Nutech Ventures. Hunt energy is primarily working for the interfacial layers, bi-tri-layered structures, and additional layers for performance enhancement of PSC, whereas IBM is working structural arrangements, tandem layered device, doubled absorber layers and using processes like sputtering for transparent conducting oxide formation. Solar-Tectic LLC is working for development of large grained to single crystal semiconductor films and Tandem solar cell with germanium. Nutech Ventures is developing methods for continuous process and for single crystal growth (Table 13.18).

Table 13.15 Japanese companies: areas of expertise in PSC development

University and private industry	Area of technology
Konica Minolta Holdings Inc. (2)Univ Shinshu	• Solar cell-doped lanthanum tantalate crystal particle • Solar cell-tantalate crystal particle having layered perovskite structure
Tosoh Corp [JP] (2)	• Method—solar cell–oxide sintered compact and sputtering target composed of a bixbyite oxide and perovskite oxide crystal phase • Conductive film–perovskite-type oxide phase
Panasonic Corp (2)	• Semiconductor device • Optically pumped semiconductor–semiconductor of a perovskite oxide
Peccell Technologies Inc. (2)	• Solid junction photoelectric conversion device • Solid junction solar cell replaceable for a DSSC with a second semiconductor layer-HP compound
Seiko Epson Corp (2)	• DSSCs composition and device • Perovskite structure containing titanium and strontium
Fujifilm Corp (1)	• Photoelectric conversion element
Hunan Province Tianci Sunshine Solar Energy Co Ltd (1)	• Perovskite based flexible film solar cell-can play the role of graphene and improve the energy conversion efficiency 27%(not claimed but mentioned in description)
Japan Science & Tech Agency (1)	• Superconducting compound
Ricoh Co Ltd (1)	• Perovskite solar battery with Perovskite compound general formula: XYM. In the general formula, X is a halogen atom, Y is an alkylamine compound, and M includes a mixture of lead and antimony in ratio of 3:1:1
Yokohama Rubber Co Ltd (1)	• Solar cell–conductive composition for solar cell electrode collector

Table 13.16 Japan universities: areas of expertise in PSC development

University and private industry	Area of technology
Nat Inst For Materials Science (2)	• Perovskite type multicomponent oxide visible light responsive photocatalyst can be used for solar cells • Nanosheet paint (perovskite octahedral crystals combined together in a planar configuration)
Nat Inst of Advanced Ind Scien [JP] & Ngk Insulators Ltd [JP]	• Phosphor microparticles–solar cell
Nat Inst of Advanced Ind Science and Technology (1)	• Perovskite crystal structure represented by formula: an R_2CuO_4 (wherein R is a rare earth element) and $XZ_2Cu_3O_6$ (wherein X is yttrium, lanthanum or praseodymium, and Z is an alkali earth metal)
Univ Shizuoka (1)	• Substrate having a perovskite layer

Table 13.17 The US companies: areas of expertise in PSC development

Companies	Area of technology
Hunt Energy Enterprises LLC (4)	• Perovskite material has the formula CMX_3 with titanate interfacial layer • Interfacial layers-Bi or Tri layer-lead halide ink composition • Preparing photoactive perovskite materials with lead halide precursor ink with different concentration • Additional perovskite and interfacial layers
IBM [United States] (4)	• Photoconversion device with grating structures arrangement • Organic tandem thin-film solar cells—layers • First absorber layer—chalcopyrite perovskite and second absorber layer-perovskite • Transparent conducting oxide formation using sputtering process
Solar-Tectic LLC [United States] (2)	• Producing large grained to single crystal semiconductor films • Tandem solar cell with germanium, perovskite-layer of CsGeSnI3
Nutech Ventures [United States] (2)	• Continuous processes—scalable perovskite device fabrication • Method—single crystal growth
Sharp Lab Of America Inc. [United States] (1)	• Thin film with adjustable band gap
Alliance Sustainable Energy [United States] (1)	• Solar cell assembly—films
Kateeva Inc. [United States] (1)	• Coating system
Los Alamos Nat Security LLC [United States] (1)	• Crystalline perovskite thin-film compositions

Table 13.18 The US universities: areas of expertise in PSC development

University	Area of technology
Univ Northwestern (5)	• Monolithically integrated perovskite–on-silicon multijunction PV device • Lead-free solid-state organic-inorganic layers • Perovskite has formula A_2MX_6 • A-Cs, Rb, K, Na, and Li, M-group consisting of a Group 4 metal, a Group 10 metal, a Group 14 metalloid, and a Group 16 metalloid—Sn, Ge, Te, Ti, Zr, Hf, Ni, Pd, and Pt • Hole-transporting materials—dopant-free polymeric • Semiconducting organic–inorganic tin halide perovskite–composition of layers
Massachusetts Institute of Technology [United States] (4)	• Method for biomimetic mineralization of perovskite nanomaterials–crystalline perovskite • Recycling car batteries-recovered lead iodide perovskite nanocrystals • PSC assembly • Mixed metal perovskite
Midwest Research Inst [United States] (1)	• Series-connected, monolithic tandem PV cells
Univ Akron [United States] (1)	• Photodetector includes an active layer—perovskite hybrid
Univ California [US] (1)	• Assembly—all solution processed metal oxide transporting layer
Univ California [United States] (1)Cosmo Chemical Co Ltd [Korea]Cosmo Am&T Co Ltd [Korea]	• PSSC—electrochemical solar cell–layers-double-wall CNTs-TiO_2 nanotubes
Univ Leland Stanford Junior [Korea] (1)	• A solar cell includes a light-absorbing layer comprising a 2D-perovskite
Univ Leland Stanford Junior [United States] (1) Massachusetts Inst Technology [United States]	• Two-terminal metal halide semiconductor-multi-junction–tunnel junction
Univ Nebraska [United States] (1)	• Assembly/Fabrication—Spiro-MeOTAD and TiO_2 Nanowire
Univ of Washington Though Its Center for Commericalization [United States] (1)Univ Washington Through Its Center for Commercialization [United States]	• Organic semiconducting compound—composition

Well-known universities like Northwestern University and Massachusetts Institute of Technology (MIT) are maintaining the status by working on newer areas like PSC development in the United States. Northwestern University is working in the area of monolithically integrated perovskite–on-silicon multijunction PV device, lead-free solid-state organic–inorganic layers also working for dopant-free polymeric hole-transporting materials and semiconducting organic–inorganic tin halide perovskite composition for active layers. MIT as usual has chosen a novel track by devising method for biomimetic mineralization of perovskite nanomaterials for crystalline perovskite, recycling car batteries to recover lead iodide for perovskite nanocrystals, and working for PSC assembly, mixed metal PSCs also. Midwest Research Institute of the United States is also working for tandem PV cells. University Leland Stanford Junior in collaboration with MIT has developed the two-terminal metal halide semiconductor multijunction–tunnel junction.

13.9.1 PV Market and Trends

Recent scenario of PV shows that PV provides 0.1% of total global electricity generation. However, rapid growth in the area is due to effective supporting policies and recent dramatic cost reductions. PV is commercially available and a reliable technology with significant potential for futuristic growth worldwide. In the IEA solar PV roadmap vision, PV is projected to provide 5% of global electricity consumption in 2030, rising to 11% in 2050 and avoid 2.3 gigatonnes of CO_2 emissions per year, which will require a selected policy support.

13.9.1.1 The Top 10 Countries in 2016

According to a report by International Energy Agency, photovoltaics development with novel PV concepts aims at achieving ultra-high efficiency solar cells via advanced materials and new conversion concepts and processes. A report IEA PVPS T1-31:2017 reported that 75 GW of PV systems have been installed and connected to the grid in the world last year, which is currently the subject of basic research. Although trends can be estimated by this projected data, the significant growth in PV market which has been found to be 74.4 GW in 2016, compared to 50 GW in 2015. It has been representing 50% growth year-on-year. This is the first time that for the global market growth happening with declining two major contributors (Japan and Europe), whereas China–Asia ranking top places efficiently making it continuously for the fourth year with 67% of total PV market last year. China reached 34.54 GW in 2016 for the PV market and also is the leader in terms of total capacity with 78 GW. Japan instead of growth was found declining by 20% with around 8.6 GW installed and connected to grid in 2016. Europe market was found declining from 7 GW in 2014, 8 GW in 2015, and 6 GW in 2016; United Kingdom with 2 GW, Germany 1.5 GW, French market 0.56 GW, whereas the US market increased to 14.7 GW, India progressed significantly to 4 GW, doubles installation from 2015; Chile 746 MW, South Africa 509 MW, Turkey 583 MW, Thailand 726 MW, Mexico 100 MW, and the Philippines 756 MW [47].

According to one report [47] for the energy market revolution, 6 of the top 10 markets for PV in 2016 have installed at least 1 GW of PV systems which were 7 in 2014. For the total installed capacity, 24 countries are entering the 1 GW club, only one more compared to 2015 [47].

13.9.1.2 Perovskite Structure

As described in detail in an article [11], Perovskites have the common formula ABX_3, where A and B are cations which reside in the corner and the body center of the pseudo-cubic unit cell and X is the anion which occupies the face center. Crystal structure of perovskite can be visualized as a corner-linked BX_6 octahedral with interstitial A cation. The crystallographic stability and apparent structure can be deduced by considering a Goldschmidt tolerance factor t and an octahedral factor μ. Here, t is defined as the ratio of the distance A–X to the distance B–X in an idealized solid sphere model ($t = (RA + RX)/\{\sqrt{2}(RB + RX)\}$, where RA, RB, and RX are the ionic radii of the corresponding ions) and μ is defined as the ratio RB/RX. For halide perovskites (X = F, Cl, Br, I), t and μ have a narrower range, i.e., 0.89–1.0 and 0.4–0.88, respectively. For ABX_3, the larger cation A is considered as organic, which is generally methyl ammonium ($CH_3NH_3^+$) with RA = 0.18 nm. Ethyl ammonium ($CH_3CH_2NH_3^+$; RA = 0.23 nm) and formamidinium ($NH_2CH = NH_2^+$; RA = 0.19–0.22 nm) also provide excellent results. The anion X is a halogen, normally iodine (RX = 0.220 nm). Br and Cl are also used in perovskites as well as mixed halide materials. By findings of research in this area for cation B, Pb is universally considered for stabilization and efficiency in solar cells. Although Sn has similar band gap (1.55 eV) and is in the same group as Pb, due to the ease of oxidation and lack of stability, it performs poorly compared to Pb in efficiency. In recent years, for improved efficiency (~20.1%), methyl ammonium lead triiodide ($CH_3NH_3PbI_3$) perovskites have been widely used for the fabrication of solar cells, and they also found that mixed halides $CH_3NH_3PbI_3 - xClx$ and $CH_3NH_3PbI_3 - xBrx$ are also important.

PSCs have already shown remarkable improvement from 2.2% efficiency to 31% in principal in theory according to the latest efficiency records provided by the National Renewable Energy Laboratory (NREL) in the United States. Perovskites have already surpassed organic PVs and DSSCs and are now, at least in terms of efficiency, in the same range as leading thin-film PVs like cadmium telluride (CdTe) and copper indium gallium selenide (CIGS).

High and rapidly improved efficiencies are not the only advantage of PSCs. Flexibility, semitransparency, tailored form factors, better low-intensity light performance, low potential processing cost, and low material cost are other value propositions of PSCs.

For commercially available silicon-based solar cells, recent advances in organic–inorganic hybrid PSCs with methyl ammonium lead iodide as the archetypal material have led to the advent of new low-cost PV technology that could be a viable competitor and have the capability to replace silicon-based solar cells market [48]. Apart from the low cost, simple device processing and manufacturability combined compatibility with roll-to-roll processing and fabrication on flexible substrates add to the merits of the perovskite PV technology. The term "perovskite" was attributed to the crystal structure of calcium titanate ($CaTiO_3$), which was discovered by the German mineralogist Gustav Rose in 1839 and named in honor of the Russian mineralogist Lev Perovski [49].

Perovskite material has special electronic structure, better charge transport, band gap tunability, easy processing, and low cost. PSCs considered as promising candidate because it has demonstrated unprecedented progress in efficiency and its architecture evolved over the period of the last 5–6 years, achieving a high PCE of about 22% in 2016 However, hysteresis, stability, and toxicity are major issues in PSCs that serve as a bottleneck for successful commercialization of this promising PV technology.

13.10 STABILITY

PSCs are a promising technology due both to the possibility of using low-cost materials and fabrication techniques and in view to their application in markets such as BIPV.

Despite the relatively high performance of PCSs, air (moisture and oxygen) stability, photo stability, and thermal stability are the three main issues that need to be investigated and improved for future commercialization. Air (moisture and oxygen) stability, photo stability, and thermal stability, UV light, thermal and electrical stress are the main sources of degradation. Moisture induces rapid degradation of the perovskite layer in PSCs. Contact with oxygen degrades the cells' constituent materials by oxidation. In addition, temperature leads to degradation of perovskite semiconductor as well as its phase change. UV light also is a source of aging, which needs to be filtered. Improvements in lifetime must be tackled in two ways: developing materials and interfaces of the cells and their architectures which are inherently more stable and via more effective encapsulation strategies.

Current products on the market typically have a warranty of 20–25 years (c-silicon solar cells), which suggest that the installation will retain 80% of its initial output after this time period, corresponding to a system loss of <1%/year [50]. The PSCs demonstrated by Li et al. [51] would pass the temperature or thermal stress test but yet to overcome the moisture ingression issues or pass the damp test. Research conducted over the past couple of years indicates that the major key factors of degradation namely moisture and heat must be addressed to improve the commercial viability of PSCs and to provide a competitive edge over the currently existing alternatives.

Apparently from the literature survey that not enough has been done on investigating the stability of large area devices. Here their new stability issues are arising because of the size, the applications of different deposition processes, interconnections, and the electrical stresses due to several unit cells being connected either in series or in parallel in a module. Even though limited, more stability investigations on modules are found in the DSSC field due to many more years of its development both with accelerated tests and outdoor installations. These studies show that at the module level there exist challenges which can be different from those at the small cell level such as encapsulation (not only of the unit cells but of interconnections and contacts), nonuniformity of the layer stacks and unit cells, and reverse bias stresses. More studies of this type must be encouraged not only for DSCs but also for the perovskite field to identify issues and criticalities related to large area modules early in the development phase of this new technology. For delving into these issues, it is important to adopt as much as possible the ISOS protocols and the IEC (61646) tests as well as testing outdoor installations. The latter two are those that ultimately determine the long-term suitability of a PV technology in particular environments. IEC qualification must be carried out on modules. It is large area modules, obviously, which represent the unit embodiment of the PV technology which will be industrialized [52].

In the review article [11] by Prasanthkumar and Giribabu, 2016, they have elaborated the improvement in the stability of perovskites for the last decade. Their study included the time from 2011 to 2016. It was found that in 2011, stability of perovskite-based DSSC was limited ($\sim$10 min) before degradation (about 80% degradation) because perovskite QDs dissolved into redox electrolyte. Further in 2012, stability of PSCs was enhanced from minutes to over 500 hours using spiro-MeOTAD as a solid-state HTM, which was solely due to solid hole conductor. For more improvement, stability over 500 hours is achieved by other method, i.e., long-term light soaking at a light intensity ($100\ mW \cdot cm^{-2}$) and a temperature of 45°C. Where TiO_2-based devices have been used for PSC and attained stability above 500 hours, Al_2O_3-based PSC with stability over 1000 hours at nearly $15\ mA \cdot cm^{-2}$ has a rapid decay in photocurrent and a small decay in photovoltage. Other researchers utilized HTMs such as spiro-OMeTAD, poly-3-hexylthiophene (P3HT), and PDPPDBTE in PSC. They found that spiro-MeOTAD-based devices showed a 28% decrease in PCE when aging was extended, which is lower when compared to the initial PCE value. Hydrophobic behavior of PDPPDBTE-based cells improved the stability remarkably even after 1000 hours, precluding water flow into the perovskite. Mixed-halide perovskites also improved the stability on nanostructured TiO_2. Inclusion of Br to mixed halide improved efficiency as well as stability, which was checked by removing Br and inclusion of Br in mixed halides. In addition, rearrangement of the 3D configuration of PSCs over time for the equimolar Br/I devices revealed a noteworthy improvement of 37% PCE. Other researcher developed a device by solution drop casting method where the device was consists of a double-layer of TiO_2 and ZrO_2 in a porous carbon film, with a significant long-term stability of 1008 hours upon direct exposure to sun. Stability of mesoporous TiO_2/$CH_3NH_3PbI_3$/C without encapsulated PSC devices achieved over 2000 hours under dark conditions. Device stability being major aspect for commercialization success for which the moisture, temperature, light, oxygen, and other concerns like intrinsic stabilities at the interface and device architecture remain major issues to practical applications. There has been significant achievement for PSC stability. For the ease of fabrication, similar design strategy of dye-sensitized and OSCs has resulted in a rapid increase in the number of researchers working in this field. During the last 5 years, the number of articles published on PSC devices has rapidly increased, from two to thousands, which can also be seen in the publication analysis given above. Therefore, PSC play a major role in commercialization of future generation of PVs because of its achievements in energy efficiencies as well as stability.

World's increasing energy needs call for the PSCs with high efficiency, low cost, high stability, non-toxicity, and reproducibility for better future. It was found that the fast progress in performance suggests that PSCs have great potential, which can surpass the traditional silicon solar cells. There is a need of research in the areas of gap in advancement like stability, toxicity, and reproducibility. Over these things chemical instability dramatically restricts the commercial development in the future. For better understanding, degradation of PSCs has three factors which have been investigated: air (moisture and oxygen) stability, photo stability, and thermal photonics, which should be taken into consideration for a high-stability PSC and better operations in energy conversion as well as storage [11].

13.11 HYSTERESIS

For long-term reliability, hysteresis and transient effects are two important aspects in optimization of device architectures and selective contact materials. Because efficiencies are found rapidly increasing, researchers speculated the slow transient effects causing hysteresis, where hysteresis in hybrid PSCs indicates the variations in the current–voltage response curves in which the corresponding PV parameters vary depending on the direction and rate of the scan. Effects like current–voltage scan direction, light and voltage bias conditions prior to measurement, measurement delay time, and also photo-induced ion migration can affect shape of IV curves and apparent device hysteresis. As reported by researchers [49], reversible hysteresis was found in groups in $MAPbI_3$-based devices of all architectures (normal and inverted). "The scanning direction is called as backward or reverse scan when the voltage is swept from positive (open circuit) to negative to negative (short circuit), whereas the scan in the opposite direction is termed as forward scan." For better quantification, it is without over or underestimation of solar cell device efficiency. It is very much indispensable to decipher the origin or mechanisms of the

I–V hysteresis to ensure reliable quantification [50]. So, every report based on device performance should include hysteresis effects.

13.12 COST COMPETITIVENESS

Fabrication of traditional solar cells in comparison to PSCs has been found complex. Perovskite, on the other hand, has been more efficient for which the cost analysis have been carried out by Cai et al., 2017. In this study they have selected two representative examples of PSCs and performed a cost analysis of their productions: one was a moderate-efficiency module produced from cheap materials, and the other was a high-efficiency module produced from expensive materials. Calculating the levelized cost of electricity (LCOE), 3.5–4.9 US cents/kWh for the efficiencies more than 12% and lifetime 15 years. Study found that both the modules have lower cost compared to traditional PVs and estimated to produce perovskite solar modules cost in the range of 0.21–0.28 US $/W. LCOE of a perovskite solar module was calculated by assuming a module cost of 0.25 US$/W and a lifetime of 15 years. The LCOEs were 4.9, 4.2, and 3.5 US cents/kWh corresponding to module efficiencies of 12%, 15%, and 20%, respectively, which were lower than that of traditional energy sources which are generally estimated 7.04–11.90 US cents/kWh, and the costs of solar PV technologies were 9.78–19.33 US cents/kWh [53]. Rather than finding a cheaper material, we should focus more on improving lifetime and efficiency (Fig. 13.10).

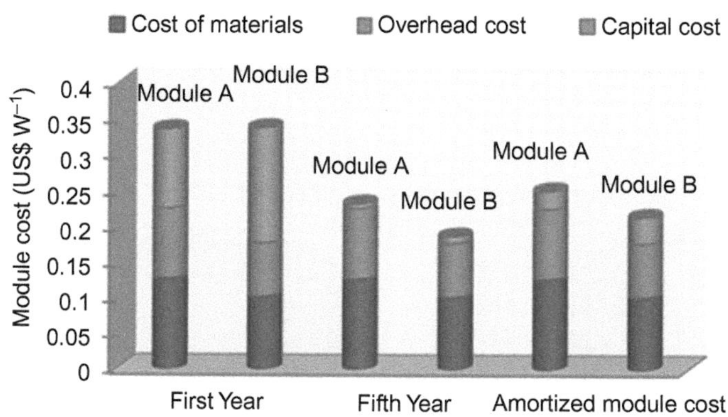

Figure 13.10 Module cost effectivity. *From M. Cai, Y. Wu, H. Chen, X. Yang, Y. Qiang, L. Han. Cost-performance analysis of perovskite solar modules. Adv. Sci. 4 (2017) 1600269, doi:10.1002/advs.201600269.*

13.13 COMMERCIALIZATION ISSUES OF PSC

In 1954, the first practical PV device based on crystalline silicon was demonstrated at Bell Laboratories. After many decades of progress, crystalline silicon technology dominates the global PV market with a 55% and 36% market share for polycrystalline- and monocrystalline-silicon modules in 2014, respectively. The remaining 9% of the market was split between a variety of other established and emerging PV technologies, including polycrystalline thin films, amorphous semiconductors, DSSCs, organics, and quantum dot solar cells. To gain market share from crystal silicon solar cells, alternative technologies have to provide a desirable combination of high PCE, low manufacturing costs, and excellent stability. Recent research suggests that organo-metal halide perovskites, with methylammonium lead iodide ($CH_3NH_3PbI_3CH_3NH_3PbI_3$ or $MAPbI_3MAPbI_3$) being the prototypical example, have the potential to meet these conditions and become competitive in the marketplace. As a result of intensive research efforts across the globe over the past 3 years, perovskite-based solar cell PCEs are now comparable to or better than most other PV technologies, and the simple device processing promises lower manufacturing costs, suggesting the potential to challenge the prevailing silicon technology in the foreseeable future (Figs. 13.11–13.13).

It is well cited in the literature that thermal, photo, and moisture stability are now the most critical issues for commercialization of PSCs [55]. The cells currently only last for months outdoors, whereas silicon solar panels are usually guaranteed to work for at least 25 years. Weather changes and extreme light levels, temperature, and humidity all cause perovskite cells to decompose. Moisture is the worst problem. Reactions with water form hydrates that alter the structures of the crystals so that they

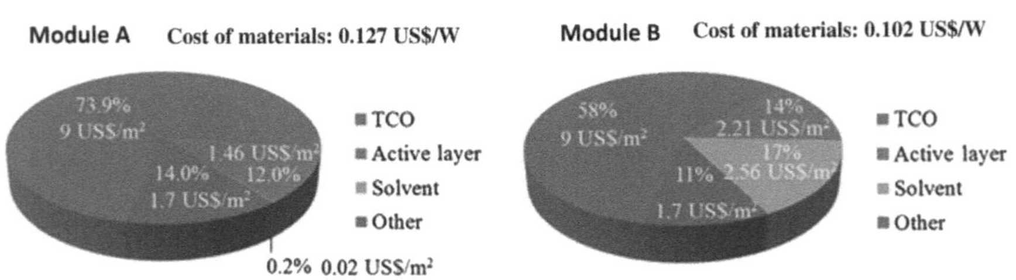

Figure 13.11 Cost of material applied for module. *From M. Cai, Y. Wu, H. Chen, X. Yang, Y. Qiang, L. Han. Cost-performance analysis of perovskite solar modules. Adv. Sci. 4 (2017) 1600269, doi:10.1002/advs.201600269.*

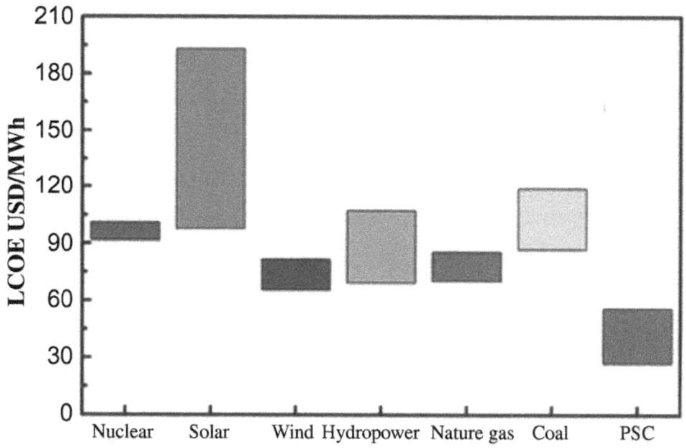

Figure 13.12 Comparison of solar cell with other sources. *From M. Cai, Y. Wu, H. Chen, X. Yang, Y. Qiang, L. Han. Cost-performance analysis of perovskite solar modules. Adv. Sci. 4 (2017) 1600269, doi:10.1002/advs.201600269.*

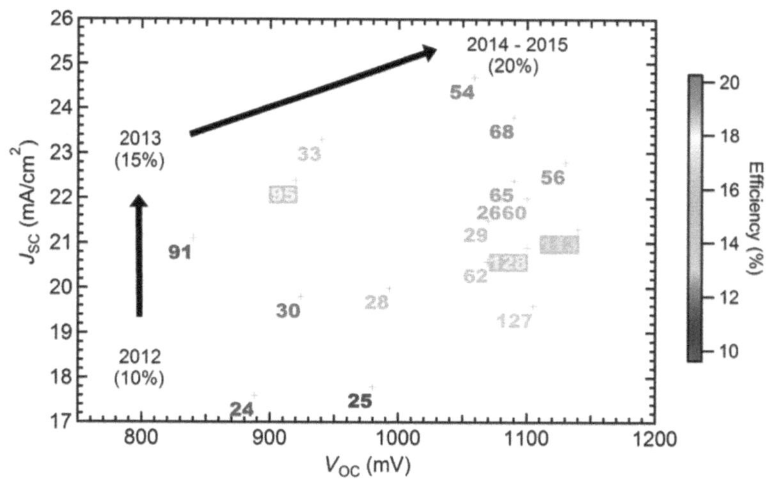

Figure 13.13 Efficiency mapping of PSCs. *From Z. Song, S.C. Watthage, A.B. Phillips, M.J. Heben. Pathways toward high-performance perovskite solar cells: review of recent advances in organo-metal halide perovskites for photovoltaic applications J. Photon. Energy 6(2) (2016) 022001, doi:10.1117/1.JPE.6.022001 [54].*

cannot absorb visible light. In the past decade, the lifetime of a cell has extended from a few minutes to 6 months. But much more work is needed.

It is claimed that the encapsulated perovskite film itself is stable at 85°C for 1000 hours under UV-free light. However, the encapsulated device exposed at 85°C with a relative humidity of 85% rapidly degraded, which was ascribed to ingression of water. A recent study on, how stability can be enhanced through the utilization of porous carbon electrode, which was attributed to the strong chemical interaction between carbon materials and perovskite. Irreversible degradation of PSCs resulted from trapped charges at the interfaces between the perovskite and selective contacts was also evidenced by Seoul National University. Stability is an important factors for successful commercialized perovskite based solar modules.

But there are major challenges to using perovskites on a commercial scale. The main one is stability: the cells currently only last for months outdoors, whereas silicon solar panels are usually guaranteed to work for at least 25 years. Weather changes and extreme light levels, temperature, and humidity all cause perovskite cells to decompose. Moisture is the worst problem. Reactions with water form hydrates that alter the structures of the crystals so that they cannot absorb visible light. In the past decade, cell lifetimes have extended from a few minutes to 6 months. But much more work is needed. Yans in his study [56] suggested five priorities to improve the stability of PSC: crystal structure, alternative designs, film quality, conducting layers and interfaces, and encapsulation.

The study explained that stability of the perovskite cell can be increased to from less than 100 hours to beyond 1000 hours through "Doping" the crystal, or introducing small imperfections with cesium ions or organic groups has also increased stability from less than 100 hours to beyond 1000 hours [56]. Stability that can also be enhanced by using bromide ion instead of iodine in perovskites solar cell has the formula ABX_3, in which A is an organic or inorganic component, B is a metal atom and X is a halide. They further explained that we believe that introducing a small number of large groups into the A site is the best way to improve the stability without sacrificing too much efficiency.

13.14 EFFICIENCY AND VARIABILITY

Efficiency considerations and variability are the crucial factor in success of any PV market. The highest efficiency is reported around 25% in case of conventional silicon solar cells; in this existing scenario, the PSC should show at least 30%–40% efficiency to translate research to real-life solution as well as cost reduction in scaling-up process. It is well cited that the percentage difference of laboratory scale reported cell efficiency and manufactured module efficiency [57,58]. Recently, the best PSC efficiency of 31% has been reported on laboratory-based research that shows perovskite as a promising candidate (Fig. 13.14).

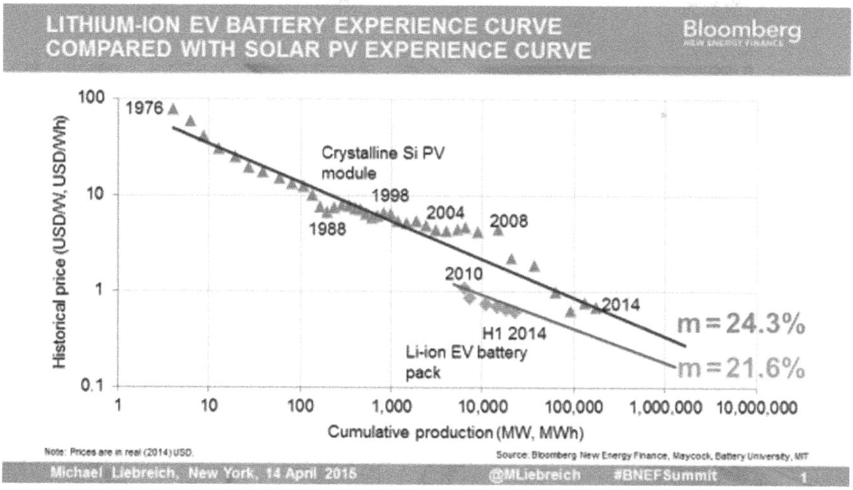

Figure 13.14 Comparison of experience curve for nuclear vs that for solar. *From A.A. Asif, R. Singh, G.F. Alapatt. Technical and economic assessment of perovskite solar cells for large scale manufacturing. J. Renew. Sustain. Energy 7 (2015) 043120, doi:http://dx.doi.org/10.1063/1.4927329 [59].*

Review article by Prasanthkumar and Giribabu (2016) [11], for organic–inorganic halide perovskite devices, it has been noticed that when researchers used methylammonium lead halide (perovskite) as a light harvesting material in excitonic solar cells, some utilized perovskites as sensitizers in DSSC, with the conversion efficiency of 3.2% for $(CH_3NH_3)PbBr_3$ and 3.8% for $(CH_3NH_3)PbI_3$. It was found that these devices were unstable due to degradation of perovskite in liquid electrolyte containing lithium halide. Subsequently, other group developed quantum dot–sensitized solar cell using perovskite $(CH_3NH_3)PbI_3$ sensitizer, perovskite nanoparticles which exhibit optical properties better than the standard N719 dye sensitizer reaching efficiencies of 6.5%. But it was found that these devices reduce performance because of dissolution of halides in liquid electrolytes. Corrosive liquid electrolytes in perovskite DSSC and others have developed solid-state electrolyte, Spiro-MeOTAD (2,2′,7,7′-tetrakis(N,N-di-p methoxyl phenylamine)-9,9′-spirobifluorene), which can be used as HTM. This advancement helped them to reach PCE of 9.7% also with long-term stability. This was achieved as demonstrated charge separation by hole transfer from the excited $(CH_3NH_3)PbI_3$ nanoparticles to the mesoscopic TiO_2 film and spiro-OMeTAD, which was confirmed by femtosecond laser studies. According to their findings, solid hole conductor as a HTM in comparison to liquid electrolyte—based conductor has improved PCE and stability. Snaith & co-workers who developed solution processable PSCs with nonconducting Al_2O_3 replacing the conducting nanoporous TiO_2 achieve high efficiency (~10%) in comparison to previous DSSC with liquid electrolyte. Additional advances count like for mixed halide $CH_3NH_3PbI_3-xCl_x$ in PSC demonstrated good stability and carrier transport; then coating thin perovskite layer onto the nanoporous TiO_2 surfaces and preparing the extremely thin absorber (ETA) cells; restoring the conducting nanoporous TiO_2 with non-conducting Al_2O_3, thus improving the open-circuit voltage (Voc) and achieving an efficiency of 10.9%. Also researchers have tried ambipolar transport in simple planar cells with the eliminated scaffolding. Also reported by other research group is the PCE of 12% using both flexible layers, scaffolding TiO2 infiltrated by perovskite. After studying several HTMs-based PSC as spiro-OMeTAD and polytriarylamine (PTA), this group highlighted that PTA-based PSC has achieved the highest efficiency in comparison to others. Similarly one group found improved PCE to 12.3% using mixed halide $CH_3NH_3PbI_3-xBrx$ perovskites by addition to low ionic radius, enhancing efficiency and stability by tetragonal to pseudo-cubic structural transition of perovskites. In the pace of development, there was also emphasis on other materials like using non-halide ($PbAc_2$) under one-sun illumination was done to achieve 15.2% prepared films using one-step coating and subsequent annealing. Also when devices were fabricated with $PbCl_2$ and PbI_2, significant improvement in efficiency was observed. This one-step annealing method, abandoned precipitation, yields morphological defects in PVs affecting PCE. Whereas two-step coating deposition method was developed to make smooth and crystalline films. The film prepared by two-step coating deposition method for which solution of PbI_2 coated on the substrate and then CH_3NH_3I in 2-propanol was spun on the film, resulting in high-quality perovskite films. For this two groups achieved 15% efficiency, one by sequential deposition and other by vapor-deposited PSC planar heterojunction thin-film architecture. From these findings, deposition approaches were found significantly influencing PCE of PSC when compared to solution-processed solar cells. Efficiency was further increased by altering the energy levels of mixed halide $CH_3NH_3PbI_3-xBrx$ with PTAA (HTM) and more by adjusting the thickness ratio of perovskite infiltrated TiO_2 scaffolding relative to the continuous perovskite layer, and also there were findings for the increased efficiency by adjusting the band alignment of HTM/ETM to the perovskite layer. They reported that careful selection of HTM, ETM, and film preparation leads to reduced carrier recombination and good carrier transport. It is theoretically reported that PSC exhibited maximum efficiency of 26% by maintaining 1 μm thickness of $CH_3NH_3PbI_3$ absorber (phase with Pm3m symmetry), which is much higher than the reported GaAs solar cell with the same thickness.

Stability improvement for PSCs: even though the PCE has increased much in the recent years, durability is still significant concern for practical applications because the existing technology has lesser durability, for commercialization of these solar cells viable extended device lifetime. Earlier in 2011 perovskite-based DSSC durability was limited ($\sim$10 min) before degradation because perovskite QDs dissolve into redox electrolyte. In 2012 stability was increased from minutes to over 500 hours using spiro-MeOTAD as solid-state HTM and also achieved the same stability by another approach, i.e., long-term light soaking at a light intensity ($100\ mW \cdot cm^{-2}$) and temperature 45°C. Al_2O_3-based PSC derived later in comparison to the TiO_2-based devices has improved stability of 1000 hours which is twice the stability attained by those TiO_2. Stability of Al_2O_3-based PSC was at nearly $15\ mA \cdot cm^{-2}$ for over 1000 hours with rapid decay in photocurrent and a small decay in photovoltage. Other researchers have tried spiro-OMeTAD, poly-3-hexyl thiophene (P3HT), and PDPPDBTE in PSC and found that 28% decrease in PCE when aging was extended for spiro-MeOTAD-based devices. In comparison, there was a remarkable increase in stability for PDPPDBTE-based cells even after 1000 hours, the reason being hydrophobic behavior of PDPPDBTE, precluding water flow into the perovskite. Technology for mixed-halide perovskites also improved the stability on nanostructured TiO_2. Studies regarding PSC for structure with Br and without Br showed that 20% drop in initial PCE and also the equimolar Br/I devices revealed a noteworthy improvement of 37% PCE which can be accounted as a result of rearrangement of the 3D configuration of PSCs over time. Devices also have played an important role in drop casting methods consisting of double-layer of TiO_2 and ZrO_2 in a porous carbon film, with significant long-term stability of 1008 hours upon direct exposure to sun. A 2000-hour stability under dark conditions was achieved by mesoporous $TiO_2/CH_3NH_3PbI_3/C$ without encapsulated PSC devices. All these technological advancements and application methods have improved PCE significantly; however, long road is waiting ahead in the future. For commercialization of a PSC solar device, stability is a major issue. Although few reports highlight controlling the device stabilities relating to moisture, temperature, light, and oxygen, other concerns like intrinsic stabilities at the interface and device architecture remain major obstructions to practical applications. Notably, these latest results show the significant improvement in PCE and stability of PSCs by considering the possibility of using one or two optional layers, three different mixed-halide perovskites, and a few diverse HTMs. Another significant aspect in PSCs is ease of fabrication, and similar design strategy of dye-sensitized and OSCs has resulted in a rapid increase in the number of researchers working in this field. During the last 5 years, the number of articles published on PSC devices has rapidly increased, from two to thousand. Therefore, PSC play a major role in commercialization of future generation of PVs.

13.14.1 Toxicity

Lead toxicity being a matter of concern in developing PSCs, it is needed to replace lead in perovskite crystal with less toxic metal. Most suitable replacements are Sn and Ge, also members of Group 14 metals but there arise chemical instability issue as the stability of 2 + oxidation state decreases. It has been demonstrated that when the Sn2 + ion is oxidized to Sn4 + , the Sn4 + acts as a p-type dopant within the material in a process referred to as "self-doping." Solar cell utilizing completely lead-free perovskite as the absorber layer needs to be found.

Development of PSCs addressing environmental concerns due to lead toxicity which is a stumbling block envisioning their large-scale applications; this was addressed by Babayigit et al. in 2016 [60].

A study by Noel et al., 2014 [61] reported a mixed metal, Sn−Pb perovskite which allowed tunability of the band gap of the perovskite absorber by varying the Sn to Pb ratio, indicating that Sn could be a good choice of metal ion, especially for lower band gap solar cells. However, the same study reported that the neat $CH_3NH_3SnI_3$ perovskite does not exhibit significant PV properties, and that a minimum content of Pb is needed to stabilize Sn in its 2 + oxidation state. Partially replacing Pb with Sn in organic−inorganic lead halide perovskites has been proven as a promising approach to reduce environmental toxicity and develop low band gap (as low as 1.20 eV) PSCs beneficial for constructing perovskite-based tandem solar cells. It was found that for the entire series of $FAPb_{1-x}Sn_xI_3$ alloys, PCEs are significantly improved after Cs incorporation. The performance of Pb—Sn binary perovskite systems will be improved.

Toxicity, recycling, and large-area technology application are important factors through the commercialization point of view. Recently the enthusiasm towards the purported environmental friendliness of Sn-based perovskites compared to Pb-based has been seen among the research community, and many proposed the replacement of Pb by tin (Sn) to render perovskite light absorbers less toxic or even nontoxic. Research work carried out by Aslihan Babayigit et al., (2016) [60,62] alarmed the community that the use of the Sn in formation of solar cell should be approached with great caution. She mentioned in her work revealed that the release of HI using the tin which introduces a new perspective in the possible routes of toxicity. Notably, the strong acidification induced by the SnI_2 is found to be more harmful than the combined effect of milder acidification and the expected Pb intoxication induced by PbI_2. These results challenge the current view in solar cell development, where Sn-based perovskites are regularly put forward as the nontoxic alternative of Pb-based ones. This leads to the recommendation that the enthusiasm towards the purported environmental friendliness of Sn-based perovskites compared to Pb-based ones should be approached with great caution. Besides that, Tsutomu Miyasaka (Toin University of Yokohama, Japan) also argued that that $\sim 0.32\ g/m^2$ of Pb in a PSC is much smaller than that used for soldering in commercial Si solar cells ($43\ g/m^2$).

13.15 DISCUSSION AND CONCLUSION

Successful commercialization of perovskite in market has generated different views, where perovskite has been considered a great replacement for Si-based solar cell capturing 90% market share. PSC efficiency has reached more than 22% in few years and can surpass standard solar cells in upcoming years. Because the increase in number of publication cannot justify the successful implementation of technology in field, products in market should be manufactured with improvised technology considering the requirements for field applications not only efficiencies but others like moisture, toxicity, durability, and also lower cost. PSCs have several advantages like simple processing, instead of expensive multistep processes, conducted at high temperature used for traditional solar cells.

Solar energy has been considered as a more attractive source compared to other renewable energy generation options. The scientists and researchers are trying their best to achieve the best out of resources they have. On the other hand, industrialists are also trying to invest in the solar power markets. The knowledge pool that has been generated in the past decade in this area by the efforts of intellects worldwide starting from first-generation semiconducting wafers developing second-generation thin films to the level of third-generation emerging thin films incorporating perovskite as the material in PVs is phenomenal. Although the efforts were there for developing low-cost material, the advancement, fabrication, and other aspects of commercialization have been considered for better growth, like hysteresis, toxicity, and generating novel absorber materials giving rise to long charge carriers' lifetimes and mobility.

REFERENCES

[1] Green Technology Report. Available from: http://www.nistads.res.in/images/ISTIP/report/green%20technology.pdf.

[2] M. Guo, W. Song, J. Buhan, Bioenergy and biofuels: history, status, and perspective, Renew. Sustain. Energy Rev. 42 (2015) 712–725.

[3] G. Xinyu, J. Boc, L. Bina, Y. Kaia, Z. Hongguanga, F. Boyuana, Study on renewable energy development and policy in China, Energy Procedia 5 (2011) 1284–1290. Available from: https://doi.org/10.1016/j.egypro.2011.03.224.

[4] Energy Statistics, Central Statistics Office, Government of India. 2017. Available from: http://www.cbgaindia.org/wp-content/uploads/2017/04/Energy_Statistics_2017.pdf.

[5] IRENA Report 2015. Available from: http://www.ren21.net/wp-content/uploads/2015/11/IRENACOP21_Fast-Facts.pdf.

[6] IRENA Report 2012. Available from: https://costing.irena.org/media/2769/Overview_Renewable-Power-Generation-Costs-in-2012.pdf.

[7] IRENA Renewable Energy Target Setting report 2015. Available from: http://www.irena.org/DocumentDownloads/Publications/IRENA_RE_Target_Setting_2015.pdf.

[8] IRENA 2017 Report. Available from: http://www.irena.org/DocumentDownloads/Publications/IRENA_RE_Capacity_Statistics_2017.pdf.

[9] IRENA 2016 Report. Available from: http://www.irena.org/DocumentDownloads/Publications/IRENA_Measuring-the-Economics_2016.pdf.

[10] D. Shi, Y. Zeng, W. Shen, Perovskite/c-Si tandem solar cell with inverted nanopyramids: realizing high efficiency by controllable light trapping, Sci. Rep. 5 (2015) 16504. Available from: https://doi.org/10.1038/srep16504. Available from: https://www.ncbi.nlm.nih.gov/pmc/articles/PMC4643242/pdf/srep16504.pdf.

[11] S. Prasanthkumar, L. Giribabu, Recent advances in perovskite-based solar cells, Curr. Sci. 111 (7) (2016). Available from: http://www.currentscience.ac.in/Volumes/111/07/1173.pdf.

[12] Reinders A, Verlinden P, Sark WV, Freundlich A, (2016) Photovoltaic solar energy: from fundamentals to applications. Wiley & Sons: London, UK 978-1-118-92746-5.

[13] H.J. Snaith, Perovskites: the emergence of a new era for low-cost, high-efficiency solar cells, J. Phys. Chem. Letters 4 (21) (2013) 3623–3630. Available from: https://doi.org/10.1021/jz4020162.

[14] Perovskite Photovoltaics 2016–2026: Technologies, Markets, Players, IDTechEx, 2016.
[15] N.G. Park, Organometal perovskite light absorbers toward a 20% efficiency low-cost solid-state mesoscopic solar cell, J. Phys. Chem. Letters 4 (15) (2013) 2423–2429. 1948-7185.
[16] R.F. Service, Perovskite solar cells keep on surging, Science 344 (6183) (2014) 458. Available from: https://doi.org/10.1126/science.344.6183.458.
[17] M.A. Green, Third generation photovoltaics: solar cells for 2020 and beyond, Physica E Low Dimens. Syst. Nanostruct. 14 (1–2) (2002) 65–70. Available from: https://doi.org/10.1016/S1386-9477(02)00361-2.
[18] P.P. Boix, T.M. Koh, S. Agarwala, G.S. Mahisalkar, Perovskite solar cell: beyond methylammoium lead iodide, J. Phys. Chem. Lett. 6 (2015) 898–907. Available from: https://doi.org/10.1021/jz502547.
[19] J.H. Rhee, C.C. Chung, E.W.G. Diau, A perspective of mesoscopic solar cells based on metal chalcogenide quantum dots and organometal-halide perovskites, NPG Asia Mater. 5 (2013). Available from: https://doi.org/10.1038/am.2013.53.
[20] P.P. Boix, K. Nonomura, N. Mathews, S.G. Mhaisalkar, Current progress and future perspectives for organic/inorganic perovskite solar cells, Mater. Today 17 (1) (2014) 16–23. Available from: https://doi.org/10.1016/j.mattod.2013.12.002.
[21] S. Gamliel, L. Etgar, Organometal perovskite based solar cells: sensitized versus planar architecture, RSC Adv. 4 (55) (2014). Available from: https://doi.org/10.1039/C4RA03981E.
[22] J.F. Yan, B.R. Saunders, Third generation solar cells: a review and comparison of polymer: fullerene, hybrid polymer and perovskite solar cells, RSC Adv. 4 (82) (2014). Available from: https://doi.org/10.1039/C4RA07064J.
[23] B. Wang, X. Xiao, T. Chen, Perovskite photovoltaics: a high efficiency newcomer to the solar cell family, Nanoscale 6 (21) (2014). Available from: https://doi.org/10.1039/C4NR04144E.
[24] S. Kazim, M.K. Nazeeruddin, M. Graetzel, S. Ahmad, Perovskite as light harvester: a game changer in photovoltaics, Angew. Chem. Int. Ed. 53 (11) (2014). Available from: https://doi.org/10.1002/anie.201308719.
[25] W.H. Zhang, B. Cai, Organolead halide perovskites: a family of promising semiconductor materials for solar cells, Chin. Sci. Bull. 59 (18) (2014). Available from: https://doi.org/10.1007/s11434-014-0259-9.
[26] M.A. Green, A. Ho-Baillie, H.J. Snaith, The emergence of perovskite solar cells, Nat. Photon. 8 (7) (2014). Available from: https://doi.org/10.1038/nphoton.2014.134.
[27] P. Gao, M. Graetzel, M.K. Nazeeruddin, Organohalide lead perovskites for photovoltaic applications, Energy Environ. Sci. 7 (8) (2014). Available from: https://doi.org/10.1039/C4EE00942H.
[28] T.C. Sum, N. Mathews, Advancements in perovskite solar cells: photophysics behind the photovoltaics, Energy Environ. Sci. 7 (8) (2014) 2518–2534. Available from: https://doi.org/10.1039/C4EE00673A.
[29] C.C. Chueh, C.Z. Li, A.K.Y. Jen, Recent progress and perspective in solution processed Interfacial materials for efficient and stable polymer and organometal perovskite solar cells, Energy Environ. Sci. 8 (4) (2015) 1160–1189. Available from: https://doi.org/10.1039/C4EE03824J.
[30] A.K. Chilvery, A.K. Batra, B. Yang, K. Xiao, P. Guggilla, M.D. Aggarwal, et al., Perovskites: transforming photovoltaics, a mini-review, J. Photon. Energy 5 (1) (2015) 057402. Available from: https://doi.org/10.1117/1.JPE.5.057402.
[31] M. Sessolo, H.J. Bolink, Perovskite solar cells join the major league, Science 350 (6263) (2015) 917. Available from: https://doi.org/10.1126/science.aad5891.

[32] H. Kim, K.G. Lim, T.W. Lee, Planar heterojunction organometal halide perovskite solar cells: roles of interfacial layers, Energy Environ. Sci. 9 (1) (2016) 12–30. Available from: https://doi.org/10.1039/C5EE02194D.
[33] P. Agarwala, D. Kabra, A review on triphenylamine (TPA) based organic hole transport materials (HTMs) for dye sensitized solar cells (DSSCs) and perovskite solar cells (PSCs): evolution and molecular engineering, J. Mater. Chem. 5 (4) (2017) 1348–1373. Available from: https://doi.org/10.1039/C6TA08449D.
[34] Heuveln V, Hendrik F, Lamberts P, Douwe EG, (2005) Method of making an electrode for an electrochemical cell. IN Patent 200782.
[35] Web of Science database.
[36] World Intellectual Property Organization WIPO.
[37] European Classification Search. Available from: https://worldwide.espacenet.com/.
[38] NREL. Available from: https://www.nrel.gov/pv.
[39] BP Statistical Review of World Energy. 2017. Available from: https://www.bp.com/content/dam/bp/en/corporate/pdf/energy-economics/statistical-review-2017/bp-statistical-review-of-world-energy-2017-full-report.pdf.
[40] R. Fang, W. Zhang, S. Zhang, W. Chen, The rising star in photovoltaics-perovskite solar cells: the past, present and future, Sci. China Tech. Sci. 59 (7) (2016) 989–1006. Available from: https://doi.org/10.1007/s11431-016-6056-8.
[41] Next Generation Photovoltaics 3. Available from: https://energy.gov/eere/sunshot/next-generation-photovoltaics-3.
[42] Z. Xiao, W. Meng, J. Wang, D.B. Mitzi, Y. Yan, Searching for promising new perovskite-based photovoltaic absorbers: the importance of electronic dimensionality, Mater. Horizon. 4 (2016) 206–216. Available from: https://doi.org/10.1039/c6mh00519e.
[43] N. Rolston, B.L. Watsonb, C.D. Bailie, M.D. McGeheeb, J.P. Bastos, R. Gehlhaar, et al., Mechanical integrity of solution-processed perovskite solar cells, Extreme Mech. Lett. 9 (2016) 353–358. Available from: https://doi.org/10.1016/j.eml.2016.06.006.
[44] N.G. Park, M. Grätzel, T. Miyasaka, K. Zhu, K. Emery, Towards stable and commercially available perovskite solar cells, Nat. Energy 1 (2016). Available from: https://doi.org/10.1038/Nenergy.2016.152. 16152.
[45] Asia Biomass Energy Cooperation promotion office. Available from: https://www.asiabiomass.jp/english/topics/1507_04.html.
[46] S.S. Shin, E.J. Yeom, W.S. Yang, S. Hur, M.G. Kim, J. Im, et al., Colloidally prepared La-doped $BaSnO_3$ electrodes for efficient, photostable perovskite solar cells, Science 356 (6334) (2017) 167–171. Available from: https://doi.org/10.1126/science.aam6620.
[47] IEA PVPS T1-31:2017 report (2016). Available from: https://www.elektrotechnik.ch/fileadmin/elektrotechnik.ch/documents/PDF/001WachstumPV_Global.pdf.
[48] X. Yin, Z. Yao, Q. Luo, X. Dai, Y. Zhou, Y. Zhang, et al., High efficiency inverted planar perovskite solar cells with solution-processed nio_x hole contact, ACS Appl. Mater. Interfaces 9 (3) (2017) 2439–2448. Available from: https://doi.org/10.1021/acsami.6b13372.
[49] N.K. Elumalai, M.A. Mahmud, D. Wang, A. Uddin, Perovskite solar cells: progress and advancements, Energies 9 (2016) 861. Available from: https://doi.org/10.3390/en9110861.
[50] M. Vazquez, I. Rey-Stolle, Photovoltaic module reliability model based on field degradation studies, Prog. Photovoltaics Res. Appl. 16 (2008) 419. Available from: https://doi.org/10.1002/pip.825.
[51] O. Schultz, S. Glunz, G.P. Willeke, Multicrystalline silicon solar cells exceeding 20% efficiency, Prog. Photovoltaics Res. Appl. 12 (2004) 553–558.

[52] S.C. Hermosa, S.K. Yadav, L. Vesce, A. Guidobaldi, A. Reale, A.D. Carlo, et al., Stability issues pertaining large area perovskite and dye-sensitized solar cells and modules, J. Phys. D: Appl. Phys. 50 (2016).
[53] M. Cai, Y. Wu, H. Chen, X. Yang, Y. Qiang, L. Han, Cost-performance analysis of perovskite solar modules, Adv. Sci. 4 (2017). Available from: https://doi.org/10.1002/advs.201600269.
[54] Z. Song, S.C. Watthage, A.B. Phillips, M.J. Heben, Pathways toward high-performance perovskite solar cells: review of recent advances in organo-metal halide perovskites for photovoltaic applications, J. Photon. Energy 6 (2) (2016) 022001. Available from: https://doi.org/10.1117/1.JPE.6.022001.
[55] I.J. Ogundana, S.Y. Foo, Improving the morphology of the perovskite absorber layer in hybrid organic/inorganic halide perovskite MAPbI3 solar cells, J. Solar Energy 2017 (2017). Available from: https://doi.org/10.1155/2017/8549847. Article ID 8549847, 9 pages.
[56] Y. Yang, J. You, Nature News and comments, Make perovskite solar cells stable, Nature 544 (2017) 155–156. Available from: https://www.nature.com/news/make-perovskite-solar-cells-stable-1.21807.
[57] J. Ayre. New CIGS solar cell record—21.7% CIGS cell conversion efficiency achieved at ZSW, *Clean Technica*, 2014. Available from: http://cleantechnica.com/2014/09/27/new-cigs-solar-cell-record-21-7-cigs-cell-conversion-efficiency-achieved-zsw/.
[58] TSMC Solar. TSMC solar commercial-size modules (1.09 m2) set CIGS 16.5% efficiency record, PR Newswire, 2015, Available from: http://www.prnewswire.com/news-releases/tsmc-solar-commercial-size-modules-109m2-set-cigs-165-efficiency-record-300072332.html.
[59] A.A. Asif, R. Singh, G.F. Alapatt, Technical and economic assessment of perovskite solar cells for large scale manufacturing, J. Renew. Sustain. Energy 7 (2015) 043120. Available from: https://doi.org/10.1063/1.4927329.
[60] A. Babayigit, A. Ethirajan, M. Muller, B. Conings, Toxicity of organometal halide perovskite solar cells, Nat. Mater. 15 (2016) 247–251. Available from: https://doi.org/10.1038/nmat4572.
[61] N.K. Noel, S.D. Stranks, A. Abate, C. Wehrenfennig, S. Guarnera, A.A. Haghighirad, et al., Lead-free organic–inorganic tin halide perovskites for photovoltaic applications, Energy Environ. Sci. 7 (9) (2014) 3061–3068. Available from: https://doi.org/10.1039/c4ee01076k.
[62] A. Babayigit, D.D. Thanh, A. Ethirajan, J. Manca, M. Muller, H.G. Boyen, et al., Assessing the toxicity of Pb- and Sn-based perovskite solar cells in model organism Danio rerio, Sci. Rep. 6 (2016). Available from: https://doi.org/10.1038/srep18721. Article number: 18721.

CHAPTER 14

Perovskite Photovoltaics: Life Cycle Assessment

Minu Mohan
Indian Institute of Science Education and Research-Thiruvananthapuram (IISER-TVM), Thiruvananthapuram, Kerala, India

14.1 INTRODUCTION

Growing global energy demands the necessity for an efficient technology that can harvest untapped energy resources with minimal environmental impacts. Currently, fossil fuels are the main fuel source that meets our present energy needs. But the availability of fossil fuels in the coming years is of great concern. Not only the availability but also the use of fossil fuels will lead to many environmental hazards by emitting poisonous gases on burning, which can lead to global warming. So the urge for an alternative source of energy is crucial. Renewable energy resources such as solar, wind, geothermal, and hydroelectricity have the potentiality to replace conventional energy resources. Compared to all these renewable resources, solar energy is more advantageous in many ways. Solar energy is more versatile, can be installed in large as well as small areas, are easily available resource in most of the places, and can be easily installed even on a roof top are some of the features that highlight solar energy among other renewable energy resources. But only a tiny fraction of these renewable energies are used effectively for energy production. Researchers in the field of solar cells date back to 19th century though Bell Laboratories developed the first practical silicon solar cell with an efficiency of 6% in 1954 [1]. But the cost of these cells was very high. Thereafter, various researches progressed with the aim to reduce the price of a silicon solar cell. Tang discovered the first heterojunction organic solar cell in 1986 [2]. This was a major breakthrough in the field of organic solar cells.

Typically, based on the material and technology used, solar cells can be broadly classified into three generations. The first generation comprises silicon wafer–based solar cells. They include both single- and

Perovskite Photovoltaics
DOI: https://doi.org/10.1016/B978-0-12-812915-9.00014-9

multi-crystal solar cells. Single-crystal solar cells are more efficient than multi-crystal solar cells, but the latter is cheaper. Though the efficiency of silicon solar cells have reached 26.33% and holds about 86% of the solar cell market, the high production cost is a drawback. Thin-film solar cells also known as the second-generation solar cells use amorphous silicon, CdTe, and Cu(In, Ga)Se_2 as the active materials. When compared to silicon wafer solar cells, they are less expensive and have lower efficiency. The third generation covers organic solar cells in which more features were introduced such as light weight and flexible solar cells. They have low production cost and enable large-scale production. But the efficiency of organic solar cells is still low when compared to its inorganic counterpart. A new era in the third-generation solar cells started with the discovery of perovskite solar cells. The research community witnessed an exponential growth in the performance of these solar cells from 3.8% [3] to 22.1% in a short period [4].

Perovskite is an organic–inorganic hybrid material. It has the goodness of both organic and inorganic parts such as long charge carrier diffusion length, high charge carrier mobility, better absorption coefficients, and direct band gap, which makes it a promising candidate in the photovoltaic family [5,6]. Active researches are going on in many laboratories across the world for developing efficient perovskite solar cells. To date, the highest reported efficiency for a perovskite solar cell is 22.1%. Theoretically, for perovskite solar cells, the power conversion efficiency (PCE) can go up to 30% [7]. Commercialization of a perovskite solar cell is not too far from reality. At this point, it is crucial to understand the environmental impacts of perovskite solar cells. Life cycle assessment (LCA) is an extensively used tool for estimating the environmental implications of a product in its life cycle, which range from raw material extraction to its disposal or recycle phase. LCA at this early development stage will be very useful as most of the product's impact can be determined at this stage [8].

14.2 PEROVSKITE

14.2.1 Material Perspective

The term perovskite is coined after the Russian mineralogist Count Lev Aleksevich von Perovski [9]. Perovskite is an organo-metal halide with the chemical structure ABX_3 where A is a cation, B anion, and X is a halide (Cl^-, Br^-, I^-) (Fig. 14.1) [11]. The band gap of the inorganic layers is modified by changing either the halogen or metal sites. By

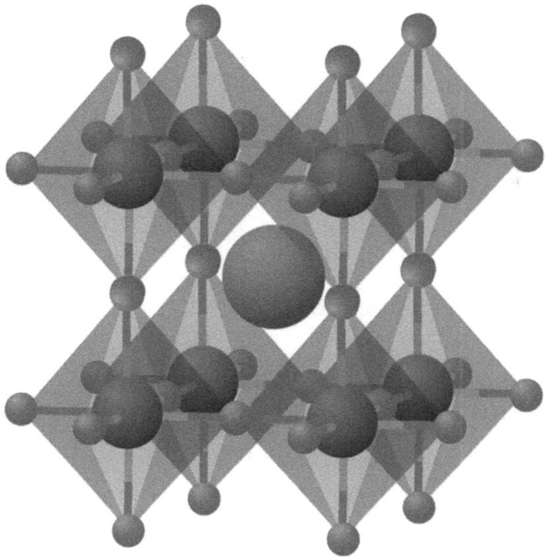

Figure 14.1 Crystal structure of cubic metal halide perovskites with the generic chemical formula ABX_3. Organic or inorganic cations occupy position A (green), whereas metal cations and halides occupy the B (gray) and X (purple) positions, respectively. *Reprinted from M. Gratzel. The light and shade of perovskite solar cells. Nat. Mater. 13(9) (2014) 838–842, with permission from Macmillan Publishers Ltd.*

varying the length of the organic cations, the width of the barrier is modified. The layers of the well are altered by adjusting the number of perovskite sheets in between each organic layer [12]. Typically $CH_3NH_3^+$ (methylammonium, MA) and $HC(NH_2)_2^+$ (formamidinium, FA) are used as organic cation A in organic solar cells. Divalent metal ions like Pb^{2+}, Sn^{2+}, Eu^{2+}, Cu^{2+}, and Ge^{2+} are used as metal cations B and Cl^-, Br^-, I^- as halide (X) [13,14]. Among these, MA lead halide perovskite ($CH_3NH_3PbX_3$/$MAPbX_3$) is more popular as solar cell active material. MAPbX3 was initially used in dye-sensitized solar cells (DSSC). But the stability of these devices was very low. Later in 2012, Kim et al. reports a solid-state perovskite device with efficiency above 9% [15]. The lead cations that are in a 6-fold coordination are surrounded by halide anions and ammonium cations in a 12-fold cuboctahedral coordination [11]. Perovskite materials possess tunable direct band gap, a wide range of absorption spectra with high absorption coefficient and high carrier mobility. Mixed halide perovskites like $MAPbI_{3-x}Cl_x$ and $MAPbI_{3-x}Br_x$ are also widely used for fabricating solar cells. Compared to $MAPbI_3$, $MAPbI_{3-x}Cl_x$ has long charge diffusion length ($L_D > 1\ \mu m$) and better stability in air [16]. The role of Cl in $MAPbI_{3-x}Cl_x$ is still under study. There are some reports regarding this. Li et al. concluded that the final perovskite film will have Cl, and it cannot change the fermi level but can bring some variations in the perovskite structure [17]. But even under inert atmosphere, $MAPbI_3$ is thermally unstable at temperatures above 85°C, whereas its FA counterpart is thermally more stable and has broad absorption spectra with reduced band gap (1.48 eV). This is due to the improved ionic radius of FA [18,19]. Perovskites have unique optical, conductive, and excitonic properties [20].

14.2.2 Perovskite Layer Formation

The final performance of the device greatly depends on the morphology and crystallinity of the perovskite film. The carrier diffusion length can be improved by controlling the film morphology. Grain size is one of the important parameters that needs to be optimized for better device performance. Deposition methods and processing conditions have great control over deciding the grain size. Grain boundaries act as recombination centers and can lead to degradation of the device. There are two ways to overcome this problem. One way is passivating the grains, thereby reducing the trap density [21]. Another way is to increase the grain size, which will reduce the grain boundaries and decrease the recombination centers. As the recombination rate reduces, the fill factor (FF), open circuit voltage (V_{oc}), PCE, series resistance (R_s), and the external quantum efficiency (EQE) increase. So a single-crystal perovskite can exhibit excellent device performance with reduced trap density and enhanced diffusion length [22]. Film composition also plays a vital role in deciding the device performance. Usually, $MAPbI_3$ precursor solution is prepared by mixing MAI and PbI_2 in the ratio 1:1, whereas $MAPbI_{3-x}Cl_x$ solution is formed by taking MAI and $PbCl_2$ in the ratio 3:1. These compositions with controlled annealing helps in fabricating pinhole-free perovskite films. Pinhole-free films with better coverage and uniformity result in devices with high PCE. Perovskite films show structure or composition dependent properties that can be varied by various processing approaches [23,24]. A number of deposition techniques are used to coat the perovskite thin film. Solution process, vapor-assisted solution process (VASP), and dual-source vacuum thermal evaporation (DSVD) are some of the commonly used methods [14]. Solution process is a low-cost technique that is commonly used for fabricating high-efficiency devices. Uniform perovskite layers are obtained using this technique. There are different types of solution process techniques, among which one-step and two-step (sequential) are the commonly used methods. In the one-step solution process, the precursor solution is prepared by mixing MAI and the metal halide (e.g., PbI_2) in a single solvent. The resultant precursor is spin cast to a film of desired thickness. Using this technique, an efficiency of ~20% has been reported [25,26]. Instead of a single solvent, mixed solvents can be used for preparing the precursor solution, which will improve the film quality. Usually, solvents like N,N-dimethylformamide (DMF) and γ-butyrolactone (GBL) are mixed in the ratio of 97:3 (vol%) to make the perovskite precursor solution. This produces perovskite films with uniform crystal domains [27]. In the two-step sequential method, the MAI and PbI_2 precursor solutions are prepared separately using different solvents. Then one of the precursor solutions, usually PbI_2 solution, is spin coated on the substrate, and the substrate is dipped into the other precursor solution for an optimized period followed by thermal annealing [28]. In VASP, the PbI_2 layer is spin coated, and the substrate is allowed to react with MAI vapors. This method provides sufficient coverage with a reported PCE of 12.1% [29]. DSVD is a costlier process where the organic and inorganic materials are taken as two sources and evaporated together. The DSVD method produces smooth uniform perovskite films. Some other methods are also used to coat the perovskite layer such as spray coating, doctor blade method, and slot-die coating [30–34]. These methods will be useful for large area coating.

14.2.3 Perovskite Device Fabrication

The elementary structure of an organic solar cell comprises a thin layer of active material sandwiched between two electrodes where one electrode is essentially transparent. For better charge extraction, there will be an electron transporting layer (ETL) and a hole transporting layer (HTL) on either side of the active layer depending upon the structure [35]. Usually, indium tin oxide (ITO) or fluorine tin oxide (FTO) on a glass substrate is used as the transparent electrode. The back electrode can be gold (Au), silver (Ag), or aluminum (Al). Titanium dioxide (TiO_2) and zinc oxide (ZnO) are widely used ETLs. Spiro-OMeTAD or PEDOT:PSS are commonly used HTLs.

Perovskites are ambipolar in nature. So they can behave as both hole and electron conductors. There are reports on ETL-free and HTL-free perovskite devices, but HTL-free devices are more acceptable as electrons have shorter diffusion length compared to holes. But it is always good to use both HTL and ETL layers for better charge extraction and reduced charge recombination rate. To enhance the efficiency of the device, many device architectures are investigated. Among these, FTO/TiO_2/Perovskite/spiro-OMeTAD/Au and ITO/PEDOT:PSS/Perovskite/PCBM/Al are the two most studied direct and inverted structures, respectively. In the initial years of perovskite solar cell development, most of the device architecture included both mesoporous TiO_2 (m-TiO_2) and a compact TiO_2 (c-TiO_2). m-TiO_2 is widely used in DSSCs where the perovskite solar cells have its origin. So the direct structure resembles the one used in DSSC. m-TiO_2 is coated on top of the c-TiO_2 layer. In direct structure, c-TiO_2 is formed by either spin coating or spray pyrolysis on the FTO substrate followed by high-temperature sintering in air. The high-temperature TiO_2 process can be replaced by a low-temperature TiO_2 coating or ZnO layer. Such alternatives can save energy during large-scale production without compromising the device performance. The m-TiO_2 is coated using different techniques such as spin coating, sputtering, and electrospinning. Then, the perovskite layer is coated on top of this mesoporous layer. Perovskite infiltrates the mesoporous layer and forms a capping over the m-TiO_2 layer. The better solubility and high hole mobility make Spiro-OMeTAD a widely used HTL in direct perovskite devices. But studies show that the dopants in Spiro-OMeTAD reduce the stability of perovskite devices [36]. PEDOT:PSS is the most commonly used HTL in the inverted perovskite architecture. The efficiency of the inverted planar perovskite device has improved up to 13% [37]. In spite of many disadvantages like acidity, PEDOT:PSS is widely used in perovskite solar cell fabrication due to its ease of availability, affordability, and low-temperature processability that make it a better candidate for flexible perovskite solar cells [38]. In the inverted structure, PEDOT:PSS is used as the HTL. But the moisture in PEDOT:PSS is harmful for the perovskite layer. In most cases, PCBM is used as the ETL.

14.2.4 Challenges

At the initial stage, researchers were working on how to improve the PCE of perovskite solar cells. They have succeeded that mission by enhancing PCE from 3% to 21% in this short duration of time. Many related issues need to be addressed for the perovskite solar cells to reach the market. Nonreproducibility of the results, instability, hysteresis, poor adhesion between layers, and the use of toxic materials are some of the issues related to perovskite solar cells. Hysteresis is a phenomenon where the efficiency of the device depends on the scan rate and the scan direction. The reason for hysteresis is still under study. Capacitive effects, defect trapping, etc., are some of the explanations for hysteresis. The degree of hysteresis can be quantified using a $J-V$ hysteresis index. Degradation of perovskite solar cells upon exposure to light, moisture, or oxygen is one of the serious problems [39,40]. Both the perovskite layer as well as the interfacial layer such as TiO_2 can contribute to the rapid degradation of the cell. TiO_2-based perovskite solar cells degrade rapidly under UV illumination. Moisture-assisted degradation is also common among MA lead halide perovskites. The hygroscopic nature of amine salt also gives rise to moisture instability. Perovskite solar cells fabricated under ambient conditions were found to be more stable upon exposing to high humidity. T_{80} time characterizes the stability of perovskite devices. It is the time taken by an unencapsulated device for its efficiency to decay to 80% of its original value [41]. For the device architecture ITO/PEDOT:PSS/$MAPbI_3$/$PC_{61}BM$/TiO_2/Al, Yu et al. achieved a T_{80} of 100 hours under ambient conditions [42]. Many research groups are working currently on this issue.

Lead is a widely used component in perovskites that are used for fabricating solar cells. Lead is highly toxic and has got worst environmental impacts. So researchers have come up with lead-free perovskite solar cells where they use tin (Sn) to replace lead. The first Sn-based solar cell reported an efficiency of 5.73% where $CH_3NH_3SnI_{3-x}Br_x$ was used as the active material [43]. Li et al. reported an efficiency of 13.6% for 50% Sn-based perovskite solar cells [44]. But Sn-based solar cells are highly sensitive to oxygen and reduce the performance of the device [43]. Studies are in progress to develop a highly efficient lead-free perovskite solar cell.

For mass production, flexible substrates are preferred over glass substrates. Flexible substrates include polymer substrates such as PET, PEN or PEI, and metal foils [45–47]. Different layers are coated on these flexible substrates by spin coating, slot-die coating, printing, etc. But polymer substrates may experience deformations upon heating above certain temperatures. High-temperature-processed perovskite solar cells are incompatible with flexible substrates. So low-temperature processed perovskite devices using PEDOT:PSS or ZnO will be preferred for large area flexible solar cell production. The flexible solar cells should also withstand repeated bending. Researches are progressing in this field also.

14.3 LIFE CYCLE ASSESSMENT

LCA is a widely used tool for calculating the impact of a product on the environment during its lifetime. LCA studies give a statistical picture of the contribution of each component towards various impact categories. This is very useful in designing a photovoltaic device with least environmental hazards. Different impacts are assessed at various stages of the product's lifetime. This ranges from raw material extraction, to processing, manufacturing, and distribution of materials and also the material maintenance and disposal. LCA studies consider all the environmental effects that can be assigned to the product. It quantifies all the materials used and its effect on the environment. It is a multistep procedure that includes four phases as mentioned in the ISO 14040 and 14044 standards (Fig. 14.2) [48,49]:

a. Goal and scope definition,
b. Life cycle inventory (LCI) analysis,
c. Life cycle impact assessment (LCIA), and
d. Life cycle interpretation.

One of the critical applications of LCA is the enhancement of goods and services. In that case, LCA studies have to be done at an early stage of the design process. Perovskite is one of the rapidly growing sectors in the photovoltaic family. Researchers all over the world are working to expand the laboratory scale perovskite fabrication to large-scale

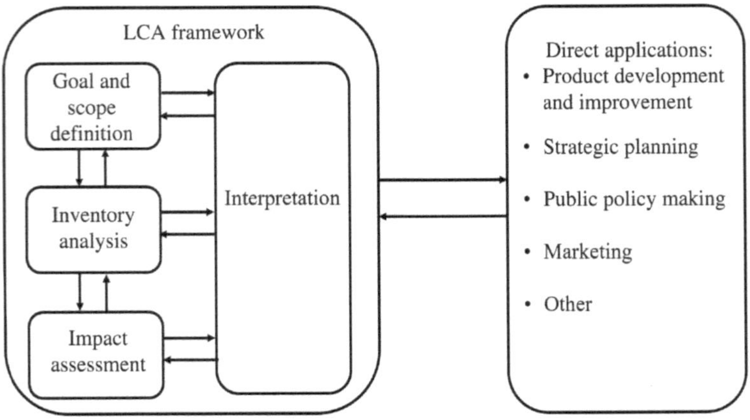

Figure 14.2 Phases and applications of an LCA (based on ISO 14040, 1997). *Reprinted from G. Rebitzer, T. Ekvall, R. Frischknecht, et al. Life cycle assessment: part 1: framework, goal and scope definition, inventory analysis, and applications. Environ. Int. 30 (5) (2004) 701–720, with permission from Elsevier.*

productions. There are many challenges for that. In order to fabricate an efficient, low-cost perovskite module, many parameters have to be considered. LCA studies at this early stage of perovskite solar cells can provide proper guidelines to researchers working in the field of device fabrication. Various research groups have performed LCA studies of perovskite solar cell for different architectures and fabrication techniques [50–54].

Espinosa et al. had performed LCA studies on solution-processed and vapor-deposited lead perovskite solar cells. Whereas Gong et al. has done studies on TiO_2-based and ZnO-based perovskite solar cells. LCA study on tin-based perovskite solar cells were done by Serrano-Lujan et al. Celik et al. analyzed SnO_2-based perovskite solar cells. All these studies give a general idea about our present perovskite research from an industrial point of view.

14.3.1 Goal and Scope Definition

LCA studies always start with a goal and scope statement. The goal and scope of the study explain the product regarding functional unit and system boundaries. Functional unit quantifies the service provided by the product. But it is not just a quantity of material; furthermore, it allows to compare and analyze alternative goods and services [55]. The system boundaries define what all processes need to be included in the framework of an LCA study [56].

The goal of all these studies is the assessment of life cycle environmental impacts from perovskite solar cell manufactured via different routes. In most perovskite LCA studies, the functional unit is defined as the supply of 1 kWh of energy generated for the entire lifetime of the device. Gong et al. described the functional unit as 1 m^2 of the perovskite solar module [52].

System boundaries provide a restriction to the processes to be incorporated in the analysis of the product. The system boundaries are usually taken as cradle-to-grave or cradle-to-gate. Cradle-to-grave covers the whole life cycle of the product ranging from resource extraction ("cradle") to the use and disposal of the product ("grave"). Cradle-to-gate considers only the partial life cycle where the study ends at the factory gate, omitting the use and disposal phase. Since perovskite solar cells are in the beginning state of development, LCA studies are mainly focused on cradle-to-gate excluding the end of life. However, Gong et al. studied a cradle-to-grave LCA assuming a lifetime of 2 years.

14.3.2 Life Cycle Inventory Analysis

The LCI denotes the appropriate inputs and outputs of a product system. The product system involves flows from and to the environment. LCI provides an inventory of these flows. The input side includes material and energy. The output side comprises products, by-products, and all the releases to air, land, and water. Using these input and output data, a flow model of the system is created. The flow model illustrates the various procedures that are to be assessed. A flow chart is used to illustrate the flow model. The flow chart contains all the activities that are to be and also provides a clear description of the system boundaries. Only activities inside the system boundaries are considered for collecting input and output data. All the data must be linked to the functional unit as described in the goal and scope definition. Perovskite solar cell fabrication includes several material synthesis steps. LCI data for most of the compounds are discussed in the literature. The data are created by following the same approach as in the EcoInvent database [57].

14.3.3 Life Cycle Impact Assessment

LCIA studies the importance of environmental impacts on the LCI flow results. LCIA provides category indicators and selection of impact categories. LCIA evaluates the product life cycle by the functional unit. There are many impact assessment methods like TRACI, commonly used in the United States; Ecoindicator, ReCiPe, and ILCD, employed in Europe; and CML method [58–61]. CML is a procedure used to estimate the measure of environmental impact that is caused by the product. This method uses various impact categories such as eutrophication, ionization radiation, aquatic ecotoxicity, land use, and human toxicity [58]. In midpoint method, the characterization factors are directly applied to the LCI results. In Ecoindicator 99, three categories are used to evaluate the environmental impact of a product [62]. These three categories are ecosystem quality, depletion of nonrenewable resources, and human health. The total environmental impact score is obtained by quantifying the impact of three categories and then normalized and weighted. There are three different types of perspectives based on the weighting step. They are Individualist, Egalitarian, and Hierarchist. The Individualist neglects the hazard of a near fossil fuels depletion. They only consider the mineral extraction. The outlook of an average scientist can be represented using the Hierarchist perspective, whereas in the Egalitarian perspective, a long time duration into the future is considered and given more importance to the ecosystem quality.

Several indicators are discussed in perovskite LCA studies. Energy payback time (EPBT) and CO_2 emission factor are two important sustainable indicators. EPBT gives the time taken by a photovoltaic device to generate equal energy that was used up during its production. Inorganic solar cells like crystalline silicon module have an EPBT of about 3–4 years. EPBT was reduced to 1.1 years using amorphous silicon thin-film solar cells. Another predominant impact indicator is the carbon footprint. It is a measure of greenhouse gas emission during the product's life cycle [63]. The CO_2 emission factor is obtained by dividing the carbon footprint by the electricity generated during the product's life cycle. For calculating the electricity generated, we need the lifetime of the solar cell. But the exact lifetime of perovskite solar cells is not known as it is in the early stage of development. So there can be fluctuations in the carbon footprint results and primary energy consumption.

14.3.4 Life Cycle Interpretation

Life cycle interpretation is the final step in the LCA study. It is a systematic method where the outcomes of the LCI and the LCIA are classified, quantified, checked, and evaluated. Interpretations will result in appropriate conclusions and recommendations. Based on the LCI and the LCIA results, interpretation should identify the important issue. Interpretation should evaluate a study taking into account the completeness, consistency, and sensitivity of the study. Most importantly, life cycle interpretation should draw a conclusion, limitations of the study, and recommendations.

14.4 DIRECT PEROVSKITE SOLAR CELL WITH TIO_2 AS ELECTRON TRANSPORTING LAYER

As discussed in the Section 14.2.3, TiO_2-based perovskite solar cells are the most popular structure with high power conversion efficiencies. A c-TiO_2 layer acts as ETL and Spiro-MeOTAD is used as HTL (Fig. 14.3). Both these layers are usually spin coated. Different perovskite techniques are used for coating the perovskite layer. Here vapor deposition, 2-step process, and dipping will be discussed in detail.

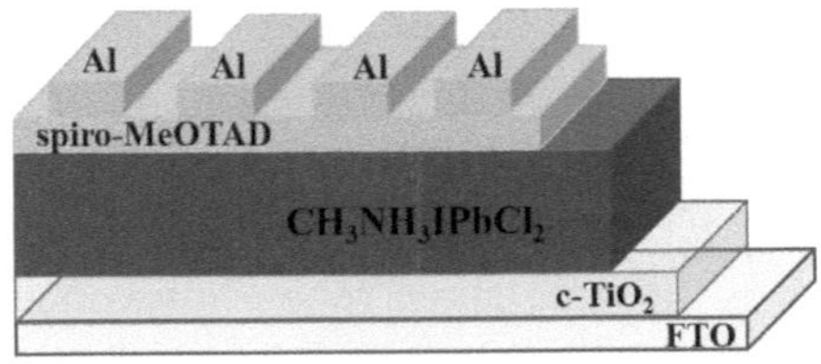

Figure 14.3 Schematic of direct perovskite solar cell with TiO_2 as ETL.

14.4.1 Vapor Deposition

The device architecture is FTO/c-TiO_2/$CH_3NH_3IPbCl_2$/spiro-MeOTAD/Ag. Here fluorine-doped tin oxide (FTO)—coated glass is used as the transparent electrode. c-TiO_2 layer is formed by spin coating and followed by sintering at 500°C. Then perovskite is co-evaporated on top of the c-TiO_2 layer. Co-evaporation is done by evaporating two sources, in this case lead chloride ($PbCl_2$) and methylammonium iodide (MAI), simultaneously under high vacuum. Spiro-MeOTAD, the HTL, is spin coated above the perovskite layer. Finally a layer of silver (Ag) is thermal evaporated, which acts as a back electrode. The reported efficiency for this structure (using vapor deposition technique) is 15.4% [64]. This was one of the highest efficiencies reported for a simple heterojunction device architecture without using any complex nanostructures. Vapor deposition method is advantageous over spin coating when large area devices are considered. Vapor deposition enables the coating of layered multi-stack films over a large area. Vapor deposition technique is widely used in liquid crystal display industries, glazing industries, and in solar cell fabrication. It helps in tuning the interfaces more efficiently. There are many concerns regarding to the commercialization of perovskite solar cells. Some of the significant challenges are the toxicity of lead, stability of the device, and cost-effectiveness. Espinosa et al. have reported a detailed LCA study on this architecture [51]. The goal of this cradle-to-gate LCA study was to assess the environmental impact of hybrid perovskite solar cells. The main agenda of this LCA study is to evaluate the overall environmental impact of lead from lead-based perovskite solar cells. They have also discussed how promising is the perovskite solar cells, when upscale, in terms of the environmental impacts. They have chosen the functional unit as the supply of 1 kWh of energy generated from the cell with a lifetime of 1 year. The light harvesting area is taken as 47.75 cm^2. The lifetime of perovskite device is under study. Espinosa et al. has considered the lifetime as 1 year. Since it is a cradle-to-gate LCA study, end of life is not taken into account. This would create large ambiguities.

Manufacturing of perovskite solar cell involves both material preparation and electricity for coating various layers. Material preparation has many chemical synthesis steps. Complex processes are involved in the synthesis of perovskite salt and Spiro-MeOTAD. Preparation of perovskite salt involves various chemicals such as iodine, lead, and nitric acid. Spiro-MeOTAD preparation involves very complex procedures. Fig. 14.4 gives the processing routes of these two compounds. The life cycle inventories of perovskite solar cells that include material extraction and energy consumption for each layer is shown in Table 14.1. The electricity consumption for coating the front electrode, TiO_2, perovskite layer, Spiro-MeOTAD, and back electrode are calculated as 4.82E − 02, 4.86E − 03, 4.34E − 02, 2.76E − 02, and 2.16E − 02 kWh, respectively.

The LCI times the input per functional unit will give the total emission of pollutant and consumption of resources. LCIA methods are used to convert them into potential impact indicators. The environmental impacts of vapor-deposited perovskite solar cell are investigated by Espinosa et al. that were mainly focused on impact categories such as climate change, human toxicity (cancer and noncancer effect), respiratory inorganics (RI), ionizing radiation (IR), freshwater eutrophication, marine eutrophication, freshwater ecotoxicity, land use, and nonrenewable resource depletion [51]. Among these, freshwater ecotoxicity has a significant impact. While considering the layer-wise impact on freshwater ecotoxicity, FTO contributes 31%, perovskite layer 28%, spiro-MeOTAD 20%, and TiO_2 17%. Human toxicity cancer effects and noncancer effects are in the second and third place. But the contribution from lead was found to be less than 5%. The contribution of silver, the back electrode, towards the nonrenewable resource depletion category is comparatively high. The least impact is for categories like IR, marine eutrophication, land use, and RIs. Table 14.2 gives the electricity contribution to all impact categories. Most of the impact categories fall within the range of 80%–95% except the mineral, fossil, and nonrenewable resource depletion category. Compared to Garcia-Valverde's work on organic photovoltaics, Espinosa et al. have estimated the carbon footprint for this vapor-deposited perovskite solar cell to be 366 g CO_2/kWh. Garcia-Valverde reported the carbon footprint of organic photovoltaics as 54.92 g CO_2/kWh [65]. This discrepancy in the carbon footprint value can be

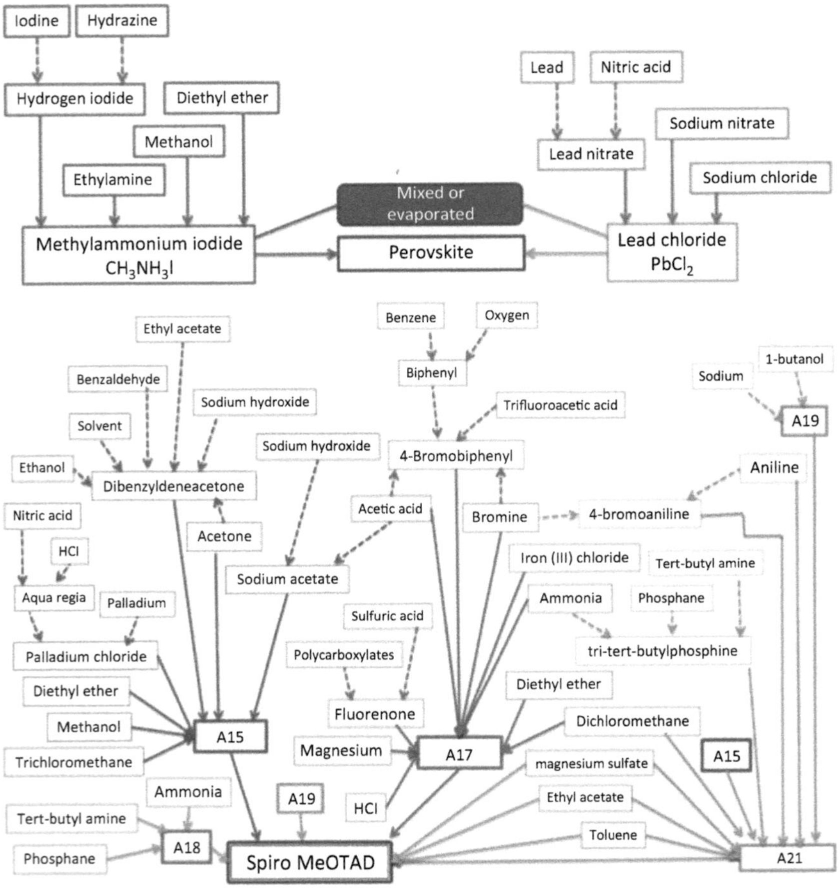

Figure 14.4 Process routes to the two components comprising the perovskite salt for solar cell devices (top) and the much more complex process route to Spiro-MeOTAD (A21:3,4′-dimethoxydiphenylamine; A15:Tris(dibenzylideneacetone)dipalladium(0); A17:2,2′,7,7′-Tetrabromo-9-9′spirobi[9H-fluorene]; A19:Sodiumtert-butoxide; A18:Tri-tert-butylphosphine). Source: *Reprinted from N. Espinosa, L. Serrano-Luján, A. Urbina, et al. Solution and vapour deposited lead perovskite solar cells: ecotoxicity from a life cycle assessment perspective. Sol. Energy Mater. Sol. Cells 137 (2015) 303–310, with permission from Elsevier.*

due to the significant energy consumption during the perovskite solar cell fabrication. The material loss in this vapor deposition method is about 70%. The LCA study shows that perovskite layer contributes more towards the environmental impacts. The electricity consumption is a dominant factor. Most importantly the role of lead (II) chloride was insignificant in creating environmental impact.

Table 14.1 Aggregated inventory for manufacturing perovskite solar cells by R1 VD

	Input materials	Amount	Unit	Comments
	PSC1.1-Front electrode	1.00	cm^2	[64]
I	Fluorine, liquid (RER)	3.21E-08	kg	Amounts estimated from [77]. Considered density 6.993/g/cm^3 [8]
I	Anti-reflex-coating, etching, solar glass/DK	1.00E + 00	cm^2	
I	Oxygen, liquid (RER)	4.29E-05	kg	Oxygen plasma treatment
I	Tin (GLO)	1.04E-06	kg	Estimated amount of material from (version 3.1., 2014)
I	Flat glass, uncoated (GLO)	1.54E-04	kg	
P	Electricity, low voltage (DK)	4.82E-02	kWh	Oxygen plasma treatment plus FTO-associated electrical process
	PSC1.2-n-type layer (TiO_2)	1.00	cm^2	TiO_2 evaporation onto glass
I*	TiO_2	3.76E-05	kg	90% lost. Thickness 330 nm. 10 nm thickness considered from [78] (average)
P	Electricity, low voltage (DK)	4.86E-03	kWh	Dried, spin coated, and sintered
	PSC1.3-Perovskite layer	1.00	cm^2	
I*	Perovskite (by vapor deposition	4.80E-07	kg	70% of material is lost during vapor deposition process
I	Nitrogen, liquid (RER)	4.91E-03	kg	Nitrogen consumed by glove box
I	Water, deionized, from tap water, at user (GLO)	3.51E-04	kg	Use of water estimated from [79]
P	Electricity, low voltage (DK)	4.34E-02	kWh	Evaporated, heated, water-cooled, annealed, and nitrogen chamber process

(*Continued*)

Table 14.1 (Continued)

	Input materials	Amount	Unit	Comments
	PSC1.4-Hole transport material (Spira-MeOTAD)	1.00	cm^2	Spiro-MeOTAD by spin coating [64]
I	Monochlorobenzene (GLO)	2.81E-06	kg	
I	Water, deionized, from tap water, at user (GLO)	4.50E-07	kg	
I*	Spiro-MeOTAD	1.88E-06	kg	
I*	Tert-butylpyridine	1.86E-07	kg	
I*	Lithium bis(trifluoromethanesulphonyl) imide	1.87E-07	kg	
I	Nitrogen, liquid (RER)	4.91E-03	kg	
P	Electricity, low voltage (DK)	2.76E-02	kWh	Spin coated, annealed, nitrogen glove box
	PSC1.5-Bak electrode	1.00	cm^2	Silver deposited by evaporation
I	Silver (GLO) market for	1.50E-07	kg	Amount estimated from [80]
P	Electricity, low voltage (DK)	2.16E-02	kWh	Evaporation and nitrogen glove box

VD, vapor deposition.
The first column describes the types of input/output, I: material input; P: process required; W: waste treatment; E: emission. When the entry is marked with an asterisk (*), the process has been created/edited.
Source: Reprinted from N. Espinosa, L. Serrano-Luján, A. Urbina, et al. Solution and vapour deposited lead perovskite solar cells: ecotoxicity from a life cycle assessment perspective. Sol. Energy Mater. Sol. Cells 137 (2015) 303–310, with permission from Elsevier.

Table 14.2 Contribution of all electricity processes to every impact categories considered in the study from ILCD methodology

Environmental impact category	Electricity impact contribution per FU (1 kWh) of PSC	
	R1 VD ($47.75\ cm^2$) (%)	R2 SC ($63.94\ cm^2$) (%)
Climate change	96	90
Human toxicity, cancer effects	94	90
Human toxicity, noncancer effects	85	90
Particulate matter	86	90
Ionizing radiation HH	83	85
Freshwater eutrophication	93	91
Marine eutrophication	93	90
Freshwater ecotoxicity	93	90
Land use	90	90
Mineral, fossil, and nonrenewable resource depletion	17	88

VD, vapor deposition.
Source: Reprinted from N. Espinosa, L. Serrano-Luján, A. Urbina, et al. Solution and vapour deposited lead perovskite solar cells: ecotoxicity from a life cycle assessment perspective. Sol. Energy Mater. Sol. Cells 137 (2015) 303–310, with permission from Elsevier.

14.4.2 Two-Step Process

Solution-processed perovskite are favorable for large-scale production mainly due to its low energy consumption [66]. Among the solution process techniques, the two-step method has better control over the morphology of the perovskite layer [67–69]. Gong et al. have performed a cradle-to-grave LCA study for the device structure FTO/BL-TiO_2/nc-TiO_2/$CH_3NH_3PbI_3$/spiro-MeOTAD/Au where the perovskite layer is formed by first spin coating PbI_2 layer followed by sintering and then dipping in CH_3NH_3I-isopropanol solution [52]. The blocking layer of TiO_2 (BL-TiO_2) is deposited via spray pyrolysis method and annealed at 450°C. A nanocrystalline mesoporous TiO_2 layer (nc-TiO2) was coated on top of the FTO/BL-TiO_2 layer by screen printing. PbI_2 dissolved in DMF was spin coated on top of the nc-TiO_2 layer. The first 20 seconds of the PbI_2 spin coating was done in an air flow of 1 bar. This will produce a smooth PbI_2 layer. This was followed by annealing at 70°C for 1 hour. Then the samples were dipped in CH_3NH_3I-isopropanol solution for 30 minutes. The HTL, spiro-MeOTAD, was spin coated above the perovskite layer. Finally, the Au back electrode was deposited under high vacuum. The reported efficiency is 9.10% [70].

Gong et al. defined the functional unit for the LCA study as 1 m^2 of the perovskite solar module [52]. This cradle-to-gate LCA study considers four stages such as raw material extraction, module production, utilization of the module, and the disposal stage. There are many disposal methods like using waste devices as landfills, incineration, or recycling. The energy inventory ratio for each layer with an active area ratio of 70% is calculated, and the corresponding energy consumption was estimated by taking the product of equipment power and operating time. Fig. 14.5 shows the pie charts of primary energy consumption. It can be seen that maximum contribution towards primary energy consumption comes from raw materials. Gold cathode contributes 56%, and FTO glass contributes to about 36%. Similarly, the cathode evaporation and TiO_2 sintering also consume a significant part of electricity. Sintering consumes 45% of the energy. Cathode evaporation contributes 43% of the total energy consumption, whereas spin coating consumes only 3%. The high-temperature annealing of TiO_2 layer demands more energy. So replacing thermal evaporation and sintering with other suitable methods can lower the primary energy consumption. The total energy consumption for developing a 1 m^2 perovskite solar cell with an active area of 70% is 7.78 kWh. The distribution of carbon footprint is also similar to the primary energy consumption. The maximum contribution is from the gold cathode (53%) followed by FTO (40%). The main reason for gold

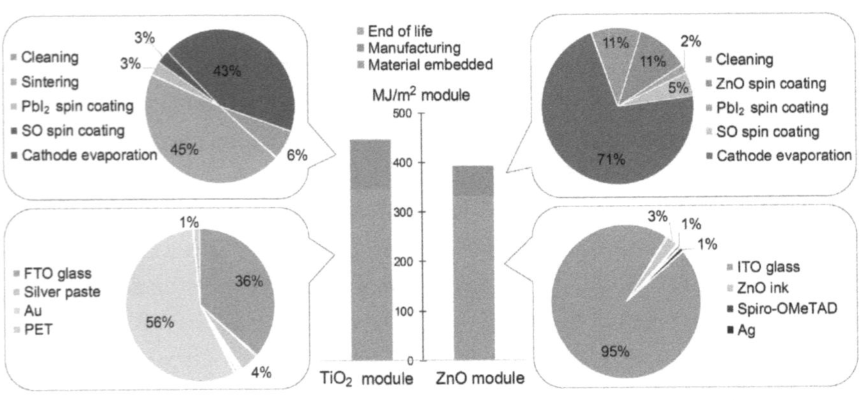

Figure 14.5 Distributions of the primary energy consumption for manufacturing two perovskite solar modules. Contributions less than 1% are not shown in the pie charts. Source: *Reproduced ("Adapted") from J. Gong, S.B. Darling, F. You. Perovskite photovoltaics: life-cycle assessment of energy and environmental impacts. Energy Environ. Sci. 8 (7) (2015) 1953–1968, with permission from The Royal Society of Chemistry.*

consuming more energy is that the production of gold from ores require huge amount of energy. Cathode evaporation and sintering contributes 43% and 45%, respectively.

Gong et al. have investigated 16 midpoint indicators such as acidification, eutrophication, human toxicity, freshwater sediment ecotoxicity, and marine aquatic ecotoxicity according to the CML method, the Intergovernmental Panel on Climate Change (IPCC) 2013 method, and the cumulative energy consumption [57,58,63]. Fig. 14.6 shows the contribution of each layer towards these indicators.

The dominant contribution of the gold cathode can be seen in most of the categories. About 93% of eutrophication, freshwater aquatic ecotoxicity, marine aquatic ecotoxicity, and freshwater sediment ecotoxicity are due to the contribution from the gold cathode. Gold cathode also has a high influence on land use (76%), human toxicity (65%), stratospheric ozone depletion (62%), depletion of abiotic resources (64%), terrestrial ecotoxicity (84%), malodorous air (43%), and acidification (40%). The contribution of gold dominates in all categories because in the production of gold, toxic mine drainage are released to rivers and lakes. This drainage contains many harmful chemicals such as mercury, sulfides, nitrates, and antimony, which are very dangerous to aquatic life and also cause eutrophication and acidification. Hence to reduce the environmental impacts, gold has to be replaced by some environmental-friendly metals. FTO glass contributes mainly to acidification potential (AP) (52%) and malodorous air (49%). During FTO glass production, gases such as sulfur oxide, carbon dioxide, and nitrogen oxide are emitted directly into the air. These gases react with water easily and form acid. Organic solvents are emitted directly to the atmosphere during drying. These emissions have major contributions in photochemical oxidation. Therefore, recycling organic solvents can decrease the photochemical oxidation effect. The EPBT and the CO_2 emission were also evaluated for the two-step perovskite solar cell. The EPBT of TiO_2 perovskite module was estimated to be $\sim$0.3 years. Sensitivity analysis shows that the performance ratio has a significant impact on EPBT. The performance ratio contributes to about 67% of the variance. Performance ratio also has a high contribution of 45.6% in the sensitivity analysis for CO_2 emission factor. In both the sensitivity analysis for EPBT and CO_2 emission factor, the contribution of insolation is only 2% and 1.4%, respectively. Lifetime also plays an important role in CO_2 emission factor (31.1%). On comparing the EPBT of perovskite with other existing photovoltaic technologies such as silicon-based

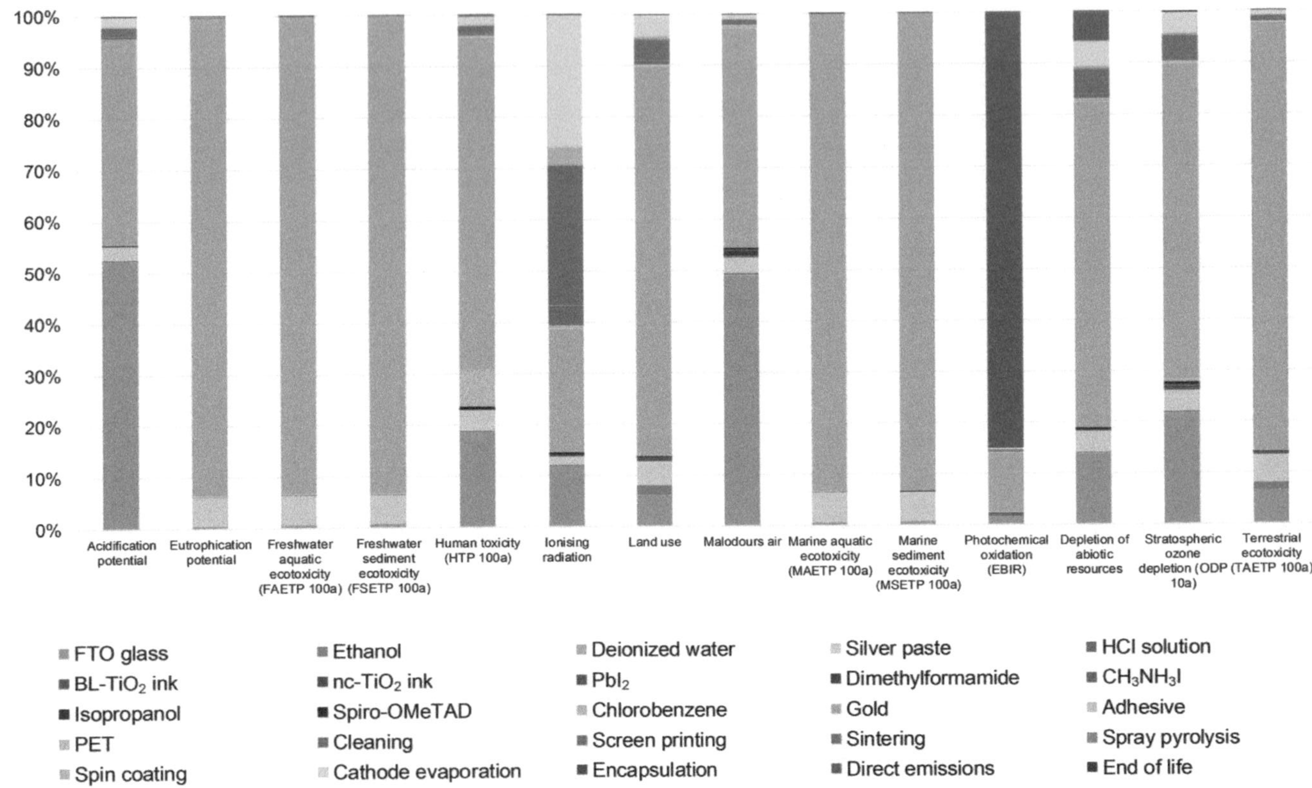

Figure 14.6 Environmental profile of 1 m^2 of the TiO_2 module. Source: *Reproduced ("Adapted") from J. Gong, S.B. Darling, F. You. Perovskite photovoltaics: life-cycle assessment of energy and environmental impacts. Energy Environ. Sci. 8(7) (2015) 1953–1968, with permission of The Royal Society of Chemistry.*

modules, CdTe modules, and organic photovoltaics, it is observed that perovskite modules have a lower EPBT value. But the CO_2 emission factor is high for perovskite solar cells. The simulation results of TiO_2 perovskite module give a mean CO_2 emission factor of 82.5 g CO_2-eq per kWh. They are less stable and sensitive to fluctuations in performance ratio, lifetime, carbon footprint, module efficiency, and insolation.

14.4.3 TiO_2 Nanotube-Based Liquid Perovskite Solar Cell

Zhang et al. has replaced the conventional m-TiO_2 by TiO_2 nanotubes (TNT) and studied the cradle-to-gate LCA of liquid perovskite solar cells. Nanotube arrays have enhanced light trapping ability thereby improving the absorption of light. This will directly reflect in the charge collection efficiency. TNTs were formed by electrochemically anodizing the titanium foil [54]. A titanium layer is first spin coated on top of the FTO glass followed by annealing at 500°C. The free-standing TNT film was transferred to the titanium layer and again annealed at 500°C. On this TNT film, a layer of PbI_2 is spin coated followed by annealing at 70°C for 30 minutes. Then the film is dipped in the CH_3NH_3I-isopropanol solution again annealed at 70°C for 30 min. The top edge of the FTO glass was dipped in platinum precursor paste. Finally, a liquid electrolyte was used that comprises of 0.9 M LiI, 0.45 M I_2, and 0.05 M urea in ethyl acetate and 0.5 M 4-tert-butyllpyridine (tBP). The optimized device exhibits a PCE of 6.5%. The major contribution towards the total mass is from FTO glass (89.61%), whereas perovskite dye, TNT anode, Pt precursor cathode, and electrolyte contributes 2.22%, 1.82%, 2.83%, and 3.22%, respectively. The active area is taken as 60% [71]. To produce a 1 cm^2 active area TNT perovskite solar cell, the total energy was calculated to be 0.27 kWh and a total of 8.58 kg raw materials are required from cradle-to-gate. About 90% of the material consumption takes place during the TNT film preparation (5.04 kg/cm^2) and perovskite dye formation (2.69 kg/cm^2). So it can be clearly seen that TNT film production requires more materials than the perovskite dye formation. The energy required to fabricate this solar cell is 0.01 kWh/cm^2. Major portion of energy consumption takes place during the perovskite dye preparation and TNT-based anode preparation. In the perovskite dye formation, significant energy consumption arises from the use of 2-propanol and diethyl ether. Apart from this energy consumption during evaporation, spin coating and cleaning will sum to about 0.025 kWh/cm^2. In TNT film formation, cleaning and etching are major energy-consuming processes. Environmental impacts of nanowastes and nanoparticle emissions are also considered in this LCA study. On analyzing the solid nanowastes from TNT electrode fabrication, the composition shows the presence of Ti and O derived from TiO_2 and C from carbon paste of the sample stage. LCIA was studied on 10 different categories like AP, eutrophication potential (freshwater EP), IR, ozone depletion potential (ODP), photochemical ozone creating potential, freshwater ecotoxicity potential (ETP), human toxicity potential (HTP), global warming potential (GWP), RI, and abiotic depletion potential (ADP). On analyzing various layers of the TNT perovskite solar cell, the major contribution is from the perovskite dye. It contributes mainly towards EP (91.14%), ODP (81.29%), ETP (79.42%), and ADP (69.54%). Solvents used for dissolving chemicals and cleaning have significant environmental impacts. Manufacturing energy also has a significant contribution towards environmental impacts. Major contributions are in categories like HTP (50.06%), ADP (43.87%), and GWP (30.24%). In manufacturing energy, preparation of TNT film and perovskite dye has a leading contribution towards environmental impacts. About 48.65% of total manufacturing energy is consumed by the perovskite dye preparation in which 97.7% is used for evaporation while synthesizing CH_3NH_3I. Diethyl ether has a high HTP of 29.35%. Zhang et al. has provided a detailed study on the role of lead in human toxicity. Perovskite dye preparation shows high HTP value. But upon analyzing it separately, it can be seen that 62.31% of HTP is due to CH_3NH_3I. Only about 3.68% of the HTP is contributed by lead iodide. A very little amount of lead is involved in the perovskite dye preparation. It is estimated that only 0.046 g of PbI_2 is required to fabricate 1 cm^2 area perovskite solar cell. Surprisingly, it was seen that the impact of CH_3NH_3I is of major concern than PbI_2. This is mainly because of the human toxic cancer effect of 2-propanol and diethyl ether which are used in the production of CH_3NH_3I. This LCA study suggests that perovskite solar cell production can be improved by minimizing the usage of solvents and by using energy-efficient manufacturing processes like recycling and reusing the solvents.

14.5 DIRECT PEROVSKITE SOLAR CELL WITH ZNO AS ELECTRON TRANSPORTING LAYER

ZnO thin film is used as the ETL. When compared to TiO_2, ZnO requires only low-temperature processing techniques, and thus reduces the energy consumption during ZnO thin-film formation. Gong et al. have performed a cradle-to-grave LCA study on the structure ITO/ZnO/$CH_3NH_3PbI_3$/spiro-MeOTAD/Ag, which has reported a lab efficiency of 15.7% [52,72]. For an active area ratio of 70%, the module efficiency is assumed to be 11%. Here $CH_3NH_3PbI_3$ layer was formed by spin coating ITO/ZnO substrate with PbI_2 and sintering for 1 hour followed by dipping in CH_3NH_3I−isopropanol solution. The HTL, spiro-MeOTAD, is spin coated on top of the perovskite layer and is followed by silver electrode deposition via thermal evaporation. The main advantage of this structure is that there is no need of any blocking layer coating and sintering. This helps to reduce the energy consumption considerably. As the functional unit of the LCA study, 1 m^2 of the perovskite module is taken. The majority of the primary energy consumption is from raw materials. In this device structure, about 95% of the primary energy consumption is contributed by ITO glass. It is to be noted that ITO substrates use 2.5 times the energy used by FTO substrate. The ZnO spin coating only consumes 11% of the total energy consumption, whereas the cathode evaporation consumes 71% (see Fig. 14.5). The carbon footprint of the perovskite solar module also exhibits similar distribution as that of the primary energy consumption. The environmental profile of 1 m^2 ZnO module is shown in Fig. 14.7. Here too, ITO glass dominates in most of the categories. ITO significantly contributes towards AP, IR, land use, malodorous air, depletion of abiotic resources, and terrestrial ecotoxicity. Silver cathode also contributes to about 52% in fresh water aquatic ecotoxicity, 40% in eutrophication, 51% in the marine aquatic ecotoxicity. But the contribution of silver cathode is less when compared to the gold cathode. Eight-five percent of the stratospheric ozone depletion is due to the ZnO ink. On replacing, solvents like chloroform from the ZnO ink, which converts ozone to oxygen, can reduce the stratospheric ozone depletion rate. According to simulation results, the EPBT of ZnO-based perovskite solar modules is about 0.193 years.

14.6 INVERTED PEROVSKITE SOLAR CELL WITH PEDOT:PSS AS HOLE TRANSPORTING LAYER

In the inverted device structure, perovskite is coated on top of the HTL. LCA studies were carried out in the inverted device architecture ITO/PEDOT:PSS/$CH_3NH_3IPbCl_2$/PCBM/Al where PEDOT:PSS and PCBM are the HTL and ETL, respectively. PEDOT:PSS was spin coated on top of the ITO substrate and annealed at 120°C for 15 minutes.

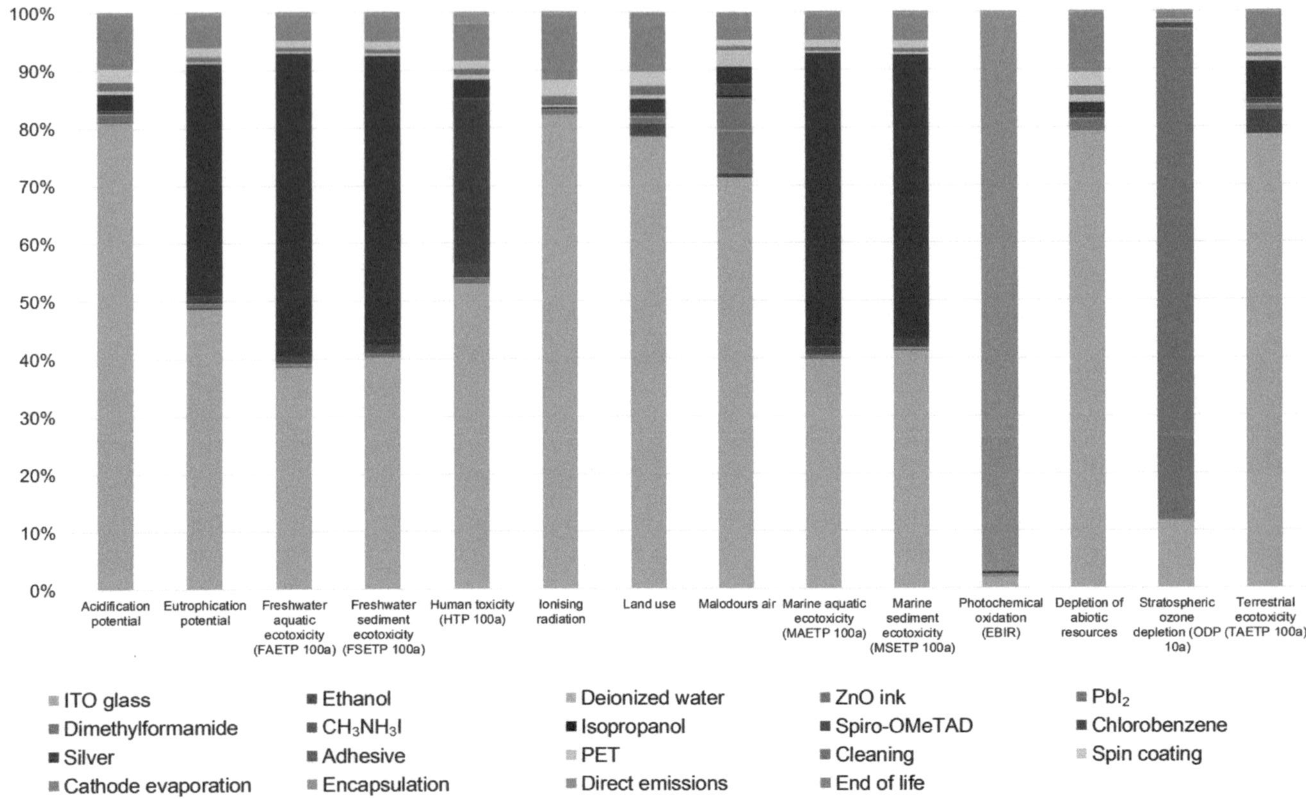

Figure 14.7 Environmental profile of 1 m^2 of the ZnO module. *Reproduced ("Adapted") from J. Gong, S.B. Darling, F. You. Perovskite photovoltaics: life-cycle assessment of energy and environmental impacts. Energy Environ. Sci. 8(7) (2015) 1953–1968, with permission of The Royal Society of Chemistry.*

$CH_3NH_3IPbCl_2$ precursor is prepared by mixing CH_3NH_3I and $PbCl_2$ in the ratio 3:1 and is spin coated on top of PEDOT:PSS layer and annealed at 90°C for 2 hours. This was followed by spin coating a 110-nm-thick PCBM layer. So this is a low temperature—processed perovskite solar cell that is compatible for flexible substrates also. You et al. has reported an efficiency of 11.5% for this structure [73]. Espinosa et al. have conducted a cradle-to-gate LCA study on this inverted structure where the functional unit was chosen to be 1 kWh of the generated energy [51]. The LCI of each layer is given in Table 14.3.

Among various impact categories studied, the principal impact is on freshwater ecotoxicity towards which the perovskite layer and the Al electrode contribute 53% and 20%, respectively. In the perovskite layer, soluble lead(II) chloride contributes less towards the impact categories compared to CH_3NH_3I. Compared to other techniques, the amount of perovskite required for solution-processed layer is less. But the electricity consumption is more in this case which accounts for about 95% and hence raises all environmental impacts. Fig. 14.8 compare the environmental impacts caused by two perovskite solar cells manufactured via different routes. Here, R1 is the TiO_2-based device where the perovskite layer is vapor deposited, and R2 is the PEDOT:PSS-based devices with solution-processed perovskite layer. It can be clearly seen that the environmental impact of solution-processed perovskite layer is higher than the vapor-deposited perovskite layer. This high impact of solution-processed perovskite layer can be accounted to the energy consumption during annealing even though the material requirement for solution-processed perovskite is less than the vapor-deposited perovskite layer. So for this structure, perovskite layer has a major contribution towards the environmental impact. The contribution from Al electrode is 26% and that of ITO is 17%.

14.7 TIN-BASED PEROVSKITE SOLAR CELLS

Lead-based perovskite solar cells have shown a PCE of above 20%, which makes it capable of competing in the solar cell market. Still, there are some problems that need to be addressed like the stability of the device and the risk posed by lead(II) chloride. LCA studies show that only 0.001% of the total environmental impact is contributed by lead in perovskite solar cell [51]. But taking into account the toxicity of lead and the serious health issues it may cause in the long run, lead-free perovskite solar

Table 14.3 Aggregated inventory for manufacturing perovskite solar cells by R2 SC (spin coated)

	Input materials	Amount	Unit	Comments and references
	PSC2.1-Front electrode	1.00	cm^2	ITO layer evaporated [73], edited from (version 3.1.2014)
I	Indium (GLO)	2.27E − 07	kg	
I	Tin (GLO)	2.27E − 08	kg	
I	Flat glass, uncoated (GLO)	1.54E − - 04	kg	
P	Electricity, low voltage (DK)	1.42E − 02	kWh	
	PSC2.2 n-type layer (PEDOT:PSS)	1.00	cm^2	PEDOT:PSS spin coated ([73]; Espinosa et al., 2013)
I*	PEDOT:PSS	7.57E − 07	kg	
P	Electricity, low voltage (DK)	9.90E − 03	kWh	
E	2-Propanol	2.27E − 03	kg	
E	Waste water/m^3	7.62E − 07	kg	
E	PEDOT:PSS	7.99E − 09	kg	
E	Propylene glycol	4.59E − 05	kg	
	PSC2.3 perovskite layer	1.00	cm^2	Perovskite by spin coating [73]
I*	Perovskite 75-25	2.55E − 07	kg	90% losses, density 3944 g/cm^3
P	Electricity, low voltage (DK)	6.21E − 02	kWh	Spin coating and annealing
	PSC2.4 Hole transport material (PCBM)	1.00	cm^2	PCBM by spin coating [73]
I*	PCBM	1.54E − 08	kg	
P	Electricity, low voltage (DK)	5.00E − 05	kWh	Spin coating
	PSC2.5 Back electrode	1.00	cm^2	Aluminum evaporated, from García-Valverde et al. [65]
I	Aluminum, primary, ingot (GLO)	9.18E − 08	kg	
P	Electricity, low voltage (DK)	2.16E − 02	kWh	Evaporated in nitrogen glove box and vacuum

The first column describes the types of input/output, I, Material input, P, Process required, W, Waste treatment, E, Emission. When the entry is marked with an asterisk (*), the process has been created/edited.

Reprinted from N. Espinosa, L. Serrano-Luján, A. Urbina, et al. Solution and vapour deposited lead perovskite solar cells: ecotoxicity from a life cycle assessment perspective. Sol. Energy Mater. Sol. Cells 137 (2015) 303–310, with permission from Elsevier.

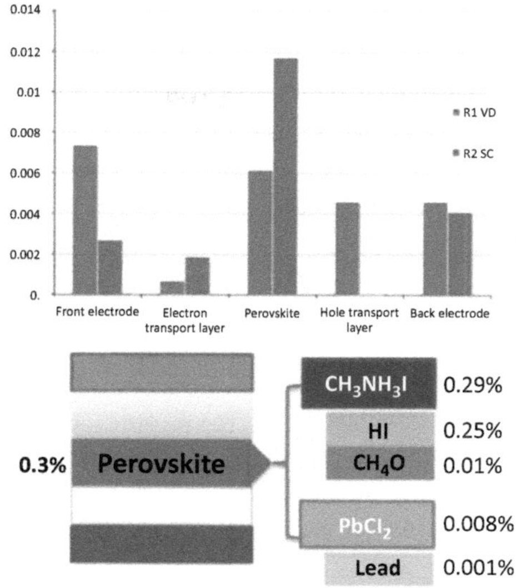

Figure 14.8 Comparison between total environmental impacts of two perovskite solar cell manufacturing routes [3,74] of 1 kWh presented for each by layer (front electrode, interface n-type, active layer, interface hole-transport material, and back electrode). Total scores have been calculated as the sum of its values when the ILCD methodology is applied (bottom). The breakdown for Climate Change impact of the perovskite material is shown in graphical form, absolute percentages in the total contribution to the FU. *Reprinted from N. Espinosa, L. Serrano-Luján, A. Urbina, et al. Solution and vapour deposited lead perovskite solar cells: ecotoxicity from a life cycle assessment perspective. Sol. Energy Mater. Sol. Cells 137 (2015) 303–310, with permission from Elsevier.*

cells are gaining more importance. Hao et al. had fabricated a tin-based perovskite solar cell with the structure FTO/c-TiO_2/$CH_3NH_3SnI_3$/spiro-OMeTAD/Au and got an efficiency of 5.73% [43]. A cradle-to-gate LCA study on this structure was done by Serrano-Lujan et al. where the functional unit was chosen to be 1 kWh of the generated energy [53]. It was seen that tin-based perovskite solar cells showed double the value of embodied energy per functional unit (124.1 MJ/kWh) and an equivalent amount of CO_2 (9.4 kg/kWh) compared to lead-based perovskite solar cells. Different environmental impacts were studied for each layer of tin-based perovskite solar cells. Contribution from the gold back electrode is maximum in all impact categories. One gram of emitted tin showed seven to eight times larger environmental impact than 1 g of emitted lead. The calculation was done using the Life Cycle Data System (ILCD) methodology. When compared to lead, tin has negligible human toxicity in both cancer and noncancer effects. But considering the high cost and limited production of tin, tin-based perovskite solar cells may not be a good option over the lead-based perovskite solar cells. The contribution of tin towards terrestrial ecotoxicity and global warming are very high (>10 times) than lead. A 1-cm^2 tin-based perovskite solar cell has a lower impact compared to a lead-based perovskite solar cell of the same area. But as the PCE of tin-based devices are smaller than lead-based device, the former requires twice the area of the latter. As tin has a negligible effect on human toxicity, the disposal of tin-based perovskite solar cells is much easier than lead-based perovskite solar cells.

Efficient disposal of lead-based perovskite solar cells is the next step to be considered. Primarily, two disposal scenarios are considered: (1) cells are landfilled; (2) cells are incinerated. The first scenario was modeled according to the study of Spalvins et al. where 70% of lead will be leached to the soil in the first year [75]. In the second case, about 98% of lead is recovered from the ashes of the incinerated cells, which was modeled according to the study of Nagib and Inoue [76]. On the assumption that the recycled materials maintain the inherent properties as that of the virgin materials, the recovered energy from the materials is considered as an "environmental" credit.

14.8 COST-EFFECTIVE PEROVSKITE SOLAR CELL

Celik et al. designed a more cost-effective and large-scale processable perovskite solar cell using SnO_2 and CuSCN as ETL and HTL, respectively [50]. The device architecture is $FTO/SnO_2/CH_3NH_3PbI_3/CuSCN/MoO_x/Al$. FTO was preferred over ITO as it is cheaper than ITO, and indium is designated as "critical metal" by the US Department of Energy's Solar Energy Technologies Program. There are four commonly used ETLs. They are TiO_2, ZnO, Al_2O_3, and SnO_2. Among these, TiO_2 is the most commonly used ETL, but its high-temperature processing is not suitable for manufacturing flexible solar cells on a large scale and it consumes more energy. Similar is the case with ZnO and Al_2O_3. These are not compatible for low-cost production. On the other hand, SnO_2 which has energy band match with perovskite is low temperature processable. Spiro-MeOTAD is the most commonly used organic HTL. But it is very expensive and unstable. Other organic HTLs are P3HT, PTAA, etc., and again the cost of these materials is very high. CuSCN is a copper-based inorganic HTL with better stability, high hole mobility, and ease of processability. Al with MoO_x interface forms a less expensive back contact compared to Au and Ag back electrodes. As discussed in Section 14.2.2, there are several methods to coat the perovskite layer such as spin coating, sequential deposition, spraying, printing, and co-evaporation technique. Among these, spray and inkjet coating would be a better option for large area coating. In this study, Celik et al. has used $CH_3NH_3PbI_3$ as the active material. HTL-free perovskite solar cells have gained popularity due to its low cost and stability. An LCA study of HTL-free perovskite is also studied using carbon paste back contact.

A cradle-to-grave LCA study was conducted for the same structure assuming an active area of 65%. Using the TRACI impact assessment model, impact categories like human toxicity, cancer and noncancer, acidification, GWP, ecotoxicity, primary energy demand, and eutrophication were modeled. Celik et al. have compared the environmental impacts of perovskite solar cell with the first- and second-generation solar cells. It is clear from Fig. 14.9 that the acidification impacts are high for solution-processed and HTL-free perovskite solar cells. But the overall impact of

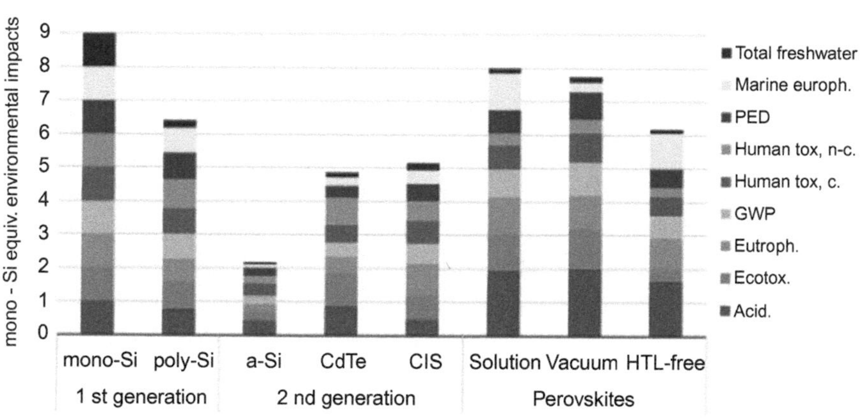

Figure 14.9 Comparison of perovskite devices with commercial PV technologies when normalized to mono-Si for selected impact categories. Note that GWP and PED stand for global warming potential and primary energy demand. Within each impact category, the impact (per m^2 of module manufactured) from each technology was divided by the impact from mono-Si. For mono-Si, the bar height is unity for each impact category. For nine impact categories, the total impact for mono-Si is given as 9 units. The data for mono-Si, poly-Si, a-Si, CdTe, and CIS are directly extracted from the EcoInvent database. *Reprinted from I. Celik, Z. Song, A.J. Cimaroli, et al. Life cycle assessment (LCA) of perovskite PV cells projected from lab to fab. Sol. Energy Mater. Sol. Cells 156 (2016) 157–169, with permission from Elsevier.*

perovskite solar cells is lower than the mono-Si solar cells. The environmental impacts of perovskite solar cells are high when compared to the existing commercial technologies. The electricity required for manufacturing perovskite solar cell is high when compared to other solar cells in market. The only way to reduce the environmental impacts is to improve the PCE and lifetime of perovskite solar cells.

14.9 CONCLUSIONS

Perovskite solar cells have gained a significant position in the photovoltaic family within this short duration of time. It has become one of the promising photovoltaic technologies for the future which can meet our energy needs at a low cost and more efficient way. LCA studies at this early stage of production will provide useful inputs for an efficient large-scale production with reduced environmental impacts. There are many discussions regarding the toxicity of lead used in perovskite devices, the stability of the device, etc. The LCA studies show that the environmental impacts caused by lead are negligible when compared to MA iodide. The electrical energy requirement for MA iodide is high. But other factors contribute significantly to various environmental impacts. One such factor is the use of gold as a back electrode. The primary energy consumption by gold cathode is very high. All the major impact categories are affected considerably. The front electrodes like FTO and ITO also contribute towards the total energy consumption. Studies show that the performance of ZnO-based perovskite solar cells is more environmental friendly than the TiO_2-based devices [52]. The electricity requirements for several processes are very high. In the case of TiO_2-based devices, sintering consumes the major portion of energy. In ZnO-based devices, ITO substrates consume more energy than FTO substrates. Perovskite solar cells were also found to have the shortest EPBT compared to current photovoltaic technologies. Hence while designing perovskite solar cells, researchers also have to focus on the environmental impacts and electricity consumption along with improving the PCE of the device. This approach will be very helpful in developing low-cost, efficient, highly stable perovskite solar cells that can create a revolution in the solar cell market. So life assessment studies will improve the probability of launching a fully developed product with fewer defects and replace the conventional solar cells.

REFERENCES

[1] D.M. Chapin, C.S. Fuller, G.L. Pearson, A new silicon p-n junction photocell for converting solar radiation into electrical power, J. Appl. Phys. 25 (5) (1954) 676–677.

[2] C.W. Tang, Two-layer organic photovoltaic cell, Appl. Phys. Lett. 48 (2) (1986) 183–185.

[3] A. Kojima, K. Teshima, Y. Shirai, et al., Organometal halide perovskites as visible-light sensitizers for photovoltaic cells, J. Am. Chem. Soc. 131 (17) (2009) 6050–6051.

[4] W.S. Yang, B.W. Park, E.H. Jung, et al., Iodide management in formamidinium-lead-halide–based perovskite layers for efficient solar cells, Science 356 (6345) (2017) 1376–1379.

[5] M. Xiao, F. Huang, W. Huang, et al., A fast deposition-crystallization procedure for highly efficient lead iodide perovskite thin-film solar cells, Angew. Chem. 126 (37) (2014) 10056–10061.

[6] B. Cai, Y. Xing, Z. Yang, et al., High performance hybrid solar cells sensitized by organolead halide perovskites, Energy Environ. Sci. 6 (5) (2013) 1480–1485.

[7] W.E.I. Sha, X. Ren, L. Chen, et al., The efficiency limit of CH3NH3PbI3 perovskite solar cells, Appl. Phys. Lett. 106 (22) (2015) 221104.

[8] B. Li, X. Gao, J. Li, et al., Life cycle environmental impact of high-capacity lithium ion battery with silicon nanowires anode for electric vehicles, Environ. Sci. Technol. 48 (5) (2014) 3047–3055.

[9] A. Navrotsky, D.J. Weidner, Perovskite: A Structure of Great Interest to Geophysics and Materials Science, American Geophysical Union, Washington, DC, 1989.

[10] M. Gratzel, The light and shade of perovskite solar cells, Nat. Mater. 13 (9) (2014) 838–842.

[11] P. Gao, M. Grätzel, M.K. Nazeeruddin, Organohalide lead perovskites for photovoltaic applications, Energy Environ. Sci. 7 (8) (2014) 2448–2463.

[12] D.B. Mitzi, Solution-processed inorganic semiconductors, J. Mater. Chem. 14 (15) (2004) 2355–2365.

[13] D.B. Mitzi, K. Chondroudis, C.R. Kagan, Organic-inorganic electronics, IBM J. Res. Dev. 45 (1) (2001) 29–45.

[14] P.P. Boix, S. Agarwala, T.M. Koh, et al., Perovskite solar cells: beyond methylammonium lead iodide, J. Phys. Chem. Lett. 6 (5) (2015) 898–907.

[15] H.-S. Kim, C.-R. Lee, J.-H. Im, et al., Lead iodide perovskite sensitized all-solid-state submicron thin film mesoscopic solar cell with efficiency exceeding 9%, Sci. Rep. 2 (2012) 591.

[16] S.D. Stranks, G.E. Eperon, G. Grancini, et al., Electron-hole diffusion lengths exceeding 1 micrometer in an organometal trihalide perovskite absorber, Science 342 (6156) (2013) 341–344.

[17] Y. Li, J.K. Cooper, R. Buonsanti, et al., Fabrication of planar heterojunction perovskite solar cells by controlled low-pressure vapor annealing, J. Phys. Chem. Lett. 6 (3) (2015) 493–499.

[18] B. Conings, J. Drijkoningen, N. Gauquelin, et al., Intrinsic thermal instability of methylammonium lead trihalide perovskite, Adv. Energy Mater. 5 (15) (2015) 1500477.

[19] G.E. Eperon, S.D. Stranks, C. Menelaou, et al., Formamidinium lead trihalide: a broadly tunable perovskite for efficient planar heterojunction solar cells, Energy Environ. Sci. 7 (3) (2014) 982–988.

[20] G.C. Papavassiliou, Synthetic three- and lower-dimensional semiconductors based on inorganic units, Mol. Cryst. Liq. Crys. 286 (1) (1996) 231–238.

[21] Y. Shao, Z. Xiao, C. Bi, et al., Origin and elimination of photocurrent hysteresis by fullerene passivation in CH3NH3PbI3 planar heterojunction solar cells, Nat. Commun. 5 (2014) 5784.
[22] B. Yang, O. Dyck, J. Poplawsky, et al., Perovskite solar cells with near 100% internal quantum efficiency based on large single crystalline grains and vertical bulk heterojunctions, J. Am. Chem. Soc. 137 (29) (2015) 9210−9213.
[23] C. Kagan, D. Mitzi, C. Dimitrakopoulos, Organic-inorganic hybrid materials as semiconducting channels in thin-film field-effect transistors, Science 286 (5441) (1999) 945−947.
[24] D.B. Mitzi, Synthesis, structure, and properties of organic-inorganic perovskites and related materials, Prog. Inorg. Chem. 48 (2007) 1−121.
[25] N.J. Jeon, J.H. Noh, W.S. Yang, et al., Compositional engineering of perovskite materials for high-performance solar cells, Nature 517 (7535) (2015) 476−480.
[26] H. Zhou, Q. Chen, G. Li, et al., Interface engineering of highly efficient perovskite solar cells, Science 345 (6196) (2014) 542−546.
[27] H.-B. Kim, H. Choi, J. Jeong, et al., Mixed solvents for the optimization of morphology in solution-processed, inverted-type perovskite/fullerene hybrid solar cells, Nanoscale 6 (12) (2014) 6679−6683.
[28] S. Aharon, B.E. Cohen, L. Etgar, Hybrid lead halide iodide and lead halide bromide in efficient hole conductor free perovskite solar cell, J. Phys. Chem. C 118 (30) (2014) 17160−17165.
[29] Q. Chen, H. Zhou, Z. Hong, et al., Planar heterojunction perovskite solar cells via vapor-assisted solution process, J. Am. Chem. Soc. 136 (2) (2013) 622−625.
[30] G. Cotella, J. Baker, D. Worsley, et al., One-step deposition by slot-die coating of mixed lead halide perovskite for photovoltaic applications, Sol. Energy Mater. Sol. Cells 159 (2017) 362−369.
[31] Y. Deng, E. Peng, Y. Shao, et al., Scalable fabrication of efficient organolead trihalide perovskite solar cells with doctor-bladed active layers, Energy Environ. Sci. 8 (5) (2015) 1544−1550.
[32] H. Back, J. Kim, G. Kim, et al., Interfacial modification of hole transport layers for efficient large-area perovskite solar cells achieved via blade-coating, Sol. Energy Mater. Sol. Cells 144 (2016) 309−315.
[33] A.T. Barrows, A.J. Pearson, C.K. Kwak, et al., Efficient planar heterojunction mixed-halide perovskite solar cells deposited via spray-deposition, Energy Environ. Sci. 7 (9) (2014) 2944−2950.
[34] T.M. Schmidt, T.T. Larsen-Olsen, J.E. Carlé, et al., Upscaling of perovskite solar cells: fully ambient roll processing of flexible perovskite solar cells with printed back electrodes, Adv. Energy Mater. 5 (15) (2015). 1500569-n/a.
[35] S. Lattante, Electron and hole transport layers: their use in inverted bulk heterojunction polymer solar cells, Electronics 3 (1) (2014) 132−164.
[36] A.D. Sheikh, A. Bera, M.A. Haque, et al., Atmospheric effects on the photovoltaic performance of hybrid perovskite solar cells, Sol. Energy Mater. Sol. Cells 137 (2015) 6−14.
[37] G. Adam, M. Kaltenbrunner, E.D. Głowacki, et al., Solution processed perovskite solar cells using highly conductive PEDOT:PSS interfacial layer, Sol. Energy Mater. Sol. Cells 157 (2016) 318−325.
[38] K. Sun, P. Li, Y. Xia, et al., Transparent conductive oxide-free perovskite solar cells with PEDOT: PSS as transparent electrode, ACS Appl. Mater. Interfaces 7 (28) (2015) 15314−15320.
[39] T. Leijtens, G.E. Eperon, N.K. Noel, et al., Stability of metal halide perovskite solar cells, Adv. Energy Mater. 5 (20) (2015) 1500963.
[40] Y. Rong, L. Liu, A. Mei, et al., Beyond efficiency: the challenge of stability in mesoscopic perovskite solar cells, Adv. Energy Mater. 5 (20) (2015).

[41] M. Jørgensen, K. Norrman, S.A. Gevorgyan, et al., Stability of polymer solar cells, Adv. Mater. 24 (5) (2012) 580–612.
[42] Y.-Y. Yu, R.-S. Chiang, H.-L. Hsu, et al., Perovskite photovoltaics featuring solution-processable TiO2 as an interfacial electron-transporting layer display to improve performance and stability, Nanoscale 6 (19) (2014) 11403–11410.
[43] F. Hao, C.C. Stoumpos, D.H. Cao, et al., Lead-free solid-state organic-inorganic halide perovskite solar cells, Nat. Photon 8 (6) (2014) 489–494.
[44] Y. Li, W. Sun, W. Yan, et al., 50% Sn-based planar perovskite solar cell with power conversion efficiency up to 13.6%, Adv. Energy Mater. 6 (24) (2016) 1601353.
[45] B.J. Kim, D.H. Kim, Y.-Y. Lee, et al., Highly efficient and bending durable perovskite solar cells: toward a wearable power source, Energy Environ. Sci. 8 (3) (2015) 916–921.
[46] J. Troughton, D. Bryant, K. Wojciechowski, et al., Highly efficient, flexible, indium-free perovskite solar cells employing metallic substrates, J. Mater. Chem. A 3 (17) (2015) 9141–9145.
[47] X. Xu, Q. Chen, Z. Hong, et al., Working mechanism for flexible perovskite solar cells with simplified architecture, Nano Lett. 15 (10) (2015) 6514–6520.
[48] ISO Technical Committee ISO/TC 207 EmSS, Life Cycle Assessment. *Environmental Management: Life Cycle Assessment: Requirements and Guidelines*: ISO, 2006.
[49] ISO Technical Committee ISO/TC 207 EmSS, Life Cycle Assessment. *Environmental Management: Life Cycle Assessment: Principles and Framework*: ISO, 2006.
[50] I. Celik, Z. Song, A.J. Cimaroli, et al., Life cycle assessment (LCA) of perovskite PV cells projected from lab to fab, Sol. Energy Mater. Sol. Cells 156 (2016) 157–169.
[51] N. Espinosa, L. Serrano-Luján, A. Urbina, et al., Solution and vapour deposited lead perovskite solar cells: ecotoxicity from a life cycle assessment perspective, Sol. Energy Mater. Sol. Cells 137 (2015) 303–310.
[52] J. Gong, S.B. Darling, F. You, Perovskite photovoltaics: life-cycle assessment of energy and environmental impacts, Energy Environ. Sci. 8 (7) (2015) 1953–1968.
[53] L. Serrano-Lujan, N. Espinosa, T.T. Larsen-Olsen, et al., Tin- and lead-based perovskite solar cells under scrutiny: an environmental perspective, Adv. Energy Mater. 5 (20) (2015) 1501119.
[54] J. Zhang, X. Gao, Y. Deng, et al., Life cycle assessment of titania perovskite solar cell technology for sustainable design and manufacturing, ChemSusChem 8 (22) (2015) 3882–3891.
[55] G. Rebitzer, T. Ekvall, R. Frischknecht, et al., Life cycle assessment: part 1: framework, goal and scope definition, inventory analysis, and applications, Environ. Int. 30 (5) (2004) 701–720.
[56] G. Finnveden, M.Z. Hauschild, T. Ekvall, et al., Recent developments in life cycle assessment, J. Environ. Manage. 91 (1) (2009) 1–21.
[57] Ecoinvent database v3.1, http://www.ecoinvent.ch/ (accessed 12.2014).
[58] J.B. Guinée, Handbook on life cycle assessment operational guide to the ISO standards, Int. J. Life Cycle Assess. 7 (5) (2002) 311–313.
[59] V. Kabakian, M. McManus, H. Harajli, Attributional life cycle assessment of mounted 1.8 kWp monocrystalline photovoltaic system with batteries and comparison with fossil energy production system, Appl. Energy 154 (2015) 428–437.
[60] C. Lamnatou, D. Chemisana, Evaluation of photovoltaic-green and other roofing systems by means of ReCiPe and multiple life cycle–based environmental indicators, Build. Environ. 93 (2015) 376–384.
[61] R. Battisti, A. Corrado, Evaluation of technical improvements of photovoltaic systems through life cycle assessment methodology, Energy 30 (7) (2005) 952–967.
[62] Spriensma MGaR. The Eco-indicator 99: a damage oriented method for life cycle impact assessment. *Methodology Report, PRé Consultants B.V.*, 2009.

[63] S. Solomon, Climate Change 2007—The Physical Science Basis: Working Group I Contribution to the Fourth Assessment Report of the IPCC, Cambridge University Press, Cambridge, UK and New York, 2007.
[64] M. Liu, M.B. Johnston, H.J. Snaith, Efficient planar heterojunction perovskite solar cells by vapour deposition, Nature 501 (7467) (2013) 395−398.
[65] R. García-Valverde, C. Miguel, R. Martínez-Béjar, et al., Life cycle assessment study of a 4.2 kW p stand-alone photovoltaic system, Sol. Energy 83 (9) (2009) 1434−1445.
[66] P.W. Liang, C.Y. Liao, C.C. Chueh, et al., Additive enhanced crystallization of solution-processed perovskite for highly efficient planar-heterojunction solar cells, Adv. Mater. 26 (22) (2014) 3748−3754.
[67] J.H. Heo, S.H. Im, J.H. Noh, et al., Efficient inorganic-organic hybrid heterojunction solar cells containing perovskite compound and polymeric hole conductors, Nat. Photonics 7 (6) (2013) 486−491.
[68] D. Bi, S.-J. Moon, L. Haggman, et al., Using a two-step deposition technique to prepare perovskite (CH3NH3PbI3) for thin film solar cells based on ZrO2 and TiO2 mesostructures, RSC Adv. 3 (41) (2013) 18762−18766.
[69] J. Burschka, N. Pellet, S.-J. Moon, et al., Sequential deposition as a route to high-performance perovskite-sensitized solar cells, Nature 499 (7458) (2013) 316−319.
[70] F. Matteocci, L. Cinà, F. Di Giacomo, et al., High efficiency photovoltaic module based on mesoscopic organometal halide perovskite, Prog. Photovoltaics 24 (4) (2016) 436−445.
[71] W.J. Lee, E. Ramasamy, D.Y. Lee, et al., Dye-sensitized solar cells: scale up and current−voltage characterization, Sol. Energy Mater. Sol. Cells 91 (18) (2007) 1676−1680.
[72] D. Liu, T.L. Kelly, Perovskite solar cells with a planar heterojunction structure prepared using room-temperature solution processing techniques, Nat. Photonics 8 (2) (2014) 133−138.
[73] J. You, Z. Hong, Y.M. Yang, et al., Low-temperature solution-processed perovskite solar cells with high efficiency and flexibility, ACS Nano 8 (2014) 1674−1680.
[74] N. Espinosa, M. Hosel, D. Angmo, et al., Solar cells with one-day energy payback for the factories of the future, Energy Environ. Sci. 5 (1) (2012) 5117−5132.
[75] E. Spalvins, B. Dubey, T. Townsend, Impact of electronic waste disposal on lead concentrations in landfill leachate, Environ. Sci. Technol. 42 (19) (2008) 7452−7458.
[76] S. Nagib, K. Inoue, Recovery of lead and zinc from fly ash generated from municipal incineration plants by means of acid and/or alkaline leaching, Hydrometallurgy 56 (3) (2000) 269−292.
[77] R. García-Valverde, J.A. Cherni, A. Urbina, Life cycle analysis of organic photovoltaic technologies, Prog. Photovolt. Res. Appl 18 (7) (2010) 535−558.
[78] V. Baglio, M. Girolamo, V. Antonucci, et al., Influence of TiO_2 film thickness on the electrochemical behaviour of dye-sensitized solar cells, Int. J. Electrochem. Sci. 6 (8) (2011) 3375−3384.
[79] N. Espinosa, R. Garcia-Valverde, A. Urbina, et al., A life cycle analysis of polymer solar cell modules prepared using roll-to-roll methods under ambient conditions, Sol. Energy Mater. Sol. Cells 95 (5) (2011) 1293−1302.
[80] N. Espinosa, F.O. Lenzmann, S. Ryley, et al., OPV for mobile applications: an evaluation of roll-to-roll processed indium and silver free polymer solar cells through analysis of life cycle, cost and layer quality using inline optical and functional inspection tools, J. Mater. Chem. A 1 (24) (2013) 7037−7049.

INDEX

Note: Page numbers followed by "*f*" and "*t*" refer to figures and tables, respectively.

A

"A" Site, 129–131
AAO. *See* Anodized aluminum oxide (AAO)
Abiotic depletion potential (ADP), 21–22
Absorber, 295–296
Absorption, 141–142
  band, 34–36
  spectra of PbI_2 and $CH_3NH_3PbI_3$ films, 304–305
ABX_3. *See* Perovskite (ABX_3)
Acetonitrile, 328
Acidification potential (AP), 19–22
Activation energy measurements (E_a measurements), 187
ADP. *See* Abiotic depletion potential (ADP)
ALD. *See* Atomic layer deposition (ALD)
Alkali halide, 197–198
  perovskite system, 9–10
All-inorganic alkali-metal lead and tin halide perovskites, 165–166
AlN. *See* Aluminium nitride (AlN)
Alumina (Al_2O_3), 28–29, 236, 332–333
  nanoparticles, 110
Aluminium nitride (AlN), 37
Aluminum (Al), 5
Aluminum doped ZnO (AZO), 15
Ambient air, 185–186
5-Aminovaleric acid (5-AVA), 131, 312
AMO_3. *See* Chalcogenide perovskite (AMO_3)
Amorphous silicon (a-Si), 1–2, 4
  solar cells, 2
Amplified spontaneous emission (ASE), 152
AMX_3. *See* Perovskite (ABX_3)
Annealing, 269–272, 305
  temperature, 3–4, 242, 269
Anode, 2–3
Anodized aluminum oxide (AAO), 206–207
Antisolvent technique, 70–71, 328
AP. *See* Acidification potential (AP)
Area-specific polarization resistance (ASR), 250–251
ARENA. *See* Australian Renewable Energy Agency (ARENA)
Argon, 269–272
a-Si. *See* Amorphous silicon (a-Si)
As-synthesized materials, 138–139
ASE. *See* Amplified spontaneous emission (ASE)
ASR. *See* Area-specific polarization resistance (ASR)
Atomic layer deposition (ALD), 4–5
Australian Renewable Energy Agency (ARENA), 21
5-AVA. *See* 5-Aminovaleric acid (5-AVA)
$(5\text{-AVA})_x(MA)_{1-x}PbI_3$ perovskite, 312–313
A–X bonds, 127
AZO. *See* Aluminum doped ZnO (AZO)

B

"B" Site, 131–132
Band diagram, 17
Band dispersion, 145
Band gap, 32
  engineering of perovskites, 73–76
  of perovskites with variation based on A site cations, 33*f*
  tunability, 25–27
  tuning, 32–33, 35*f*
Band structure, 145–146
Batch process, 1, 18–19
Bathocuproine (BCP), 57
BCP. *See* Bathocuproine (BCP)
Bending cycles, 12–13
Benzene, 328

Benzonitrile, 328
BHJ films. *See* Bulk heterojunction films (BHJ films)
Bibliometric analysis, 11–18
Biomass technology, 1–3
Bis(trifluoromethane) sulfonamide lithium salt (Li-TFSI), 317
Blocking layer of TiO_2 (BL-TiO_2), 17
Bonding, 27
Bourgoin–Corbett mechanism, 173–174
Bridgmanite, 232
Bromide (Br), 133–134
  bromide-based perovskite, 150
Bulk heterojunction films (BHJ films), 104
Burstein–Moss band filling model, 146
B–X bonds, 127
BX_6 octahedral unit, 123–124, 127–128

C

C_{60} (nonoxide material), 295–296
C_{60}-SAM, 101–102
Cadmium indium diselenide (CIS), 4
  solar cell, 4
Cadmium telluride (CdTe), 1–2, 4, 20–26, 43
  modules, 19–21
  solar cell, 4
Calcium titanate ($CaTiO_3$), 43, 197–198
Calcium titanium oxide ($CaTiO_2$), 6, 232
Capacitive hysteresis, 126
Carbon
  black/graphite composite material, 3–4
  electrodes, 8–9
  footprint, 11
Carbon counter electrodes (CCEs), 260–262
Carbon dioxide emission (CO_2 emission), 1, 11
Carbon nanotubes (CNTs), 3–4
  and graphene-based perovskite solar cells, 4–8
Carbon-based HTM–free perovskite solar cell, 308–317. *See also* Hybrid organic–inorganic perovskites (HOIPs)
  cross-sectional SEM images of carbon black, 314*f*
  J–V curves, 311*f*, 313*f*, 316*f*
  microscopy images, 312*f*
  SEM image from cross-section, 310*f*
  solar cell parameters, 315*t*
  structure of fully printable mesoscopic PSC, 309*f*
Carbon-based materials, 3–4
Carbon-based PSCs, 313–315
Carrier diffusion length (L_D), 142
Carrier dynamics, 147
Carrier mobility, 142–147
  carrier dynamics, 147
  defects and band structure, 145–146
  density of states, 144
  effective mass, 144–145
  electronic structure, 143–144
Cathode, 2–3
CBM. *See* Conduction band minimum (CBM)
CCEs. *See* Carbon counter electrodes (CCEs)
CdTe. *See* Cadmium telluride (CdTe)
CE. *See* Counter electrode (CE)
Ceramic dielectric laminated supercapacitor, 37
Cesium (Cs), 131
  Cs-based perovskite-like $CsSnI_3$, 131
Cesium bromide (CsBr), 267–268
Cesium lead halide ($CsPbX_3$), 197–198, 202
  NWs, 224–225
$CH_3NH_3^+$. *See* Methyl ammonium (MA)
CH_3NH_3I. *See* Methylammonium iodide (MAI)
$CH_3NH_3PbBr_3$. *See* Methylammonium lead bromide ($MAPbBr_3$)
$CH_3NH_3PbI_3$ perovskite solar cells. *See* Graphene-based perovskite solar cells ($CH_3NH_3PbI_3$ perovskite solar cells)
$CH_3NH_3PbI_3$. *See* Methylammonium lead iodide ($MAPbI_3$)
$CH_3NH_3PbX_3$. *See* Colloidal halide perovskite ($CH_3NH_3PbX_3$)

Chalcogenide perovskite (AMO_3), 197–198
Charge dissociation efficiency, 234–235
Charge transfer process, 316–317
Charge transport mechanism, 52
Charge-transfer resistance (R_{CT}), 184–185
Charge–dipole interaction, 30
Chemical capacitance element (C_μ), 184–185
Chemical deposition techniques, 232–234
Chemical-assisted spin coating, 232–234
Chemical-vapor deposition (CVD), 4–6, 140–141, 141*f*, 206, 232–234
Chemical-vapor transport (CVT), 206
China, PSC development in, 28–43
  Chinese companies' areas of expertise, 29*t*
  Chinese universities areas of expertise, 30*t*
  PV market and trends, 41–43
  US companies areas of expertise, 39*t*
  US universities areas of expertise, 40*t*
Chlorine (Cl), 133
Chlorobenzene, 328
Chloroform, 328
CIGS. *See* Copper indium gallium selenide (CIGS)
CIS. *See* Cadmium indium diselenide (CIS)
Class-based searches, 15
Climate change, 1–2
CML method, 10–11, 19
CNTs. *See* Carbon nanotubes (CNTs)
Coal, 5–6
Coevaporation, 232–234, 246–248, 254–255
Colloidal $CsPbX_3$ QDs, 218–219
Colloidal halide perovskite ($CH_3NH_3PbX_3$), 218
  nanoparticle-based LEDs, 218
  organometallic halide perovskite, 134–136
  single crystal, 137
Colloidal metal halide perovskite QDs, 218
Colorless PI substrate (CPI substrate), 24
Compact TiO_2 (c-TiO_2), 5–6, 47–48
Compositional engineering techniques, 44–45
Conduction band minimum (CBM), 143
  of alumina, 335–336
Confocal fluorescence lifetime imaging microscopy, 126
Constant phase element (CPE), 183–184
Contact materials engineering, 71–73
Convention thin-film technology, 4
  cadmium indium diselenide solar cell, 4
  cadmium telluride solar cell, 4
Conventional energy sources, 1
Conventional hydrothermal process, 238–241
Conventional silicon-based PVs, 237
Copper indium gallium selenide (CIGS), 4, 43
  solar cells, 20–26
Cost competitiveness, 47
Cost-effective perovskite solar cell, 28–30
Counter electrode (CE), 232–234, 260–262
CPE. *See* Constant phase element (CPE)
CPI substrate. *See* Colorless PI substrate (CPI substrate)
Cradle-to-gate LCA study, 12–13, 25–28
Cradle-to-grave LCA study, 17, 22–23, 29–30
Crystalline silicon (c-Si), 2, 6
  solar cells, 2, 18, 20–26, 44, 289–290
Crystallization process, 37, 324
Crystallographic analysis, 128–129
CsBr. *See* Cesium bromide (CsBr)
c-Si. *See* Crystalline silicon (c-Si)
$CsPbX_3$. *See* Cesium lead halide ($CsPbX_3$)
c-TiO_2. *See* Compact TiO_2 (c-TiO_2)
Cubic geometry, 31–32
Cubic perovskites, 31–32, 31*f*
  ABX_3 crystal structure, 123, 124*f*
Cubic symmetry, 127
Cuprous iodide, 17
CuSCN, 28–29, 72–73, 268–269
CVD. *See* Chemical-vapor deposition (CVD)
CVT. *See* Chemical-vapor transport (CVT)

D

Debye forces, 30
Degradation, 15, 267
Dendrite crystal growth, 202
Density functional theory (DFT), 20, 144
 calculation, 126
Density of states, 144
density-voltage hysteresis (*J*– *V* hysteresis), 166–167
Deposition
 methods, 36
 process engineering, 69–71
Depth profile SIMS measurements, 176–177
Device architecture, 3–17, 4*f*, 26*t*, 45–59
 inverted planar p–i–n heterojunction structure, 57–59
 liquid electrolyte DSSC structure, 47–49
 mesosuperstructured structure, 51–52
 planar n–i–p heterojunction structure, 53–57
 regular structure, 52–53
 solid-state mesoscopic structure, 49–51
Device stability, 179–180
DFT. *See* Density functional theory (DFT)
Diethyl ether, 21–22, 330
Diffusion constant measurement in HOIPs, 186–187
Dimensionality of material, 211–212
2,9-Dimethyl-4,7-diphenyl-1,10-phenanthroline, 154–155
Dimethylformamide (DMF), 4–5, 36–37, 60, 133, 200–201, 203, 244–248, 326–328
Dimethylsulfoxide (DMSO), 36–37, 60, 136, 330
Dip coating, 12–13, 97, 232–234
Direct perovskite solar cell
 with TiO_2 as ETL, 11–22, 12*f*
  TiO_2 nanotube-based liquid perovskite solar cell, 21–22
  two-step process, 17–21
  vapor deposition technique, 12–16
 with ZnO as ETL, 22–23
Dissolution recrystallization, 203
Divalent metal ions, 2–4
DMF. *See* Dimethylformamide (DMF)
DMSO. *See* Dimethylsulfoxide (DMSO)
Doped Perovskite System, 9
Double-layer capacitance (C_{DL}), 184–185
Double-step solution process, 136–137
Down shifting (DS), 268–269
DS. *See* Down shifting (DS)
DSSCs. *See* Dye-sensitized solar cells (DSSCs)
DSVD. *See* Dual-source vacuum thermal evaporation (DSVD)
Dual source coevaporation technique, 294
Dual-source vacuum thermal evaporation (DSVD), 4–5
Dye-sensitized solar cells (DSSCs), 1, 2–4, 5, 5–6, 165–166, 200, 231, 289–290, 323

E

Ecoindicator, 10–11
EDS technique. *See* Electrostatic spray deposition technique (EDS technique)
EDX spectroscopy. *See* Energy-dispersive X-ray spectroscopy (EDX spectroscopy)
Effective mass, 144–145
Efficiency, 20–21, 50–55
 charge dissociation, 234–235
 luminescence, 217
 photoconversion, 231–232
EIS. *See* Electrochemical impedance spectroscopy (EIS)
EISA method. *See* Evaporation-induced self-assembly method (EISA method)
Electric field, 28, 29*f*
Electrical conductivity of graphene film, 6–8
Electricity consumption, 13–14
Electrochemical impedance spectroscopy (EIS), 180–181, 315–316
 equivalent circuit models, 181–186, 183*f*, 185*f*
Electrochemically exfoliated graphite, 106–109

Electron transport layer (ETL), 5, 9–10, 91, 142–143, 234–235, 255–260, 259*f*, 289–291, 332–333
direct perovskite solar cell
with TiO_2 as, 11–22
with ZnO as ETL, 22–23
ETL-free cells, 258–260
Electron transport materials (ETMs), 49–50, 232–234
effect, 273–274
Electron-acceptor, 295–296
Electron-transporting layer, 94–95
Electronic
band structure analysis, 127–128
charge transport/recombination resistance, 184–185
conductivity, 187
dimensionality, 27
wavefunction, 27
Electrospinning, 5–6
Electrostatic spray deposition technique (EDS technique), 249–250
Energetic carriers, 125
Energy payback time (EPBT), 11
Energy technology, 4
Energy-dispersive X-ray spectroscopy (EDX spectroscopy), 168
Energy-intensive manufacturing process, 18
Environmental hazards, 1
Environmental impact, 25–28
of hybrid perovskite solar cells, 12–13
of nanowastes and nanoparticle emissions, 21–22
of vapor-deposited perovskite solar cell, 13–14
EP. *See* Eutrophication potential (EP)
EPBT. *See* Energy payback time (EPBT)
EQE. *See* External quantum efficiency (EQE)
Espacenet database, 11
ETA. *See* Extremely thin absorber (ETA)
Ethanol, 328
Ethyl ammonium ($CH_3CH_2NH_3^+$), 42–43
Ethylene glycol, 328
ETL. *See* Electron transport layer (ETL)
ETMs. *See* Electron transport materials (ETMs)
ETP. *See* Freshwater ecotoxicity potential (ETP)
Eutrophication potential (EP), 21–22
Evaporation-induced self-assembly method (EISA method), 202–203
Excitonic PV technologies, 5–6
Exfoliation method, 207–208
External quantum efficiency (EQE), 4–5, 238–241
Extremely thin absorber (ETA), 51–52
cells, 51–53

F

FA. *See* Formamidinium (FA)
Fabrication, 2–3
of perovskites, 36–37
of PSCs, 99, 234–262
crystal structure of ABX_3 perovskite, 233*f*
fast deposition crystallization, 328
methods to deposit perovskite layers, 234*f*
performance, 263–277
sequential deposition method, 328–329
single-step spin coating method, 326–328
solution processed adduct method, 329–330
thermal evaporation, 330
vapor-assisted solution process deposition, 331
Fabry–Perot wavelength-filter, 248–249
$FAPbBr_3$, 149, 225, 265–267
$FASnI_3$. *See* Formamidinium tin iodide ($FASnI_3$)
Fast deposition crystallization, 328
solvent engineering method, 328*f*
Fast ions, 187
Fermi level, 146
Ferroelectric materials, 18
FET. *See* Field-effect transistor (FET)
FF. *See* Fill factor (FF)
Field-effect transistor (FET), 147–148, 153–154

Fill factor (FF), 3–5, 10, 48, 236–237, 293–294, 332–333
Film composition, 4–5
Film deposition methods evolution, 60–66
  single-step solution process, 60–63
  thermal vapor deposition, 65–66
  two-step solution deposition, 63–64
  vapor assisted solution process, 64–65
Film formation process, 324–325
Film morphology, 202
First-generation PV technology. *See* Crystalline silicon (c-Si)
First-generation solar cells, 6
Flexible all-solid state array type solar cells, 37
Flexible PSCs, 2–3, 2*f*, 20. *See also* Hybrid organic–inorganic perovskites (HOIPs)
  bending and stability test of, 22*f*
  device architecture, 3–17, 4*f*, 26*t*
    inverted planar flexible PSCs, 13–17, 16*f*
    mesoporous flexible PSCs, 4–9, 5*f*
    planar flexible PSCs, 9–13
  manufacturing, 18–19
    batch processing, 18–19
    roll-to-roll processing, 19
  modern applications, 25–27, 28*f*
  stability, 20–25
Flexible solar cell, 20–21
Fluoride, 134
Fluorine-doped tin oxide (FTO), 1–2, 4–6, 12–13, 89–90, 236–237, 289–290, 335
Fluorine-free perovskite LED, 150–151
Force of inertia, 251–252
Formamidinium (FA), 2–4, 124–125, 164–165, 171–173, 242–244
  FA-based perovskite, 129–131
Formamidinium tin iodide ($FASnI_3$), 215–216
Formamidinium-based perovskite films, 306–307
Forward scanning, 46–47
Fossil fuels, 1, 5–6
  transformation to fossil-fuel civilization, 1–2
Four-terminal optical coupling tandem structure, 77
Freshwater ecotoxicity, 13–14
Freshwater ecotoxicity potential (ETP), 21–22
FTO. *See* Fluorine-doped tin oxide (FTO)
FTO/$CH_3NH_3PbI_3$/MWCNT device, 315–316
Full-width-at-half-maxima (FWHMs), 149

G

Gallium arsenide (GaAs), 6
γ-butyrolactone (GBL), 4–5, 60, 136
Geothermal energy, 1
Germanium (Ge), 33, 54
  Ge-based hybrid perovskite, 131–132
Gibbs free energy, 329
Glass substrate, 47–48
Glass-based PSCs, 6–7
Global energy demands, 1
Global fossil-fuel energy demand, 1–2
Global Status Report of Renewables (2015), 1–2
Global warming, 1–2
Global warming potential (GWP), 21–22
Gold (Au), 5, 256–258, 260–262
Goldschmidt's tolerance factor, 42–43, 127–128
Graphene
  materials, 6
  nanoflakes, 238–241
Graphene-based perovskite solar cells ($CH_3NH_3PbI_3$ perovskite solar cells), 4–8, 5*f*
Gravitational force, 251–252
Green technologies, 1–3
Growth mechanism, 329
GWP. *See* Global warming potential (GWP)

H

Halide perovskite ($CH_3NH_3PbX_3$/$MAPbX_3$), 2–4, 3*f*, 25, 127, 197–198
  halide-based hybrid perovskites, 34
  system, 8–10

alkali halide perovskite system, 9–10
organometal perovskite system, 10
type perovskite material, 127
Halides, 43–44
anion, 188
HBL. *See* Hole blocking layer (HBL)
Heterojunction perovskite solar cells
fabrication techniques of perovskite solar cells, 326–331
images of one-step solution processed $MAPbI_{3-x}$, 326*f*
solar cell device architectures, 332–334
types of perovskite solar cells, 334–336
High ionicity, 27
High-efficiency single-junction PSCs, 27
High-temperature sintering process, 253
Highest occupied molecular orbital (HOMO), 50–51
HOIPs. *See* Hybrid organic–inorganic perovskites (HOIPs)
Hole blocking layer (HBL), 255–256
Hole conductor, 295–296
Hole conductor–free perovskite solar cells, 305–306. *See also* Hole transport material (HTM); HTM–free perovskite solar cell
Hole transport layer (HTL), 5–7, 12–13, 213–214, 219–220, 234–235, 255–260, 259*f*, 333–334
effect, 273–274
energy levels for materials, 333*f*
HTL-free cells, 258–260
HTL-free perovskite solar cells, 28–29
Hole transport material (HTM), 44–45, 51–53, 72–73, 89–90, 289–290, 290*f*
HTM-free mesoscopic P-N solar cells, 336
HTM-free planar perovskite cell, 297
HTM-free PSC devices, 289–290
HTM–free perovskite solar cell, 317
carbon-based, 308–317
inverted planar type, 300–301
mesoporous-type, 302–308
planar-type, 291–300
stability, 317
structure of hole conductor–free $TiO_2/CH_3NH_3PbI_3$, 291*f*
working mechanism, 290–291
HOMO. *See* Highest occupied molecular orbital (HOMO)
Hot injection method, 200, 202
HPs. *See* Hybrid perovskites (HPs)
HTL. *See* Hole transport layer (HTL)
HTM. *See* Hole transport material (HTM)
HTP. *See* Human toxicity potential (HTP)
Human toxicity, 13–14
Human toxicity potential (HTP), 21–22
Humid air, 265–267
HX. *See* Hydrogen halide (HX)
Hybrid 3D perovskites, 32
Hybrid halide perovskites, 34
Hybrid organic–inorganic metal halides. *See* Perovskite (ABX_3)
Hybrid organic–inorganic perovskites (HOIPs), 43–44, 123, 163. *See also* Carbon-based HTM–free perovskite solar cell
applications, 147–157
FET, 153–154
lasing, 151–153
LECs, 149
LED, 149–151
perovskite solar cells, 148–149
photodetector, 154–157
background, 164–166
basic synthesis, 134–141
solution-processed method, 136–140
vapor-evaporation methods, 140–141
complicating factors in measurement of ion diffusion in, 179–180
crystal structure, 123–126
cubic perovskite ABX_3 crystal structure, 124*f*
diffusion constant measurement in, 186–187
first HOIP-based PV device, 44–45
ion transport through, 167
long timescale phenomena in, 166–167
observed phenomena in, 164*f*

Hybrid organic–inorganic perovskites (HOIPs) (*Continued*)
optoelectronic properties, 141–147
photovoltaic cells, 166–167
publication statistics with respect to perovskites ABX_3, 135*f*
symmetry and phase, 127–134
timeline of research into optoelectronic properties and discoveries, 135*f*
unanswered questions on the nature of ion transport in, 188–190
Hybrid organometallic perovskites-based material, 152
Hybrid perovskites (HPs), 5–6, 33, 34*f*, 141
degradation, 38
HP-based photodetectors, 155
material, 148
stability, 38
Hydroelectricity energy, 1
Hydrogen bonding, 36–37
Hydrogen halide (HX), 241
Hydrothermal autoclaving, 238–241
Hydrothermal methods, 105–106
Hydrothermal/solvothermal synthesis, 96–97
Hysteresis, 6–7, 46–47, 335

I

ILCD methodology. *See* Life Cycle Data System methodology (ILCD methodology)
Impedance
of CPE, 183–184
spectroscopy, 180–181
In situ XRD, 140
In-operando studies of HOIP-based devices, 179
Incident light, 221–222
Incident photon conversion efficiency (IPCE), 7–8, 96–97, 98*f*
Indium tin oxide (ITO), 3–5, 8–9
Induced–dipole interaction, 30
Industrial revolution, 1
Infrared energy (IR energy), 149–150
Infrared-emitting diode (IRED), 149–150
Inorganic
compounds, 36
framework, 30
oxide perovskite system, 8
p-type semiconductor materials, 15–17
solar cells, 11
thin-film, 6–8
Inorganic perovskite, 25, 26*f*
materials, 223
perovskite system, 9
doped perovskite system, 9
intrinsic perovskite system, 9
QDs, 202
Integrated devices, 104
Intellectual property laws, 15
Interface engineering techniques, 150–151
Intergovernmental Panel on Climate Change (IPCC), 1–2, 19
Intermediate $MAI-PbI_2-DMSO$ phase, 69–70
Intermolecular interaction, 28–29
International patent classification (IPC), 11, 15
International Renewable Energy Agency, 3
Intrinsic perovskite
aggregated inventory for manufacturing perovskite solar cells, 26*t*
solar cell with PEDOT, 23–25
system, 9
Inverted planar
architecture, 13
flexible PSCs, 13–17, 16*f*
HTM-free PSC, 300–301
J–V curves of champion planar heterojunction solar cells, 302*f*
performance parameters of solar cells, 302*t*
photoluminescence spectral and impedance characterization, 301*f*
p–i–n heterojunction structure, 57–59
structure, 336
Inverted structure, 45–47, 57
Iodide (I), 132–133, 171–173
Iodine-rich halide perovskite $MAPbBrI_2$, 155

Iodoplumbate
complexes, 36
finger print, 36
Ion diffusion, direct techniques for studying, 174
Ion migration in hybrid perovskites, 163
background on HOIPs, 164–166
complicating factors in measurement of ion diffusion in HOIPs, 179–180
computational studies of ion transport, 171–174
computationally determined activation energy values, 172*t*
diffusion constant measurement in HOIPs, 186–187
direct techniques for studying ion diffusion, 174
EIS equivalent circuit models, 181–186
impedance spectroscopy, 180–181
long timescale phenomena in HOIPs, 166–167
NMR, 177–179
qualitative evidence of ion transport, 168–171
SIMS, 176–177
tracer experiments, 174–175
unanswered questions on the nature of ion transport in HOIPs, 188–190
Ion transport, computational studies of, 171–174
Ionic conductivity, 187
Ionizing radiation (IR), 13–14, 21–22
IPA. *See* Isopropanol (IPA)
IPC. *See* International patent classification (IPC)
IPCC. *See* Intergovernmental Panel on Climate Change (IPCC)
IPCE. *See* Incident photon conversion efficiency (IPCE)
IR. *See* Ionizing radiation (IR)
IR energy. *See* Infrared energy (IR energy)
IRED. *See* Infrared-emitting diode (IRED)
Isopropanol (IPA), 63–64
Isothermal annealing, 175
ITO. *See* Indium tin oxide (ITO)

J

Jahn Teller effect, 30–31, 127–128
Japan
Japanese companies areas of expertise, 38*t*
universities areas of expertise, 39*t*
Japan, solar cell development, 28
Keyword-based searches, 15

L

$La_{1-x}Sr_xCo_{1-y}Fe_yO_3$ (LSCF), 249–250
$La_{1-x}Sr_xMn_{1-y}Co_yO_{3\pm\delta}$ perovskite oxide system, 176–177
Langevin theory, 147
Large-scale perovskite solar energy technology commercialization
China, 28–43
commercialization issues of PSC, 48–50
cost competitiveness, 47
efficiency and variability, 50–55
toxicity, 54–55
hysteresis, 46–47
Japan, 28
in market, 55
natural resources depletion, 2*f*
through patenting activity, 20
PSC
architecture, 6–11
landscape of, 11–18
PV technology status, 19–20
renewable energy generation options and their contribution, 3*t*
short-term industrialization prospects, 26–27
solar energy, 55–56
solar PV generations, 7*f*
South Korea, 28
stability, 43–46
technology focus of top countries, 20–26
Japan patents filed, 24*t*
renewable energy consumption *vs.* generation statistics, 25*t*
UK patents filed, 23*t*
US patents filed, 25*t*
LARP. *See* Ligand-assisted reprecipitate (LARP)

Laser, 224–225
laser-assisted crystallization, 241
scribing, 18–19
Lasing, 151–153
threshold, 224–225
Layered perovskites, 26, 27*f*, 199
optical properties, 211
structure synthesis, 210–211
LCA. *See* Life cycle assessment (LCA)
LCI analysis. *See* Life cycle inventory analysis (LCI analysis)
LCIA. *See* Life cycle impact assessment (LCIA)
LCOE. *See* Levelized cost of electricity (LCOE)
LD perovskite. *See* Low-dimensional perovskite (LD perovskite)
LE. *See* Light emitting (LE)
Lead (Pb), 7, 30, 232
lead-based perovskites, 154
lead-based PSCs, 25–28
toxicity, 54
Lead chloride ($PbCl_2$), 12–13
Lead halides, 232
coordination chemistry of lead halide perovskite precursors, 34–36
octahedral cage, 27–28
Lead lanthanum zirconate titanate (PLZT), 18–19
LECs. *See* Light-emitting electrochemical cells (LECs)
LED. *See* Light-emitting diodes (LED)
Levelized cost of electricity (LCOE), 18, 47
Lewis acid, 330
Lewis acid-base adduct method, 329–330
Li-TFSI. *See* Bis(trifluoromethane) sulfonamide lithium salt (Li-TFSI)
Life cycle assessment (LCA), 2, 8–11
goal and scope definition, 9
LCI, 10
LCIA, 10–11
life cycle interpretation, 11
Life Cycle Data System methodology (ILCD methodology), 10–11, 17*t*, 25–28
Life cycle impact assessment (LCIA), 8, 10–11
Life cycle interpretation, 8, 11
Life cycle inventory analysis (LCI analysis), 8, 10
Ligand-assisted reprecipitate (LARP), 200
Light emitting (LE), 216
Light-emitting diodes (LED), 149–151, 197–198, 216–220
Light-emitting electrochemical cells (LECs), 147–149
Light-harvesting layer, 1
Light-induced degradation, 264–265
Liquid electrolyte DSSC structure, 47–49
Lithium-treated titania film, 265–267
LN-P3HT-based device, 23–24
LO phonos. *See* Longitudinal optical phonos (LO phonos)
Long timescale phenomena in HOIPs, 166–167
Long-lived photogenerated carriers, 154
Longitudinal optical phonos (LO phonos), 216–217
Loparite, 232
Low-dimensional material synthesis
one-dimensional material, 202–207
structure synthesis and of layered perovskites, 210–211
optical properties of layered perovskites, 211
two-dimensional perovskites, 207–210
zero-dimensional materials, 200–202
Low-dimensional perovskite (LD perovskite), 198–199
applications
light-emitting diodes, 216–220
photodetector, 221–223
photovoltaic devices, 211–216
hybrid perovskite layered materials, 211
laser, 224–225
LEDs, 151
low-dimensional material synthesis, 200–211
materials, 198–199
drawbacks, 223
Low-pressure (LP), 254–255
vapor-assisted process, 107*f*

Low-temperature
process, 289–290
solution deposition, 232–234
solution processability, 1–2
solution-processed crystallization, 203–206
LP. *See* Low-pressure (LP)
LSCF. *See* $La_{1-x}Sr_xCo_{1-y}Fe_yO_3$ (LSCF)
Luminescence efficiency, 217

M

MA. *See* Methyl ammonium (MA)
Magic angle spinning nuclear magnetic resonance (MAS NMR), 178
MAI. *See* Methylammonium iodide (MAI)
$MAPbBr_3$. *See* Methylammonium lead bromide ($MAPbBr_3$)
$MAPbI_3$. *See* Methylammonium lead iodide ($MAPbI_3$)
MAS NMR. *See* Magic angle spinning nuclear magnetic resonance (MAS NMR)
$MASnI_3$ perovskite, 76, 131–132, 141–142, 144–145
Massachusetts Institute of Technology (MIT), 41
Material toxicity, 16–17
MAX_3. *See* Methylammonium halide (MAX_3)
$MA_xFA_{1-x}PbI_3$ mixed cation, 131–132
Mesoporous
architecture, 91–99
ETM, 142
flexible PSCs, 4–9, 5*f*
scaffold, 245–246, 255–256
Mesoporous TiO_2 (mp-TiO_2), 5–6, 254, 303–304
based perovskite solar cells, 335
micron thick, 47–48
mp-TiO_2-free mesosuperstructured solar cells, 267–268
scaffold, 253
Mesoporous-type HTM–free perovskite solar cell, 302–308
HR-SEM images of perovskite compositions, 306*f*
HR-SEM of TiO_2 thicknesses, 303*f*
normalized short circuit current density, 307*f*
PV performance, 304*t*
Mesoscopic perovskite solar cell, 100*f*
Mesoscopic sensitized solar cells, 335
Mesosuperstructured architecture, 109–112, 111*f*
Mesosuperstructured cells, 52
Mesosuperstructured solar cells (MSSC), 89–90, 238–241, 335–336
device architecture, 109, 109*f*
device optical response, 110
Mesosuperstructured structure, 51–52
Metal foils, 7
metal foil–based flexible PSCs, 8–9
Metal halide perovskites, 10, 25–27, 197–198
Metal ion migration, 20
Metal oxide perovskites, 26–27
Methanol, 328
Methyl ammonium (MA), 2–4, 28, 32, 42–44, 124–125, 143, 149, 232, 245–246
cation, 129
cations, 164–165
lead trihalide, 147
Methyl ammonium lead triiodide ($CH_3NH_3PbI_3$), 42–43
Methylammonium halide (MAX_3), 197–198
Methylammonium iodide (MAI), 1–2, 12–13, 21–22, 25, 91–92, 168, 203, 208–210, 212–213, 242–244, 247*f*, 292–293, 304, 324–325, 325*f*
increase of vapor pressure, 254–255
stoichiometric ratio, 36
two-step spincoating, 238–241
Methylammonium lead bromide ($MAPbBr_3$), 44–45
hybrid material, 129
nanorod arrays, 203–206
symmetries, 128–129
Methylammonium lead iodide ($MAPbI_3$), 3–4, 44–45, 48, 51–53, 62*f*, 63*f*, 141–142, 164–166, 197–198

Methylammonium lead iodide ($MAPbI_3$) (*Continued*)
  $CH_3NH_3PbI_3$-based light-emitting FET, 153
  device, 48
  film, 36
  NWs, 213–214
  perovskites, 10–11, 27–28, 154
    hybrid perovskites, 32
    layer, 268–269
    QDs-sensitized solar cells, 212–213
  photovoltaic devices, 179–180
  quantum dots, 89–90
  symmetries, 128–129
Methylammonium triiodide, 323
Micrometer-sized grain morphology, 297–298
Microwave irradiation, 241
Microwave-assisted technique, 238–241
Microwires (MWs), 206
Miniaturized light sources, 224
Miscellaneous architectures, 112–115, 116*f*, 117*f*
MIT. *See* Massachusetts Institute of Technology (MIT)
Mixed halide perovskite, 242–244
MLCC. *See* Multilayer ceramic capacitor (MLCC)
Moisture, 48–50, 263
  effect, 263–267
    decomposition pathway of hybrid halide perovskites, 264*f*
    photographs of 3D TiO_2/ $CH_3NH_3Pb(I_{1-x}Br_x)_3$ bilayer nanocomposites, 265*f*
  instability, 323–324
  moisture-assisted degradation, 6–7
Molecule octahedral framework interaction, 30
Monocrystalline silicon, crystal structure of, 2–3
Monocrystalline silicon solar cell, 2–3
Monovalent A cations, 127
Mott–Schottky analysis, 291
mp-TiO_2. *See* Mesoporous TiO_2 (mp-TiO_2)
MSSC. *See* Mesosuperstructured solar cells (MSSC)
Multi-crystal solar cells, 1–2
Multi-crystalline silicon, 6–8
Multilayer ceramic capacitor (MLCC), 18
Multilayer deposition technique, 13
Multiwalled carbon nanotubes (MWCNTs), 313–315
MWCNTs. *See* Multiwalled carbon nanotubes (MWCNTs)
MWs. *See* Microwires (MWs)
MX_6 octahedra, 43

N

^{14}N MAS NMR study, 178
N-2-methyl pyrrolidone, 60
n-doped $MAPbI_3$, 146
N-type semiconductors, 149–150, 238–241
Nano-imprinting technique, 20–21
Nanocomposite, 238–241
Nanocrystalline mesoporous TiO_2 layer (nc-TiO_2), 17
Nanocrystals (NCs), 198–199
Nanofiber, 238–241
Nanoparticle (NP), 4–5
Nanorods (NRs), 92, 238–241
Nanotube, 238–241
Nanowires (NWs), 198–199, 206–207
National Institute for Materials Science, 37
National Renewable Energy Laboratory (NREL), 28, 43
nc-TiO_2. *See* Nanocrystalline mesoporous TiO_2 layer (nc-TiO_2)
NCs. *See* Nanocrystals (NCs)
Near-infrared region (NIR), 123–124
Nickel oxide (NiO), 92, 92*f*
NIR. *See* Near-infrared region (NIR)
NMR. *See* Nuclear magnetic resonance (NMR)
Noncapacitive hysteresis, 126
NP. *See* Nanoparticle (NP)
NREL. *See* National Renewable Energy Laboratory (NREL)
NRs. *See* Nanorods (NRs)
Nuclear magnetic resonance (NMR), 177–179

Nucleation, 324
NWs. *See* Nanowires (NWs)
Nyquist plots, 181–182, 182*f*

O

Octahedral factor (μ), 42–43
Octahedral framework, 27–28, 30–31
ODP. *See* Ozone depletion potential (ODP)
Oleamide doping, 104–105
Oleic acid, 200–201
OMHP. *See* Organometal halide perovskite (OMHP)
One-dimension (1D)
  material, 202–207
  nanostructures, 96–97
  perovskites, 213–214
    nanorods or NWs, 220
One-step method, 328
  EISA method, 202–203
One-step precursor solution spin-coating process, 254
One-step sintering process, 236–237
Open circuit voltage (V_{oc}), 4–5, 51–53, 329
Optical absorbance, 141–142
Optoelectronic properties, 242
  HOIP, 141–147
    absorption, 141–142
    carrier diffusion length, 142
    carrier mobility, 142–147
OPVs. *See* Organic photovoltaics (OPVs)
Organic
  compounds, 36
  p-type semiconductors, 13
  solvents, 19–21
Organic photovoltaics (OPVs), 1–2, 19–21. *See also* Perovskite photovoltaics
  cell, 289–290
  configuration, 13
Organic solar cells (OSCs), 1, 5–6, 231
Organic–inorganic
  halide perovskite, 218–219
  hybrid perovskites, 25–26, 26*f*, 89–90
  lead halide perovskite materials, 224–225
Organic–inorganic perovskite, 289–290
  absorbers, 89–90
  nanoplatelets preparation, 207–208
  self-assembles, 252–253
Organolead
  halide perovskite, 27–28
  metal perovskites, 30
Organolead triiodide perovskites (OTP), 222–223
Organometal halide, 197–198
Organometal halide perovskite (OMHP), 1–2, 234–235
  nanoparticles, 200
  solar cells, 148
Organometal lead halide perovskite, 25
  composition of perovskites, 26–27
  coordination chemistry of lead halide perovskite precursors, 34–36
  distortion of perovskite geometry, 30–32
    change in composition of ABX_3, 32
    Jahn Teller effect, 30–31
    size effect, 31–32
  fabrication of perovskites, 36–37
    perovskite thin films, 36–37
  metal halide perovskites, 25–26
  nature of bonding in, 27–30
    intermolecular interaction, 28–29
    lead halide octahedral cage, 27–28
    molecule octahedral framework interaction, 30
  optical properties of perovskites, 32–34
  stability of hybrid perovskites, 38
Organometal perovskite system, 10
Organometal trihalide perovskite, 3–4
Organometallic hybrid perovskites, 153
  LED, 150
Organometallic lead halide perovskite solar cells, 323
OSCs. *See* Organic solar cells (OSCs)
OTP. *See* Organolead triiodide perovskites (OTP)
Oxide perovskites, 127
Oxidized state, 47–48

Oxygen molecules, 23–24
Oxygen-induced degradation, 264–265
Ozone depletion potential (ODP), 21–22

P

P3HT. *See* Poly-(3-hexylthiophene) (P3HT)
Passivation of PSC, 113–114
Patenting activity, commercialization through, 20
  China's main universities and companies, 22*t*
  Korean patents filed by companies and industries, 23*t*
Patents, 19–20
  analysis, 15–18
    classification codes using Espacenet, 17*t*
    country-wise number of published patents and collaborative patent, 18*f*
    patent distribution in top five countries, 19*f*
Pb-halide-based PSCs, 27
PbI_2 lattice, 36–37, 245–246
PbI_6 octahedra, 164–165
$PC_{61}BM$. *See* Phenyl-C_{61}-butyric acid methyl ester ($PC_{61}BM$)
PCEs. *See* Photoconversion efficiencies (PCEs)Power conversion efficiency (PCE)
PCPDTBT. *See* Poly {2,6-[4,4-bis-(2-ethylhexyl)-4H-cyclopenta(2,1-b; 3,4-b′)-dithiophene]-alt-4,7-(2,1,3-benzothiadiazole)} (PCPDTBT)
PDMS film. *See* Polydimethylsiloxane film (PDMS film)
p-doped $MAPbI_3$, 146
PDPPDBTE, 45–46, 53–54
PEA. *See* Phenylethylammonium (PEA)
Peak splitting, 177
PEDOT. *See* Poly (3,4-ethylenedioxythiophene) (PEDOT)
PEDOT:PSS. *See* Poly (3,4-ethylenedioxythiophene) polystyrene sulfonate (PEDOT: PSS)
Perfluorinated ionomer (PFI), 219–220
Perovskite (ABX_3), 2–7, 25, 28, 31, 43, 47, 73, 79, 93, 164–165, 197–198, 232, 323
  band-gap engineering, 73–76
  challenges, 6–7
  change in composition, 32
  commercialization, 20–21
  composition, 26–27
  crystals, 312
  cubic perovskite ABX_3 crystal structure, 123, 124*f*, 233*f*
  deposition and/or postdeposition processes, 70
  device fabrication, 5–6
  distortion of perovskite geometry, 30–32
  fever, 44–45
  layer formation, 4–5
  LEDs, 218
  material perspective, 2–4
  mineral, 6–7
  nanorods, 198–199
  NWs, 202, 205*f*, 206
  optical properties, 32–34
  perovskite-based cells, 13
  perovskite-based DSSC durability, 53–54
  perovskite-based LED, 149–151, 219–220
  perovskite-based powered cells, 16
  perovskite-based solar cells, 2–3, 128–129
  perovskite-MWCNT junction, 313–315
  perovskite-structured material, 20–21
  perovskite–perovskite tandem solar cells, 78
  powered cells, 15
  properties and applications, 17–20
    MLCC, 18
    of perovskite materials, 19*t*
    storage and switching devices, 18–20
  publication statistics, 135*f*
  research, 12
    top universities in most prolific countries, 16*t*
  solar modules, 18–19

structure, 6, 32, 42–43, 232
sunlight-based cells, 20–21
system classification, 8–10
halide perovskite system, 9–10
inorganic perovskite system, 9
tandem solar cells, 76–78, 77*f*
Perovskite absorber
film, 324
layer, 334
preparation methods for, 61*f*
Perovskite films, 4–5, 64–65, 70, 324, 328–329, 329*f*, 331*f*
annealing, 109–110
Perovskite layer, 241–255
advantages and disadvantages of fabrication methods, 256*t*
interfacial layer, 291
one-step solution process based on mixture PbI_2, 243*f*
Perovskite market, 19–20
commercialization through patenting activity, 20
solar cell efficiency chart, 21*f*
Perovskite materials, 43, 197–198, 211–212, 216–217, 220
based LED, 217–218
structural dimensionality, 214–215
Perovskite photovoltaics. *See also* Organic photovoltaics (OPVs)
cost-effective perovskite solar cell, 28–30
direct perovskite solar cell
with TiO_2 as ETL, 11–22
with ZnO as ETL, 22–23
inverted perovskite solar cell with PEDOT, 23–25
LCA, 8–11
tin-based perovskite solar cells, 25–28
Perovskite solar cell evolution
advanced device engineering techniques, 67–78
band-gap engineering of perovskites, 73–76
contact materials engineering, 71–73
deposition process engineering, 69–71
perovskite tandem solar cells, 76–78, 77*f*
solvent engineering, 67–69, 68*f*
cubic perovskite crystal structure, 44*f*
device architecture, 45–59
inverted planar p–i–n heterojunction structure, 57–59
liquid electrolyte DSSC structure, 47–49
mesosuperstructured structure, 51–52
planar n–i–p heterojunction structure, 53–57
regular structure, 52–53
solid-state mesoscopic structure, 49–51
film deposition methods, 60–66
Perovskite solar cells (PSCs), 1–2, 5–7, 19–26, 30, 43–45, 46*f*, 89–90, 148–149, 231–234, 263*f*, 289–290, 323–324, 330
advantages, 12–14
dip coating, 12–13
spin coating, 12
spray coating, 12
thermal evaporation, 13–14
architecture, 6–11
mesoporous architecture, 91–99
mesoscopic PSCs, 100*f*
mesosuperstructured architecture, 109–112, 111*f*
miscellaneous architectures, 112–115, 116*f*, 117*f*
photovoltaic parameters of PSCs, 96*t*
planar architecture, 99–109, 103*f*
review papers related to PSC research development, 9*t*
challenges, 14–17
degradation, 15
material toxicity, 16–17
stability, 16
characteristic features, 10–11
commercialization issues, 48–50
fabrication, 234–262, 326–331
counter electrode, 260–262
electron and hole transport layer, 255–260
perovskite layer, 241–255

Perovskite solar cells (PSCs) (*Continued*)
photoanode, 236–241
future trends, 20–21
perovskite commercialization, 20–21
research and development of perovskite solar cell, 21
history, 7–8
landscape, 11–18
patents analysis, 15–18
publication analysis, 12–15
performance, 263–277
effect of ETM and HTL, 273–274
moisture effect, 263–267
other factors, 274–277
temperature effects, 269–272
ultraviolet radiation effect, 267–269
types, 334–336, 334*f*
HTM-free mesoscopic P-N solar cells, 336
inverted planar structure, 336
meso-superstructured solar cells, 335–336
mesoporous TiO_2 based perovskite solar cells, 335
mesoscopic sensitized solar cells, 335
planar pin heterojunction solar cells, 336
working principle, 7
Perovskite thin films, 36–37
optoelectronic and photovoltaic devices, 198
solar cell materials, 36
Perovskite/graphene based hybrid device, 155–156
PFI. *See* Perfluorinated ionomer (PFI)
PFN. *See* Poly(9,9-bis(3′-(*N,N*-dimethylamino) propyl)-2,7-fluorene)-alt-2,7-(9,9-dioctylfluorene) (PFN)
Phenyl-C_{61}-butyric acid methyl ester ($PC_{61}BM$), 9–10, 57
Phenylethylammonium (PEA), 214–215
Photoanode, 236–241
Photochemical ozone creating potential, 21–22
"Photocleaning" effects, 126
Photoconductivity, 154
Photoconversion efficiencies (PCEs), 231–232, 262*f*
Photodetectors, 154–157, 221–223
Photodiodes, 221–222
Photoluminescence (PL), 126, 329
Photoluminescence quantum yield (PLQY), 202
Photons, 310
Photothermal-induced resonance (PTIR), 168
Phototransistors, 221–222
Photovoltaic (PV), 1–5, 231
conversion, 323
development, 41–42
devices, 43–44, 168–171, 211–216, 323
market and trends, 41–43
perovskite structure, 42–43
top 10 countries in 2016, 41–42
parameters, 303–304, 307–308
of PSCs, 96*t*
performance
of layered perovskites, 199
of PSC, 293–294
related applications, 32
solar cell, 8–11
technologies, 2–5, 19–21
thin-film technology, 3–5
wafer-based technology, 2–3
technology status
perovskite market, 19–20
Physical deposition techniques, 232–234
PI substrates. *See* Polyimide substrates (PI substrates)
"Pillared structures", 53
PL. *See* Photoluminescence (PL)
Planar architecture, 99–109, 103*f*
Planar device architecture, 9–10, 101–102
Planar flexible PSCs, 9–13
Planar heterojunction, 256–258
Planar n–i–p heterojunction structure, 53–57
Planar pin heterojunction solar cells, 336
Planar-type hole transport material-free perovskite solar cell, 291–300
four passes of perovskite spray, 299*f*

J–*V* curves of planar HTM-free $CH_3NH_3PbI_3$ solar cells, 294*f*
J–*V* response of $MAPbI_3/C_{60}$ solar cell, 297*f*
preparation processes for $CH_3NH_3PbI_3$ perovskite films, 292*f*
PV parameters of cells, 299*t*
SEM images of $CH_3NH_3PbI_3$ films, 293*f*
spray deposition technique, 298*f*
Plastic substrate, 5–6, 10
Platinum (Pt), 260–262
Platinum-coated FTO glass, 89–90
PLQY. *See* Photoluminescence quantum yield (PLQY)
Plumbates, 34–36
PLZT. *See* Lead lanthanum zirconate titanate (PLZT)
P–N junction, 149–150
Polarons formation, 125
Poly (3,4-ethylenedioxythiophene) (PEDOT), 5, 13–17, 22–23
inverted perovskite solar cell with, 23–25
Poly (3,4-ethylenedioxythiophene) polystyrene sulfonate (PEDOT: PSS), 9–11, 256–258, 336
Poly {2,6-[4,4-bis-(2-ethylhexyl)-4H-cyclopenta(2,1-b; 3,4-b′)-dithiophene]-alt-4,7-(2,1,3-benzothiadiazole)} (PCPDTBT), 333–334
Poly-(3-hexylthiophene) (P3HT), 45–46, 53–54, 72–73, 333–334
Poly(9,9-bis(3′-(*N*,*N*-dimethylamino) propyl)-2,7-fluorene)-alt-2,7-(9,9-dioctylfluorene) (PFN), 154–155
Poly[bis(4-phenyl)(2,4,6-trimethylphenyl) amine poly(triarylamine)] (PTAA), 53, 256–258
Polycrystalline silicon solar cell, 3
Polydimethylsiloxane film (PDMS film), 112–113, 260–262
Polyethylene terephthalate, 222–223
Polyimide substrates (PI substrates), 24
Polymer
solar cells, 6
substrates, 7
Polytriarylamine (PTA), 51–53
Porous carbon film, 308
Porous $CH_3NH_3PbBr_3$ nanowires, 155
Porous NWs, 206–207
Powder X-ray diffraction (PXRD), 214–215
Power conversion efficiency (PCE), 1–2, 4–6, 44–45, 89–90, 131, 165–166, 234–235, 289–290
Precursor flux, 140
"Prior art" searches, 15
Pristine perovskites, 155
Process gas–assisted method, 241
Processing technique, 60, 77–78
2-Propanol, 328
PSCs. *See* Perovskite solar cells (PSCs)
PTA. *See* Polytriarylamine (PTA)
PTAA. *See* Poly[bis(4-phenyl)(2,4,6-trimethylphenyl) amine poly (triarylamine)] (PTAA)
PTIR. *See* Photothermal-induced resonance (PTIR)
P-type
NiO, 333–334
semiconductors, 149–150, 238–241
Publication analysis, 12–15
PV. *See* Photovoltaic (PV)
PXRD. *See* Powder X-ray diffraction (PXRD)

Q

QDs. *See* Quantum dots (QDs)
Qualitative evidence of ion transport, 168–171
Quantum dots (QDs), 48–49, 198–202
solar cells, 5, 289–290
Quantum effects, 214–215

R

R&D. *See* Research and development (R&D)
R2R. *See* Roll-to-roll (R2R)
Radiant efficiency (RE), 216
Radiotracer method, 174–175
Ramping rate, 242
Randles circuit model, 182

RE. *See* Radiant efficiency (RE)
ReCiPe, 10–11
Red photons, 245–246
Regular structure, 45–47, 52–53
Relative humidity (RH), 238–241
Relaxation, 177–178
Renewable energy, 3
resources, 1
sources, 1–2
technologies, 4–5
Renewable power capacity, 4
Reprecipitation, 200–201
Research and development (R&D), 11–12
Respiratory inorganics (RI), 13–14, 21–22
Retrograde solubility, 37
Reverse scanning, 46–47
RH. *See* Relative humidity (RH)
RI. *See* Respiratory inorganics (RI)
Roll-to-roll (R2R), 19
fabrication of PSCs on plastic substrate, 19, 20*f*
processing, 19
Room temperature (RT), 202–203
RT. *See* Room temperature (RT)

S

S&T. *See* Science and technology (S&T)
Scanning electron microscopy (SEM), 168
Scanning-probe microscopy, 126
Science and technology (S&T), 11–12
Second-generation solar cells, 6
Secondary ion mass spectroscopy (SIMS), 174–177
"Self-doping" process, 54
SEM. *See* Scanning electron microscopy (SEM)
Semiconductor, 6–7
materials, 153
NW lasers, 224
QDs, 212–213
Semitransparent perovskite solar cells, 3–4
CNT and graphene-based perovskite solar cells, 4–8
employing graphene-based transparent top electrode, 7*f*
perovskite reference device, 10*f*
silver nanowires, 8–11, 9*f*
Sensitivity analysis, 19–21
Sequential deposition method, 328–329
Sequential-vapor deposition process (SVD process), 294, 295*f*, 296*f*
Series resistance (R_s), 4–5
Shockley–Queisser limit, 126, 166, 231
Short-circuit current density (J_{sc}), 236–237, 325
Short-term industrialization prospects, 26–27
Signal-to-noise ratio, 19
Silicon
silicon-based cells, 5–6
silicon-based modules, 19–21
silicon-based solar cells, 43, 55
wafer-based solar cells, 1–2
Silver (Ag), 5, 12–13, 253, 256–258, 260–262
Silver nanowires (Ag-NWs), 3–4, 8–11, 9*f*, 253
SIMS. *See* Secondary ion mass spectroscopy (SIMS)
Single-crystal solar cells, 1–2
Single-junction HOIPs, 166
Single-layer CVD graphene, 6–8
Single-source evaporation, 140
Single-step solution deposition, 60–62
Single-step solution process, 60–63, 136
Single-step spin coating method, 326–328
fabrication of perovskite film with methods, 327*f*
Single-walled carbon nanotubes (SWNTs), 272
Sinter-less spin-coating processes, 273–274
"Slow ions", 187
SOC-GW model. *See* Spin–orbit coupling-GW model (SOC-GW model)
Solar cell(s), 1–2, 148–149, 231, 329
active layer, 141–142
bandgap, open-circuit voltage, and energy loss of, 11*t*
characteristics, 11*t*
comparison, 6*t*
devices, 38
ETL, 332–333

HTL, 333–334
technologies, 6
Solar energy, 1–2, 5–6, 55–56
Solar PV module, 4–5
Solar technology, 1–3, 6–8
Solar-Tectic LLC, 37
Solid state
HOIP active material photovoltaics, 165–166
mesoscopic
solar cells, 237
structure, 49–51
precursors, 36
samples, 178
Solution process(ing), 12
adduct method, 329–330, 330*f*
materials, 124–125
method, 134–140, 254–255
anion exchange reaction and color change, 139*f*
setup for growth of large MAPbI single crystal, 138*f*
NWs, 221–222
PbI_2 film, 331
perovskite films, 17, 89–90
techniques, 4–5, 210–211, 246–248, 300
thin-film perovskite devices, 234–235
TiO_2, 102
Solvent engineering techniques, 44–45, 67–69, 68*f*
South Korea, PSC development in, 28
Spin coating, 12
method, 236–237
technique, 5–6, 248–249
Spin-orbit coupling, 144
Spin–lattice relaxation, 177–178
Spin–orbit coupling-GW model (SOC-GW model), 144
Spin–spin relaxation, 177–178
Spray coating method, 12, 249–250
Spray deposition method, 297
Spray pyrolysis process, 12, 17, 241
Sputtering process, 5–6, 18–19, 232–234
Stability, 16, 20–21
of flexible PSCS, 20–25
of hole transport material–free PSC, 317
and life time, 232
Standard N719 dye sensitizer, 51–53
Stoichiometric precursors, 210–211
Storage devices, 18–20
Sunlight-based cells, 20
Surface tension, 251–252
SVD process. *See* Sequential-vapor deposition process (SVD process)
Switching devices, 18–20
SWNTs. *See* Single-walled carbon nanotubes (SWNTs)

T

Tandem solar cell, 76, 77*f*
TBP. *See* 4-Tertbutylpyridine (TBP)
TCO. *See* Transparent conducting oxide (TCO)
Technology development, 11
in PSCs, 20
Temperature, 263
annealing, 18–19
effects, 269–272
temperature-dependent ^{1}H NMR, 178–179
4-Tertbutylpyridine (TBP), 21–22, 317
2,2′,7,7′-Tetrakis-(*N*,*N*-di-*p*-methoxyphenyl-amine)-9,9′-spirobifluorene (Spiro-MeOTAD), 12–13, 28–29, 45–46, 49–50, 53–54
HTL, 256–258
2,2′,7,7′-Tetrakis-[*N*,*N*-di(4-methoxyphenyl)-amino]-9,9′-spirobifluorene (Spiro-OMeTAD), 5, 7–8, 45–46, 89–90, 256–258, 273–274, 300, 336
layer, 274–277
process, 148
spiro-OMeTAD/PEDOT:PSS graphene, 6–8
TGA. *See* Thermo gravimetric analysis (TGA)
Thermal annealing, 4–5, 70–71, 242–244
effect, 269–272

Thermal annealing (*Continued*)
short rapid, 109–110
Thermal evaporation method, 13–14, 65, 242–244, 252–253, 330, 331*f*
Thermal oxidation (TO), 236–237
Thermal stability, 20
Thermal vapor deposition, 65–66
Thermo gravimetric analysis (TGA), 269–272
THF, 328
Thin-film generation, 4–5
DSSC, 5
organic solar cell, 5
PSC, 5
quantum dot solar cell, 5
Thin-film solar cells, 1–4, 231
Thin-film technology, 3–5. *See also* Wafer-based technology
convention thin-film technology, 4
emerging thin-film generation, 4–5
Third generation
PV technology, 231, 233*f*
solar cells, 6
Three-dimension (3D), 323
hybrid perovskite materials, 143
perovskites, 215
structure, 323
THz photoinduced absorption transient spectroscopy, 147
Time, 242
sensitivity, 180
time-integrated photoluminescence spectroscopy, 300
time-of-flight mass spectrometry, 126
time-resolved photoluminescence studies, 92, 300
Time-of-flight secondary ion mass spectroscopy (TOF-SIMS), 176
Tin (Sn), 54
Sn-based materials, 131–132
Sn–Pb perovskite, 54–55
tin-based perovskites, 154
solar cells, 25–28
tin-lead perovskites, 123–124
Tin oxide (SnO_2), 28–29, 97–98
TiO_2 nanotubes (TNT), 21–22
arrays, 7–8
TiO_2 nanotube-based liquid perovskite solar cell, 21–22
Titanium (Ti), 236–237
Titanium dioxide (TiO_2), 5, 28–29, 236, 277
direct perovskite solar cell with TiO_2 as ETL, 11–22
TiO_2-based devices, 45–46
TiO_2-based perovskite solar cells, 11
TiO_2-free cells, 267–268
$TiO_2/CH_3NH_3PbI_3$ interface, 93
TiO_2/perovskite interface, 291
TiO_2/ZrO_2 mesoporous layers, 310–311
TNT. *See* TiO_2 nanotubes (TNT)
TOF-SIMS. *See* Time-of-flight secondary ion mass spectroscopy (TOF-SIMS)
Tolerance factor, 73–74, 129, 131, 210
Toluene, 328
Toxicity, 54–55, 323–324
TPA. *See* Triphenylamine (TPA)
Tracer experiments, 174–175
TRACI
impact assessment model, 29–30
method, 10–11
Transparent conducting oxide (TCO), 4–5, 47–48, 93
Transparent conductive adhesive–coated PET substrate (TCA–coated PET substrate), 8–9
Triphenylamine (TPA), 258–260
Tris [2-(1H-pyrazol-1-yl) pyridine] cobalt (III), 333–334
Tunability of optical properties, 202
Two-dimension (2D), 323–324
$(C_4H_9NH_3)_2PbBr_4$ nanocrystal, 139
$CH_3NH_3PbX_3$ perovskite nanosheets, 208–210
hybrid perovskite materials, 143
layered structure, 323–324
perovskite, 207–210, 213–214
materials, 207–208
platelets, 220
synthesis, 210–211
2D-perovskites, 214–215
Two-step deposition technique, 304
Two-step fabrication method, 208–210

Two-step sequential method, 4–5
deposition method, 308
Two-step solution deposition, 63–64
Two-step spin-coating method, 213–214

U

Ultrasonic spray deposition, 249–250
Ultrastable performance of system, 152
Ultrathin metal film (UTMF), 6–7
Ultraviolet (UV), 232
irradiation, 4–6
light, 44
radiation effect, 267–269
treatment, 18–19
Uniform layer perovskite film, 246–248

V

Vacancy diffusion, 171–173, 173*f*
Vacuum-flash assisted solution process, 70–71
Valence band maximum (VBM), 143
Valence band shift, 144
Vapor assisted solution process, 64–65
Vapor deposition technique, 12–16
aggregated inventory for manufacturing perovskite solar cells, 15*t*
contribution of all electricity processes to every impact categories, 17*t*
process routes, 14*f*
Vapor-assisted solution processing (VASP), 4–5, 241
process deposition, 331, 332*f*
technique, 297
Vapor-evaporation method, 134–136
Vapor-phase method, 208
Variability, 50–55
VASP. *See* Vapor-assisted solution processing (VASP)
VBM. *See* Valence band maximum (VBM)
Viscous drag, 12–13, 251–252
Volmer-Weber growth, 324

W

Wafer-based silicon solar cells, 231
Wafer-based technology, 2–3. *See also* Thin-film technology
monocrystalline silicon solar cell, 2–3
polycrystalline silicon solar cell, 3
Wall plug efficiency (WPE), 216
Warburg diffusion element, 185–186
Warburg element (W_s element), 183–187
Waste to energy technologies, 1–3
Water-soluble fullerene (WS-C_{60}), 102
Web of Science database, 8–12
Wind
energy, 1
technology, 1–3
WO_x, 97–98, 98*f*
WPE. *See* Wall plug efficiency (WPE)

X

X-ray diffraction (XRD), 168, 180, 297, 310
X-ray photoemission spectroscopy (XPS), 168, 180
"X" Site, 132–134
Xylene, 328

Y

Yttrium-doped TiO_2 (Y-TiO_2), 148–149, 256–258

Z

Zero-dimensional materials, 200–202
Zinc oxide (ZnO), 5, 10, 28–29, 92, 236
direct perovskite solar cell with ZnO as ETL, 22–23
nanoparticles, 9–11
nanoparticles–based thin interfacial layer, 8–9
nanorods, 213–214
thin film, 22–23
Zirconium (Zr), 277
Zn_2SnO_4, 2–3, 9–10, 94–95
ETL PSC, 12
ZrO_2, 45–46, 53–54, 312

FSC
www.fsc.org
MIX
Papier | Fördert gute Waldnutzung
FSC® C083411

Kolibri 360 GmbH
Ferdinand-Jühlke-Straße 7
99095 Erfurt, Deutschland
produktsicherheit@kolibri360.de